47.70
M4
1992

Networks of Power

THOMAS P. HUGHES

Networks of Power

Electrification in Western Society, 1880-1930

The Johns Hopkins University Press
Baltimore and London

This book has been brought to publication with the
generous assistance of the Burndy Library and the
National Endowment for the Humanities.

© 1983 by The Johns Hopkins University Press
All rights reserved. Published 1983
Printed in the United States of America

Second printing, 1988

The Johns Hopkins University Press, 701 West 40th Street,
Baltimore, Maryland 21211
The Johns Hopkins Press Ltd., London

Library of Congress Cataloging in Publication Data
  Hughes, Thomas Parke.
    Networks of power.

    Includes index.
    1. Electric power systems—United States—History.
  2. Electric power systems—Great Britain—History.
  3. Electric power systems—Germany—History.   I. Title.
  TK1005.H83   1983   363.6′2   82-14858
  ISBN 0-8018-2873-2

Title-page illustration: *Ralston Crawford's* Electrification
(1936). Reproduced by permission of the Hirshhorn Museum and
Sculpture Garden, Smithsonian Institution.

For Agatha

Contents

Preface

THE roots of this book reach back to my undergraduate years as an engineering student. In those days, engineers seemed uninterested in comprehending the order in the technological world. An exception was the course of study in electrical engineering. The solution of problems in this field demanded an ability to see relationships, use mathematics imaginatively, and draw on generalizations. Especially satisfying was a course in electric power plant design taught by Professor Frederick Morse, who had the intellectual strength to use elegant electrical science in solving problems within a context of economic, political, and geographical factors. He solved not by excluding variables but by bringing to bear powerful and complex analysis and order.

Despite some disappointment in the way engineering was taught, I never gave up the search for traces of order in man-made complexities. In graduate school I turned to history and continued the search for expressions, however tremulous, of man's constructive power in a chaotic world. I sought the patterns that some engineers had imposed in the past on the material world. These were the engineers who, in the words of their nineteenth-century biographer Samuel Smiles, "were strong-minded, resolute, and ingenious," and who were "impelled to their special pursuits by the force of their constructive instincts."

My historian-mentors, Oron J. Hale and Julian Bishko, sought patterns in political, economic, and diplomatic history. They delighted in finding coherence in Bismarck's foreign policy or in the seasonal rhythms and geographical migrations of the *mesta* in medieval Spain. I sought the economy, efficiency, and system imposed by inventors, engineers, managers, and entrepreneurs of technological change. Early on I found that the most impressive of the patterns imposed on the world by men impelled by the force of constructive instincts were systems, coherent structures comprised of interacting, interconnected components that ranged from relatively simple machines to regional electric supply networks. Complex systems became my unit of study.

This study of systems has been in progress for more than a decade. Therefore, it has been conceived, researched, and written within the context of rapidly changing perspectives on, and assumptions about, the nature of history, especially the history of technology. Initially, the essential problem seemed to be to delineate a history of evolving ideas and the artifacts that gave the ideas material representation. The history of a given machine or process was often written, then, without reference to context. I found that there was interest in the impact of technology on society, but that with rare exceptions the impact of society, or culture, on the shape of technology had been virtually ignored. Dissatisfied with the internalist's approach, I turned to an exploration of a broad range of factors, events, institutions, men, and women involved in complex networks of power—technical, economic, political, and social. My exploration of the cultural context was reinforced when I became a member of a history of science department that used the contextual approach.

As my study of the history of technology continued, I found not only that technological systems in context were the units for study, but also that the history of these systems extended beyond national borders. A study of U.S. electric light and power systems alone would not do. Causal links are no respecters of political boundaries. I discovered that inventors, engineers, and managers often drew on foreign technologies—even went abroad in search of them—as they pursued solutions to their problems. I found that the networks of evolving technologies often linked Germany with the United States because both were industrializing rapidly, and that England often provided a contrast to events and trends observed in the other two countries. (Limitations of time, resources, and language prevented exploration of the sources pertaining to France, Italy, Sweden, the Benelux countries, Russia, Japan, and other industrializing regions of the world.)

In time I found that a comparative approach linking three national areas needed to be modified. In fact, the interaction between region and technology was more notable than that between nation and technology. Influences at the national level, such as legislation, affected evolving technological systems, but local geographical factors, both natural and man-made, were more direct and discernible determinants of the shape of the systems. Stated differently, the factor endowments shaping electric power systems were more regional than national.

Taking the comparative approach was made more manageable by the decision to focus on decisions made by inventors, engineers, managers, and financiers who were system builders. This emphasis on the inner workings of electric supply systems and the dynamics of their growth did not, however, produce a study that is of interest only to professionals presiding over technological change. Consumers of technology can more effectively influence the output of technological systems if they, too, understand the functioning of systems and the nature of the critical decisions made by those who direct them.

As a historian traditionally trained, I am reluctant to suggest a definitive model for the evolution of electric power systems. Nevertheless, I have proposed a loosely structured model because the history I explored was mostly untouched, and I want to provide some landmarks by which other historians can chart their explorations. I expect my findings to be revised, my map to be redrawn, and my themes to be redefined as the archives are explored far more thoroughly in the future. I also anticipate that historians who reflect on the same material that I have used will find different patterns.

Those to whom I wish to acknowledge my indebtedness for assistance, encouragement, and support are numerous. Persons who read and commented on chapters or sections of the book are Wayne Astley, James Beard, Robert Belfield, Richard Bettinger, John Brainerd, James Brittain, Harvey Brooks, Bernard Carlson, Alfred Chandler, Robert Friedel, Leslie Hannah, John Heilbron, Richard Hore, Samuel Insull, Jr., Paul Israel, Daniel Kevles, Wilmer Kleinbach, Forrest McDonald, Judith McGaw, John A. Maneatis, Merritt Roe Smith, John Staudenmaier, Rosemary Stevens, Edmund Todd, George Vanderslice, W. C. Watson, and George Wise.

Those who have aided me in researching and illustrating the book include Arthur Abel, Brian Bowers, Kathleen Bramley, J. Church, William Clough, Field Curry, Dorothy Ellison, Ted Fedder, Bernard Finn, Hugh Gibb, Janet Halder, Barbara Kelly, Reese Jenkins, Kurt Mauel, Lynn Nyhart, Corwin Overton, Ruth Pengel, Samuel Sass, Fred Schoch, Ruth Shoewalker, Patricia Sikes, Helen Slotkin, C. J. Somers, E. Symons, Diane Taylor, F. Thoma, Rudolf von Miller, Sigfrid von Weiher, Richmond Williams; students in my seminars at the University of Pennsylvania; and Agatha H. and Lucian P. Hughes.

For typing and for editorial suggestions, I am appreciative of the excellent craftsmanship and careful attention to detail of Eleanore Kurtz. Also involved in the typing of the manuscript were my friends in the office of the University of Pennsylvania's Department of the History and Sociology of Science, Marthenia Perrin, Sylvia Dreyfuss, and Pat Johnson.

For their advice and encouragement throughout the publication process, I am indebted to Jack Goellner, Jim Johnston, Henry Tom, Lisa Mirski, Susan Fillion, and especially my editor, Penny Moudrianakis, all of The Johns Hopkins University Press. Ellen Koch of the University of Pennsylvania assisted expertly in the preparation of the index.

Grants in support of my research and writing were awarded by the American Council of Learned Societies, the American Philosophical Society, the Center for Interdisciplinary Research at the University of Bielefeld, the Fulbright Commission, the John M. Glenn Fund of Washington and Lee University, the Institute of Electrical and Electronics Engineers, the Massachusetts Institute of Technology, the National Science Foundation, the Research Council of the University Center in Virginia, The Rockefeller Foundation, the Smithsonian Institution, the Social Science Research Council, the Southern Fellowship Fund, Southern Methodist University, and the University of Pennsylvania.

Dr. Bern Dibner, a connoisseur of books, science, and history, generously supported my endeavor to make this book an aesthetic as well as scholarly statement.

Networks of Power

CHAPTER I

Introduction

Of the great construction projects of the last century, none has been more impressive in its technical, economic, and scientific aspects, none has been more influential in its social effects, and none has engaged more thoroughly our constructive instincts and capabilities than the electric power system. A great network of power lines which will forever order the way in which we live is now superimposed on the industrial world. Inventors, engineers, managers, and entrepreneurs have ordered the man-made world with this energy network. The half-century from 1880 to 1930 constituted the formative years of the history of electric supply systems, and from a study of these years one can perceive the ordering, integrating, coordinating, and systematizing nature of modern human societies. Electric power systems demanded of their designers, operators, and managers a feel for the purposeful manipulation of things, intellect for the rational analysis of their nature and dynamics, and an ability to deal with the messy economic, political, and social vitality of the production systems that embody the complex objectives of modern men and women. Robert Venturi, the contemporary architect, has asked architects to embrace the complexity and contradictions of the modern world and to make of that world a habitable environment.[1] Leading engineers and managers have also recognized that their drive for order must be tempered by tolerance of messy vitality. Modern electric systems have the heterogeneity of form and function that make possible the encompassing complexity.

Man's making of the complex modern world is an appropriate subject for the twentieth-century historian. Creation of the material environment shaped by—and shaping—mankind is not a peripheral subject that can be left to narrow specialists. To direct attention today to technological affairs is to focus on a concern that is as central now as nation building and constitution making were a century ago. Technological affairs contain a rich texture of technical matters, scientific laws, economic principles, political forces, and social concerns. The historian must take the broad perspective to get to the root of things and to see the patterns. Scientists and

---

[1] Robert Venturi, *Complexity and Contradiction in Architecture* (New York: The Museum of Modern Art, 1966), pp. 22–23.

1

engineers analyze the technical systems they build, but historians are needed to comprehend the complex, multifaceted relations of these systems and the changes that take place in them over time.[2]

For historians, the study of complexity and change is engaging. Edward Gibbon sat in the ruins of the Capitol in Rome and reflected on the contrast between what he saw before him and the earlier glory that was Rome. Upon seeing drums of oil being unloaded from an American ship in an African port, the American scholar Perry Miller asked how a civilization as new as the American one could already be exporting the products of its technology to remote areas of the world that had been settled centuries earlier. Other historians, taking bareboned statistics from widely separated times, have sought to explain quantitative change by means of qualitative analysis. The drama of change provides the historian with an emphasis that sets him or her apart from the social scientist, who often dissects situations without including a time dimension.

How did the small, intercity lighting systems of the 1880s evolve into the regional power systems of the 1920s? In this case, the change is not the decline that fascinated Gibbon; it is the expansion that attracted Miller. The focus is not on contrasting data; it is on contrasting physical configurations. The problem of this book is to explain the change in configuration of electric power systems during the half-century between 1880 and 1930. Such change can be displayed in network diagrams (see Fig. I.1), but the effort to explain the change involves consideration of many fields of human activity, including the technical, the scientific, the economic, the political, and the organizational. This is because power systems are cultural artifacts.

Electric power systems embody the physical, intellectual, and symbolic resources of the society that constructs them. Therefore, in explaining changes in the configuration of power systems, the historian must examine the changing resources and aspirations of organizations, groups, and individuals. Electric power systems made in different societies—as well as in different times—involve certain basic technical components and connections, but variations in the basic essentials often reveal variations in resources, traditions, political arrangements, and economic practices from one society to another and from one time to another. In a sense, electric power systems, like so much other technology, are both causes and effects of social change.

Power systems reflect and influence the context, but they also develop an internal dynamic. Therefore, the history of evolving power systems requires attention not only to the forces at work within a given context but to the internal dynamics of a developing technological system as well. This book is not simply a history of the external factors that shape technology, nor is it only a history of the internal dynamics of technology; it is a history of technology and society.

Scientists have done much to enlighten us about the nature of dynamics of the structures of the natural world, but historians have as yet only barely

---

[2] In his essays, Harvey Brooks, a scientist and engineer, addresses the multifaceted complexity of contemporary technosocial systems. See, for example, Brooks, "A Framework for Science and Technology Policy," *IEEE Transactions on Systems, Man, and Cybernetics*, SMC–2, no. 5 (1972): 584–88.

FIGURE I.1. *THE CONFIGURATION OF EVOLVING SYSTEMS: INNER CITY (1885); CENTRALIZED URBAN (1906); AND REGIONAL (1930)*

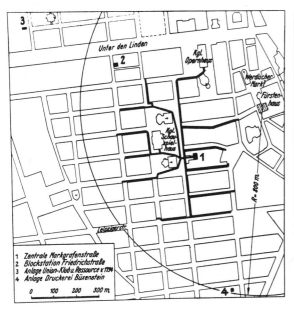

Distribution system, Berlin, 1885. From Matschoss et al., 50 Jahre, p. 13.

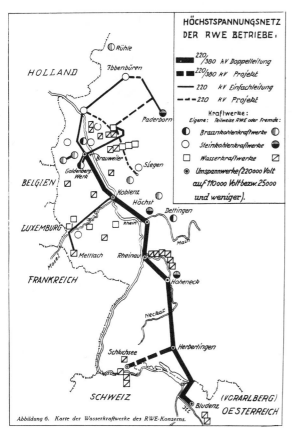

Distribution system, Chicago, 1906. Courtesy of Commonwealth Edison Co., Chicago, Ill.

Transmission system from the Dutch border to Switzerland, 1930. From Koepchen, "Das RWE" (see Chapter XIV, note 17, below).

FIGURE I.2. *THE STATISTICS OF EVOLVING SYSTEMS IN THREE COUNTRIES*

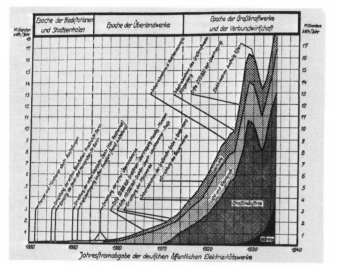

Above: *Electricity generated by public utilities in Germany, 1880–1940. From Rudolf von Miller's article in* Technikgeschichte 25 *(1936): 112.*

Right: *Electricity generated by utilities in the United States, 1882–1921. From* Electrical World 80 *(1922): 546.*

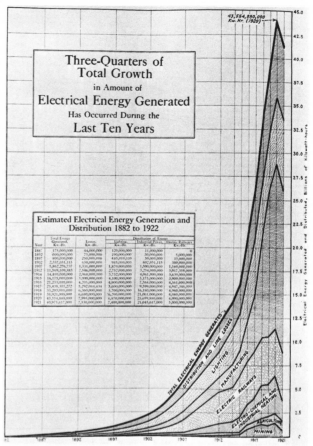

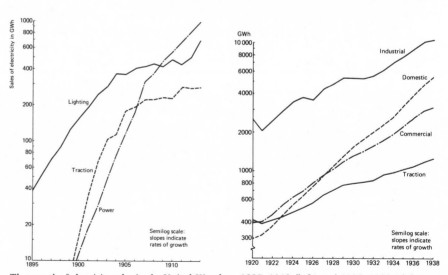

The growth of electricity sales in the United Kingdom, 1895–1913 (left) and 1920–1938 (right). Data from I. C. R. Byatt, "The British Electrical Industry, 1875–1914" (D. Phil. thesis, Oxford University, 1962), p. 111; and Hannah, Electricity Before Nationalisation, *table A.1.*

penetrated the surface of the highly organized and evolving systems of the man-made world. Historians interested in technology have written only a few monographs that concentrate on the evolution of the massive, extensive, vertically integrated production systems of the modern industrial world. Although the public senses the strong organizing forces that originated in these systems and that today influence their lives, they only dimly perceive the nature of these forces. The technological, or man-made, world awaits a Darwin to explicate the origins and dynamics of the forces that pervade it. Quoting Paul Valéry, historian Marc Bloch chides traditional historians for not taking up the task of explicating "the conquest of the earth" by electricity, one of those notable phenomena that have "greater possibilities of shaping our immediate future than all the political events combined."[3]

How might the technological systems that increasingly structure our material environment—specifically, electric power systems—be defined? Because these systems have varied over time and from place to place, the historian's definition cannot be as precise as the scientist's. Ludwig von Bertalanffy, one of the most articulate of systems theorists, needed a book, not a paragraph, to define "system."[4] Thus, an inadequate approximation must serve here as an introduction to the concept of systems. Some characteristics of systems are so general that they transcend time and place. A system is constituted of related parts or components. These components are connected by a network, or structure, which for the student of systems may be of more interest than the components. The interconnected components of technical systems are often centrally controlled, and usually the limits of the system are established by the extent of this control. Controls are exercised in order to optimize the system's performance and to direct the system toward the achievement of goals. The goal of an electric production system, for example, is to transform available energy supply, or input, into desired output, or demand. Because the components are related by the network of interconnections, the state, or activity, of one component influences the state, or activity, of other components in the system. The

[3] Marc Bloch, *The Historian's Craft* (New York: Knopf, 1959), p. 66.

[4] Ludwig von Bertalanffy, *General System Theory: Foundations, Development, Applications* (New York: Braziller, 1968). The literature on systems in general is extensive, and even a selected bibliography is beyond the scope of—and inappropriate for—a history of a particular kind of system. The interested reader might first consult the bibliography in Bertalanffy's *General System Theory* and the article by Talcott Parsons, "Social Systems," in *Encyclopedia of the Social Sciences*, 1968, 15: 458–72. More specific as an introduction to technological systems is Günter Ropohl, *Eine Systemtheorie der Technik: Zur Grundlegung der Allgemeinen Technologie* (Munich and Vienna: Hanser, 1979), which has an extensive bibliography.

In 1978, Bertrand Gille, the French historian of technology, published an extensive historical survey in which he used a systems model of the development of technology to place the history of technology in the context of general history. His model involved technology as structures, technical ensembles, and technical concatenations. These correspond roughly to machines, processes, and civil-engineering works (structures); production, communication, and transportation systems (ensembles); and such systems interrelated vertically and horizontally by general production or output (concatenations). Gille used the model to range over human history and explain technical progress. I am indebted to Professor Cecil Smith for sharing his unpublished English translation (1981) of Bertrand Gille's "A Systems History of Technology" (*Histoire des techniques* [Paris: Gallimard, 1978]). Gille's model was brought to my attention after I completed my manuscript, but I find no cause to revise my own model, which interprets a relatively circumscribed case history in fine detail.

network provides a distinctive configuration for the system. For example, a system can have its components arranged vertically or horizontally.

According to widespread usage, a horizontally arranged system interconnects components of the same kind or function, though not necessarily of the same magnitude, while a vertical system interconnects components joined in a functional chain. For example, an electrical system of the horizontal kind combines power plants under central control, while a production system of the vertical kind might link a coal mine to an electric power plant through a central control facility coordinating the supply of coal and the output of electricity. Systems are also arranged hierarchically, with small systems yielding to the overriding control of a large encompassing system. Systems also interact with one another through the coordination of semiautonomous controls, but without yielding to an overriding control. Although it is customary to define systems as technical, economic, political, or social, the centralization of at least a loose control over systems of these different kinds makes possible the conceptualization of sociotechnical systems and the like.

Those parts of the world that are not subject to a system's control, but that influence the system, are called the environment. A sector of the environment can be incorporated into a system by bringing it under system control. An open system is one that is subject to influences from the environment; a closed system is its own sweet beast, and the final state can be predicted from the initial condition and the internal dynamic. Some systems are planned to their full extent, while others grow by increments and by confluence with other systems over time. All of the kinds and conditions of systems noted in this abstract definition will be illustrated and described in detail in the history that follows.[5]

Usually in this study, "system" refers to a technical system, such as an electric transmission system. Sometimes reference is, as noted, to a system with interacting components, some of which are not technical. Centrally directed, interacting institutions and technical components comprise such a system. On occasions, however, the concept of system is used much more loosely. "System" then means interacting components of different kinds, such as the technical and the institutional, as well as different values; such a system is neither centrally controlled nor directed toward a clearly defined goal. This usage is similar to that of the historian who writes of a system of nation-states. Such a loosely structured system is similar to the concept embodied in "syndrome." All of the systems, it is important to stress, share the characteristic of interconnectedness—i.e., a change in one component impacts on the other components of the system.

[5] An interesting discussion of electrical systems and their management is found by Georg Boll, *Entstehung und Entwicklung des Verbundbetriebs in der deutschen Elektrizitätswirtschaft bis zum europäischen Verbund* (Frankfort on the Main: VWEW, 1969), pp. 13–15. Articles on interconnections and electric systems were frequently published in technical periodicals during the latter half of the period covered by the present study, and many are cited in the various chapters of this book. More recent helpful discussions of electrical systems include: Hans Glavitsch, "Computer Control of Electric-Power Systems," *Scientific American*, November 1974, pp. 34–44; Wallace Brand, "Northeast Electric Bulk Power Supply, *Public Utilities Fortnightly*, 9 June 1966, pp. 65–88; and U.S., Federal Power Commission, *The 1970 National Power Survey*, pts. 1–4 (Washington, D.C.: GPO, 1971).

Electric power systems of the technical kind consist of power generation, transformation, control, and utilization components and power transmission and distribution networks. (The primary distinction between transmission and distribution is the greater distance covered, and therefore the higher transmission voltages used, in transmission.) During the half-century 1880–1930, power generation components included coupled prime movers such as reciprocating steam engines and steam and water turbines. Various types of generators were coupled to the prime movers. Transformers became the principal mode of changing the characteristics of electric supply during transmission and distribution. Energy utilization components included lamps, motors, both stationary and traction (moving), and heating and electrochemical devices. The system incorporated a multitude of applications (see Fig. I.3). Power transmission increased in extent from a few city blocks to regions comprising tens of thousands of square miles. Power distribution networks carried the electric supply from the transmission network to the power utilization machinery and appliances. Control components regulated the supply system in accordance with established standards such as voltage and frequency and directed the system for optimum performance as measured by goals, including efficiency and economy. The most difficult challenge in defining an electric supply system arises at the extreme supply and demand ends of the system. For instance, should the mechanical prime mover be included in the definition of a system? Should the various loads be included, considering that they were usually outside the control of the system? In this study the prime movers have been included in the definition of the system because the inventors, engineers, and industrial scientists treated them as such and because the characteristics of the prime mover were coordinated with the other components of the electric system. Furthermore, the prime movers were under the system's control. The invention and development of motors have been treated in this study because inventors and designers matched the characteristics of motors to those of the electric supply system. Such complications will be clarified by historical example.

The rationale for undertaking this study of electric power systems was the assumption that the history of all large-scale technology—not only power systems—can be studied effectively as a history of systems. It is hoped, therefore, that this history of a particular kind of system will be of some assistance to other historians who wish to study other systems. The assumption of similarity is based in part on an analysis of studies of large systems by other historians who have used the concept of the system to organize, analyze, and draw conclusions from disparate materials.[6]

[6] In another source, I have discussed at some length the use of the systems approach by Lynn White, Jr., *Medieval Technology and Social Change* (New York: Oxford University Press, 1962); Karl Marx, *Capital: A Critique of Political Economy*, ed. Friedrich Engels, 3 vols. (Chicago: Kerr, 1932–33); and Alfred D. Chandler, Jr., *The Visible Hand: The Managerial Revolution in American Business* (Cambridge, Mass.: Harvard University Press, 1977). See Thomas P. Hughes, "The Order of the Technological World," in *History of Technology, 1980*, ed. A. Rupert Hall and Norman Smith (London: Mansell, 1980), pp. 1–16.

Among recent books in which other historians discuss technology as systems are Hugh Aitken, *Syntony and Spark: The Origins of Radio* (New York: Wiley, 1976); Edward W. Constant II, *The Origins of the Turbojet Revolution* (Baltimore: The Johns Hopkins Unversity Press, 1980);

Figure I.3.   *Universal supply system, Berlin, c. 1930. from Matschoss et al., 50 Jahre, p. 90.*

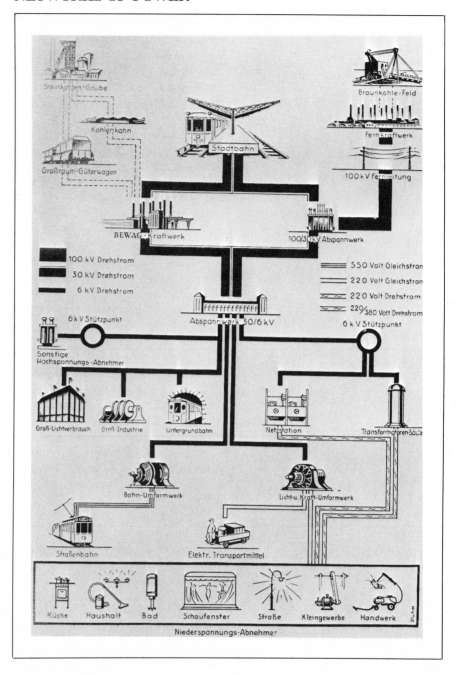

John Enos, *Petroleum Progress and Profits: A History of Process Innovation* (Cambridge, Mass.: M.I.T. Press, 1962); Louis C. Hunter, *Steamboats on the Western Rivers: An Economic and Technological History* (Cambridge, Mass.: Harvard University Press, 1949); idem, *Waterpower: A History of Industrial Power in the United States, 1780–1930* (Charlottesville: University Press of Virginia, 1979); Arthur Johnson, *The Development of American Petroleum Pipelines: A Study in Private Enterprise and Public Policy* (Ithaca, N.Y.: Cornell University Press, 1956); David Landes, *The Unbound Prometheus: Technological Change and Industrial Development in Western Europe from 1750 to the Present* (Cambridge: At the University Press, 1972); Otto Mayr, *Feedback Mechanisms in the Historical Collections of the National Museum of History and Technology* (Washington, D.C.:

Key to Figure I.3. *Translation of terms*

GERMAN	ENGLISH
Steinkohlen-Grube	Hard-coal mine
Braunkohle-Feld	Brown-coal open-face mine
Kohlenkahn	Coal barge
Stadtbahn	City railway
Fernkraftwerk	Distant power station
Grossraum-Güterwagen	Long-distance coal transport
Fernleitung	Long-distance transmission lines
BEWAG-Kraftwerk	Berlin power station
Abspannwerk	Step-down transformer station
Drehstrom	Polyphase current
Gleichstrom	Direct current
Stützpunkt	Distribution center
Sonstige Hochspannungs-Abnehmer	Special high-voltage consumer
Gross-Lichtverbrauch	Large-scale light consumer
Gross-Industrie	Heavy industry
Untergrundbahn	Subway
Netzstation	Dispatching center
Transformatoren-Säule	Distribution transformer
Bahn-Umformwerk	Motor-generator converter (traction load)
Licht-u. Kraft-Umformwerk	Motor-generator converter (light and power load)
Strassenbahn	Streetcar
Elektr. Transportmittel	Electric truck
Küche	Kitchen appliances
Haushalt	Household appliance
Bad	Hot water
Schaufenster	Display window
Strasse	Street lighting
Kleingewerbe	Commercial consumer
Handwerk	Craftsman

The emphasis on, and delineation of, technological systems by historians and social scientists; the drive of inventors, engineers, appliers of science, managers, and financiers to create systems;[7] and the obvious systematic character of electric power systems have all stimulated the organization of this study as a history of systems. The study is complex because it is a

Smithsonian Institution, 1971); Bruce Mazlish, ed., *The Railroad and the Space Program: An Exploration in Historical Analogy* (Cambridge, Mass.: M.I.T. Press, 1965); Elting Morison, *Men, Machines, and Modern Times* (Cambridge, Mass.: M.I.T. Press, 1966); Nathan Rosenberg, ed., *The American System of Manufacturers* (Edinburgh: Edinburgh University Press, 1969).

There are interesting parallels between this study of electrical systems and the history of growth and technological change in the chemical industry as periodized and organized by L. F. Haber in *The Chemical Industry, 1900–1930* (Oxford: Clarendon Press, 1971).

[7] One such inventor and engineer, Elmer Sperry (1860–1930), devoted a lifetime to inventing and developing technological systems. His systems were mostly electromechanical in nature and involved highly complex feedback controls. The study of hundreds of his patents for these systems has influenced my own concepts of technological systems. See Thomas P. Hughes, *Elmer Sperry, Inventor and Engineer* (Baltimore: The Johns Hopkins Press, 1971).

DEVELOPMENT OF ELECTRIC SYSTEMS
END OF YEAR 1900

Transmission systems served by Pennsylvania Power & Light, 1900

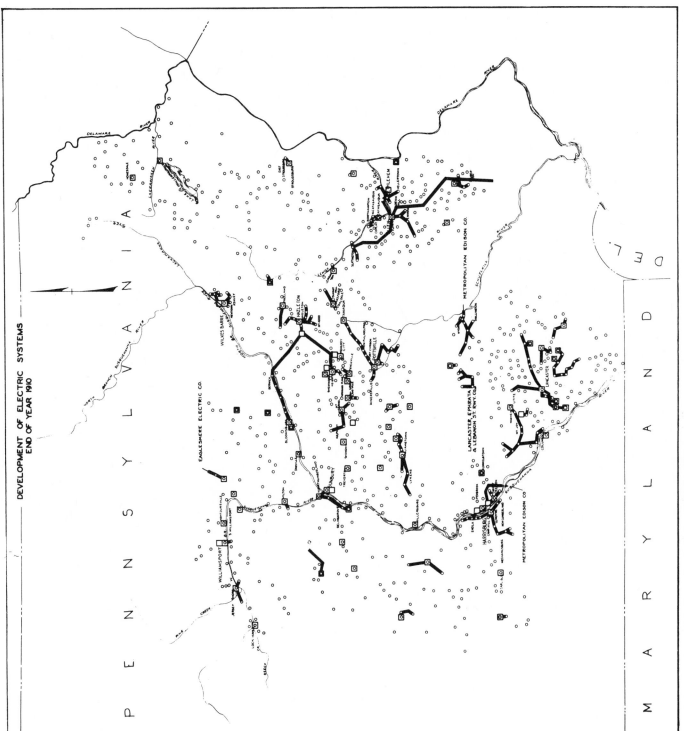

Transmission systems served by Pennsylvania Power & Light, 1910

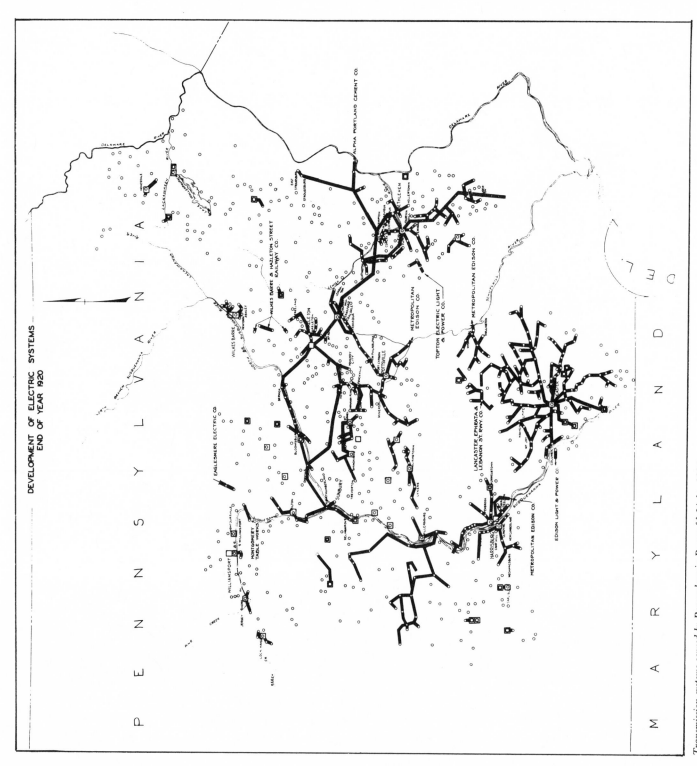

Transmission systems served by Pennsylvania Power & Light, 1920

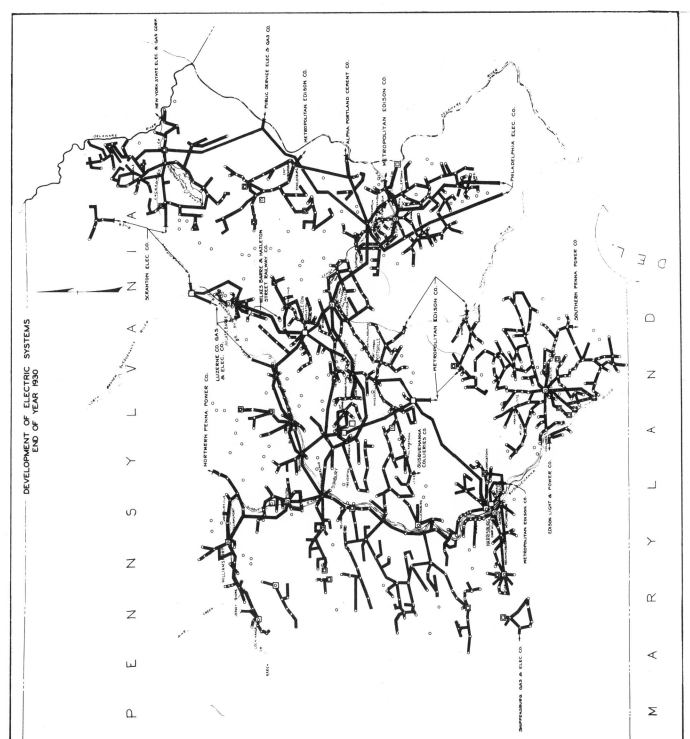

DEVELOPMENT OF ELECTRIC SYSTEMS
END OF YEAR 1930

Transmission systems served by Pennsylvania Power & Light, 1930

comparative one involving developments in three different countries over a period of fifty years. The problem of organization was further complicated by the necessity of selecting representative power systems from different regions for different phases of the history. There were thousands of independent utility systems to choose from. An explanation of each selection will be given in the body of the text; here the overall structure of the history will be outlined.

Although the electric power systems described herein were introduced in different places and reached their plateaus of development at different times, they are related to one another by the overall model of system evolution that structures this study at the most general level. The model has phases, and dominant characteristics are shown to emerge during each; in addition, the model identifies the particular capabilities and interests of the professionals who presided over system growth in each of the phases. In the first phase, the invention and development of a system are considered. The professionals playing a predominant role during this phase are inventor-entrepreneurs, who differ from ordinary inventors in that the former preside over a process which extends from the inventive idea through development to the time when the invented system is ready to be used. Engineers, managers, and financiers also are involved in this first stage, but they do not preside over the system's growth until later phases.

The second phase of the model directs attention to the process of technology transfer from one region and society to another. The transfer of the Edison electric system from New York City to Berlin and London is a case in point. The sites are specific, but general observations about the transfer process can be made. During this phase the agents of change are numerous; they include inventors, entrepreneurs, organizers of enterprises, and financiers.

The essential characteristic of the third phase of the model is system growth. As noted earlier, the historian is responsible for analyzing growth, and analyzing the growth of systems is a particularly interesting and difficult challenge. The method of growth analysis used in this study involves reverse salients and critical problems. Because the study unit is a system, the historian finds reverse salients arising in the dynamics of the system during the uneven growth of its components and hence of the overall network. In labeling such areas of imbalance "reverse salients," the author has borrowed from military historians, who delineate those sections of an advancing line, or front, that have fallen back as "reverse salients." The metaphor is appropriate because an advancing military front exhibits many of the irregularities and unpredictable qualities of an evolving technological system. In the case of a technological system, inventors, engineers, and other professionals dedicate their creative and constructive powers to correcting reverse salients so that the system can function optimally and fulfill system goals.

Having identified the reverse salients, the system tenders can then analyze them as a series of critical problems. Defining reverse salients as critical problems is the essence of the creative process. An inventor or applier of science transforms an amorphous challenge—the backwardness of a system—into a set of problems that are believed to be solvable. Engineers in particular are known for their ability to define solvable problems. The inventor's or engineer's confidence that the reverse salient can be corrected increases dramatically once the problems are defined, because the articulation of a problem often implies its solution.

When engineers correct reverse salients by solving critical problems, the system usually grows if there is adequate demand for its product. On occasion, however, a critical problem cannot be solved. For instance, the first of the major types of electric systems, direct current, had a reverse salient in that it was uneconomical to transmit. Despite precise definitions of the problem, the direct-current inventors and engineers could not in the 1880s find a solution. As a result, other inventors found a solution outside the d.c. system, and for a time the two systems were in conflict. After a compromise was worked out, the two systems existed in a complementary way until the newer system became the dominant one. Thus, this study offers an explanation not only of the evolution of systems as reverse salients are identified and solved, but also of the occasional emergence of new systems out of the failure to solve critical problems in the context of the old.

As a system grows, it acquires momentum. The fourth phase of the system model is characterized by substantial momentum. A system with substantial momentum has mass, velocity, and direction. In the case of technological systems, as defined in this study, the mass consists of machines, devices, structures, and other physical artifacts in which considerable capital has been invested. The momentum also arises from the involvement of persons whose professional skills are particularly applicable to the system. Business concerns, government agencies, professional societies, educational institutions, and other organizations that shape and are shaped by the technical core of the system also add to the momentum. Taken together, the organizations involved in the system can be spoken of as the system's culture. A system with such mass usually has a perceptible rate of growth or velocity. Often the rate accelerates. A system usually has a direction, or goals. The definition of goals is more important for a young system than for an old one, in which momentum provides an inertia of directed motion.

In the case of electric power systems, the institutions that presided over and were influenced by them most directly were the utilities, both public and private. From about 1890 until World War I, the major electric power utilities in the United States, Germany, and England concentrated on supplying the most heavily populated and industrialized urban centers. The decisions made by the utilities' managers during this period shaped the character of the systems more obviously than did the decisions of inventors and engineers, whose solutions to critical problems of a technical kind had cleared the way for growth through the creation of a universal system of supply. With increasing frequency during the two decades before the Great War, the utilities found themselves confronting other institutional contenders for authority over economic development and social change. The tension between the utilities and political institutions such as local governments was high during this phase of systems development. In this instance, however, a *modus vivendi*, if not a lasting arrangement, was found by the contending powers. Three chapters in this study have thus been devoted to an examination of the evolution of the electric power systems in three major cities: Berlin, Chicago, and London.

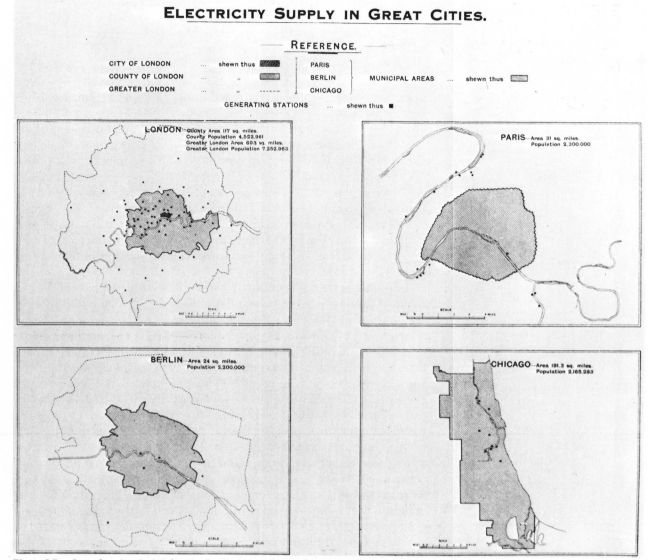

Figure I.5. *Central stations in London, Berlin, Paris, and Chicago, c. 1920. From the County of London Electric Supply Co., Ltd.,* Public Inquiry Held by the Electricity Commissioners in Connection with Application for Consent for the Erection of a Power Station at Barking. *Courtesy of NESCO, Newcastle upon Tyne, England.*

Despite the momentum of systems and the inertia of motion, however, contingencies push systems in new directions. To demonstrate this phenomenon, this study explores the impact of World War I on electric power systems. The engineers and managers who presided over these systems were persuaded by political and military leaders and by public pressure to attenuate their customary drive for autonomous growth and profit and to emphasize the cooperative production of energy. Assigning energy production a higher priority than either profit or organizational autonomy led to new managerial and engineering policies for the duration, and some of these survived the war. The essential point, however, is not the particular instance of war as a contingent and shaping force; rather, it is the possibility of external forces redirecting high-momentum systems.

The last phase of system history delineated by this study is characterized by a qualitative change in the nature of the reverse salients and by the rise of financiers and consulting engineers to preeminence as problem solvers. Managers played the leading role during the phase characterized by an increase in momentum. In the newer phase, which involved planned and evolving regional systems, major reverse salients became essentially problems of funding extremely large regional systems and clearing political and legislative ground. Financiers and associated consulting engineers responded effectively to problems of this kind and scale. The phase was also characterized by an increased capability on the part of engineers and managers, especially consulting engineers and managers, to plan new systems and the growth of old ones. In some cases planned systems were financed by government agency entrepreneurs drawing on public funds.

This loosely structured model has been used to bring order and comprehensibility to the myriad events in the history of electric power systems. In fact, utility systems did not evolve according to one strict pattern. Chapter XIV, which describes the different styles of three mature regional systems, demonstrates variations. All three had the same pool of technology to draw from, but because the geographical, cultural, managerial, engineering, and entrepreneurial character of the three regions differed, the power systems were appropriately varied as well (see Fig. I.5). The concept of style suggests that there was—and probably is—no one best way of supplying electricity. Embodied in the different power systems of the world is a complex variation on major themes that keeps the technology from becoming homogeneous and dull and that provides the historian with the challenging task of description and interpretation.

CHAPTER II

Edison the Hedgehog: Invention and Development

QUOTING the Greek poet Archilochus, Isaiah Berlin wrote in *The Hedgehog and the Fox:* "The fox knows many things, but the hedgehog knows one big thing." Hedgehogs, according to Berlin, are those "who relate everything to a single central vision, one system less or more coherent or articulate." Foxes, in contrast, pursue many ends, ends that are "often unrelated and even contradictory." Berlin counted Dante, Plato, Lucretius, Pascal, Hegel, Dostoyevsky, Nietzsche, Ibsen, and Proust among the hedgehogs.[1] Thomas Edison's name should be added to the list.[2]

Edison invented systems, including an electric light system that took form as the Pearl Street generating station and distribution network of the Edison Electric Illuminating Company of New York, now known as the Consolidated Edison Company. Edison focused on one level of the process of technological change—invention—but in order to relate everything to a single, central vision, he had to reach out beyond his special competence to research, develop, finance, and manage his inventions. Because of this organizational, system-building drive, he is known as an inventor-entrepreneur.[3]

Edison was a holistic conceptualizer and determined solver of the problems associated with the growth of systems. The history of Edison system building, therefore, is also a history of ideas and a study of problem solving. Edison's concepts grew out of his need to find organizing principles that were powerful enough to integrate and give purposeful direction to diverse factors and components. The problems emerged as he strove to fulfill his ultimate vision.

As an inventor-entrepreneur, Edison presided over the process of technological change from problem identification to innovation and technology transfer. Creative fulfillment, however, came to him mostly from the in-

---

[1] Isaiah Berlin, *The Hedgehog and the Fox: An Essay on Tolstoy's View of History* (New York: Simon & Schuster, 1953), p. 1.

[2] Parts of this chapter are drawn from Thomas P. Hughes, "The Electrification of America: The System Builders," *Technology and Culture* 20 (1979): 124–61.

[3] For a discussion of the concept of an entrepreneur as one who presides over invention, development, and innovation, see Thomas P. Hughes, *Elmer Sperry, Inventor and Engineer* (Baltimore: The Johns Hopkins Press, 1971), pp. 63–70, 241, 290–95.

ventive act, not from the other phases of technological development. He counted his patents more than his money, at least until his later years, when he began to look to industrialists like Henry Ford as status models. Edison flourished as an inventor-entrepreneur in the late 1870s and early eighties, the period when he was presiding over the invention and introduction of his system of electric lighting. His historical peers were other inventor-entrepreneurs, such men as Robert Fulton, Samuel Morse, and Cyrus Hall McCormick, who, like himself, did not rest until companies (usually those they established) were manufacturing their inventions. Edison formed a number of companies to organize his invention and the introduction of the lighting system: a company for research and development, others for manufacturing components, and another to preside over the operation of the system. In each case, he allied himself with men whose interests and capabilities complemented his own. Persons with legal and financial experience, for instance, compensated for his lack of experience and special aptitude for the complexities of organization and financing. Despite their presence, however, it was Edison, as inventor-entrepreneur, who pulled most of the strings of the complex system. Later in the history of electric lighting and power systems, other entrepreneurs—manager-entrepreneurs and financier-entrepreneurs—took center stage because the most difficult problems blocking the growth of the system became managerial and financial. Inventors and engineers still had roles to play in the history of the evolving light and power systems, but the inventor-entrepreneurs moved on to other newly emerging fields of technology.

Edison's genius lay in his ability to direct a process involving problem identification, solution as idea, research and development, and introduction into use. These phases of change need to be defined, but because the process was, and is, so complex, and because there are so many variations on the central theme, an encompassing, general definition will suffice here. In problem identification, an inventor perceives a situation that can be defined as a problem. The ability to define the situation as a problem implies that a solution is likely to be found. Experienced inventors recognize that many situations cannot be defined as problems, because the state of the technology, availability of funding, or some other factor is not favorable. Idea response is the inventor's effort—active and passive (subconscious perhaps)—to formulate concepts that will solve in his imagination his definition of the problem. An imaginary device is functioning in an imaginary environment. Usually the inventor gathers information as he pursues—or even awaits—ideas. The idea response will become an invention after the idea has been given form. The inventive concepts of Edison and other inventors are often, perhaps usually, visual rather than verbal or mathematical. For this reason, the first expression of an idea often appears as a drawing in a notebook or on a scrap of paper. Subsequently the idea is given form as a mechanical and electrical device or as a chemical process. This invention is then brought by research and development to the stage at which it can be introduced to the market. Research is an information-gathering exercise and can be done by literature search or by scientific experimentation. Development, an important part of the innovation process, often involves the redefinition of the problem, new ideas, and research as the invention is tried in environments that are increasingly like the real-use environment within which the innovation must function. The invention is no longer an imaginary device functioning in the inventor's mind. It is important to add that the innovation process is not straightforward; it involves backtracking to identify new subproblems, elicit additional ideas, and make new subinventions.[4]

The identification of a problem by experienced inventor-entrepreneurs like Edison usually involved bridging the gap between resources and demand. The professional identified a demand, either existing or potential, and the available resources that might fill it. The resources included available endowments such as existing technology, capital, labor, and land (natural resources). Having identified the problem of using the resources to meet the demand, the inventor then created the technology, or the idea for the technology, that would make the resources usable in filling the demand. An excellent invention used the available resources efficiently and economically to respond to the demand precisely. The less-than-excellent invention needed to be refined to meet the demand. Not every invention was a response to a demand, actual or anticipated, however; many that were not demand oriented were ingenious utilizations of available resources, including existing technology. The response to available endowments, especially technological ones, is sometimes identified as "technological push" in contrast to "market pull." Edison, like so many professional inventors, acted in response to a combination of the two.[5]

Edison preferred to invent systems rather than components of other persons' systems. During his long career as a professional inventor-entrepreneur, he turned to the invention of systems to such an extent that preference for systems can be identified as a salient characteristic of his approach. The history of several of his major inventions—the quadruplex telegraph, the telephone, the incandescent electric lighting system, magnetic-ore separation, Portland cement, and the storage battery—illustrates the spectrum of his methods.[6] Some of these ventures were successful,

---

[4] These definitions are developed further with illustrative examples in Thomas P. Hughes, "Inventors: The Problems They Choose, the Ideas They Have, and the Inventions They Make," in *Technological Innovation: A Critical Review of Current Knowledge*, ed. P. Kelly and M. Kranzberg (San Francisco, Calif.: San Francisco Press, 1978), pp. 168–82.

[5] The literature on the nature of invention is voluminous, and much of it is written by economic historians, sociologists, and historians of technology. Among the most useful books are Jacob Schmookler, *Invention and Economic Growth* (Cambridge, Mass.: Harvard University Press, 1966); S. C. Gilfillan, *The Sociology of Invention* (Cambridge, Mass.: M.I.T. Press, 1970); and the revised edition of Abbott P. Usher, *A History of Mechanical Invention* (Cambridge, Mass.: Harvard University Press, 1954). An annotated listing of many articles and books on innovation (and invention) can be found in S. H. Cutcliffe, J. A. Mistichelli, and C. M. Roysden, *Technology and Values in American Civilization* (Detroit, Mich.: Gale Research Co., 1980).

[6] Edison's inventive activities are described in detail in many biographies, the quality of which varies greatly. The most recent are Robert Conot, *A Streak of Luck* (New York: Seaview Books, 1979); Ronald W. Clark, *Edison: The Man Who Made the Future* (New York: Putnam, 1977); and Matthew Josephson, *Edison* (New York: McGraw-Hill, 1959). The most thorough on technical matters and adulatory in tone are Frank L. Dyer and Thomas C. Martin, *Edison: His Life and Inventions*, 2 vols. (New York: Harper & Bros., 1910); and the 1929 edition of that work, which was written in collaboration with William H. Meadowcroft and also published by Harper & Bros. The most intimate study of the inventor is the account by Francis Jehl, *Menlo Park Reminiscences*, 3 vols. (Dearborn, Mich.: Edison Institute, 1937–41). Wyn Wach-

some were not. Edison's method was not always the same; it varied with time and according to the problem, as one would expect from a professional. The history of his electric lighting system, however, reveals the essential characteristics of his systems approach.[7]

Edison is most widely known for his invention of the incandescent lamp, but it was only one component in his electric lighting system and was no more critical to its effective functioning than the Edison Jumbo generator, the Edison main and feeder, or the parallel-distribution system. Other inventors with generators and incandescent lamps and comparable ingenuity have been forgotten because they did not carry the process further and introduce a system of lighting.[8]

Why did Edison so often choose to work on systems? If the inventor created only a component, he remained dependent on others to invent or supply other components. The inventor of components could not have the control over innovation that Edison wanted. An apt example of an inventor of components, but not systems, is Joseph Swan (1828–1914), the British inventor of the incandescent lamp. Swan's lamps were incorporated with components invented by others into a system, but in private conversation Swan acknowledged the superiority of Edison's system.[9] Swan cannot com-

---

horst's *Thomas Alva Edison: An American Myth* (Cambridge, Mass.: M.I.T. Press, 1981) relates Edison the cultural hero to American values. See also the exhibit catalogue by Bernard Finn and Robert Friedel, *Edison: Lighting a Revolution* (Washington, D.C.: Smithsonian Institution, 1979). For a brief study of Edison, see Thomas P. Hughes, *Thomas Edison, Professional Inventor* (London: HMSO, 1976).

[7] Biographies of Edison usually include discourses on the Edison method. The chapter "Edison's Method in Inventing" in Dyer and Martin's *Edison* (2: 596–628) is informed and considered, if not critical. G. P. Lathrop ("Talks with Edison," *Harper's New Monthly Magazine* 80 [1889–90]: 425–35) rehearses the familiar, but adds occasional helpful items as he tries to perceive how an inventor invents. M. A. Rosanoff ("Edison in His Laboratory," *Harper's Magazine* 165 [1932]: 401–17) provides a scientist's appraisal which is not sentimental. Richard H. Schallenberg ("The Alkaline Storage Battery: A Case History of the Edison Method," *Synthesis* 1 [1972]: 1–13) generalizes about the Edison method from the case of the alkaline battery. H. M. Paynter of M.I.T. kindly provided me with a copy of his Thurston Lecture "Edison in Retrospect: Experimental Physicist and Systems Engineer," which he delivered at the American Society of Mechanical Engineers' annual meeting in Detroit, Mich. on 13 November 1973. I have also found helpful a paper on Edison's method which was presented by his son Theodore to the M.I.T. Club of Northern New Jersey on 24 January 1969. See also Conot, *Streak of Luck*, pp. 455–72; and Thomas P. Hughes, "Edison's Method," in *Technology at the Turning Point*, ed. William B. Pickett (San Francisco, Calif.: San Francisco Press, 1977), pp. 5–22.

[8] This analysis of Edison's method of inventing systems is taken in part from Hughes, "Edison's Method," pp. 5–22.

[9] See G. P. Lowrey to Edison, 23 October 1881, and C. Batchelor to Edison, 4 October 1881, Edison Archives, Edison National Historic Site, West Orange, N.J. (hereafter cited as EA). The question of priority in the invention of the incandescent lamp has troubled many historians. Among the leading contenders for that priority were Thomas A. Edison and Joseph Swan (of Newcastle upon Tyne). Unfortunately, some of the controversy has swirled around a nebulous concept of what actually was invented. In a recent essay, "Swan's Way: Inventive Style and the Emergence of the Incandescent Lamp," George Wise sensibly focuses on the invention of the carbon filament and discusses the importance of Edison's systems approach to the invention. Wise points out that Edison's estimates of the cost of his system were grossly in error, but contributed conceptually to his invention. I am indebted to Wise for allowing me to see his pre-publication manuscript, scheduled to appear in *IEEE Spectrum* in April 1982.

pare with Edison as an inventor-entrepreneur, but his claim to have invented the practical carbon-filament lamp is comparable to Edison's.

Another reason for Edison's inclination to invent systems was more subtle: he sought the stimulation for inventive ideas which comes from seeing inadequacies in some components revealed by improvements made in others. Imbalances among interacting components pointed up the need for additional invention. By the time each system was ready for use, therefore, it involved many patents. Edison, who rarely articulated his method, said of his invention of an electric lighting system:

It was not only necessary that the lamps should give light and the dynamos generate current, but the lamps must be adapted to the current of the dynamos, and the dynamos must be constructed to give the character of current required by the lamps, and likewise all parts of the system must be constructed with reference to all other parts, since, in one sense, all the parts form one machine, and the connections between the parts being electrical instead of mechanical. Like any other machine the failure of one part to cooperate properly with the other part disorganizes the whole and renders it inoperative for the purpose intended.

The problem then that I undertook to solve was stated generally, the production of the multifarious apparatus, methods and devices, each adapted for use with every other, and all forming a comprehensive system.[10]

The interactions provided structure, or guidelines, for inventive activity. Other inventors also used the systems approach, having, like Edison, experienced its stimulating effect.[11]

Reflection on Edison's method suggests that he used the systems approach in order to employ the reverse salient–critical problems method, but since Edison did not analyze and articulate his approach and method, the historian must interpret the record carefully. In fact, the record shows that other inventors and engineers used the reverse salient–critical problems method during the half-century covered by this study; thus, attributing that method to Edison as well does not seem a far-fetched conclusion (see pp. 33–37 below). As noted earlier, reverse salients are obvious weak points, or weak components, in a technology which are in need of further development. A reverse salient is obvious, and creative imagination is not needed to define it. As will be shown, the nondurability of experimental lamp filaments before 1878 was a reverse salient in incandescent-lamp systems. Edison and many others were aware of the need for inventive activities in this area. In contrast to a reverse salient, then, the definition of critical problems by an inventor does require creative imagination. Critical problems result from the inventor's defining the reverse salient as a problem, or set of problems, that, when solved, will correct the reverse salient. (As will be seen, in Edison's work the reverse salients were often economic in nature; the critical problems, technical.) A systems approach facilitates the use of the reverse salient–critical problems method because reverse salients

[10] The quotation is taken from a photoreproduction of Edison's public testimony. The reproduced pages are numbered 3128–34. The item is on file at the Edison Archives in a folder labeled "Electric Light Histories Written by Thomas A. Edison for Henry Ford, 1926." Edison archivist A. R. Abel is unable to identify the original source of this item.

[11] Elmer Sperry, also a professional, independent inventor, introduced guidance and control systems for ships and airplanes prior to World I. His approach is similar to Edison's. See Hughes, *Sperry*, esp. pp. 51–53, 63–70, 159–61, 290–95.

Figure II.1. Associates in the Edison system: Francis Upton, John Kreusi, and Charles Batchelor. Courtesy of the Edison Archives, Edison National Historic Site, West Orange, N.J.

are observably weak in relationship to other system components, and because, as Edison himself wrote, the improvement of one component in a system will reverberate throughout the system and cause the need for improvements in other components, thereby enabling the entire system to fulfill its goal more efficiently or economically. In other words, the systems approach facilitated the conceptual formulation of Gestalt patterns and the visualization of the incomplete parts of those patterns.

The availability of assistants with a variety of knowledge and skills also stimulated Edison to choose problems that involved a system of components. There were superb mechanics, electricians, chemists, glass blowers, and other skilled persons in the Menlo Park community. After acquiring funding for his electric lighting project in the fall of 1878, Edison employed additional men whose talents were particularly well suited for the project. Of special importance among them was Francis Upton, the mathematician and physicist. Others, however, had been at Edison's side for years. Charles Batchelor, for instance, was an ingenious master craftsman, dexterous and sharp-eyed, and his wide-ranging experimental techniques and mechanical aptitude kept him at Edison's right hand. Batchelor was so closely involved with Edison in all of his work "that his absence from the laboratory is invariably a signal for Mr. Edison to suspend labor."[12] John Kreusi, who was in charge of the Menlo Park machine shop, also played a major role in building the Edison system. Trained in Switzerland as a fine mechanic, he could deftly construct Edison's various designs from nothing more than rough sketches and cryptic instructions. He, like Batchelor, had been with Edison in Newark, New Jersey, before the establishment of the Menlo Park laboratory (see Fig. II.1).[13]

When the electric lighting project entered the development phase, others at Menlo Park worked on various components of the system. Dr. Hermann Claudius, a former officer in the Austrian Telegraph Corps, built simulations of the system with batteries for generators, fine wires for the distribution system, and resistors for the load. Francis Jehl reported that Claudius had at his fingertips Kirchhoff's laws of conductor networks.[14] The names of some of the other pioneers who made it possible for Edison to invent and develop an entire system include John "Basic" Lawson, J. F. Ott, D. A. "Doc" Haid, William J. Hammer, Edward H. Johnson, Stockton Griffin, George and William Carman, Martin Force, and Ludwig Boehm (see Fig. II.2).

These varied talents were supported by a broad array of expensive machine tools, chemical apparatus, library resources, scientific instruments, and electrical equipment.[15] A major reason for the establishment of the Edison Electric Light Company, the patent-holding enterprise, in October 1878 was to acquire funds for additional laboratory equipment. The story

[12] *New York Herald*, 21 December 1879, quoted in Jehl, *Reminiscences*, 1: 393.

[13] Jehl, *Reminiscences*, 1: 54.

[14] Ibid., 2: 545.

[15] For Jehl's description of the scientific instruments, see ibid., esp. 1: 257–70. Robert Friedel, director of the Center for the History of Electrical Engineering, Institute of Electrical and Electronics Engineers, has found no evidence in the Edison record of Edison's having borrowed the Sprengel pump. The surviving record suggests that the pump was developed at Menlo Park. Robert Friedel, personal communication, 1 March 1982.

Figure II.2. Creators of the Edison system, Menlo Park, 1879. Back row, right to left: *A. "Doc" Haid (chemist); Francis Upton (mathematician); Francis Jehl; and Charles Batchelor (master mechanic).* Third row, third from right: *Thomas Edison. Courtesy of the Edison Archives, Edison National Historic Site, West Orange, N.J.*

(perhaps apocryphal) of Edison's borrowing a Sprengel pump from Princeton University to achieve the vacuum needed in his incandescent-bulb experiments is well known. It may have led to the erroneous conclusion, however, that Edison and Upton did not sufficiently appreciate the importance of scientific instruments in experimentation, at least not enough to invest heavily in them. It may also have led to the equally false conclusion that Edison's laboratory was not as well equipped as the laboratories of major universities. In fact, the Menlo Park laboratory probably built a vacuum pump of the latest design. Furthermore, Menlo Park had galvanometers, static generators, Leyden jars, induction coils (including a Rühmkorff coil capable of a 20-cm. spark), batteries, and condensers. Wooden-boxed condensers "were strewn everywhere," for they, along with variable-resistance boxes, were essential apparatus for telegraphy experiments.[16] In addition, the laboratory was equipped with a standard ohm, a Wheatstone bridge, Thomson high- and low-resistance reflecting galvanometers, an astatic galvanometer, and a Helmholtz-Gaugain tangent galvanometer. Edison's experimentation also led to the purchase of other inventors' and manufacturers' apparatus, such as generators, for testing purposes. The Edison laboratory at Menlo Park was probably one of the best electrical laboratories in the world. Moreover, Edison also equipped it, at great expense, as a chemical research laboratory. The expenses and wages for equipment and personnel were substantial. Edison reported in January 1879, about six months after commencing the electric lighting project, that he had expended $35,000 on the project and that operating

[16] Jehl, *Reminiscences*, 1: 228–29, 257–70.

expenses continued at the rate of about $800 each week. By November 1879 more than $50,000 had been spent on the project.[17]

There was a feedback relationship in all this. Edison assembled a community of craftsmen and appliers of science and the tools and scientific instruments they needed in order to work on problems of a systemic nature, and the presence of these men and their apparatus further stimulated him, even constrained him, to choose to invent and develop systems. The existence of a community at Menlo Park and later at Edison's larger laboratory at West Orange, New Jersey, did, however, lead to controversy about the significance of Edison's role and about his dependence on his staff. Francis Jehl insisted in his *Memoirs* (published years later) that Edison was the Napoleon with the master plan.[18] Other historians have interpreted Edison's contribution and method in light of Jehl's memoirs. As a result, Edison emerges as the profoundly imaginative formulator of a grand design that was fulfilled in detail by knowledgeable and skilled assistants. Edison even used insights drawn from scientific law—for example, Ohm's law—to his own advantage. On the other hand, Jehl, in a confidential memorandum, assigned the role of master conceptualizer to Francis Upton, the user of science who invented ingenious solutions to technical problems.[19]

Upton, a young "scholar and gentleman" who was nicknamed "Culture" by Edison, joined Edison's staff at Menlo Park in December 1878 on the recommendation of Grosvenor P. Lowrey, Edison's counsel and business and financial adviser. He became Edison's mathematician and physicist. He had been educated at Phillips Andover Academy, Bowdoin College, Princeton University, and Berlin University, where he attended lectures by the eminent scientist Hermann von Helmholtz. Notes on Helmholtz's lectures on physics during the winter semester of 1877 survive among Upton's papers at the Edison Archives in West Orange, New Jersey.

The Edison who emerges in Jehl's memorandum is remarkably different from the Edison depicted in Jehl's *Reminiscences*.[20] In private, Jehl described Edison as a "pusher" who gave his financial backers confidence that he would find a way to solve the problems. Jehl reluctantly acknowledged that Edison was a genius, but he regarded Upton as the thinker and the conceptualizer of systems. As he put it:

When an abnormal man can find such abnormal ways and means to make his name known all over the world with such rocket-like swiftness, and accumulate such wealth with such little real knowledge, a man that cannot solve a simple equation, I say, such a man is a genius—or let us use the more popular word—a wizard. So was Barnum! Edison is and always was a shrewd, witty business man

[17] Edison to Theodore Puskas, 28 January 1879, in Letter Book E-3407, EA; Lowrey to Edison, 13 November 1879, in folder labeled "Electric Light—General," EA.

[18] Jehl, *Reminiscences*, 1: 216, 362–63; 2: 852–54.

[19] The memorandum is in the William Hammer Collection, Museum of American History, Smithsonian Institution, Washington, D.C. It was enclosed in a letter from Francis Jehl to F. R. Upton, 22 April 1913. In a note on the memorandum, Hammer wrote that years earlier Jehl had written a similar letter to him from Budapest. I am indebted to Robert Friedel of the History Institute of the Institute of Electrical and Electronic Engineers for calling my attention to this item.

[20] Jehl wrote his *Reminiscences* years after these events while employed at the Menlo Park laboratory at Dearborn, Michigan, which had been restored by Henry Ford, a friend and admirer of Edison's.

without a soul, an electrical and mechanical jobber, who well understood how to "whoop things up," whose only ambition was to make money and pose as a sort of fetich for great masses of people that possess only a popular notion of an art, and who are always ready to yap in astonishment at some fire-work display that is blown off for the benefit of mankind. Yellow journals and hedge-writers claim him for their own, while the work of Upton is hidden in the lore of progressive science and research.[21]

Jehl insisted in private, as he did in his memoirs, that the work at Menlo Park was not haphazard experimentation, but in confidence he declared that the group's theoretical insights came from the "proficient and eruditive mind" of Upton. It was Upton who coached Edison about science and its uses in solving technical problems; it was Upton who taught Edison to comprehend the ohm, volt, and weber (ampere) and their relation to one another.[22] Upton, for instance, laid down the principles that governed the project's commercial and economic distribution system and solved the equations that rationalized it. His tables, his use of scientific units and instruments, made possible the design of the system.

Upton's interpretation of his relationship with Edison was far more appreciative of Edison's role. In a memorandum on the Edison method, Upton portrayed Edison as the director of a research and development laboratory.[23] He stressed Edison's power of concentration and single-minded pursuit of an objective. Whatever lay at hand was seized upon and molded to his purpose; occasionally, quite expensive apparatus was ruined because it was available and could be made to serve an immediate purpose. The expense mattered little to Edison during the heat of the quest. His power of concentration also showed itself in his ability to follow a single line of thought as he read through pages of densely packed information. Edison also impressed Upton with his talent for asking original questions. "I can answer questions very easily after they are asked," Upton lamented, "but find great trouble in framing any to answer." Edison posed questions that could be translated into hypotheses, which in turn established the strategy and tactics of experimentation. His questions were often drawn from his doubts about accepted explanations and procedures. According to Upton, Edison never took anything for granted; he always doubted what others thought possible. Sometimes the result was that he found a new way. The questions sometimes flowed from Edison as if he had no control over his thoughts but was intuitively penetrating to the essence of a complex and confusing situation. While waiting for the leading questions to form in his mind or the right experiment to present itself, Edison often passed the time, Upton observed, by idly doing experiments in the general area of his concern, which kept his attention focused on the general problem. If such efforts proved fruitless, he would shift to another subject area and work on another project for a time.

Upton was well versed in calculus, and Edison was notorious for his weakness in mathematics, but the way in which Edison grasped the essence of quantifiable relationships impressed Upton. Upton's task was to reduce

[21] Memorandum in Jehl to Upton, 22 April 1913, pp.10–11, Hammer Collection.
[22] Ibid., p. 4.
[23] Folder labeled "Biographical—Upton, Francis," item E-6285-11, EA.

Edison's notions to equations. Edison's ability to find metaphors that allowed him to draw on what he knew and impose order on what he did not know also was a gift. Upton admired the way in which Edison formulated general objectives, or solutions, and then worked ingeniously toward the end in view.

Further research will not ultimately resolve the complexity and contradictions in the surviving testimony of active, ambitious men like Edison and his lieutenants. The relationships were as multifaceted and involuted as Jehl's analogy of Napoleon-and-staff suggests. Pioneers like Upton enhanced the Edison legend; a few others, like Jehl, in private, cast doubts. Among those who raised questions openly were Nikola Tesla, the inventor of a polyphase power system; W. K. L. Dickson, the inventor of motion picture apparatus; and Frank J. Sprague, a pioneer in electric traction.[24] Their, or their advocates', criticisms generally focus on the alleged failure of Edison, or history, to ascribe credit due them for work done, or inventions made, while they were employed by Edison. Laboratory notebooks and other records do suggest that he was often stimulated by the ideas and achievements of others, inside the laboratory and out. He drew on the ideas of others by means of literature and patent searches. It is also true that Edison seldom singled out assistants to attribute to them particular inventive ideas, but he often acknowledged their general assistance in newspaper interviews and he gave them responsible positions in his manufacturing enterprises.

The tangled connections and contradictory evidence can be clarified somewhat, however, if we view the Edison group as an organic community whose members were functionally, or systematically, related. Edison's role had to be played, as did Kreusi's, Batchelor's, and Upton's. Jehl might have been covertly jealous of the publicity given Edison, even of his stature as a hero, but the myth helped attract financiers who would never have supported Jehl or even Jehl's hero Upton. No doubt Edison's immense prestige also influenced those who were involved in patent litigation. And it is doubtful that the others could have provided the verve and intellectual style that gave Edison the power to inspire most of the Menlo Park community to work long hours at fever pitch. It is understandable, however, that persons functioning effectively in a community would at times reject the ego-constraining role the community asked them to play.

The controversy over Edison's role is clarified somewhat when attention is directed to the systems approach informally used in his laboratory. As entrepreneur and manager of research and development, Edison was responsible for the output of his research-and-development community. Evaluating him simply as an independent inventor is inappropriate. The question should not be Did Edison invent? but rather How did he preside over inventive activity? Biographers have been led astray by focusing on Edison's patents and their priority. In truth, assigning patents to Edison was probably in many instances partially a tactical device used to take advantage of his fame and prestige to impress potential financial backers and even judges

[24] Gordon Hendricks, *The Edison Motion Picture Myth* (Berkeley: University of California Press, 1961); Harriet Sprague, *Frank J. Sprague and the Edison Myth* (New York: William Frederick Press, 1947).

Figure II.3.   Edison's laboratory, Menlo Park, the winter of 1880/81. From Insull, Public Utilities, *facing p. 40.*

in patent litigation. Because many of his inventions were undoubtedly the result of collective endeavors, it would have been fair to assign the patents to several persons in the laboratory including Edison. This, however, is an issue separate from the question of the role he played as a manager and entrepreneur of invention and development. It is interesting to note in this regard that the American inventor Eli Whitney has recently been described as a manager and entrepreneur.[25]

The systems Edison chose were not simply technical ones; they involved economics as well. For example, from the start of his sustained concentration on the electric lighting project in the fall of 1878, Edison and his laboratory assistants analyzed the costs of generating and distributing electricity. Various notebook entries, though they do not provide a chronology and complete record of Edison's inventive ideas and development activities, show the focus of his thoughts on, for instance, the cost of operating the Gramme and Wallace arc-light generators that were acquired for test purposes.[26] From the available literature, Edison and his assistants also ascertained the cost of operating a Jablochkoff arc-lighting system, the invention that had created so much excitement during its display in Paris and other European cities. Other items in the notebooks for the early months of the project also reveal Edison's concern about the cost of copper needed for

[25] See Merritt R. Smith, *Harpers Ferry Armory and the New Technology* (Ithaca, N.Y.: Cornell University Press, 1977); and, especially, idem, "Eli Whitney and the American System of Manufacturing," in *Technology in America*, ed. Carroll W. Pursell, Jr. (Cambridge, Mass.: M.I.T. Press, 1981), pp. 45–61.

[26] Menlo Park Notebook no. 6 (4 December 1878–30 January 1879), pp. 22–30. Edison's laboratory notebooks are part of the Edison Archives on file at the Edison National Historic Site, West Orange, N.J. Hereafter, if only one date is given for a notebook, it is for the first dated entry in the notebook; if two dates are given, the second is for the last dated entry.

distribution and generator wiring.[27] On 8 September 1878 Edison visited the Ansonia, Connecticut, plant of William Wallace, the brass manufacturer, to see his arc-light dynamo. While there, Edison made rough calculations, including fuel costs and transmission losses.[28] He also decided in the early days that a successful lighting system would be one that produced light at a price lower than the cost of gas.[29] Spurred by his vision of an electric lighting system that would be analogous to existing gaslight systems, he began—in 1878, according to his own recollections—collecting

every kind of data about gas; bought all the transactions of the gas engineering societies, etc., all the back volumes of gas journals. Having obtained all the data and investigated gas-jet distribution in New York by actual observations, I made up my mind that the problem of the subdivision of the electric current could be solved and made commercial.[30]

Edison could not reduce general statements about the comparative costs of gas lighting and electric lighting to quantitatively more precise ones until he had developed some of his inventions and made a detailed analysis of the cost of gas lighting as supplied in a particular locality. From the start, he clearly realized, however, that his system would have to be economically competitive, and thus he conceived of the problem to be solved by invention as inseparably technical and economic. He did not set out to invent a lighting system the cost of which would not be considered until it was built. As an economic historian later observed, "From the economist's viewpoint, the most significant aspect of Edison's activities in electric lighting was his concern at every step with economic factors."[31] It would perhaps be more correct to say that he defined problems as econotechnical.

Because technological change involved economic, legal, and legislative factors as well as technical and scientific ones, Edison needed a Grosvenor Lowrey to help him fulfill his objectives as an inventor-entrepreneur. Lowrey guided Edison in matters involving Wall Street, New York City politicians, and patent applications. Edison, however, did not step back, immerse himself in technological and scientific problems, and leave the "politics" to Lowrey; the surviving correspondence shows that Edison played a prominent role in the financial and political scenarios concerning his inventions.

Born in Massachusetts, Lowrey took up the practice of law in New York

[27] See Menlo Park Notebook no. 1 (28 November 1878–24 July 1879), section on wire calculations; and Notebook no. 12 (20 December 1878), pp. 174–75, 232–33. On the cost of operating Jablochkoff candles, see Notebook no. 6, p. 57.

[28] Christopher S. Derganc, "Thomas Edison and His Electric Lighting System," *IEEE Spectrum*, February 1979, p. 50.

[29] Harold C. Passer, *The Electrical Manufacturers, 1875–1900* (Cambridge, Mass.: Harvard University Press, 1953), p. 83.

[30] Testimony of Edison quoted in Thomas C. Martin, *Forty Years of Edison Service, 1882–1922* (New York: New York Edison Co., 1922), p. 9. Contrary to Edison's claim that he undertook investigation of the gaslight industry in 1878, notebook records and other original sources show that no systematic research about the industry was done until the end of 1879 and during 1880, when Edison was planning his first central station in New York City. I am indebted to Paul Israel of the Thomas Edison Papers Project at Rutgers University for this information. Israel is investigating Edison's method in the Menlo Park period. Paul B. Israel, personal communication, 16 February 1982.

[31] Passer, *Electrical Manufacturers*, p. 83.

City and rose to prominence there. He acted as counsel to the U.S. Express Company, Wells Fargo & Company, and the Baltimore & Ohio Railroad. He was also legal adviser to the financial entrepreneur Henry Villard, who became closely associated with Edison. In 1866 Lowrey became general counsel for the Western Union Telegraph Company, a position which brought Edison and Lowrey together in connection with telegraph patent litigation. Lowrey was one of those who persuaded Edison to turn to electric lighting.[32] Having observed the sensational publicity given to the introduction of the Jablochkoff arc light in Paris in 1878, Lowrey urged Edison to enter the field and offered to raise the money Edison needed to expand Menlo Park. Not only did he advise Edison, he often encouraged the inventor. Lowrey promised in 1878 that the income from electric-lighting patents would be enough to fulfill one of Edison's dreams: it would "set [him] up forever . . . [and] enable [him] . . . to build and formally endow a working laboratory such as the world needs and has never seen."[33] (At that time, the only buildings in the Menlo Park complex were the laboratory building, the carpentry shop, and the carbon shed; still to come were a machine shop, library, and office buildings.) Shortly afterward, Edison gave Lowrey a free hand in negotiating the sale of forthcoming electric-lighting patents and establishing business associations and enterprises at home and abroad: "Go ahead, I shall agree to nothing, promise nothing and say nothing to any person leaving the whole matter to you. All I want at present is to be provided with funds to push the light rapidly."[34]

Lowrey had close contacts with the New York financial and political world. His law offices were on the third floor of the Drexel Building— Drexel, Morgan and Company occupied the first floor. Working closely with his long-time friend Egisto P. Fabbri, "an Italian financial genius"[35] and partner of J. Pierpont Morgan's, he obtained the funds for Edison from the Vanderbilts and Drexel, Morgan and Company. His skill and effectiveness in dealing with politicians and political problems is conveyed by the following episode. In December 1880 Lowrey arranged a lobbying extravaganza. The objective was to obtain a franchise allowing the Edison Illuminating Company to lay the distribution system for the first commercial Edison lighting system in New York City. Behind the opposition of some New York City aldermen lay gaslight interests and lamplighters who might be thrown out of work by the new incandescent light. A special train brought the mayor and aldermen to Menlo Park. Arriving at dusk, they saw the tiny lamps glowing inside and outside the laboratory buildings. After a tour and demonstration by Edison and his staff, someone pointedly complained of being thirsty, which was a signal for the group to be led up to a darkened second floor of the laboratory. Lights suddenly went on to disclose a lavish "spread" from the famous Delmonico's. After dinner, Lowrey presented Edison and Edison's case, and in due time the franchise was

[32] Payson Jones, *A Power History of the Consolidated Edison System, 1878–1900* (New York: Consolidated Edison Co., 1940), p. 27; on Lowrey, see p. 161.
[33] Lowrey to Edison, 10 October, 1878, EA.
[34] Edison to Lowrey, 2 October, 1878, EA.
[35] Lewis Corey, *The House of Morgan* (New York: G. H. Watt, 1930), p. 23, quoted in Jones, *History of the Consolidated Edison System*, p. 162.

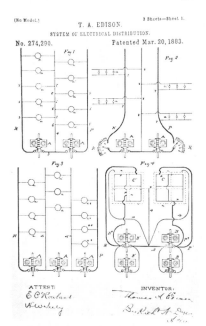

Figure II.4. *System diagrams: Generators in series, lamps in parallel, and three-wire distribution. Patent no. 274, 290 (20 March 1882). Three-wire distribution was not introduced until 1883.*

granted.[36] The franchise was as necessary for commercial success as a well-working dynamo.

Turning to the simple chronological history of the invention and development of Edison's electric lighting system will serve as a reminder that even though Edison's concepts were holistic and his approach was essentially systematic, his day-to-day activities were analogous to the hunting and backtracking of a goal-seeking organism that does not know precisely how to proceed. Edison had taken time off from some of his other projects to experiment with incandescent lamps before the fall of 1878, but it was not until 1878 that he envisaged his goal and organized the electric-lighting project. To suggest that Edison's approach to the invention of electric lighting was systematic does not imply, however, that from the start of the project he conceived of the system in all of its precise technical and economic relationships and quantitative characteristics. What is clear is that he intended from the start of the project to invent not only an incandescent lamp but also related components, such as a distribution network.

The components of the system about which Edison first had inventive ideas in 1878 may have been the distribution network and the incandescent lamp. These ideas, as is usually the case with inventions, were improvements and new combinations. Until Edison's numerous notebooks with their often incomprehensible and scrambled entries are more thoroughly analyzed,[37] the answer to the question of what exactly he discovered or invented in the early fall of 1878 will remain in doubt, but circumstances suggest that the inventions may have been a parallel system of distribution and a filament lamp with a circuit-interrupting device (see Fig. II.4). The design of a circuit-interrupting device in each lamp may have been suggested to him by the design of a tasimeter that he had recently invented to measure the heat of the sun's corona during eclipse.[38] He believed that these ideas could be developed into practical devices. As he wrote to an associate, "Have struck a bonanza in electric light—indefinite subdivision of light."[39]

Edison's early problem identification and inventive ideas were not, however, remarkably original. The subdivision of light had been considered a critical problem ever since it became obvious that arc lamps were too intense for small-space lighting. Subdivision refers to the division of light, or electrical energy, into smaller units. The use of low-intensity, incandescent-filament lamps was seen as a likely answer to this problem because a number of inventors before Edison had designed incandescent lamps of varied characteristics. These lamps, however, were not durable. One historian of the incandescent electric light cites the invention of twenty types of incan-

---

[36] Jehl, *Reminiscences*, 2: 778–85.

[37] At the time of this writing the Thomas A. Edison Papers Project, headed by Professors Reese Jenkins and Leonard Reich, is engaged in the editing and analysis of the Edison collection at West Orange. In cooperation with this project, Dr. Bernard Finn of the Smithsonian Institution and Dr. Robert Friedel of the Institute of Electrical and Electronic Engineers' history center are making an in-depth study using the Edison papers on the invention of the incandescent lamp. Both studies should provide fresh information, understanding, and insight into Edison's method and activities.

[38] Conot, *Streak of Luck*, pp. 116, 120.

[39] Telegram from Edison to Puskas, 22 September 1878, EA. (Edison had written William Wallace to this effect on 13 September.)

descent lamps from 1809 to 1878 (what has been called the precommercial period). One of the inventors, Moses G. Farmer, was the former partner of Wallace, whom Edison visited in September 1878.[40] Farmer had used a platinum-strip incandescent lamp as early as July 1859 and returned to experimentation in 1877; in 1878 he proposed connecting the incandescent lamps in parallel. He also used a circuit-breaking device to cool the filament.[41] Edison may have learned of this on his Ansonia visit; in any event, he had had the opportunity to consider parallel circuits, as contrasted to series connections, in the telegraph and in arc-light and electrochemical battery systems.[42]

His notebooks for the last three months of 1878 and his early electric-light patents support the conclusion that his bonanzas were the durable filament and parallel circuitry. He had decided even earlier to use a platinum filament.[43] Early in October he was experimenting with a platinum filament and a thermostatic device to cool the platinum before it fused. At the same time, he had in mind a parallel circuit.[44] He applied for a patent on a platinum filament with a thermostatic regulator on 5 October 1878 and was granted the patent (no. 214,636) on 22 April 1879.[45] Edison's notebooks and patents suggest that the invention of a thermostatic regulator and the parallel circuit were inextricably linked because the regulator briefly interrupted the flow of current to cool the filament and because, if the lamps had been wired in series, all the lamps would have been extinguished when the regulator of any one of them operated. Thus the interaction of components served as the stimulus for invention.[46]

With their inventive ideas, Edison and his associates embarked on the long and tedious research and experimentation that was needed to develop their general notions into practical devices. Not all of Edison's experiments involved physical apparatus. Many were simply calculations; by using data available in the technical literature and elementary science, such as Ohm's law, Edison and his staff could anticipate phenomena. Because the simple circuit equations and economic calculations used were quantitative, the experiments, both physical and symbolic, were econotechnical (though Edison and his assistants would never have employed such jargon).

Edison announced his brainchild prematurely, with fanfare, in the *New York Sun* on 20 October 1878. He told reporters of his plans for underground distribution in mains from centrally located generators in the great cities; predicted that his electric light would be brought into private houses and substituted for gas burners at a lower cost; and confidently asserted that his central station "[would furnish] light to all houses within a circle

[40] Arthur A. Bright, Jr., *The Electric-Lamp Industry: Technological Change and Economic Development from 1800 to 1947* (New York: Macmillan, 1949), pp. 39–40.

[41] Conot, *Streak of Luck*, p. 120; Bright, *Electric-Lamp Industry*, p. 46; George F. Barker to Edison, 22 November 1878, EA.

[42] Frank L. Pope, *Modern Practice of the Electric Telegraph* (New York: Van Nostrand, 1877), p. 153; Edwin Houston, *A Dictionary of Electrical Words, Terms, and Phrases* (New York: Johnston, 1888), p. 131; Memorandum in Jehl to Upton, 22 April 1913, p. 8, Hammer Collection.

[43] Charles Stowell to Edison, 31 May 1878, EA.

[44] Drawing labeled "Caveat no. 4" and dated 8 October 1878, EA.

[45] Bright, *Electric-Lamp Industry*, pp. 61–64; and Jehl, *Reminiscences*, 1: 235–36.

[46] Edison also saw the possibility of using a shunt circuit for each lamp to make the lamps independent. Robert Friedel, personal communication, 1 March 1982.

of half a mile." He spoke not only of his incandescent lamp but of other envisaged components of his system, such as meters, dynamos, and distribution mains. In fact, he had no generator, no practical incandescent lamp, much less a developed system of distribution—these were at least a year away. He did however, have the concept. As he wrote in private to an associate,

I have the right principle and am on the right track, but time, hard work and some good luck are necessary too. It has been just so in all of my inventions. The first step is an intuition, and comes with a burst, then difficulties arise—this thing gives out and [it is] then that 'Bugs'—as such little faults and difficulties are called—show themselves and months of intense watching, study and labor are requisite before commercial success or failure is certainly reached.[47]

Others also reported in 1878 that Edison had ideas for a lamp and for a parallel wiring system as well as a general outline for his project. Francis Jehl, who joined Edison as a laboratory assistant early in 1879 and who later published reminiscences of the Menlo Park days, believed that in October 1878—twelve months before the construction of a practical incandescent lamp and the announcement of his basic generator design—"Edison had his plans figured out, as a great general figures out his battle strategy before the first cannon is fired."[48] The key to Edison's success, according to Jehl, "lay in his early vision, far in advance of realization."[49]

Assuming that Edison was thinking systematically helps explain why he soon directed his attention away from the durability of the filament to a combination of durability and high resistance. Analysis of system costs led him to identify the need for a high-resistance filament. Ultimately this became the critical problem and the essence of his lamp as a patentable invention. It is not clear when the search for high resistance began, but in a 1926 essay that has been attributed to him, Edison stated that in the fall of 1878 he experimented with carbon filaments, the major problem of which was their low resistance. As he explained it, "In a lighting system the current required to light them in great numbers would necessitate such large copper conductors for mains, etc., that the investment would be prohibitive and absolutely uncommercial. In other words, an apparently remote consideration (the amount of copper used for conductors), was really the commercial crux of the problem."[50] Further evidence for the dating of his high-resistance concept is his statement that "about December 1878 I engaged as my mathematician a young man named Francis R. Upton. . . . Our figures proved that an electric lamp must have at least 100

---

[47] Edison to Puskas, 13 November 1878, EA.
[48] Jehl, *Reminiscences*, 1: 216.
[49] Ibid., p. 217.
[50] Thomas A. Edison, "Beginnings of the Incandescent Lamp and Lighting System" (typescript), p. 4, EA. The typescript is dated 1926 and identified as an item sent to Henry Ford at his request; it should be used cautiously, however, because by 1926 Edison and his patent lawyers had organized history with their own priorities in mind. Also, it follows in many ways the account of the incandescent lamp given in the authorized biography of Edison; see Dyer and Martin, *Edison*, 1: 234–66. The 1926 typescript was written with the help of W. H. Meadowcroft, who later collaborated with Dyer and Martin in the publication of the 1929 edition of the Edison biography. I am indebted to Robert G. Koolakian for this information.

ohms resistance to compete commercially with gas."[51] Edison said that he then tried various metals, including platinum, in order to obtain a filament of high resistance, and that he continued along these lines until about April 1879, when he developed a platinum filament of great promise because the occluded gases had been driven out of it, thereby increasing its infusibility. In his *Reminiscences*, Jehl maintains that Edison wanted a high-resistance lamp as early as October 1878, and that he had reached this conclusion by thinking through his envisaged system of electric lighting.[52] Jehl also states that by applying Joule's and Ohm's laws, Edison arrived at the essentials of his system.

Edison seems to have been refining his inventive ideas for components and his more general ideas for a system throughout the fall of 1878 (see Fig. II.5). His early emphasis on the durable filament and parallel wiring was specific, but his notion of a system involving generators, lamps, and distribution from a central station was vague. He probably first conceived of it as an analogue of central-station, illuminating gas supply. Experimenting with filaments and calculating costs from available data about existing generator and arc-light systems allowed him, and then Upton, to introduce the costs into the system. This perspective revealed that a durable filament and parallel wiring might yield a technically workable system, but not an economically feasible one. As a result of a systematic analysis of technical and economic factors, the high-resistance filament then emerged as the critical problem. Joule's and Ohm's laws, equations that are considered simple today, were of great value to the inventors, for they allowed the characteristics of the components to be introduced as variables and related systematically. The effect of an intended change in one variable (component) on other variables (components) could be anticipated, and actual experiments would test the assumptions and predictions.

Jehl's stress on the importance of Ohm's law, and items in the laboratory notebooks and secondary literature, suggest more about the nature of Edison's reasoning and that of his assistants, especially Francis Upton.[53] As

[51] Edison, "Beginnings," p. 5. Upton stated that Edison had high resistance in mind when Upton joined Edison's staff; see Francis R. Upton's 1918 speech to Edison Pioneers, item X–EG285–13, folder labeled "Biographical—Upton, Francis," EA.

[52] Contrary to his recollection, however, Jehl did not come to Menlo Park until the end of February 1879, "at the earliest." Paul B. Israel, 16 February 1982, personal communication (see note 30 above). Therefore, Jehl's statement should be taken as secondhand evidence.

[53] Jehl (*Reminiscences*, 1: 362–63 and 2: 852–54) stresses Edison's reliance upon Ohm's law. The stress suggested a clue to Edison's reasoning. In a reference often used at Menlo Park (the essay "Electricity" in the *Encyclopaedia Britannica*, 9th ed. [1878], p. 41), Ohm's law is stated as R = E/C (resistance equals electromotive force divided by current). In a misleadingly titled article, Harold C. Passer argues cogently that Edison's reasoning was as Jehl suggested ("Electrical Science and the Early Development of the Electrical Manufacturing Industry in the United States," *Annals of Science* 7 [1951]: 382–92). Passer offers no evidence, however, from the notebooks or other original Edison sources. Passer similarly discusses Edison's reasoning and concepts in *Electrical Manufacturers* (pp. 82, 84, and 89). Dyer and Martin's *Edison* (1: 244–60), parallels the memorandum attributed to Edison in 1926. Josephson's *Edison* (pp. 193–204, 211–220) stresses Edison's use of Ohm's law but does not note the importance of Edison's having used Ohm's law in conjunction with Joule's law in order to conceptualize his system. Josephson also dates Edison's first high-resistance lamp as January 1879 (a platinum filament), but gives no source for the statement (ibid., p. 199); nor does he provide a source for the statement that Edison came to the idea for high resistance in a "flash of inspiration"

Figure II.5. *Creation of the Edison system, Menlo Park. From* Leslie's Weekly *(1880), as reproduced in* Forty Years of Edison Service, 1882–1922 *(New York: Press of the New York Edison Co., 1922), facing p. 21.*

noted, Edison and Upton saw that the cost of copper, especially that which depended on the cross-sectional area and the length of the conductors, was a major variable in the overall cost equation. Large and long conductors would raise the price of electric light above that of gas. To keep conductor length down, Edison sought a densely populated consumer area; to keep the cross-sectional area small, he had to reason further, using the laws of Ohm and Joule. The notebooks show that Edison and Upton used Joule's law (heat or energy = current^2 × resistance = voltage × current) to calculate the amount of energy that would be expended by the incandescent filaments.[54] They also used an adaptation of that law to show energy loss in the conductors. Energy loss was taken to be proportional to the current^2 times the length of the conductor times a constant dependent upon the quality of the copper used, all divided by the cross-sectional area of the conductor (energy loss proportional to C^2La/S).[55] The formula posed an enigma, for if Edison increased the cross section of the copper conductors to reduce loss in distribution, he would increase copper costs, which was to be avoided. Obviously, a trade-off, to use the jargon of the engineering profession, was in order. There was, however, another variable—the current—to consider. If current could be reduced, then the cross-sectional area of the conductors need not be so large. But current was needed to light the incandescents, so how was one to reduce it?

To solve the dilemma, Edison and Upton could have reasoned as follows. Wanting to reduce the current in order to lower conductor losses, they realized that they could compensate and maintain the level of energy transfer to the lamps by raising the voltage proportionately (H = C × V). Then they brought Ohm's law into play (resistance equals voltage divided by current). It was a eureka moment, for they realized that by increasing the resistance of the incandescent-lamp filament they would raise the voltage in relationship to the current. (Resistance was the value of the ratio.) Thus began the time-consuming search for material suitable for a high-resistance filament. The notable invention was this logical deduction; the filament was a hunt-and-try affair.

Edison's reasoning can be illustrated with a simple example using approximate, rounded-off values. In 1879 he obtained a carbonized-paper filament whose resistance ranged from 130 ohms cold to about 70–80 ohms heated.[56] In order to arrive at a lamp whose candlepower was equivalent to that of gas, he found, this filament would require—in present-day units—

---

after 8 September 1878 (ibid., p. 194). Bright's *Electric-Lamp Industry* does not offer any additional information on the way in which Edison conceived of his system. Jehl's *Reminiscences* remains the most suggestive published source, despite the book's lack of organization; see 1: 214–15, 243–45, 255–56, and 2: 820–21, 852–54.

[54] Menlo Park Notebook no. 3 (21 November 1878), p. 107, and Notebook no. 9 (15 December 1878–10 March 1879), p. 41, contain some of the many early entries involving Ohm's law. In Notebook no. 6, pp. 11 ff., Joule's law is mentioned; the dates in this notebook (4 December 1878–30 January 1879, indicate that Edison was using Joule's law early in his electric lighting project. In Notebook no. 10 (December 1878–January 1879), the word "Joule" is jotted down where the equation H = C^{2}R is used.

[55] Menlo Park Notebook no. 12, pp. 174–76.

[56] Menlo Park Notebook no. 52 (31 July 1879), p. 229; the entry is dated 15 December 1879.

the equivalent of about 100 watts. This meant that the product of the voltage across the lamp and the current must equal 100 watts. Since the resistance was 100 ohms, the current had to be 1 amp, for according to Joule's law the heat energy was equal to the product of C^2 and the resistance ($100 = C^2 \times 100$; $C = 1$). It then followed that the voltage must be 100 in order to fulfill the energy need ($H = C \times V$; $V = 100/1 = 100$). Therefore, the approximate specifications of the lamps in Edison's system in today's terminology were 100 watts, 100 volts, 1 amp, and 100 ohms.[57]

Another possible explanation for the decision to seek high resistance, and a complementary one, was the search for a platinum filament with a small surface of radiation. The objective was to maintain the incandescent material of the filament at a high temperature with a low expenditure of energy. To do this Edison shaped the filament into a spiral, thereby reducing the radiating surface as well as the size of the filament.[58] The experimentation with small surfaces of radiation resulted in higher-resistance filaments. With Francis Upton's help, Edison could then have ascertained the effect of higher resistance on the entire circuit. They could have serendipitously discovered the advantages of high resistance.

The funds raised in part by Lowrey for personnel and equipment allowed the simultaneous invention and development of other components of the system besides the incandescent lamp. Upton, for instance, was especially active in the design and construction of an appropriate generator. The generators on the market—for instance, the Siemens, Wallace, and Gramme— were designed for arc lights wired in series. A network of feeder wires and distribution mains for the distribution network also received considerable attention. The inventive ideas embodied in the various versions of the components of the system resulted in a large family of patents covering the Edison system.

In the fall of 1879, about the same time that Edison and his assistants obtained a filament with the high resistance and durability desired, Menlo Park announced impressive progress in the design of an incandescent-lamp generator. On 18 October 1879 *Scientific American* carried an article—unsigned, but written by Upton—describing the generator. In contrast to the earlier, high-resistance generators of other inventors, the Edison generator had low internal resistance. The article claimed that it would obtain nearly double the foot-pounds of energy from the same input that earlier generators had achieved, or an efficiency of 90 percent. Although original in terms of the specifics that were designed for the incandescent system, the generator was based on experiments done at Menlo Park with the gener-

[57] In January 1881 Edison conducted an economy test of the electric lighting system installed at Menlo Park to demonstrate its practicability. This was a prelude to his installation of a full-size system at the Pearl Street station in New York City in 1882. In the test he used two lamp sizes, 16 c.p. and 8 c.p. The 16-c.p. lamp depended on an electromotive force of 104.25 volts and a resistance of 114 ohms; therefore, the current in the 16-c.p. lamp was .9 amps. See C. L. Clarke, "An Economy Test of the Edison Electric Light at Menlo Park, 1881," dated 7 February 1881, and published for the first time in Committee on St. Louis Exposition of Association of Edison Illuminating Companies, *Edisonia: A Brief History of the Early Edison Electric Lighting System* (New York: Association of Edison Illuminating Companies, 1904), pp. 166–78.

[58] Derganc, "Edison and His Electric Lighting System," pp. 55–56.

ators of Gramme, Wallace, and others. In fact, the publicity for the Edison generator exaggerated its excellence compared with other inventors' generators. The Edison generator was designed for a system of incandescent lamps connected in parallel and, therefore, a system with low resistance. It correctly had a low internal resistance to match the low external resistance. To compare it with a high-resistance arc-light generator designed for a system of arc-lights connected in series, however, was not appropriate. The Edison generator was not designed for arc lights in series. Still, the claims made about the generator excited much professional and public reaction, as was usually the case with the well-publicized Edison inventions, and Edison should be given credit for seeing the need to design components systematically in relationship to others in the system—a truth other inventors failed to grasp.[59]

Believing late in 1879 that the essential design of the incandescent lamp and generator would prove practical, Edison and his associates shifted their emphasis to development of the components and the system. This meant increased stress on economic factors. As early as 1880 they were calculating the cost of a central station of 10,000 lamps, possibly in anticipation of the planned central station for New York City's Pearl Street. They had experimented enough to estimate that 1 h.p. from a steam engine and generator could supply eight 16-c.p. lamps. Therefore they needed about 1,200 h.p. for their system. The calculations that followed this assumption involved fixed capital, operating expenses, and other costs (see Fig. II.6). Finally, assuming that the 10,000 lamps would be used five hours out of every twenty-four, and knowing what gas companies charged for five hours of equivalent light, they calculated the surplus income of an electric utility charging the same price for light. Edison and the parent Edison company holding the patents would receive payment from a surplus income of $90,886.[60]

As development proceeded, Edison advanced the level of experimentation from components to laboratory-scale models of the system and then to a small, pilot-scale system for lighting Menlo Park in December 1879. Because the components were new and because Edison wanted to control their manufacture, Lowrey assisted in the formation of several companies. Edison had to draw heavily on his own resources to fund these manufacturing companies. Edison and Lowrey also founded an operating company to build a central-station system that would both function commercially and serve as a demonstration for potential franchise purchasers.

The organization and management of the companies formed by Lowrey and Edison in connection with the electric lighting system have been de-

---

[59] For the reaction of the scientific community to Edison, Edison's inventions, and especially the claims made for the generator, see David Hounshell, "Edison and the Pure Science Ideal in Nineteenth-Century America," *Scientific American*, February 1980, p. 614. More general reactions are analyzed in Jean-Jacques Salomon, "Public Reactions to Science and Technology: The Wizard Faces Social Judgment," in *Science, Technology and the Human Prospect*, ed. C. Starr and P. Ritterbush (New York: Pergamon Press, 1980), pp. 77–93. I am indebted to Dr. Thomas Simpson of St. John's College, Sante Fe, New Mexico, for his views in comparing arc-light and incandescent-lamp generators.

[60] For a discussion of this cost analysis, see Hughes, "Electrification of America," pp. 133–35.

Capital Investment:			Depreciation	
Power plant building		$ 8,500	2%	$ 170
Boilers and auxiliary equipment		30,180	10%	3,018
Steam engines and dynamos		48,000	3%	1,440
Auxiliary electrical equipment		2,000	2%	40
Conductors		57,000	2%	1,140
Meters		5,000	5%	250
Total		$150,680		$6,058
Operating and Other Expenses:				
Labor (Daily):				
Chief engineer	$ 5.00			
Assistant engineer	3.00			
Wiper	1.50			
Principal fireman	2.25			
Assistant fireman	1.75			
Chief voltage regulator	2.25			
Assistant voltage regulator	1.75			
Two laborers	3.00			
Total	$20.50			
Labor (Annual)		$ 7,482		
Other:				
Executive wages (annual)		$ 4,000		
Rent, insurance and taxes		7,000		
Depreciation		6,058		
Coal (annual)		8,212		
($2.80/ton; 3#/h.p. hour; 5 hours daily; 1,200 h.p.)				
Oil, waste, and water		2,737		
Lamps (30,000 at 35c each)		10,500		
Total		$45,989		
Estimated Minimum Income from 10,000 Installed Lamps			$136,875	
Expenses			− 45,989	
			$ 90,886	

scribed well elsewhere.[61] Here it is important to stress that the pristine character of the companies manifested Edison's determination to create a coherent system. The first company—the Edison Electric Light Company (EELC)—was formed on 15 October 1878 essentially to fund Edison's invention, research, and development projects and to bring a return on his investment through the sale or licensing of patents on the system throughout the world. The Edison Electric Illuminating Company of New York, a utility, or operating, company, was incorporated in December 1880 as a licensee of the parent Edison Electric Light Company. Under Edison's personal supervision, the EEIC would build the central generating station on Pearl Street in New York City which began commercial operation in September 1882. In 1881 Edison also established in New York City the Edison Machine Works to build dynamos and the Edison Electric Tube Company to manufacture underground conductors. The Edison Lamp Works, organized in 1880, turned out incandescent lamps in quantity (see Fig. II.7). Edison also entered into a partnership with Sigmund Bergmann, a former Edison employee, to set up a company to produce various acces-

[61] Passer, *Electrical Manufacturers.*

Figure II.7.   *The incandescent-lamp factory, Menlo Park, 1880, part of the Edison manufacturing system. Executive staff* (front row, left to right): *Philip Dyer, William Hammer, Francis Upton, and James Bradley. Courtesy of the Edison Archives, Edison National Historic site, West Orange, N.J.*

sories. As engineer and manager, Edison was the pivotal figure in all of these companies during the early years, but for him the focus and the commitment remained invention (see Fig. II.8).

The history of the design and construction of the Pearl Street Station of the Edison Electric Illuminating Company has been told by many.[62] One especially serious engineering problem encountered in the construction of the station was the laying down of the distribution system underground. As noted earlier, the cost of the distribution system was a critical problem

---

[62] Of the secondary works, Matthew Josephson's *Edison* tells the story with unusual clarity and interest and includes an extensive bibliography of articles about the historic station. The Edison Pioneers have recounted their experiences and other authors have celebrated the achievement on various anniversaries of the station's opening. Essays by Edison's contemporaries can be found, for example, in the Association of Edison Illuminating Companies' *Edisonia*. Especially valuable because of its extensive reprinting of early source material and information on the Pearl Street station and the Edison Electric Illuminating Co. is Payson Jones's *History of the Consolidated Edison System*. The Consolidated Edison Co. of New York is a successor company to the EEIC.

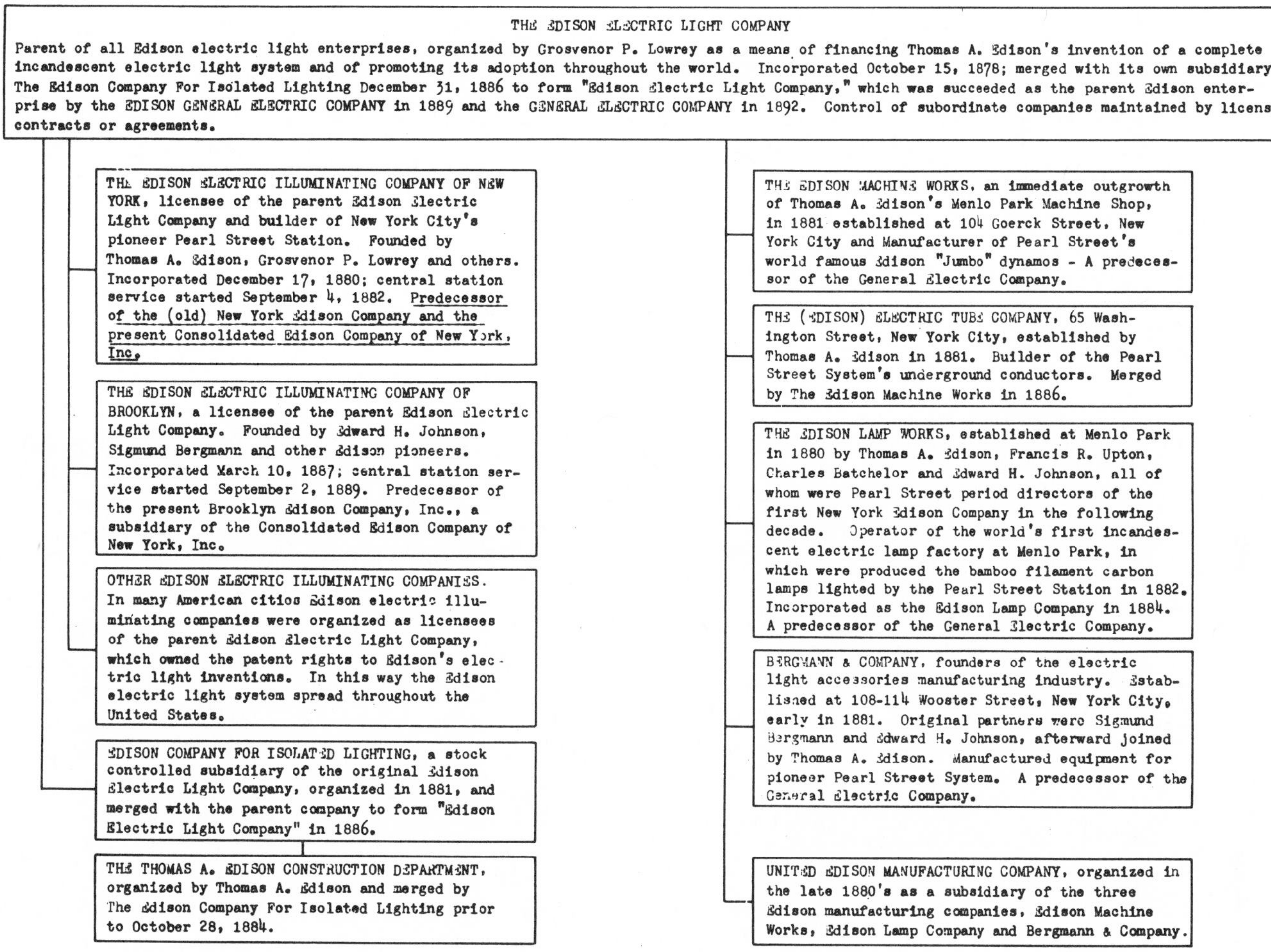

THE EDISON ELECTRIC LIGHT COMPANY

Parent of all Edison electric light enterprises, organized by Grosvenor P. Lowrey as a means of financing Thomas A. Edison's invention of a complete incandescent electric light system and of promoting its adoption throughout the world. Incorporated October 15, 1878; merged with its own subsidiary The Edison Company For Isolated Lighting December 31, 1886 to form "Edison Electric Light Company," which was succeeded as the parent Edison enterprise by the EDISON GENERAL ELECTRIC COMPANY in 1889 and the GENERAL ELECTRIC COMPANY in 1892. Control of subordinate companies maintained by license contracts or agreements.

THE EDISON ELECTRIC ILLUMINATING COMPANY OF NEW YORK, licensee of the parent Edison Electric Light Company and builder of New York City's pioneer Pearl Street Station. Founded by Thomas A. Edison, Grosvenor P. Lowrey and others. Incorporated December 17, 1880; central station service started September 4, 1882. Predecessor of the (old) New York Edison Company and the present Consolidated Edison Company of New York, Inc.

THE EDISON ELECTRIC ILLUMINATING COMPANY OF BROOKLYN, a licensee of the parent Edison Electric Light Company. Founded by Edward H. Johnson, Sigmund Bergmann and other Edison pioneers. Incorporated March 10, 1887; central station service started September 2, 1889. Predecessor of the present Brooklyn Edison Company, Inc., a subsidiary of the Consolidated Edison Company of New York, Inc.

OTHER EDISON ELECTRIC ILLUMINATING COMPANIES. In many American cities Edison electric illuminating companies were organized as licensees of the parent Edison Electric Light Company, which owned the patent rights to Edison's electric light inventions. In this way the Edison electric light system spread throughout the United States.

EDISON COMPANY FOR ISOLATED LIGHTING, a stock controlled subsidiary of the original Edison Electric Light Company, organized in 1881, and merged with the parent company to form "Edison Electric Light Company" in 1886.

THE THOMAS A. EDISON CONSTRUCTION DEPARTMENT, organized by Thomas A. Edison and merged by The Edison Company For Isolated Lighting prior to October 28, 1884.

THE EDISON MACHINE WORKS, an immediate outgrowth of Thomas A. Edison's Menlo Park Machine Shop, in 1881 established at 104 Goerck Street, New York City and Manufacturer of Pearl Street's world famous Edison "Jumbo" dynamos - A predecessor of the General Electric Company.

THE (EDISON) ELECTRIC TUBE COMPANY, 65 Washington Street, New York City, established by Thomas A. Edison in 1881. Builder of the Pearl Street System's underground conductors. Merged by The Edison Machine Works in 1886.

THE EDISON LAMP WORKS, established at Menlo Park in 1880 by Thomas A. Edison, Francis R. Upton, Charles Batchelor and Edward H. Johnson, all of whom were Pearl Street period directors of the first New York Edison Company in the following decade. Operator of the world's first incandescent electric lamp factory at Menlo Park, in which were produced the bamboo filament carbon lamps lighted by the Pearl Street Station in 1882. Incorporated as the Edison Lamp Company in 1884. A predecessor of the General Electric Company.

BERGMANN & COMPANY, founders of the electric light accessories manufacturing industry. Established at 108-114 Wooster Street, New York City, early in 1881. Original partners were Sigmund Bergmann and Edward H. Johnson, afterward joined by Thomas A. Edison. Manufactured equipment for pioneer Pearl Street System. A predecessor of the General Electric Company.

UNITED EDISON MANUFACTURING COMPANY, organized in the late 1880's as a subsidiary of the three Edison manufacturing companies, Edison Machine Works, Edison Lamp Company and Bergmann & Company.

Figure II.8. *Edison's manufacturing system. From Jones,* History of the Consolidated Edison System, *p. 13. Courtesy of the Consolidated Edison Co. of New York.*

that Edison recognized early in the development of the system, and thus, inventive talent and developmental skills were heavily invested in the problem. Subsequently, other utilities also found that distribution and transmission were problems of the first order, the popular attention given to lamps and generators notwithstanding. The electrical network is, after all, the essence of the system.

Edison's ultimate objective was to introduce central-station supply. A central station would distribute electric light to the public, in contrast to generating plants, or isolated stations, which would be used only by their owners. The steam boilers, steam engines, generators, and auxiliary equipment would be housed in the station building itself; from the central station the distribution system would, in the case of New York City, fan over an area of one square mile. Edison chose 257 Pearl Street, in the financial district, as the location of the New York City station. So located, the central station would supply restaurants and shops that could afford the new light as a way of attracting customers and would illuminate workplaces and offices that could afford it as an unusually effective light without hazardous

Figure II.9.   *Thomas Edison, 1882.*
Courtesy of the Edison Archives, Edison
National Historic Site, West Orange, N.J.

and noxious fumes. In the Wall Street district, the station would also catch the attention of financiers and the investing public, persons who were needed to fund Edison stations elsewhere.

In 1882 Edison was in New York supervising construction of the Pearl Street system. As chief engineer, he spent long hours, week after week, not in an office, but out with the workers sweating in the hot summer sun as they wrestled with the new and difficult task of laying the underground cables for the Pearl Street station. Edison did not stop at supervision; he worked in the dug-out trenches with the laborers, responding to the most minute problems, many of which arose from the difficulties of maintaining adequate insulation. Often after a frantic day unregulated by the clock, he would sleep for only a few hours on piles of tubes stacked in the station building, his bed place softened only by his overcoat. Invention continued as problems arose during the development and engineering phases of construction. Edison applied for 60 patents in 1880, 89 patents in 1881, and 107 in 1882. Most covered inventions pertaining to the electric lighting system. Only once again in his lifetime, in 1883, a year in which he applied for 64 patents, would the total number of his patents reach the level applied for during the three years he dedicated to electric lighting.[63]

On 4 September 1882 John W. Lieb, an Edison electrician, threw the switch that fed current from a Jumbo generator in the Pearl Street station to the first of the lamps installed in the district. The generators bore the name of the great elephant brought to America by P. T. Barnum. Edison switched on the lamps in 23 Wall Street, the offices of Drexel, Morgan and Company. The single generator was driven by a directly connected Porter-Allen engine supplied with steam from Babcock & Wilcox boilers (the two components of the system which were unusual in that they had not been built by Edison). Initially, only about one-third of the one-square-mile district to be supplied from the station had current and only four hundred lamps were lit. In place, but not operating the first day, were five other Jumbos, steam engines, and boilers. The noncondensing Porter-Allen steam engines were rated at 125 nominal h.p., weighed about 6,500 pounds, took steam at 120 pounds per square inch of pressure, and ran at 350 revolutions per minute. They were designed to light twelve hundred 16-c.p. lamps. When cooled with an air blast, the machines could supply 850 amps at 115–120 volts. The generators weighed about seven times more than the steam engines driving them.

Control and regulation, vital functions in this and subsequent supply systems, have been overlooked by many historians, but they were not overlooked by Edison and his associates (see Fig. II.10). Voltage regulation was obtained by inserting and removing resistance from the field circuit of the generators. An automatic indicator utilizing an electromagnet connected across the main circuit indicated the voltage level. If the voltage was within one or two volts of the desired amount, the indicator arm, the armature of the electromagnet, remained in a neutral position in balance between the pull of the electromagnet and the pull of a delicately adjusted spring. If the line voltage rose, the armature would swing under the influence of

[63] Arthur E. Kennelly, "Thomas Edison," in National Academy of Sciences, *Biographical Memoirs* (Washington, D.C., 1943), p. 301.

the stronger magnetic field toward the electromagnet and close a contact, thereby lighting a red lamp; if the line voltage dropped, the attraction of the electromagnet weakened and the spring exerted the greater pull, causing the armature to reverse its swing and to close the low-side contacts and light a blue lamp. The attendant accordingly turned a handwheel to control the amount of field resistance. To devise such a control to match output and load was simple for Edison, for electromagnetic devices, like relays, were the essence of the telegraph he knew so well.[64]

Edison's station was noted for its handling of material and for its economical engineering. An uninterrupted flow of material and energy would later characterize power plant engineering. A 20-h.p. engine drove a screw conveyor for lifting coal from the vault under the building to a position above the boilers where the coal dropped by gravity into the stoke hole of the boiler. Another screw conveyor took ashes from beneath the grates and discharged them into a container under the sidewalk. The exhaust from the steam engines passed through feed-water heaters and then into the atmosphere; a fan forced combustion air into the furnace and cooling air onto the generator commutator. Under full headway, the four boilers, rated at 240 h.p. each, would consume about five tons of coal and 11,500 gallons of water per day (see Fig. II.11).

Initially, the most serious problem arose when two generator sets were connected in parallel. The governors of the two steam engines began to hunt, first cutting off and then providing steam for full power. Edison found an ingenious means of temporarily correcting the anomaly by mechanically connecting the governors of the steam engines. In November 1882, however, two Armington & Sims engines were substituted for the Porter-Allen engines, and when these operated successfully with generators connected in parallel, all of the Porter-Allen engines were replaced. It is interesting to note that this difficulty involved a component that was not designed and built by Edison for the system.

Countless other problems plagued the station during the breaking-in period. Current leaked from conductors and junction boxes under the streets. In addition, there were fires from faulty wiring. Slowly, however, service increased. On 1 October 1882, a few weeks after service began, wiring was in place for 1,626 lamps in the district and 1,284 were in use; on 1 October the following year there was wiring for 11,555 lamps and 8,573 were in use. The Pearl Street station remained in service until 2 January 1890, when a disastrous fire shut down the original system. Then, on 12 January of that year, a reconstructed station began supplying current from Pearl Street. On 1 April 1894 the station was retired and the building was sold. Then only a plaque marked the historic site where Edison demonstrated so effectively his system for generating, controlling, distributing, measuring, and utilizing electricity for lighting.

If the Pearl Street station had not been part of a business as well as a technological system, however, it might not have survived until the fire of 1890. A similar station in London was abandoned several years after its founding, for financial and political reasons (see pp. 55–62 below). The

---

[64] This account of the station and its equipment is based in part on Hughes, *Thomas Edison*, pp. 29–36.

Figure II.10. *Wiring system for the Pearl Street central station. From Association of Edison Illuminating Companies,* Edisonia, p. 70.

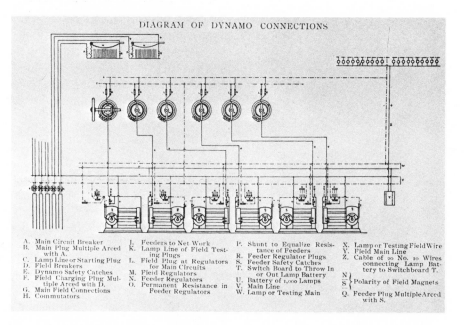

DIAGRAM OF DYNAMO CONNECTIONS

A. Main Circuit Breaker
B. Main Plug Multiple Arced with A.
C. Lamp Line or Starting Plug
D. Field Breakers
E. Dynamo Safety Catches
F. Field Charging Plug Multiple Arced with D.
G. Main Field Connections
H. Commutators

J. Feeders to Net Work
K. Lamp Line of Field Testing Plugs
L. Field Plug at Regulators for Main Circuits
M. Field Regulators
N. Feeder Regulators
O. Permanent Resistance in Feeder Regulators

P. Shunt to Equalize Resistance of Feeders
R. Feeder Regulator Plugs
S. Feeder Safety Catches
T. Switch Board to Throw In or Out Lamp Battery
U. Battery of 1,000 Lamps
V. Main Line
W. Lamp or Testing Main

X. Lamp or Testing Field Wire
Y. Field Main Line
Z. Cable of 20 No. 10 Wires connecting Lamp Battery to Switchboard T.

$\frac{N}{S}$ } Polarity of Field Magnets

Q. Feeder Plug Multiple Arced with S.

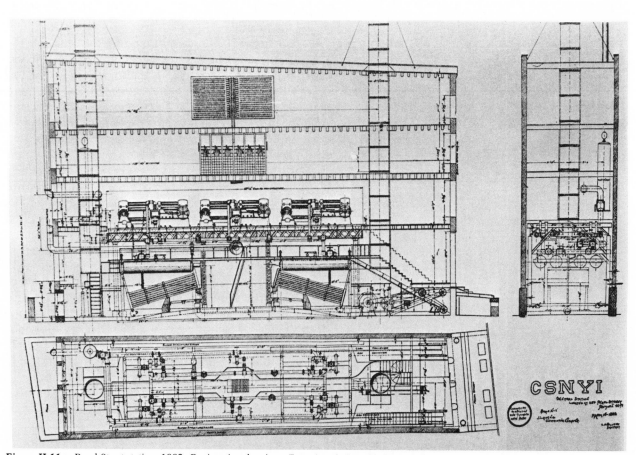

Figure II.11. *Pearl Street station, 1882: Engineering drawings. From Association of Edison Illuminating Companies,* Edisonia, p. 64.

Pearl Street station and the Edison Electric Illuminating Company (EEIC) were sustained not only because of the capital invested in the plant and the company but also because the other companies of the Edison system depended on them as showpieces and inducements for others to purchase Edison franchises and Edison equipment. This part of the early Edison history is rarely mentioned in the Edison lore; Edison and his associates were reluctant to release any detailed operating figures at the time.[65] The annual reports of the parent Edison Electric Light Company and correspondence now available, however, provide evidence of the EEIC's near collapse.

The EEIC did not charge customers for electricity in 1882 and showed a loss of over $12,000 for the first two quarters of 1883, followed by a net loss for the year.[66] In 1885 the company reported a net income of 6 percent on its capital investment of $828,800 and declared a 4 percent dividend. The latter may have been intended to encourage investors, who were needed for the construction of a new central station uptown. The company predicted that the new station would be more profitable because the newer Edison three-wire distribution system would lower costs substantially and because the load in the new district would be more evenly distributed to accommodate use at all hours of the day or night. The relationship between regularity of load and per-unit cost (load factor) had been recognized as a critical problem.

Nevertheless, Edison encountered difficulties in funding the expansion. The Edison Electric Light Company, the owner of one-quarter of the EEIC's operating stock, appropriated $171,200 of this stock as a trust fund to assist in raising funds. Despite this, the promise of larger profits from the new station, and the further inducement of a 20 percent bonus on the new expansion stock to EEIC stockholders, the company still experienced difficulties. A private canvassing of stockholders revealed that not one-tenth of the million and a half dollars authorized could be raised from this source. The prominent financier Henry Villard, an investor in Edison stock who was then residing in his native Germany after a reverse of his American fortunes, heard from his American correspondent that informed opinion was disgusted with the entire Edison business and believed that Edison officials were getting rich at the expense of EEIC stockholders. Villard's correspondent advised him to ignore the rosy, vague promises of the company's officials and not to purchase the expansion stock. Faced by the apathy of such stockholders, the company then formed a guarantee syndicate with the promise of profit for syndicate participants in the form of a large bonus of EEIC stock. News of this venture was passed on to Villard in February 1886 with the advice that he not "touch it with a ten-foot pole."[67] Finally, a "Drexel Morgan Syndicate" with heavy investments in the Edison system "[squeezed] out some money somewhere."[68]

---

[65] See p. 71 below.

[66] Edison Electric Light Co., *Annual Report, 1883* (New York: EEIC, 1884); and Edison Electric Illuminating Co., *Annual Report, 1885* (New York: EEIC, 1886).

[67] C. A. Spofford to Henry Villard, 12 January, 18 January, 25 January, and 26 February 1886, Box 124, Villard Papers, Harvard University Library, Cambridge, Mass.

[68] Spofford to Villard, 26 February 1886, Villard Papers.

Not only surviving but expanding, the EEIC, a critical component of Edison's system, seems to have stimulated the sale of Edison franchises by the parent company and of equipment by the manufacturing companies. The sale of franchises in major cities proceeded briskly at the same time that, behind the scenes, Edison's backers were having difficulty financing their showpiece. By 1888, large Edison companies or utilities were located in Detroit, Michigan; New Orleans, Louisiana; St. Paul, Minnesota; Chicago, Illinois; Philadelphia, Pennsylvania; and Brooklyn, New York. Abroad, central lighting stations had been established in Milan and Berlin. In 1890 the New York utility reported dramatic growth since 1888: the number of its customers had risen from 710 to 1,698; the number of lamps (16 c.p.) used had increased from 16,377 to 64,174; and the company's net earnings had nearly doubled, jumping from $116,235 to $229,078.[69] Indicative of the approach of a new era of electric power, in 1889 the company reported the first "motorload" (470 h.p.).

---

[69] Edison Electric Illuminating Co., *Annual Report, 1890*. (New York: EEIC, 1891).

CHAPTER III

Edison's System Abroad: Technology Transfer

EDISON and his associates invented, developed, and introduced into use at the Pearl Street station an electric lighting system that was designed for the New York City site. The technology of the Edison central-station system incorporated both the natural geography (coal and water supply) and the human geography (load or demand) of New York City. In addition, Edison, Lowrey, and others took advantage of the financial resources and organizational forms that were available in New York City. The legislative, or regulatory, constraints of New York little troubled the innovators once they obtained a franchise from city council. The Edison technology also manifested the experience and skills of craftsmen and appliers of science who had been drawn mostly from the native and immigrant populations of New York City and the industrial Northeast. Thus, without articulating the intent to do so, Edison and his associates designed a site-specific technology. Undoubtedly, however, they believed that a system designed for New York City would function well in the other great cities of the Western world.

In fact, the Edison interests were not content to introduce the technology in New York City or even to limit it to the urban centers of the United States. Edison and his associates also actively engaged in the diffusion and transfer of their central-station technology to other American cities and to the cities of Europe. The history of the transfer of Edison's system to London and Berlin is instructive, for it reveals different modes of technology transfer and the fate of a site-specific technology in two new environments.

Grosvenor P. Lowrey promoted the Edison enterprises not only in the United States but in England and on the Continent as well. His activities provide an outstanding example of modes of technology transfer. Lowrey's associates, members of the great banking and investment house of Drexel, Morgan and Company, had the financial resources, the foreign contacts, and the organizational wherewithal to move technology across national boundaries. From the start of the Edison electric lighting project, Lowrey anticipated the business that could be developed abroad. In October 1878 he told Edison that the way to fulfill his dream of building a working laboratory "such as the world needed and had never seen" was to sell patents, including foreign ones.[1] It was not unusual in the 1870s for Americans to look abroad for financing and for a major market for their technology. Before the invention of his electric lighting system, for instance, Edison had promoters representing his telegraph and telephone patents in England and on the Continent.[2]

For the telephone patents, Edison's representatives were Colonel George E. Gouraud in England and Theodore Puskas in Paris. Puskas also handled Edison's phonograph business in Paris. Gouraud, a transplanted American of the "boomer" type, often represented American bankers in London and had valuable contacts among British capitalists; Puskas, a native of Budapest, was known among inventors as the first to suggest a telephone exchange.[3] After Lowrey and Edison agreed that Drexel, Morgan and Company should dispose of Edison's inventions in England and Europe, however, Gouraud's and Puskas's call upon Edison because of earlier services became awkward.

When Edison first believed he had struck a bonanza by solving the problem of subdividing the electric light, he confided in Puskas and Gouraud. He promised Gouraud that he would be "counted in," even though the light was being developed with the economic support of financiers whose interests would have to be taken into account in making such an arrangement.[4] Gouraud enthusiastically informed Edison that news of his discovery had spread throughout the British press and constituted the best, as well as free, publicity. Gouraud saw more than a million pounds "trembling in the balance," pounds that could come down on Edison's side if the facts of his invention were made known in detail; there was no time like the present to launch a great Edison company in England. Furthermore, he reported, the scientific community waited for more news as eagerly as did the financiers.[5]

After communicating the news of the subdivision bonanza to Gouraud and Puskas with the unmistakable implication that they would be beneficiaries, Edison gave Grosvenor Lowrey the power of attorney to sell his financial interest in the electric light in Great Britain. He authorized Lowrey to select a suitable banking house with correspondents in England and to form a partnership between Edison and the bankers for the purpose of taking out patents, organizing exhibitions of the inventions, and disposing of the rights. Edison was to receive one half of the net proceeds; Lowrey and the banking house the other. Edison took pains, however, to add that he expected Lowrey to satisfy "any reasonable claims made by Colonel Gouraud for any services which he may have rendered, desiring you to be

[1] G. P. Lowrey to Edison, 10 October 1878, Edison Archives, Edison National Historic Site, West Orange, N.J. (hereafter cited as EA).

[2] For a recent account of the Edison telegraph and telephone in England, see Robert Conot, *A Streak of Luck: The Life and Legend of Thomas Alva Edison* (New York: Seaview Books, 1979), esp. pp. 53–55, 145–47. See also Matthew Josephson, *Edison: A Biography* (New York: McGraw-Hill, 1959), pp. 102–3, 149–55; and Frank L. Dyer and Thomas C. Martin, *Edison: His Life and Inventions*, 3rd ed., 2 vols. (New York: Harper & Bros., 1930), 1: 148–52.

[3] Josephson, *Edison*, p. 149; Dyer and Martin, *Edison*, 1: 179.

[4] Edison to T. Puskas, 22 September and 5 October 1878, and Edison to G. E. Gouraud, 14 October 1878, EA.

[5] Gouraud to Edison, 24 October 1878, EA.

not only just to him, but to save me from any fair course of complaint by him: this is to be treated as one of the expenses."[6]

Edison also wrote to Puskas, expressing "great delicacy" in informing him about the arrangements that would be made for the European business. Explaining that he had turned to "friends" in America for a large sum to develop the invention, Edison revealed that they had immediately raised the question of his European patents. He had told them about his agreements with Puskas for telephone and phonograph business. They wanted the rights to dispose of the electric lighting patents in Europe, but were willing to discuss with Puskas his acting as their agent in all European countries except Italy; there, according to Egisto Fabbri, the Morgan partner, Fabbri's brother's firm, Messrs. Fabbri & Chauncey of New York, would be the most effective agent. Edison attempted to console Puskas by predicting that while Drexel, Morgan and Company would have general control over promotional matters and negotiations, Puskas's income as agent, though a reduced share, would actually be larger.[7]

Lowrey had persuaded Edison that the American banking house and its overseas partners could transfer Edison's technology more effectively than "Continental people," especially those in Paris. From his and his associates' experience, Lowrey had concluded "that Americans are the honestest and most straightforward, as well as the best hearted people in the world." The English, he believed, "are next." In Paris, Lowrey reported, "[there] was no chicanery or trick considered to be beneath a great banker." Therefore, he urged, Edison should be represented by a strong American banking house that was capable of dealing with chicanery and tricks. Furthermore, since the offices of Drexel, Morgan and Company were only twenty miles from Menlo Park, Edison could remonstrate quickly if his affairs were not handled to his liking. In addition, Drexel, Morgan and Company's British and European network could effectively exhibit the Edison system in London, Paris, Vienna, Berlin, and elsewhere in Europe in order to draw attention to the technology and to dispose of the patent rights. The bankers, with their highly developed negotiating skills, could define the best arrangements, sometimes selling patents outright, other times reserving partial interest in them. Lowrey added that he wanted Puskas to settle for, perhaps, 10 percent as an agent, unless Puskas was able to raise $50,000 or more in cash to invest in Edison's development of the electric light. The American bankers, whom Lowrey characterized as "a new class of men" in Edison's experience, were not, it seems, prepared to invest more money for development until more progress had been made at Menlo Park, despite the "highest confidence in your [Edison's] ability and respect for your character."[8] Therefore, Puskas's cash could alter the arrangements.

The kind of technology transfer agreement that the bankers were ready to negotiate with the inventor is revealed in a document of 31 December 1878 signed by Edison; Drexel, Morgan and Company; Egisto Fabbri; and Lowrey. Fabbri and Lowrey signed as trustees.[9] The banking house agreed

[6] Edison to Lowrey, 29 October, 1878, EA.
[7] Edison to Puskas, 13 November 1878, EA.
[8] Lowrey to Edison, 10 December 1878, EA.
[9] Agreement of 31 December 1878, EA.

to aid Edison in obtaining electric lighting, power, and heating patents in Great Britain, Ireland, and portions of the dominions and to manage and exhibit the inventions described in the patents. Drexel, Morgan and Company was to reimburse Edison for prior expenses in taking out British patents and to defend these and patents for inventions in the subject area over the next five years (1879–83). Edison agreed to turn over to the trustees title to all of these patents so that the trustees could assign them as directed by Drexel, Morgan and Company. If after three years Drexel, Morgan and Company had not disposed of the principal patents, Edison could serve notice demanding their return to him at the end of six months. Expenses arising from obtaining and disposing of the patents, as well as the cost of exhibiting the inventions, would be deducted from the income of the joint enterprise before the profits were divided between Edison and the banking house. Payments to Lowrey, Gouraud, Puskas, or other financial agents and associates would be made by Drexel, Morgan and Company, presumably because the agents would relieve the company of some of the responsibilities assigned to it by the agreement.

Drexel, Morgan and Company did use the international exhibitions effectively. In so doing, they upheld a tradition established in 1851, when Cyrus McCormick showed his reaper and Samuel Colt exhibited his revolver to great advantage at the Great Exhibition in the Crystal Palace, the first in a series of major international technological and industrial exhibitions.[10] Although it was presented on a smaller scale than the other international industrial exhibitions, the International Electrical Exhibition of 1881 in Paris drew the attention of European scientists, engineers, inventors, financiers, and entrepreneurs to electrical technology, especially electric lighting, and Edison and his associates used this opportunity to great advantage (see Fig. III.1).

A. P. Trotter, a leading British electrical engineer, wrote that "electrical engineering was born at the . . . [Paris] Exhibition . . . , a lusty child of science and machinery."[11] The British journal *Engineering* judged the four-month exhibition as "literally brilliant" and as a powerful attraction not only for technical and scientific visitors but for the public as well. Professionals were drawn together for the first time by the international congress for electricians that was held during the exhibition. Altogether, there were at least fifty different displays of arc and incandescent lighting, including the incandescent lamps of Swan, Lane-Fox, Maxim, and Edison.[12]

[10] On international exhibitions of technology and industry, see Kenneth W. Luckhurst, *The Story of Exhibitions* (London: Studio Publications, 1951). On the Crystal Palace exhibition of 1851 as a microcosm of technology and social reaction at mid-century, see Thomas P. Hughes, "Industry through the Crystal Palace" (Ph.D. diss., University of Virginia, 1953). Following the Paris exhibition of 1881, there were electrical exhibitions in Munich (1882), Vienna (1883), and Philadelphia (1884). On the Philadelphia exhibition, see Jane Mork Gibson, "The International Electrical Exhibition of 1884," *IEEE Transactions on Education* E-23 (1980): 169–76.

[11] A. P. Trotter, "Early Days of the Electrical Industry and Other Reminiscences" (typescript at the library of the Institution of Electrical Engineers, London), pp. iii–xiii.

[12] *Engineering* 32 (1881): 534. For a listing of publications about the International Electrical Congress, Paris, 1881, see *Catalogue of the Wheeler Gift of Books, Pamphlets, and Periodicals in the Library of the American Institute of Electrical Engineers*, ed. William Weaver (New York: AIEE, 1909), vol. 2, items 5422–39.

Figure III.1. *Edison exhibit at the Paris Exhibition of 1881. From* Edison: Lighting a Revolution *(Washington, D.C.: Smithsonian Institution, 1979), p. 56.*

Edison's incandescent-lamp system was received enthusiastically at the Paris exhibition. The display persuaded Europeans that Edison's system was preeminent in the field and this led entrepreneurs to seek exclusive rights to manufacture and market the system. The outstanding instance of technology transfer involved Emil Rathenau, the founder of the German Edison Company, later known as German General Electric (Allgemeine Elektrizitäts-Gesellschaft). A mechanical engineer, Rathenau returned to Germany from the exhibition determined to obtain an Edison franchise. Young Oskar von Miller visited the Paris exhibition and returned to his native Munich to help organize an international electrical exhibition there. He described the exhibition as "overwhelming," and the Edison incandescents as most impressive.[13] Frank Sprague, who was to become the outstanding pioneer of electric traction in the United States, visited the Paris exhibition as a young naval ensign. He was persuaded that a new era of lighting had begun, and he believed that the public and investors had been greatly influenced.[14] The combined effect of motivated young engineers and stimulated investors streaming away from Paris with a favorable impression of Edison and his works is an informal but important example of technology transfer.

Edison sent Charles Batchelor, one of his principal technical assistants at Menlo Park, to install and superintend the display at the fair grounds on the Champs-Elysées. An Edison 200-h.p. generator capable of lighting twelve hundred 16-c.p. lamps dominated the exhibit (see Fig. III.2). Some of the success of the lamp exhibit was attributed to the use of copper wiring

[13] Walther von Miller, *Oskar von Miller Nach Eigenen Aufzeichnungen, Reden und Briefen* (Munich: Bruckmann, 1932), p. 15; Theodore Heuss, *Oskar von Miller und der Weg der Technik* (Munich: Deutsches Museum, 1950), p. 7.

[14] Frank J. Sprague, *Report on the Exhibits at the Crystal Palace Electrical Exhibition, 1882* (Washington, D.C., 1883), p. 33.

Figure III.2.   Edison generator at the Paris exhibition of 1881. From Telegraphic Journal and Electric Review *9 (1881): 437.*

Figure III.3.   Incandescent lamps illuminating paintings at the Paris exhibition of 1881. From La Lumière électrique *4 (1881): 227.*

Figure III.4.   Edward H. Johnson, Edison's representative in London. Courtesy of the Edison Archives, Edison National Historic Site, West Orange, N.J.

of unusually low resistance, wiring that was specially made by William Wallace of the Ansonia Copper Company. Plans for New York's Pearl Street station also were on exhibit and stimulated many technical and economic questions. The French architect C. Garnier asked some of the exhibitors to light the Paris Opera House, and to the Edison group he assigned the responsibility for lighting the building's great foyer. This, too, attracted attention to Edison's system.[15]

After the Paris exhibition, Drexel, Morgan and Company supported the organization of an Edison display at an electrical exhibition in London. Drexel, Morgan and Company and Edison had placed Edward H. Johnson (1846–1919) in charge of promoting the Edison system in England. Johnson, once a telegraph operator, had assisted Edison in his telegraph enterprises and subsequently became a close associate in the electric lighting business. He was an original partner in the Edison lampworks, and in 1886 he would become president of the Edison Electric Light Company. In 1882, he presided over the Edison exhibit at the International Electric Exhibition held in the Crystal Palace in Sydenham, near London. Assisting him was the electrician William Hammer, another of Edison's close associates.

Johnson assured Edison that his display at the Crystal Palace had elicited "the admiration of every Engineer." As he described it, "All the other work in the Palace is of the most temporary character. . . . Swan's lights, for instance, are just as likely to go out as they are to burn." Johnson stressed that the crowds left the exhibition persuaded of the superiority not only of the Edison system but of Johnson's explanation of electric lighting as

[15] Etienne de Fodor, "The Edison System in Europe Forty Years Ago," *Electrical World* 80 (1920): 547–48. Fodor was one of the installers of the Edison exhibit.

Figure III.5. *The Edison isolated plant at the Crystal Palace exhibition of 1882. Courtesy of the National Museum of American History, Smithsonian Institution, Washington, D.C.*

well.[16] (See Fig. III.5.) The *London Daily News* judged Edison's system more advanced than all its rivals: "[It] is the wonder of the show, and his representative is certainly the prince of all showmen. There is but one Edison, and Johnson is his prophet."[17]

Judging by his letters to Edison, Johnson was especially effective in informing and impressing British aristocrats and scientists. The emphasis he placed on influencing them indicates the importance of the role the aristocrats had in financing entrepreneurs, and that scientists had in authenticating the new technology. By contrast, scientists in America remained in the background as Edison invented, developed, and innovated. Even before Johnson took charge in London, Colonel Gouraud had approached Sir William Thomson, one of Great Britain's most distinguished physicists, to ask him to act as a consultant for the Edison interests in Britain. Thomson in turn wanted to make an "unprofessional," comparative test of Edison and Swan lamps, between which he found little difference, especially as to economy.[18]

One of Johnson's efforts involved guiding members of the Society of Arts through the Edison exhibit and afterwards dining with them in the Crystal Palace. Lord Alfred Churchill and Sir Frederick J. Bramwell, the

[16] E. H. Johnson to Edison, 19 March 1882, William Hammer Collection, Museum of American History, Smithsonian Institution, Washington, D.C. I am indebted to Dr. Robert Friedel for guidance in using the Hammer Collection.

[17] *London Daily News*, 8 April 1882, quoted in Payson Jones, *A Power History of the Consolidated Edison System, 1878–1900* (New York: Consolidated Edison Co., 1940), p. 49.

[18] Gouraud to W. Thomson, 15 March 1881, and Thomson to Gouraud, 18 March and 30 April 1881, in folder labeled "Electric Light 1880, Foreign," EA.

leaders of the delegation, admired the display and Johnson openly, and their praise was such as "to elicit from Lord Alfred at the dinner the one toast of the Evening."[19] Johnson envisaged that the scientists and aristocrats would both invest in and testify to the excellence of the Edison system. Their support was needed both for the central-station ventures and for the small, isolated plants Johnson anticipated would be purchased for the great homes and country estates. The duke of Sutherland saw the Crystal Palace display and was so delighted that he could not stay away. After his fourth visit he invited Johnson to dinner at his London residence, Stafford House, to discuss installation of the system there. Only because of a prior investment made in the Brush Electric Light Company before he learned of the Edison system did the duke refrain from associating with the Edison company in England.[20] The duke of Westminster and the duke of Edinburgh also liked the display and expressed their intention to install the Edison system.

The Crystal Palace display was only part of Drexel, Morgan and Company and Edison's master scheme, however. Although the banking house did not want its name publicly associated with the undertaking, the Crystal Palace exhibition was to be followed by the opening of a central station at 57 Holborn Viaduct in the heart of London.[21] Holborn Viaduct would serve as an "exemplification" for the central stations that the English Edison company expected to license. Holborn Viaduct was to the English company what the Pearl Street station was to the Edison Electric Light Company of New York, with the exception that the English company obtained its rights to the English patents from America.

On 15 March 1882 the English Electric Light Company, Ltd., registered and held its first board meeting. The major stockholders and directors included men of scientific reputation, aristocratic connections, and financial means. (In New York, the directors of the Edison Electric Illuminating Company were men of business and finance.) Among the board members characterized by Johnson as "scientific representatives" were Sir Frederick J. Bramwell (1818–1903), an expert on municipal engineering and a well-qualified technical witness on legal matters; William Preece (1834–1913), chief electrical engineer for the Post Office Department; Sir John Lubbock (1834–1913), a banker, sometime member of Parliament, and chairman of the London County Council; and Sheeford Bidwell, who advised Johnson on Edison patents in Britain. The board initially included Lubbock, Bidwell, Viscount Anson, Richard B. Wade, and E. Plydell Bourverie.[22] With £1,000,000 in authorized capital, the company's purpose was threefold: to acquire all of Edison's British patents in electric lighting, heat, and power; to purchase the lamps, dynamos, and other Edison equipment then on display in England; and to lease the property at 57 Holborn Viaduct. The company agreed to pay £20,000 for the patent rights and equipment and to allot one fully paid share of "B" classification stock for every £10 of "A" share stock paid up. There were 50,000 £5 "A" shares and 50,000 £10 "B" shares.

[19] Johnson to Edison, 19 March 1882, Hammer Collection.
[20] Ibid.
[21] Ibid.
[22] Ibid.

Thus Edison shipped two Jumbo generators from New York to the newly formed English company, and early in 1882 these were installed in the Holborn Viaduct station. Holborn Viaduct thus had generators of the same kind as those at New York's Pearl Street station, and its components—Edison lamps, some underground cable, and other electrical equipment—were similar to those found in the New York station. Moreover, Edison-trained electricians (including Hammer) and managers presided over both the London and the New York station. Yet the New York station survived and the Holborn station did not. This failure of Edison technology to adapt and put down roots in London calls for other than strictly technical explanations.

The station formally opened for inspection on 11 April 1882. Equipped with 1,000 incandescent lamps (but with a capacity for 2,200), the station lit streetlights from Holborn Circus along the Viaduct and Newgate Street to the vicinity of the General Post Office. Incandescent lamps in hotels, stores, restaurants, and other private buildings fronting along the way enhanced the effect. The four hundred lamps that lit the General Post Office and the lamps in Dr. Joseph Parker's City Temple excited comment among the passers-by and the distinguished and financially affluent visitors who were given a tour of the station (see Fig. III.6). Johnson, who remained in London for a time after the opening and who was a major shareholder in the English company, later recalled that the station elicited extremely favorable comment.[23] The publicity release of the Edison Electric Light Company in America described the Holborn station as "a successful exhibition on a large scale of the practical working of the Edison system of incandescent lighting."[24] (See Fig. III.7.)

The Holborn Viaduct station was much more than a "try out . . . in the provinces"[25] for the American company and more than a pilot plant for the English company.[26] The station was both a commercial-technological venture and an advertisement for the English company.[27] With rights to the Edison patents within the United Kingdom, English Edison had reason to expect a substantial profit from the sale of concessions or licenses to prospective central-station companies throughout the United Kingdom. As early as 20 May, the London *Economist* reported the sale of local licenses by English Edison.[28]

The English patent-owning company established rates for Holborn consumers which were calculated to create good will and favorable publicity.

[23] Thomas C. Martin, *Forty Years of Edison Service, 1882–1922* (New York: New York Edison Co., 1922), p. 29. Some of the account of the Holborn Viaduct station and the following analysis of the effect of the Electric Lighting Act of 1882 is taken from Thomas P. Hughes, "British Electrical Industry Lag: 1882–1888," *Technology and Culture* 3 (1962): 27–44.

[24] Edison Electric Light Co., *Bulletin* (New York) 10 (1882): 6.

[25] Committee on St. Louis Exposition of Association of Edison Illuminating Companies, *Edisonia: A Brief History of the Early Edison Electric Lighting System* (New York: Association of Edison Illuminating Companies, 1904), p. 159.

[26] Dyer and Martin, *Edison*, 1: 336. Edison did, however, use information from Holborn at Pearl Street; see ibid., 1: 417.

[27] At a meeting of the shareholders of the Edison & Swan United Electric Light Co., Ltd., successor to English Edison, the directors offered this interpretation of the intent of Holborn's founders.

[28] *Economist* (London) 40 (1882): 604.

Figure III.6. Plan of illumination of the
Holborn Viaduct, 1881–82. Courtesy of
the National Museum of American History,
Smithsonian Institution, Washington, D.C.

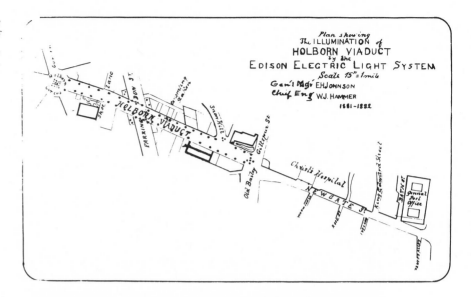

Figure III.7. On the left, the Edison
central station at 57 Holborn Viaduct,
London. Courtesy of the National Museum
of American History, Smithsonian
Institution, Washington, D.C.

From April until July 1882, the station supplied street lighting without charge to the city authorities; for the next six months the rates were the same as those for gas. Individual arrangements were made with private consumers, but the plan for Holborn kept the price of electricity near that of gas, even if it meant no profit.[29] The supply of electricity to customers at rates comparable to gas rates undoubtedly cost the station at least twice the selling price in 1882.[30] Company planners predicted that increased and more regular loads, along with further technical developments, would reduce operating costs. Time would test the wisdom of English Edison's investment in the Holborn Viaduct station, but in the spring of 1882 the station seemed to be proof of the technical workability of the Edison system. Holborn served as a reminder of the international character of technology.

Holborn was a technical, business, and economic venture. Engineers awaited the technical results; businessmen and financiers wanted the balance sheet. Thus English Edison commissioned two of the country's rising electrical engineers, Dr. John Hopkinson and Dr. J. A. Fleming, to analyze the costs of the Holborn station. After careful record keeping over several months and after making assumptions about long-term operations and load, the two engineers concluded: "It will be possible with careful management at present prices of plant to supply electricity to the public at the cost of equivalent gas, and to make a small profit if a fairly full consumption is maintained for 7 or 8 hours per day."[31] (See Tables III.1 and III.2.) They were certain, however, that use of the lamps for less than the whole night would be unprofitable if customers were charged a rate equivalent to the price of gas.

The interest created by the Paris and London exhibitions and the optimism generated by the Holborn Viaduct station contributed to heavy speculation in the spring of 1882 in shares in various electrical manufacturing and supply (utility) companies. During two weeks in May, sixteen new companies appeared, and the London *Economist* recalled earlier bubbles in railway shares (1845–46) and the submarine telegraph.[32] Then, in mid-May, the stocks fluctuated wildly. The shares of the Anglo-American Brush Electric Light Corporation, makers of the Brush arc lamp and the Lane-Fox incandescent lamp, dropped £600,000 in total value in three days of trading but remained far above their face value. The press generally advised caution.[33] The *Economist* offered sober counsel: "However great a revolution [in electric light] may possibly be in store for us, the inevitable competition amongst the real electrical systems will effectually prevent the shareholders from becoming millionaires."[34] At year's end, however, the speculation in electrical shares would stand as Britain's chief security mania of the year.[35]

---

[29] Testimony of Sir John Lubbock, M.P. and major shareholder and director of English Edison, before a select committee of the House of Lords in 1886; see p. 63 below.

[30] See p. 64 below.

[31] John Hopkinson and J. A. Fleming, "Report on the Central Station, Holborn Viaduct, to the Directors of the Edison Electric Light Company, Limited" (typescript, Hammer Collection), pp. 22–23.

[32] *Economist* (London) 40 (1882): 604.

[33] *Electrical Review* 10 (1882): 343.

[34] *Economist* (London) 40 (1882): 605.

[35] "Commercial History and Review of 1882," ibid. 41 (1883).

TABLE III.1. COST OF THE HOLBORN VIADUCT CENTRAL STATION

Cost Breakdown	£	s.	d.	£	s.	d.
Messrs. Hayward, Tyler & Co. Tanks, steampumps, waterheater, inspector, traveler pipes, and other fittings				660	1	11
Builder Erecting shaft				1,301	7	5
Drexel, Morgan & Co.						
Engine and boiler	737	3	3			
Two 1,200 light dynamos at £1,830	3,660	0	0	4,397	3	3
Installations—Cost of materials						
1. Negretti & Zamba	16	7	5			
2. Sharp & Co.	8	1	0			
3. Vaseline & Co.	7	4	7			
4. Hodge & Essex	11	9	1			
5. Holland & McConnell	8	14	6			
6. Steel & Garland	12	16	8			
7. W. D. & H. O. Wills	10	4	7			
8. Terrey & Co.	14	0	3			
9. Birch	14	9	1			
	103	7	2	6,358	12	7

Source: Hopkinson and Fleming, "Report on the Central Station, Holborn Viaduct."

In an epoch characterized as "the end of laissez-faire" and in a nation whose government showed increasing concern for the welfare of the growing body of the electorate, it is not surprising that Parliament, the central bureaucracy, and the local authorities reacted to the intense activity and optimism in the electric lighting industry. Within two weeks of the formal opening of the Holborn station a select committee of the House of Commons was established, and from 25 April to 12 June 1882 it heard testimony on proposed central-station legislation.

Newly formed central-station companies needed permission to break open the streets in order to lay the electrical lines of the underground distribution systems. In the United States this authorization came from the local government or authority, but in Britain the problem moved quickly to the national level, where Parliament and the Board of Trade sought a general solution in legislation. Thomas Edison had wined and dined New York City aldermen prior to the erection of the Pearl Street station; but in Britain, Joseph Swan, J. E. H. Gordon, and R. E. B. Crompton, all pioneer entrepreneurs in electric lighting, gave a dinner for the Speaker of the House of Commons and other leading M.P.'s in a bid to win passage of the Electric Lighting Act of 1882.[36]

The government's electric lighting bill came from the Board of Trade. Founded in 1786, this government department "grew like Topsy" during the industrial era. It held far-reaching regulatory powers not only over transportation, communications, and various industries but also over the

[36] Rookes E. B. Crompton, *Reminiscences* (London: Constable & Co., 1928), pp. 111–12.

TABLE III.2. ESTIMATED ANNUAL COST OF ONE UNIT OF THE HOLBORN VIADUCT STATION

One steam dynamo and boiler, maintained at an average
of six hundred 16-c.p. lights per night

Coal at £32 per month	384.0.0
Oil at £11 per month	132.0.0
Waste	4.0.0
Wages at £50 per month	600.0.0
Lamps at 2s. ld. (say, 1,500)	156.0.0
Share of rent for one dynamo	100.0.0
Depreciation (10%) on engines, boilers, and other materials, excluding the value of copper in the mains	500.0.0
Share of management for one dynamo per annum	200.0.0
	£2,076.0.0

Source: Hopkinson and Fleming, "Report on the Central Station, Holborn Viaduct."

surging electric light industry.[37] Heading the Board of Trade in the Liberal Gladstone's government was Joseph Chamberlain, a man who had effectively socialized the public utilities as mayor of Birmingham and who would long be remembered for his republicanism in the Victorian era.

Chamberlain's activities as president of the Board of Trade from 1880 to 1885 amounted to a fierce onslaught on the abuses of property against human rights.[38] Chamberlain and the Board of Trade drew upon precedent as they moved to regulate a public utility. The Tramways Act of 1870 had limited the tenure of private ownership; the Public Health Act of 1875 had authorized municipal ownership of waterworks;[39] and an extensive body of legislation had regulated private gas companies. This legislation had come to be thought of as protection for the public against the tendency of public utilities to abuse their monopolistic powers. On the other hand, the government had been tolerant of regulated, or natural, monopolies if a duplication of services raised costs exorbitantly.[40] The expense of breaking open the streets to lay electrical distribution lines would be heavy.

Government ownership, as well as government regulation, was viewed sympathetically in the twilight of the Victorian era. However, government ownership meant ownership by the local governing authority.[41] As mayor of Birmingham, Joseph Chamberlain had strengthened the movement for municipal socialism considerably. He had reduced the operating costs and the price to consumers while improving the service of the public utilities—

[37] George S. Bower and Walter Webb, *The Law Relating to Electric Lighting, Being the Electric Lighting Act, 1882* (London: S. Low, Marston, Searle & Rivington, 1882), p. 100.

[38] J. L. Garvin, *The Life of Joseph Chamberlain*, 4 vols. (London: Macmillan & Co., 1932–), 1: 413.

[39] Sir Henry Self and Elizabeth Watson, *Electricity Supply in Great Britain* . . . (London, 1952), p. 14 ff.

[40] Gas companies had not begun as monopolies, but had assumed that character when competing distribution systems involved ruinous expense. The Metropolitan Gas Company Act of 1860 confirmed the monopolistic districting of London that the companies had taken up on their own initiative. H. H. Ballin, *The Organisation of Electricity Supply in Great Britain* (London: Electrical Press, 1946), pp. 1–3.

[41] Local authorities were those bodies to whom the public health of the United Kingdom was entrusted—e.g., metropolitan district boards and vestries and urban and rural sanitation authorities. *Engineering* 35 (1883): 374.

including the gasworks—he had bought for Birmingham.[42] Permanent Secretary of the Board of Trade Farrer told a parliamentary committee holding hearings on the proposed legislation that it was "much too late" to argue that public enterprise could not compete with private enterprise or to question bringing the public purse to bear in the public-utility area.[43]

The proposed legislation for governing central stations clearly manifested its sponsor's sympathy with government regulation and tender feeling toward municipal ownership. Yet the claims of private enterprise upon the economy also were recognized. The bill provided a regulated monopoly with limited tenure for private enterprise (or a regulated monopoly for a local governing authority). Private enterprise had the opportunity to risk capital in an area of the economy that was a technological frontier. After the time allotted under the tenure had elapsed, the local authority would have the power to purchase the central station and distribution system (now having stood the test of time) with ratepayers' (taxpayers') money. In the atmosphere of enthusiasm and sanguine speculation of the spring of 1882, Chamberlain, Parliament, and the Board of Trade had every reason to expect private enterprise to venture capital under restricted conditions.

As a matter of fact, the counselor for six private light companies gave his general approval to the Board of Trade bill, excepting several provisions.[44] A trade publication characterized the bill as equitable and in the public interest, and the London *Economist* predicted that the monopoly given the private central station would allow rates to rise high enough to bring a return on the investment within the time of tenure.[45] Spokesmen for private enterprise singled out for criticism the bill's provision of a seven-year tenure. Sir Frederick J. Bramwell, of English Edison, told the parliamentary committee the companies needed a minimum of twenty-one years.[46] Edward H. Johnson probably summed up the attitude of American investors when he told the committee he would prefer seventy years, or better still, no limit, but English businessmen accepted the bill with equanimity upon its enactment on 18 August 1882. Clause 27 of the bill provided for a twenty-one-year tenure.

Another of the companies' criticisms was directed against the so-called "scrap iron" provision of clause 27. The Board of Trade intended that if the local authority made a compulsory purchase after the period of tenure had elapsed, the sale price would be based upon the value, as isolated elements, of dynamos, wire, bricks, mortar, and other components of the operating central-station system—bricks as bricks and mortar as mortar, Board of Trade Secretary Farrer explained.[47] Yet the large number of applications by companies for authorization[48] under the Electric Lighting

[42] Garvin, *Joseph Chamberlain*, 1: 190 ff.

[43] Great Britain, Parliament, House of Commons, *The Electric Lighting Act 1882: Minutes of Evidence Given before the Select Committee of the House of Commons* (London, 1882), p. 13 (hereafter cited as *House of Commons Hearing, 1882*).

[44] *Electrical Review* 10 (1882): 283.

[45] *Economist* (London) 40 (1882): 907.

[46] Letter to the *Times* (London), 21 July 1882.

[47] *House of Commons Hearing, 1882*, p. 28.

[48] See p. 62 below.

Act of 1882[49] indicated that in the year of optimism the companies expected to make a profit, clause 27 notwithstanding.

Other sections of the act would subsequently prove significant, especially the clauses designed to prevent what were envisaged as monopolistic abuses. Undertakers (utilities) were not to tyrannize customers by imposing the system components of only one manufacturer. The central station, for example, could not prescribe any special form of lamp (clause 18). Potential customers could demand that supply mains be laid to their premises under conditions that could prove costly to the central station (clause 19), for restrictions all but outlawed the cheaper, overhead distribution system. In short, no British engineer could design a central-station system solely in terms of economic and technological considerations; instead, he had to design for an environment permeated by legislation.[50] Edison had designed his technology and his American business venture under legislative conditions that were vastly different from these.

After passage of the Electric Lighting Act, in April, and after the legislated conditions had been defined, Great Britain—still considered the world's greatest industrial nation and one distinguished by her scientists and engineers—seemed ready to move ahead with her central-station industry. Economic conditions were favorable: 1882 fell within a limited period of recovery during the "great depression in Britain, 1873–1896."[51] By 1 July, however, it was apparent that the electric lighting companies and the manufacturers of electrical components were not making preparations for the anticipated winter demand.[52] In September 1882 disgruntled Brush shareholders questioned company policy.[53] By the end of December the value of electric-light company shares had dropped dramatically from their spring highs.[54]

[49] 45 & 46 Vict., ch. 56.

[50] Under the act, three means were provided for bestowing powers on the electrical undertakings: a special act of Parliament, a Board of Trade license, and a provisional order. The last-named proved to be used most frequently. The powers bestowed pertained chiefly to the breaking up of public streets, for if a company could manage to supply electricity without doing this, there were no legal fetters. The Board of Trade bestowed provisional orders subject to subsequent approval by Parliament. Private companies and individuals and local authorities could obtain powers. A provisional order defined the area of supply; the responsibility of the utility (undertaker) to supply would-be consumers in the area (who might not always be in the immediate area of supply lines); the maximum rates and the method of assessing rates (satisfactory meters were not available); maximum voltages, in the interest of public safety; and inspection procedures to determine if undertakers had fulfilled their obligations. The act is summarized, and the provisional orders explained, in *Engineering* 35 (1883): 374.

[51] See H. L. Beales, "The 'Great Depression' in Industry and Trade," *Economic History Review* 5 (1934): 65–75; and A. E. Musson, "The Great Depression in Britain, 1873–1896: A Reappraisal," *Journal of Economic History* 19 (1959): 199–228.

[52] *Electrical Review* 10 (1882): 473.

[53] Ibid. 11 (1883): 209–10. At a stockholders' meeting of the Devon and Cornwall (Brush) Electric Light & Power Co. in September 1882, a small stockholder challenged the directors' purchase of the Lane-Fox rights from the parent company. The exchange that followed was interrupted by another small stockholder, who offered to bring a mine lamp to the company to make good on any losses that might result from the Lane-Fox blunder—he added that he saw no reason to wind up the company. The chairman characterized these remarks as a little beside the mark, as he had heard no hint about winding up the company.

[54] Stock market quotations in *Economist* (London) 40 (December 1882).

The parent patent-holding and franchise-granting Brush Electric Light Company had to face not only a decline in the value of its shares but a legal test of its claim to an incandescent-lamp patent. Apparently St. George Lane-Fox, inventor of Brush's lamp, had assigned the exclusive right of using and granting licenses for his patents to the British Electric Light Company on 20 July 1881, and only nine days later had made the same arrangement with Brush.[55] Problems for Anglo-American Brush increased in 1883 when subsidiary companies, "who [had been] all eagerness to obtain a high-priced concession, . . . [were] only too glad to resell it to the original grantor on almost nominal terms."[56] A year after the bright promise of the spring of '82, the Anglo-American Brush Corporation stood revealed as a patent-holding and manufacturing company which had been founded on an arc-lighting system that was no longer outstanding in the field and on an incandescent-lamp patent of doubtful value.[57]

Caught in the sweeping reaction, the parent English Edison Company sought to ease its difficulties. In October 1883 the company merged with its principal competitor in the manufacture and sale of incandescent lamps, Swan Electric Light Company, which had been founded on the basis of the patents of Joseph Swan of Newcastle upon Tyne, the inventor of an incandescent lamp that was said to be comparable to Edison's.[58] Thus the Holborn Viaduct station became the property of Edison and Swan United Electric Light Company, Ltd. James Staat Forbes, director of the new enterprise, told stockholders in November 1884 that Holborn had been a costly affair and had caused the directors a great deal of anxiety, having been operated at a serious annual loss.[59] The station then passed from public view and was abandoned by the company in 1886.

On 20 October 1884 Edison & Swan also informed the Board of Trade that none of the provisional orders granted the company under the Electric Lighting Act of 1882, orders that enabled the company to erect central-station systems in London, would be carried out. In the opinion of the company's directors, the "vexatious and harassing" legislation of 1882—especially the limited tenure and "scrap iron" clause—condemned to failure such ventures on the part of private business. This action indicated a trend: in 1883 the Board of Trade granted sixty-nine provisional orders for central-station systems and sixty-two were subsequently revoked; in 1884 only four were granted and all were revoked; and in 1885 none was issued.[60]

By the end of 1884, the doldrums had set in. Bad times had not come upon the British electrical industry alone, however. Innovators in the electric light and central-station industry found their melancholia echoed by

[55] *The Electrician* 10 (1882) (2 and 9 December). In a subsequent settlement, Lane-Fox repurchased his patents from Anglo-American Brush and then granted, as part of the settlement, a full and free license to the corporation to sell and use all of his inventions that were connected with electricity. *Electrical Review* 14 (1884): 95–96.

[56] *Economist* (London) 42 (1884): 1480.

[57] *Electrical Review* 11 (1883): 452–53.

[58] Kenneth R. and Mary E. Swan, *J. W. Swan: A Memoir* (London: Benn, 1929); Brian Bowers, "Joseph Swan and the Invention of the Incandescent Filament Lamp," *Papers Presented at the Fourth Weekend Meeting on the History of Electrical Engineering, Durham, England, 2–4 July 1976* (London: Institution of Electrical Engineers, 1976), pp. 68–70.

[59] Report of shareholders' meeting, *Electrical Review* 15 (1884): 397.

[60] Emile Garcke, *Manual of Electrical Undertakings* (London, 1896), p. 393.

financiers and managers from the older industries of shipbuilding and iron and steel.[61] Eighteen eighty-two had marked the end of the recovery cycle within the "great depression"; 1883 and 1884 marked the beginning of a slump.[62] Yet spokesmen for the private-enterprise sector of the electrical industry were reluctant to view the problems of the industry in the broader perspective of the overall economy; they preferred to limit their analysis to government activity. Because investment in electric lighting had been heavy and because influential persons were interested in the electrical industry, in 1886 Parliament considered amendments to the 1882 Electric Lighting Act.[63]

A committee of the House of Lords heard representatives of private enterprise speak as one in attributing the doldrums to government legislation. Sir Frederick J. Bramwell stated unequivocally that the Electric Lighting Act of 1882 had set the British electrical industry back four years.[64] Lionel L. Cohen, financier and stock exchange official, testified that capital could not be found for the industry under the conditions of the act.[65] A director of the Bank of England and leading engineers and scientists swelled the chorus of complaints that focused upon the tenure and "scrap iron" clause of the act.[66]

The committee also learned of other provisions within the act, and provisional orders granted under the act which worked technical and economic hardships on the engineers and station managers.[67] Committee members heard as well that foreign states had not enacted legislation constraining the electric lighting industry.[68] Spokesmen for private capital complained not only of government regulation but also of the stimuli the act had given municipal socialism. James Staats Forbes, head of Edison & Swan, argued that the local authorities would not innovate in the electrical lighting field nor would they allow private capital to innovate. He asked rhetorically, "Is that in the interests of the public?" and replied categorically, "I am stupid enough to believe that it is not."[69]

Spokesmen for the central government and the local authorities, however, called the attention of the Committee of Lords to the adverse effects of the speculation mania of 1882 and to the competition offered by cheap

[61] The real depression of the eighties began in 1883 in shipbuilding, engineering, and the iron and steel industry. See J. H. Clapham, *An Economic History of Modern Britain*, 3 vols. (Cambridge: At the University Press, 1931–38), 3: 5–6.

[62] See p. 61, n. 51, above; and Walt Rostow, *British Economy of the Nineteenth Century* (Oxford, 1949), pp. 49, 50.

[63] Great Britain, Parliament, House of Lords, *Report from the Select Committee of the House of Lords on the Electric Lighting Act (1886)* . . . (London, 1886) (hereafter cited as *Lords Committee Hearing, 1886*).

[64] Ibid., pp. 46–47.

[65] Ibid., p. 4.

[66] Ibid., pp. 3–4, 15, 33.

[67] Regulations regarding service for would-be customers enabled the customer to choose the type of lamp and to limit the number of overhead wires.

[68] Where local (not national) legislation existed, it was in the nature of safety regulations and was intended to prevent interference with telegraph and telephone installations; see *Lords Committee Hearing, 1886*, p. 218. In the opinion of one British engineer, the Americans were indifferent to accidents and fires; ibid., p. 634.

[69] Ibid., p. 106.

gas.[70] Government electrical engineer William H. Preece spoke of a "money making world" that was eager to "reap the benefit of the patient student, the ingenious inventor, and the deserving engineer."[71] The town clerk of Chamberlain's Birmingham testified that the ratio of the price of gas to that of electricity was 1:2 in Britain, but only 2:3 in the United States. This, he contended, had more to do with the progress of electric lighting than with government legislation.[72] Electric light in Britain was a luxury, he argued, and thus was not suitable for the investment of the ratepayer's taxes. Recent analysis of the relative cost of gas and electric lighting at that time supports the town clerk's explanation.[73]

Other explanations for the decline of the British central-station industry were heard during the time of its doldrums, but in the end the case made by private enterprise was most effective: the Electric Lighting Act of 1882—government legislation—had paralyzed the central-station industry. The spokesmen for the electrical industry chose to ignore the more complex questions of national inertia, technology and technological education, and business cycles. They concentrated on a more immediate problem, the amendment of the Electric Lighting Act of 1882. Changes in government, the Irish Home Rule bill, and changes at the Board of Trade all contributed to the delay in amending the act. The act had passed during Gladstone's rule; the amending bill passed in 1888 under Lord Salisbury.[74] In 1887 private investors had anticipated passage when capital once again began to flow into the electrical industry. The London Electric Supply Corporation registered in August 1887 with an authorized capital of £1,000,000 and bold plans to erect the world's greatest central station in London (see pp. 238–47 below).[75]

On 28 June 1888 the amending bill became law. Most significant among its provisions were the extension of tenure for private companies from twenty-one to forty-two years and a redefinition of the compulsory-purchase terms to the advantage of the private enterprise. The municipalities or local authorities, on the other hand, received a limited veto on the granting of a provisional order by the Board of Trade.

Just as private enterprise had been reluctant to interpret the doldrums in the industry within the general context of the economic slump, now, with the amendment of the "iniquitous Electric Lighting Act," the correlation between the passage of the amendments and the general economic recovery went unnoticed. The Jubilee Year, 1887, marked the beginning of a "brisk revival."[76] An atmosphere of optimism and enthusiasm now prevailed among private entrepreneurs. In the House of Lords the earl of Crawford predicted an outpouring of capital and forty thousand new jobs

[70] In 1885 the competitive position of gaslight had been even further strengthened by the introduction of the Welsbach mantle, which gave a more efficient and agreeable light. Philip Chantler, *The British Gas Industry* (Manchester: Manchester University Press, 1938), p. 9.

[71] *Journal of the Society of Arts* 32 (1883–84): 347.

[72] *Lords Committee Hearing, 1886*, p. 562.

[73] I. C. R. Byatt, *The British Electrical Industry, 1875–1914* (Oxford: Clarendon Press, 1979), pp. 21–28.

[74] 51 & 52 Vict., ch. 12.

[75] See *Engineering* 94 (1912): 163–64, for an account of the Deptford station.

[76] Clapham, *Economic History of Modern Britain*, 3: 5–6.

in the newly surging sector of the economy.[77] Engineers believed that the technological foundations had been laid.[78] The editors of British technical journals were cautious but sanguine. As one of them wrote, "At last . . . the air clears, and activity begins to show itself all round."[79] Britain would move ahead sure-footedly, not plunge ahead with the reckless abandon of the American central-station managers who hung wire in festoons and the manufacturers who countenanced imminent obsolescence. The British businessman would not drum up demand frenetically, as did his American cousin.[80] Cold statistics supported warm sentiments as the value of electrical shares turned upward late in 1888, and the trend continued in 1889.[81] The best indication of renewed optimism was a dramatic upswing in the number of provisional orders granted: after 1884 there had been only one; in 1889 there were twelve; and in 1890 there would be seventy-four.[82]

Edison's central-station technology, the kind that had been transferred to Holborn Viaduct, was not revived, however. During the six years that separated the original and amended electric lighting acts, new technology took root in England. The attention of the professionals and the public shifted to Sebastian Z. de Ferranti, a young British engineer (see pp. 97–98 below). His plans for electric lighting were as bold as Edison's, and he soon became known as the British Edison. English Edison's Holborn station was remembered by only a few and as a history of the failure of technology transfer. But in fact it was not the technology that had failed. The disillusionment of private investors in 1882 after the collapse of stock market speculation together with the passage of the Electric Lighting Act discouraged risk taking. The adverse effects of the lighting act may have been exaggerated by opponents of regulation and municipal socialism, but prudent investors realized that a Parliament that could pass a constraining act in 1882 had the authority and possibly the will to frustrate development. Furthermore, in the 1880s, although local governments had the authority to develop utilities and were often committed to municipal socialism, they did not believe that ratepayers', or taxpayers', money should be invested in a technology which had not yet been proven. Also, municipal governments had invested heavily in gas lighting. The relatively low price of gaslight discouraged investment in an electrical system. It should be noted, however, that in the United States the Edison company (and later its competitors) was able to sell central-station light as a luxury.

The relatively slow development of English electric supply systems, especially in London, is easier to probe with hindsight than it was for analysts in the 1880s. The mid-decade doldrums were thought by most to be a transitory phase, but historical perspective reveals that failure of the Holborn Viaduct station and other projects was symptomatic of a trend. As this study will show, the Holborn failure was soon followed by the Deptford failure (see pp. 238–47 below). Moreover, as late as 1920 London's

[77] *Electrical Engineer* (London) 1 (1888): 380.

[78] *Electrical Engineer* (New York) 8 (1889): 478.

[79] *Electrical Engineer* (London) 2 (1883): 245.

[80] *Electrical Review* 23 (1888): 329–30.

[81] *Economist* (London) 47 (1889): 1269.

[82] Garcke, *Manual of Electrical Undertakings*, p. 394.

Figure III.8.   Emil Rathenau. Courtesy of Berliner Kraft-und Licht AG (formerly BEW).

electric lighting and power facilities were judged backward when compared to the systems employed in other metropolitan centers of the Western world (see p. 260 below). With the exception of a power system on the northeast coast, the general situation in England was also judged to be negative when evaluated in terms of the prevailing criteria of electrical development.

Believing that the economic stagnation of the mid-1880s was an abnormal condition, advocates of the development of electric lighting identified only its immediate causes, such as the 1882 legislation and the competitive price of gas lighting. If they could have foreseen the events of the next three decades, they would have known that after revising the offending legislation, lowering the price of electric lighting, and introducing forms of electric power with which gas could not compete, London and much of England would continue to be labeled backward by electrical engineers and managers when compared to Berlin and Chicago, Germany and the United States. Knowing this, they too would have probed more deeply for long-run, or underlying, causes. With hindsight, the frustrating influence of a series of legislative acts, especially those that were parliamentary in origin, looms large as an underlying cause. These acts, however, reflected the pervasively conservative attitudes of economic, political, and technological interest groups that were quite content with their status and future in a world without electric power systems. For example, electric power supply threatened to overrun and weaken the authority of local governments, to displace the established gaslight technology, and to devalue investments in that older technology and its institutions. The power of these and other conservative interests was revealed clearly during the Deptford episode and during the efforts to bring a unified electric supply system to London (see pp. 247 and 260 below).

The history of the transfer of Edison technology to Berlin is notably different, despite the fact that the system exported to Germany was similar to that installed at Holborn Viaduct. Again, however, the explanation of how the system fared is not simply technical.

The history of the transfer begins at the Paris International Electrical Exhibition of 1881. Among the hundreds of thousands of visitors to the Paris exhibition were three men who are now recognized as outstanding figures in the history of electrical technology, and all three were influential in bringing Edison technology to Germany. Werner von Siemens (1816–1892) already had the reputation of a leading German electrical inventor and manufacturer of telegraph apparatus, and his company, Siemens & Halske, was moving into arc lighting. Emil Rathenau (1838–1915), a relatively unknown mechanical engineer, visited the exhibition hoping to find new opportunities for technological innovation. Having studied in Hannover and Zurich, he was a member of the new generation of engineering-school-trained engineers. Before buying and operating a small machine factory in Berlin, he had worked as an engineer with the Borsig Locomotive Works in Berlin and in the John Penn engineering shops in Greenwich, England. In 1882 Rathenau sold his factory and traveled extensively abroad seeking entrepreneurial opportunities. The third German visitor of note was young Oskar von Miller (1855–1934), who also had been trained at an engineering school (Munich) and would later become the head of a leading Bavarian electrical engineering consulting firm and a pioneer in the de-

Figure III.9.   *Decorative incandescent lamps surrounding statue of Virgin and child, Munich exhibition, 1882. From* Offizieller Bericht . . . München 1882, *p. 65.*

velopment of water power and regional power systems (see pp. 334–50 below).

After viewing the Edison display in Paris, Rathenau decided to buy the rights to the Edison patents in Germany. He negotiated with the Compagnie continentale Edison, which had been established to dispose of Edison patent rights in Europe. In addition to this company, the Edison interests had established the Société électrique Edison to market small, isolated plants and the Société industrielle et commerciale Edison, located in Ivry, near Paris, to manufacture Edison inventions.[83] Rathenau turned to several German banking houses for financing and in July 1882 acquired the patent rights from the French companies. The parties to the agreement were, on the one hand, the banking house of Gebrüder Sulzbach of Frankfort on the Main, Jacob Landau of Berlin, and the National-Bank für Deutschland; and, on the other hand, Compagnie continentale Edison and Société électrique Edison of Paris. By agreement the banking houses (and Rathenau) established a Société d'étude (an exploratory venture to ascertain the technical and commercial feasibility of the Edison system in Germany) with the intent of establishing a permanent enterprise after a stipulated period of months.[84] The Société d'étude proceeded to try the technology and explore the market in Berlin, Munich, and Stuttgart by making several small installations in private clubs and theaters and by displaying the system at an electrical exhibition organized in 1882 in Munich by Oskar von Miller (1855–1934). Rathenau and von Miller came to know each other through the exhibition and von Miller joined Rathenau in the new enterprise.

Persuaded of favorable developments, on 13 March 1883 the three German banking houses founded a permanent company headed by Rathenau. The Deutsche Edison Gesellschaft für angewandte Electricität (German Edison Company for Applied Electricity) was established to exploit "industrially" all patents granted to Edison in Germany related "to measurement, distribution, and application of Electricity for lighting and the transmission of energy." To exploit the patents and the license acquired from Compagnie continentale, German Edison had the right to manufacture or have manufactured in the factories of the foreign Edison companies the apparatus of the Edison system and the right to erect lighting installations or cede this right to others. Cash payment to the Paris company for the rights and license was 350,000 marks. There would also be royalties on incandescent lamps sold and used in Germany and on the installations according to horsepower. German Edison was to deal exclusively in the Edison system of incandescent lighting, but could employ any system of its choice for arc lighting. The company also had the authority to establish subsidiary enterprises. Its capital of five million marks was divided into ten

[83] Conrad Matschoss, "Die geschichtliche Entwicklung der Allgemeinen Elektricitäts-Gesellschaft in den ersten 25 Jahren ihres Bestehens," *Beiträge zur Geschichte der Technik und Industrie* 1 (1909): 55–56.

[84] The 1882 agreement is summarized in an undated (1883?) "Contract, Memorandum of Agreement" entered into by the German banks named above and by the Edison companies in Paris. This document in the Edison Archives is not dated, but is an amendment to the 1882 agreement. On German banks and the electrical industry, see Hugh Neuburger, "The Industrial Politics of the *Kreditbanken,* 1880–1914," *Business History Review* 51 (Summer 1977): 192.

thousand shares. By resolution of the board and a general shareholders meeting, common stock could be increased and preferred stock issued. Besides stock, there were founders' certificates that entitled the holders as a group to take 35 percent of the net profits after a 6 percent dividend had been declared on the paid-up stock. The company issued 2,500 founders' certificates. These, however, did not convey a share, or equity, in the company. One thousand certificates were issued to the first subscribers to stock on the basis of one certificate for ten shares paid up. The Compagnie continentale and the Société électrique in Paris received the remaining fifteen hundred. The Compagnie continentale also had the power to appoint two permanent "commissioners" to guard its rights and interests in the company. The commissioners and directors of the company would get 10 percent of the company surplus remaining after the 6 percent dividend to stockholders, the 35 percent share of the net profits to the holders of founders' certificates, and the 5 percent of the surplus to a reserve fund had been paid.[85]

The Compagnie continentale, with the sanction of Thomas Edison and the Edison Electric Light Company of Europe, Ltd., of New York, not only transferred patent rights and licenses for the manufacture and erection of lighting installations but also agreed to keep German Edison apprised of all Edison inventions until 15 November 1886. In addition, the Paris company would aid and assist German Edison in the technical development of manufacturing facilities and would furnish to German Edison suitable "instructors."[86]

In Germany, prudence dictated that the German Edison Company reach an understanding with the dominant company in electrical technology and manufacturing. In March 1883, therefore, German Edison negotiated a "memorandum of agreement" with the firm of Siemens & Halske. The parties to the agreement besides Siemens & Halske were Thomas Edison; the Edison Electric Light Company of Europe, Ltd.; Compagnie continentale; and the banking houses Gebrüder Sulzbach, Jacob Landau, and the National-bank für Deutschland. According to this agreement, Siemens & Halske, which had tentatively begun manufacturing incandescent lamps with a license from the Swan Company, Edison's competitor, agreed not to oppose Edison incandescent-lamp patents through litigation and to manufacture Edison lamps under license and with royalty payment. Both German Edison and Siemens & Halske, as a result, had Edison manufacturing rights for incandescent lamps. The latter firm, however, agreed not to install Edison lighting in Germany, but to leave this work to German Edison. In turn, German Edison would purchase an Edison type of generator and other equipment for these installations (with the exception of the lamps) from Siemens & Halske. The license to Siemens & Halske to manufacture Edison equipment was to run for ten years.[87]

[85] "Articles of Association of the German Edison Company for Applied Electricity" (1883); a copy is in the Edison Archives.

[86] "Articles of Association," EA.

[87] "Memorandum of Agreement . . . Siemens & Halske," 16 March 1883, EA. See Georg von Siemens, *History of the House of Siemens*, trans. A. F. Rodger, 2 vols. (Freiburg/Munich: Alber, 1957), 1: 92–93; and Conrad Matschoss et al., *50 Jahre Berliner Elektrizitätswerke, 1884–1934* (Berlin: VDI Verlag, n.d.), p. 10.

William Hammer, the experienced Edison electrician, left English Edison to join Rathenau's new company for eighteen months as chief engineer and expert for the transfer of the Edison technology. Hammer left the Edison company because of his differences with Arnold White. Thomas Edison wrote to Hammer, saying, "I am perfectly well satisfied that everything [about your departure] is explained by the name 'Arnold White.' "[88] Years later, according to Hammer, Edison confided that Hammer must have been up against "the two meanest men I ever met . . . Arnold White

[88] Edison to W. Hammer, 4 August 1883, Hammer Collection.

Figure III.11.   *Thomas Edison and Emil Rathenau. Courtesy of Berliner Kraft-und Licht AG (formerly BEW).*

in England & Emil Rathenau in Germany."[89] (See Fig. III.11.) But in the summer of 1883, Hammer was acting as liaison for Rathenau with Edison. Hammer informed Edison that the German company was looking beyond the isolated installations of the exploratory enterprise to two central stations in Berlin: one to supply a block of buildings, including the Café Bauer, a well-known Berlin establishment, the lighting of which would attract considerable attention; and another to light central Berlin with fifty thousand incandescent lamps. The latter would function as a model for central stations throughout Germany, just as Pearl Street had done for the United States and Holborn Viaduct had been expected to do for Britain. Hammer made it clear that Rathenau preferred to obtain his technical information directly from Edison rather than from Paris. Rathenau wanted plans, blueprints, and detailed specifications for central stations based on Edison's experience. The layout of a central-station distribution system especially concerned Rathenau, so Hammer wanted to know if a small-scale analog model of distribution should be built, such as the one done at Menlo Park in preparation for Pearl Street, or whether Edison could meet the requirements sent by the German company by supplying a detailed plan. As agreed, Rathenau also wanted word of the newest Edison inventions disclosed to German Edison immediately. Siemens & Halske, anticipating future orders from German Edison, was already disassembling Edison generators to analyze construction details. In addition, Rathenau wanted a well-informed Edison man to come to Germany to assist in the establishment of an incandescent-lamp factory. In 1883, Rathenau, though obviously dependent on Edison in electrical matters, was already planning modifications and adaptations in steam-engine design, an area in which he had expertise and in which Edison had relied on others.[90]

By the fall of 1883 the bloom was off the rose. Rathenau himself wrote to Edison telling him that he and his Paris friends had gotten German Edison into "a very sad dilemma." The German company had cultivated the soil well for "the grain we meant to sow," but the lead in the European lighting business had passed to others. Among the problems were these. Because Thomas Edison had told Rathenau that the Edison type-H generator would run five hundred lamps, he had proceeded to order fifteen generators from Siemens & Halske and to take orders from customers. The engineers at Siemens & Halske had opined that the generator would supply only four hundred lamps, but, relying on Edison, Rathenau had gone ahead selling the machine for five hundred. When so tried, the armatures of the machines burned through the first day. At considerable expense Rathenau had to compensate for these losses. Also, Edison had not sent the detailed plans for generators that would supply a hundred lamps, and these were the machines for which competitors were finding a lively market in Germany. Scarcely any reliable help for these and other problems was forthcoming from Paris, and the state of affairs was becoming unbearable as the business went to others, despite the effective preparations and advertising of German Edison.[91]

[89] Note by Hammer about his conversation with Edison in 1889, Hammer Collection.
[90] Hammer to Edison, 16 July 1883, EA.
[91] Rathenau to Edison, 13 October 1883, EA.

Another problem was the status of the Edison incandescent-lamp patents in Germany. These were not secure, in part because of the Swan patents. Edison was advised by Joshua Bailey, a representative for the Edison Electric Light Company of Europe and the Compagnie continentale, to come to an understanding with Swan in Germany. Bailey also advised Edison to send, as requested, a thoroughly competent person, like Francis Upton, to Germany to help Rathenau place an incandescent-lamp factory in operation. German Edison was prepared to pay him up to $300. The alternative, persuading German Edison to purchase lamps from the United States, was not a viable option because German nationalism would bring Germans to buy lamps made in Germany: "You may think," Edison was told, "that national feeling does not weigh against economy, but that is a complete error. Any German interloper would have all the national sympathies, and would hold the German market against a foreign lamp that should be both cheaper and better." Furthermore, German patent law required that within three years from its granting, the patent should be worked to an extent adequate to supply the market. Since Siemens was manufacturing only one hundred lamps a day, German Edison needed to place a factory into production. The purchase of lamps from the Paris factory was impractical because the quality of the lamps was uneven. Also, James Hipple, who advised the Paris company on lamp manufacture, was believed to be out of touch with newly developed techniques in the United States. Thomas Edison was also urged to provide an economic analysis of the costs of the Pearl Street station since it began operating in the fall of 1882. Despite publicity from the United States to the effect that large profits were accumulating, Bailey had been unable to confirm this. Investors were reluctant to invest in central stations in Europe until credible figures were available. They had heard rumors that the New York utility was operating at a loss, but even a statement of loss might have been preferable to the rumors.[92] In fact, as internal correspondence revealed, the Edison Illuminating Company was losing money, but apparently Edison and his advisers were not inclined to share the information (see p. 45 above).

Nevertheless, German Edison went ahead with its plans to build two central stations. Before the technical problems could become the center of attention, however, political obstacles had to be overcome. Rathenau, Oskar von Miller, and the other managers of German Edison had probably learned of the frustrating political situation in Britain and anticipated government reaction with trepidation. The Berlin government had among its legislators and officials persons whose views paralleled the views expressed in Britain. Berlin owned its gas utility, and there were those who wanted the city to own and operate an electric utility as well. The central, or national, government deferred to the city government because, as in the United States, constitutional arrangements provided local authorities independence in the regulation of utilities.[93] The government of the state of Prussia expressed some interest in framing special provisions to govern the incorporation and operation of electrical enterprises, but serious moves in this direction

[92] J. Bailey to Edison, 21 October 1883, EA.
[93] G. Siegel, *Die Elektrizitätsgebung* (Berlin: VDI, 1930), pp. 1, 39–40.

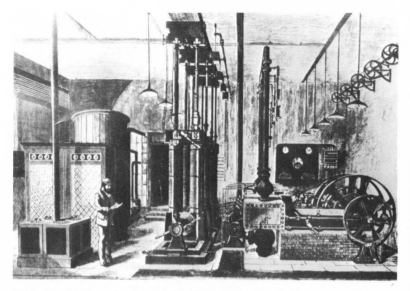

Figure III.12.   *Central station at Friedrichstrasse 85 and Unter den Linden, Berlin, 1884. From Matschoss et al., 50 Jahre, p. 11.*

Figure III.13.   *Laying an underground supply system. From Matschoss et al., 50 Jahre, p. 104.*

did not come until after the turn of the century.[94] In the Berlin municipal council the prevailing argument—also advanced in London—was that private enterprise should take the risks of innovation, probably because the granting of a charter to the German Edison company provided for detailed regulation of supply and a share of profits for the municipal government. As Burgomaster Duncker summed it up, "All risks fall on the private company; all financial advantage to the city."[95]

The city of Berlin and German Edison reached agreement in February 1884 (see p. 185 below).[96] The company obtained authority to supply electricity to a defined area in the heart of Berlin using the city's streets for the distribution system. German Edison then proceeded to establish an operating utility much in the manner that the Edison Electric Light Company had established the Edison Electric Illuminating Company of New York. On 8 May 1884 the Berlin utility, Städtische Elektrizitäts-Werke (StEW), was founded with an authorized capital of three million marks.

In September the company began operating a small (100-kw.) lighting plant at Friedrichstrasse 85 to supply the Café Bauer and the block of buildings in which it was located (see Fig. III.12). There were numerous problems, the character of which is suggested by such stories as that of

[94] Georg Dettmar, *Die Entwicklung der Starkstromtechnik in Deutschland* (Berlin: ETZ Verlag, 1940), pp. 139, 293–94.

[95] Conrad Matschoss, "Geschichtliche Entwicklung der Berliner Elektricitäts-Werke von ihrer Begründung bis zur Übernahme durch die Stadt," *Beiträge zur Geschichte der Technik und Industrie* 7 (1916): 9.

[96] "Vertrag zwischen der Stadtgemeinde Berlin und der Deutschen Edison-Gesellschaft Berlin," 6 and 19 February 1884; a copy is in the library of the Deutsches Museum, Munich.

Oskar von Miller in the basement tending the generators while Rathenau soothed cafe customers when the lights suddenly went out. The Markgrafenstrasse central station, the first constructed by StEW, opened in August 1885. This, too, had problems, but not as many as the Pearl Street station, where, the Germans took pains to point out, "in the early days it had been impossible to operate one of the generators an entire day without repairs."[97] This opinion contrasts remarkably with the memoirs of the Edison pioneers who operated the New York station and who reported that, with the exception of one three-hour interruption, the eight Jumbo generators produced current without a break from September 1882 until fire destroyed the station in 1890.[98]

The generators in Markgrafenstrasse were the Edison type built by Siemens & Halske, and the capacity of the station was 754 kw., of which 492 kw. were for incandescent lamps. Pearl Street had used 540 kw. for incandescents in 1882. In the case of Markgrafenstrasse, however, there were twelve 41-kw. incandescent-lamp generators instead of six 90-kw. Jumbos as installed at Pearl Street. The German company also used single steam engines to belt-drive two incandescent-lamp generators and one 27-kw. arc-lamp generator instead of one engine for one generator as at Pearl Street. Current was distributed at 100 volts by a two-wire network.[99] (See Figs. III.14–III.16.) Because of the use of the basic Edison design and because of the relatively small size of the machines, the German station was not a bold technical venture. In 1886 the company opened another station at 80 Mauerstrasse, but its capacity of 285 kw. was smaller than that of its predecessor.

The incandescent-lighting load carried by the two central stations in the heart of Berlin constituted luxury lighting. In 1886 the largest percentage of the load was for "theaters" (24 percent), followed by banks (20 percent), eating and drinking establishments (20 percent), shops (17 percent), hotels and guesthouses (8 percent), street lighting for the city (7.5 percent), industry and trade (2 percent), residences (1 percent), and miscellaneous (.5 percent).[100] As in New York and London, electric lighting in Berlin was not for the poor.

German Edison and StEW began training technical staff and relying less than at the start on foreign Edison men and information from abroad. In 1884, before either of the German central stations was completed, William Hammer left Berlin. He had had serious disagreements with Rathenau "about certain Englishmen," but the antipathy seems to have been deep and not localized to a particular episode. In explaining his departure to Thomas Edison, in whose good graces he wished to remain and to whom he had explained earlier misunderstandings with the English company, Hammer besmirched the reputation of Rathenau. Referring to him as that "Jew Emil Rathenau, a man utterly unfit to be at the head of any organization," Hammer reported that because of numerous past transactions Rathenau bore anything but a good name in Berlin. Hammer believed that,

---

[97] Matschoss et al., *50 Jahre*, p. 100.

[98] Jones, *History of the Consolidated Edison System*, p. 203.

[99] Matschoss et al., *50 Jahre*, pp. 101, 104, 138, 146, 172, 192, 209.

[100] Arthur Wilke, *Die Berliner Elektrizitäts-Werke* (Berlin, 1890).

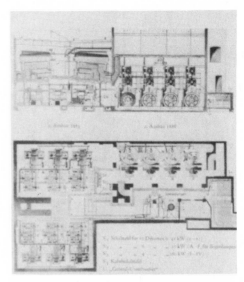

Figure III.15. *The Markgrafenstrasse central station: Stages of expansion, 1885 and 1888. From Matschoss et al.,* 50 Jahre, *p. 101.*

Figure III.14. *Interior of the Markgrafenstrasse central station, 1887. From Matschoss et al.,* 50 Jahre, *p. 104.*

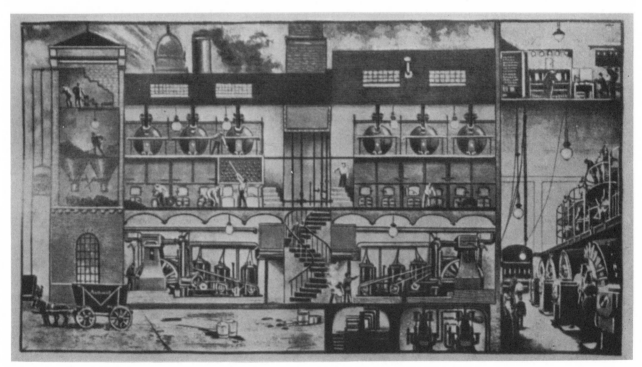

Figure III.16. *Drawing of the interior of the Markgrafenstrasse station, 1886. Courtesy of Rudolf von Miller.*

among other unfortunate characteristics, Rathenau harbored prejudice against foreigners (a characteristic obviously shared by Hammer). To further justify his own difficulties, Hammer listed for Edison other employees who had recently left German Edison because of Rathenau.[101]

As the growing pains of Rathenau's enterprise attenuated, the Germans became increasingly restless with the ties that bound them to the Edison company in Paris and with the Siemens company in Berlin. The technology had been transferred and adapted and the Germans were ready to develop their own style. A letter signed by both Rathenau and Oskar von Miller, but probably written by von Miller, whose good nature and diplomatic skills were legendary, informed Edison of the success of German Edison during its first two years of operation. Having installed 42,000 incandescent lamps, not including those of its licensees, the German company had done more business in Europe than any other European Edison company and more than the Edison companies in the United States, considering the shorter time it had been in existence. Edison was told he should drop his prejudices against his European associates, for the Germans would "glorify" his name and "honor" his inventions.[102] The glory might be short-lived, however. Edison's own sources warned of an impending stalemate and possible rupture between the German company and the Edison company in Paris. The Paris Company was in poor condition, and its contractual links with German Edison frustrated the innovations of the latter company. Unless strong and resourceful measures were soon taken, Edison would lose not only income from the Paris company but his income from Germany as well. A strong entrepreneur was needed to clarify and reorganize the Edison enterprises in Europe.[103]

Henry Villard (1835–1900) promised to do this—and much more—for Edison. Villard, who was born in Rhenish Bavaria of an upper-middle-class family (Hilgard), emigrated to America in 1853. There he achieved recognition as a correspondent for the *New York Tribune* during the Civil War and married Fanny Garrison, daughter of William Lloyd Garrison, the abolitionist. Drawing on his entrepreneurial skills, his friendship with J. Pierpont Morgan, and his good financial connections in Germany, Villard organized a railway empire in the U.S. Northwest, serving as president of the Northern Pacific Railroad (with Morgan's backing) from 1881 to 1884. During these eventful years, he also purchased controlling interest in the *New York Evening Post* (1881) and took an active role in the affairs of the Edison Electric Illuminating Company of New York while the Pearl Street station was being built. He was a director of the EEIC for almost a decade, and he also served on the board of the Edison Electric Light Company during its early years. During the depression of 1883–88, however, the financial exigencies of the Northern Pacific Railroad forced him to resign from that company and he returned to Germany, where he renewed his association with German banking houses.

Villard first discussed his plans for reorganizing the Edison enterprises

---

[101] Hammer to Edison, 16 June 1884, EA.

[102] E. Rathenau and O. von Miller to Edison, 29 June 1885, EA.

[103] Francis R. Upton (in Berlin) to Edison, 3 June 1886, and Upton to Samuel Insull, 7 June 1886, EA.

Figure III.17. *Deutsche Edison*
Gesellschaft administration building,
Markgrafenstrasse 44 (middle building).
Note stack for central station in the rear.
From Matschoss et al., 50 Jahre, *p. 14.*

in Europe with Francis Upton, Edison's assistant at Menlo Park, who had become head of lamp manufacture in the United States. Upton, who was touring Europe in 1886, was persuaded that Villard had the resources, the intelligence, and the entrepreneurial ability to achieve his objectives. These objectives were to raise money in Germany to stimulate central-station construction in Europe and in the United States, to resolve the differences between German Edison and the Paris company, and to negotiate a new relationship between German Edison and Siemens & Halske. He also wanted to stimulate the central-station business of German Edison, possibly by reorganizing that company.[104] Among other results promised was a substantial increase in the returns to Thomas Edison and the Edison Electric Light Company of Europe.

Villard returned to the United States in 1886, but the plan he had put into motion bore fruit. In May 1887 the German Edison Company was reorganized and took the new name Allgemeine Elektrizitäts-Gesellschaft (General Electric Company); the contract of German Edison with Compagnie continentale Edison was abrogated after substantial payment was made to the Paris company; the contractual arrangement with Siemens & Halske was revised; and the capital of Allgemeine Elektrizitäts-Gesellschaft (AEG) was raised from five million to twelve million marks to provide for expansion of the two Berlin central stations and the building of two additional stations in the city. The AEG took over management of the Städtische Elektrizitäts-Werke Berlin and changed its name to Berliner Elektricitäts-Werke (BEW) (see pp. 183 ff. below for the company's subsequent history). Thomas Edison was not entirely pleased with these changes, how-

[104] Upton to Edison, 7 and 10 June 1886, and H. Villard (by C. A. Spofford) to Edison Electric Light Co. of Europe, Ltd., New York, 29 June 1886, EA.

ever. AEG was no longer an "Edison" company, and the new companies did not even carry his name.[105]

From the perspective of technology transfer, the initiative taken by AEG and BEW to develop their own distinctive style of central-station technology is of particular interest. Indicative of this emerging trend was the design of the extension of the Markgrafenstrasse station in 1888 and the opening of the first of the new central stations at Spandauerstrasse 49 in the fall of 1889. The Markgrafenstrasse extension had directly coupled steam engines and generators, and the Spandauerstrasse station had unprecedentedly large generators driven by 1,000-h.p., slow-running steam engines directly coupled to specially designed multipolar generators built by Siemens. Thomas Edison disapproved of such large units, preferring, in the face of erratic loads, the flexibility offered by smaller ones. Edison also used high-speed steam engines that were belt-coupled to generators. The steam engines for Spandauerstrasse and for the station that would open at Schiffbauerdamm in 1890 were ordered from the Belgian factory van der Kerkhoven.[106] John Beggs, vice-president of the Edison Illuminating Company in New York, described the German engines as an advance unmatched in the United States. Edison viewed them with skepticism during his visit to Berlin in 1889.[107]

Villard's involvement with electrical affairs was by no means at an end. He returned to the United States in 1886 with a commission from the Deutsche Bank to negotiate with Drexel, Morgan and Company about the investment of German money in American enterprises, including electrical companies. In 1889 Villard took the lead in reorganizing the Edison enterprises in the United States into a new company named the Edison General Electric Company. It included the old Edison Electric Light Company, Edison Lamp Company, Edison Machine Company, Bergmann and Company, and the manufacturers of other components. More than half of the capital of Edison General Electric was furnished by German banks and other German sources.[108] Villard then proceeded to take an active part in promoting the merger of Edison General Electric and the Thomson-Houston Electrical Manufacturing Company to form the General Electric Company, a merger that was sponsored in part by Drexel, Morgan and Company. Again, Edison was not pleased to see his name eliminated.

The Edison style was indelibly impressed upon the history of electrical manufacturing and supply in Germany and the United States, however. Only in England did the technology transfer fail. The most penetrating explanation for the failure in London and the success in Berlin is neither technological nor economic; it is political. The comparative price of illuminating gas and electric lighting seems to have been about the same in

[105] Allgemeine Elektrizitäts-Gesellschaft to Edison, 19 February 1889, EA.

[106] Felix Pinner, *Emil Rathenau und das elektrische Zeitalter* (Leipzig: Akademische Verlag, 1918), p. 141.

[107] W. von Miller, *Oskar von Miller*, p. 52; comments of John Beggs quoted in Jones, *History of the Consolidated Edison System*, pp. 330–33.

[108] Henry Villard, *Memoirs of Henry Villard: Journalist and Financier*, 2 vols. (Boston: Houghton Mifflin, 1904), 2: 320–26. See also Villard to Charles Coffin, 11 March 1889 (129-60-61) and 3 April 1889 (130-62-3), Houghton Library, Harvard University, Cambridge, Mass.; the numbers in parentheses refer to box, folder, and letter number, respectively.

both cities, and financing was available in London until a combination of speculation and restrictive laws dampened investors' enthusiasm. On the other hand, in the realm of political power, the situations in the two cities were different. Rathenau and von Miller, in alliance with the investment banks and industrial interests (Siemens), had sufficient power to persuade the local government to clear the way for their electrical ventures. In London, the alliance of aristocratic money and influence, scientific eminence, and technical competence contended with a host of conflicting interests in Parliament. The result was confusion, contradiction, and parliamentary constraint. The next decade revealed even more clearly the primacy of politics in Britain and a coherent political economy in industrial Germany.

CHAPTER IV

Reverse Salients and Critical Problems

HAVING explored the nature of the invention, development, and transfer of Edison's direct-current system, we now turn to the endeavor of inventors and engineers to maintain the momentum by solving the critical problems that frustrated growth of the system. "Evolution" is an apt metaphor for an electric supply system like Edison's because internal forces alone did not direct its growth; the system grew within a context of geographical, economic, political, and organizational factors. This chapter will focus on the inventions and developments that allowed and promoted the evolution of the Edison system within this context. Growth was not foreordained; it was usually promoted. Special note will be taken of the process by which a new system emerged as a result of the failure to solve a major problem in the old system. At the end of this process, the old and new systems existed for a time in a relationship of dialectical tension, a struggle called "the battle of the systems."

Several terms that will be used in discussing evolving systems need further definition. The first is "reverse salient." The term is customarily used to identify that section of an advancing battle line, or military front, which is continuous with other sections of the front, but which has fallen behind or been bowed back. "Reverse salient" became a household expression during World War I because of the protracted struggle of the Germans to eliminate the reverse salient along the western front at Verdun. This concept is preferable to "disequilibrium" or "bottleneck," which some economists and economic historians use, because the concept of a reverse salient refers to an extremely complex situation in which individuals, groups, material forces, historical influences, and other factors have idiosyncratic, causal roles, and in which accidents as well as trends play a part. "Disequilibrium" suggests a relatively straightforward abstraction of physical science, and "bottleneck" is geometrically too symmetrical.

The idea of a reverse salient suggests the need for concentrated action (invention and development) if expansion is to proceed. A reverse salient appears in an expanding system when a component of the system does not march along harmoniously with other components. As the system evolves toward a goal, some components fall behind or out of line. As a result of the reverse salient, growth of the entire enterprise is hampered, or thwarted, and thus remedial action is required. The reverse salient usually appears as a result of accidents and confluences that persons presiding over or managing the system do not foresee, or, if they do foresee them, are unable to counter expeditiously. The causes of the lag can arise from within the system; from its environment, or context; or from some complex combination thereof. The reverse salient will not be seen, however, unless inventors, engineers, and others view the technology as a goal-seeking system.

An evolving system moves in any given direction for complex reasons. The factors that cause some components to fall behind or out of line vary according to time and place, as the episodes in this study show. Movement often results when those who are presiding over the system want growth to extend their authority. On other occasions growth improves the economy of the system. Economy and efficiency—the first cherished by managers; the second, especially by engineers—also give direction to the movement of a system. As noted, analysis of a growing system often reveals the inefficient and uneconomical components, or reverse salients.

Innumerable (probably most) inventions and technological developments result from efforts to correct reverse salients. Outstanding inventions and developments in electric lighting and power during the two decades after 1880 were responses to reverse salients. Independent inventors—and later, inventors and industrial scientists hired by business enterprises—applied themselves to solving the problems inherent in reverse salients. They approached the challenge as inventors and engineers usually do—by defining the reverse salient as a set of "critical problems," problems whose solution would bring the system back into line—i.e., correct the reverse salient. Outstanding inventors, engineers, and entrepreneurs usually have a record of defining and solving such problems.[1]

How did the system builders of the 1880s identify critical problems? As observed in the case of Edison, a person conceptualizing a technology systematically or holistically often recognized inadequacies in the patterns formed by the system's components and networks. Those who did not observe system growth firsthand surveyed the publications of others who did. Professionals knew that the competition congregated at reverse salients and critical-problem sites. A major reason for the simultaneity of invention, or the simultaneous solution of critical problems, was that inventors inferred from their familiarity with the state of the art the sites, figurative and literal, of critical-problem-solving activity and the nature of the problems. The repeated clustering of patents from different inventors within common subject categories was one indication of the collective focus. Other inventors' patent applications, as recorded regularly in technical journals, were especially revealing.[2]

[1] Elmer Ambrose Sperry (1860–1930) was a remarkably adept solver of critical problems. His record includes almost 250 patents, and these, whether actually used or not, were solutions to critical problems, some relatively trivial, some of major importance. See Thomas P. Hughes, *Elmer Sperry, Inventor and Engineer* (Baltimore: The Johns Hopkins Press, 1971), esp. pp. 64–70.
[2] Scholars in various fields have attacked the problem of simultaneity of invention. See, for instance, a recent essay by Arthur P. Harrison, Jr., "Single-Control Tuning: An Analysis of an Innovation," *Technology and Culture* 20 (1979): 314–15; and an earlier and often-cited

Inventors like Edison also set out to invent and develop new systems. Sometimes—and in this chapter such circumstances will be encountered—they invented the nucleus of a new system while responding to a reverse salient in an existing one. The solution to the critical problem, in other words, would not harmonize with the existing system's components, so the inventor went on to invent or use other components that did harmonize with his newly invented component.

The electric system introduced by Edison and his associates at the Pearl Street station in 1882 was a direct-current, limited-area-of-distribution system. It suited heavily populated urban areas, but through adaptations it continued to evolve in the twentieth century. Because it was a loser in history, however, giving way over a period of decades to the polyphase, universal system, most reminiscing engineers and managers and most historians have tended to ignore the steady improvements that were made in it and to focus instead upon the dramatic rise of the victorious current in "the battle of the systems." Such bias obscures the fact that as late as the first decade of the twentieth century, some leading engineers still argued that direct current was better for the city than the polyphase system when measured against conventional yardsticks of profitability, reliability, and efficiency.

The direct-current system is an excellent example of a system that evolved successfully, up to a point. Reverse salients were defined as critical problems, and with an exception of surpassing importance—the high cost of transmission—they were solved. There were many categories of improvements. Edison and Upton increased the efficiency of the d.c. generator, and hundreds of inventors subsequently improved on their work. Magnetic fields were wound for higher efficiency, the tendency of armatures to heat up was reduced, sparking brushes were dampened, fluctuations in output under changing load were regularized, and the transfer of energy of the magnetic field to the armature was improved. Countless patents in the industrial countries are evidence of the efforts that were made to solve these critical problems.

In Germany, Friedrich von Hefner-Alteneck (1845–1904), an associate of Werner von Siemens's and a pioneer in the design of arc-light systems, learned that central-station operators like Emil Rathenau were dissatisfied with the belt-driven, small-output, high-speed generators built by Siemens & Halske, an Edison licensee. In 1886, working with Karl Hoffmann, Hefner-Alteneck then developed a large, slow-speed, internal-pole generator suited for direct coupling to an efficient slow-speed steam engine.[3] Rookes E. B. Crompton (1845–1940), an English electrical inventor and engineer, introduced in England a directly coupled generator that was matched to

article by Robert K. Merton, "Singletons and Multiples in Science," in Merton's *The Sociology of Science*, ed. Norman W. Storer (Chicago: University of Chicago Press, 1973), pp. 343–82. For an analysis of the phenomenon in science, see Thomas S. Kuhn, "Energy Conservation as an Example of Simultaneous Discovery," in *Critical Problems in the History of Science*, ed. Marshall Clagett (Madison, Wis.: University of Wisconsin Press, 1959), pp. 321–56.

[3] Georg von Siemens, *History of the House of Siemens*, trans. A. F. Rodger, 2 vols. (Freiburg/Munich: Alber, 1957), 1: 96–98.

the British high-speed Willans engine.[4] To increase generator capacity, inventors and manufacturers, such as the German firms Helios and C. & E. Fein, increased the number of field poles. In the eighties, Wilhelm Lahmeyer, a German inventor and manufacturer, introduced generators that were designed to reduce magnetic and electrical losses. The firm of Pöge in Germany introduced carbon brushes to take the current from the commutator.[5] Carbon brushes, like many of the other improvements made in the system, were introduced at about the same time in the United States, the United Kingdom, and other industrial countries.

When improvements reduced the costs of generating electricity, engineers and managers then sought to achieve a motor load that would compete with steam and gas engines. Improvements in d.c. electric-motor design then became a critical-problems category. Inventors made improvements in motors that were suited for a particular type of mechanical load. Motors for streetcars had to stop and accelerate often and had to take the abuse of poorly laid rails and the strain of frequent turns. Motors for elevators had to respond to even more frequent accelerations and needed to be unusually reliable to avoid stalling and trapping passengers. Motors for small home appliances such as fans and sewing machines had to be designed to operate without skilled maintenance. Early models of these appliances were battery powered; later they were designed for connection to central-station systems.

In the United States, the young inventor Frank J. Sprague (1857–1934) won a reputation for the excellence of his inventions and developments related to the problems of d.c. motors. In 1884 he exhibited his first motor; two years later he took a 15-h.p. central-station motor operating at 220 volts and installed it in a freight elevator in Boston, Massachusetts; in 1887–88, using his motors, he constructed a practical streetcar system in Richmond, Virginia.[6] He also responded to the complex problem of controls for the elevators and coupled electric cars used in subways and elevated transportation sytems. His controls for the latter were known as the multiple-unit system. Sprague is a superb example of a critical-problem-solving inventor and engineer who was obviously informed about reserve salients.

Sprague is remembered primarily for his work with large motors. Other inventors found ways of solving the problems of the small motors used in home appliances. In 1884, in the United States, Philip Diehl (1847–1913) invented a variable-speed d.c. motor for dental machines, and later it was adapted for use on sewing machines. In 1887 Schuyler Skaats Wheeler (1860–1923), who is credited with inventing the electric fan, was, with Charles Curtis and Francis Crocker, manufacturing motors designed to

[4] Brian Bowers, *R. E. B. Crompton* (London: Science Museum, 1969), p. 12; Percy Dunsheath, *A History of Electrical Power Engineering* (Cambridge, Mass.: M.I.T. Press, 1962), p. 148.

[5] For improvements in d.c. generators, see Georg Dettmar, *Die Entwicklung der Starkstromtechnik in Deutschland* (Berlin: ETZ Verlag 1940), pp. 36–60. Dettmar is exceptional in his attention to the improvements made in the d.c. system.

[6] On Sprague and the development of industrial and traction motors, see Harold C. Passer, *The Electrical Manufacturers, 1875–1900* (Cambridge, Mass.: Harvard University Press, 1953), pp 237–40; on his construction of the famous Richmond streetcar system, see ibid., pp. 242–55.

operate on incandescent-light circuitry.[7] Other inventors addressed similar problems in Europe and the United Kingdom.[8] Following Edison's and Swan's nearly simultaneous introduction of the high-resistance filament, countless inventors improved upon its performance, thus lowering the initial and operating costs of these lamps. The long and important history of improvements in the incandescent lamp cannot be chronicled here, but it should be noted that the solution of problems related to this lamp was especially important for the growth of central-station systems during the first decade and a half of their history, when incandescent-lamp load was much greater than the industrial power and traction load of most systems.[9]

In the 1880s inventors and engineers steadily improved the generators, motors, and other components of the direct-current system. With one outstanding exception, the problems of the system were being solved reasonably well; the exception was the high cost of distribution, especially long-distance transmission. Edison had identified the cost of distributing electricity by wire or cable as a major reverse salient in the system he was designing. Distribution costs continued to be about the same as generation costs and rose prohibitively higher in the 1880s when the radius of distribution from a d.c. station was extended more than a mile. The majority of engineers and station managers did not believe that long-distance "transmission" (as opposed to short-distance "distribution") of direct current was feasible without storage batteries.

In 1883 Edison responded, if not entirely successfully, by introducing his three-wire system. The reverse saliency of distribution costs for electric systems is reflected in the nearly simultaneous invention of three-wire distribution by at least three inventors. The Pearl Street distribution network of 1882 incorporated Edison's feeder-and-main arrangement, which reduced costs below what was required to run a simple two-wire layout. Yet Edison and his associates realized that central stations would not find a market unless the cost of distribution was further reduced. This was especially true in less densely populated districts. Edison's official biographer writes that Edison, "being firmly convinced that there was a way out . . . pushed aside a mass of other work, and settled down to his problem."[10] Surely this was a reverse salient. On 20 November 1882, only two months after the Pearl Street station began operating on a regular basis, Edison executed a patent application entitled "System of Electrical Distribution" (U.S. Patent no. 274,290) (see Fig. IV.1). In the trade it came to be known

[7] On the manufacturers of d.c. motors in the 1880s, see Malcolm MacLaren, *The Rise of the Electrical Industry during the Nineteenth Century* (Princeton: Princeton University Press, 1943), pp. 92–95.

[8] See Dettmar, *Starkstromtechnik*, pp. 36–61; Dunsheath, *History of Electrical Power Engineering*, pp. 182–87; and E. J. Holmyard, A. R. Hall, and T. I. Williams, eds., *History of Technology* (Oxford: Clarendon Press, 1958).

For developments in France, see Robert Moïse and Maurice Daumas "L'Electricité industrielle," in *Histoire générale des techniques*, ed. M. Daumas et al. (Paris: Presses Universitaires, 1978), pp. 337–66.

[9] The history of incandescent lamps is given in detail in Arthur A. Bright, Jr., *The Electric-Lamp Industry: Technological Change and Economic Development from 1800 to 1947* (New York: Macmillan, 1949). Bright includes the history of British and European developments.

[10] Frank L. Dyer and Thomas C. Martin, *Edison: His Life and Inventions*, 3rd ed., 2 vols. (New York: Harper & Bros., 1930), 1: 424.

Figure IV.1. Drawings from Hopkinson's patent for a three-wire distribution system.

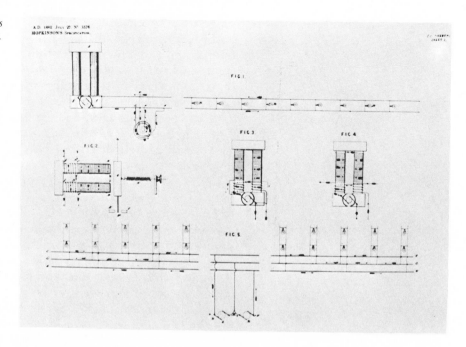

Figure IV.2. John Hopkinson. Courtesy of the Institution of Electrical Engineers, London.

as the three-wire system, and it continues to be used for distribution today. The three-wire arrangement saved 60 percent of the copper needed to operate Edison's two-wire, feeder and main, network.

John Hopkinson (1849–1898), British inventor, engineer, and scientist, patented a three-wire system a few months earlier (British Patent no. 3,576 dated 27 July 1882). The patent application provides a lucid description of the invention:

> For the purpose of economising the cost of main conductors I place two dynamo machines in series and place two systems of lamps or other appliances consuming electricity of approximately equal capacity also in a series. A main conductor is taken from each extreme pole of the two dynamos to points between the two systems of lamps, the intermediate conductor serving to bring back to the central station any electricity required for one system of lamps in excess of that required by the other system of lamps.[11]

Edison's representative in London, E. H. Johnson, asked the young American naval officer, Frank J. Sprague, who was then reviewing the Crystal Palace exhibition of 1882, to make a report on the Hopkinson system. When Sprague returned to America in 1883, he went to work for Edison, who assigned him to the installation of the pioneering three-wire system in the small town of Sunbury, Pennsylvania (see pp. 432–33 below).

In Germany, meanwhile, Wilhelm von Siemens, on behalf of Siemens & Halske, also applied for a three-wire-distribution patent, but his application was rejected because of the priority of Hopkinson's patent.[12] Nevertheless, the three-wire system is an excellent example of the near simultaneity of perception of a reverse salient.

[11] Quoted in James Greig, *John Hopkinson: Electrical Engineer* (London: Science Museum, 1970), p. 33.
[12] Dettmar, *Starkstromtechnik*, p. 117.

Another ingenious attempt to solve the distribution-transmission problem involved storage batteries. Storage batteries were used to improve the load factor; in addition, batteries were utilized for d.c. transmission. Direct current taken from generators at over a thousand volts was transmitted over substantial distances (at reasonable cost because of the high voltage) and was then fed into batteries connected in series; each battery took an appropriate increment of the high voltage in a manner similar to the connecting of arc lamps in series in a high-voltage circuit. After charging, the batteries were disconnected from the generators and connected in parallel to obtain a low voltage appropriate for the distribution network.

Because alternating-current transmission ultimately displaced direct-current and storage-battery substations, historians have tended to ignore the battery installations. R. H. Parsons, however, historian of the first generation of central power stations, has succinctly recounted the history of the use of storage batteries by the Colchester, England, undertaking and by the Chelsea Electricity Supply Company in London. The South Eastern Brush Electric Light Company, Ltd., built the Colchester system, which began operating in 1884. The Brush Company, British counterpart of American Edison, pioneered in the introduction of arc-light systems, and the Colchester plan was drawn up to accommodate Edison's incandescent electric lighting system. High-voltage Brush arc-light generators, which enjoyed a good reputation, were installed in the central station at Colchester and transmitted current to five battery substations located in cellars beneath the town's shops. Batteries that were charged in series discharged in parallel at 60 volts. The organizers of the plan intended to supply power to two thousand incandescent lamps, but they failed to do so because the batteries caused unending and insurmountable problems.[13]

On the other hand, for about forty years the Chelsea Electricity Supply Company successfully used batteries as part of its transmission-distribution system. It used Brush-Victorian generators to send out direct current at more than 1,000 volts. When the load on the system was relatively low, the generators charged the batteries, which were connected in series, as at Colchester; when the load was relatively high, the batteries, which automatically reconnected in parallel, discharged to the load at about 100 volts. The company also used motor generators in the substations to act as "continuous-current transformers." After 1893 the company supplied virtually all of Chelsea. Only in 1928 when standardization was introduced throughout England, did it convert its system to alternating current.[14]

The simultaneous invention of three-wire distribution systems (later, as many as five wires were tried), use of the complex combination of generators and imperfect storage batteries, and experiments with high-voltage, direct-current transmission (see pp. 131 and 234 below) testify to the recognition that the high cost of low-voltage distribution and transmission

[13] R. H. Parsons, *The Early Days of the Power Station Industry* (Cambridge: At the University Press, 1940), pp. 52–55. For exhaustive accounts of the neglected history of the storage battery, see Richard H. Schallenberg, *Bottled Energy: Electrical Engineering and the Evolution of Chemical Energy* (Philadelphia: American Philosophical Society, 1982); and idem, "The Anomalous Storage Battery: An American Lag in Early Electrical Engineering," *Technology and Culture* 22 (1981): 725–52.

[14] Parsons, *Early Days of the Power Station Industry*, pp. 58–65.

constituted a reverse salient. Edison, it should be recalled, identified the cost of distribution as a critical problem when inventing his system. In the early 1880s various inventors and engineers defined the critical problem differently in an attempt to improve the situation. The definition that ultimately resulted in a practical solution came from a Frenchman working with an Englishman in London. However, neither their contemporaries nor historians have given Lucien Gaulard and John Gibbs the recognition they deserve. Their history is one of litigation, counterclaims for priority, frustrating technical problems, and personal tragedy.

More than any other inventors, the Frenchman Lucien Gaulard and his British business partner, John D. Gibbs, are responsible for the series of inventions and developments that culminated in the solution of the transmission and distribution problem. Gaulard and Gibbs demonstrated that by using alternating current and transformers, high voltages could be employed for the economic transmission of electricity and low voltage for distribution at the point of consumption.

The history of Gaulard and Gibbs and their transformer system is extremely complicated. It provides insight into, and raises questions about, the nature of the invention-and-development process, reverse salients and critical problems, the emergence of new systems, and priority claims and patent controversy. Before turning to the particular critical problems that Gaulard attacked in response to perceived reverse salients, we will consider briefly the overall state of the technology.

In 1831, almost half a century before Gaulard and Gibbs obtained their first patent on their transformer system of distribution, Michael Faraday (1791–1867) discovered the principle of induction. In the intervening decades, many inventors applied the principle of induction, which stated that from a practical inventor's point of view, a variation in the electric current in a coil of conducting wire will induce a current in a coil that is in close proximity. Many of the applications that preceded Gaulard's were various forms of the spark inducer, or induction coil, developed by the German physicist Heinrich Daniel Rühmkorff (1803–1877) in 1848, and introduced in a more primitive design by Charles Grafton Page (1812–1868) in the United States in 1836 on the basis of Joseph Henry's independent discovery of induction. The object of these devices, which came to be known as Rühmkorff coils, was to induce a high electromotive force (voltage) in the secondary from a low electromotive force in the primary. The resulting voltage, or tension as it was then called, was transformed, but it was stepped up, not down.[15]

As the use of arc lamps increased in the late 1870s, several inventors addressed the problem of preventing the extinguishing of one lamp, either intentionally or by accident, from breaking the entire circuit and extinguishing the remainder of the lamps. To do this, the primaries of a number of induction coils were connected in series in a circuit supplied by a generator, and individual arc lamps were connected to the induction coils' secondaries. Because the secondaries were not in the primary electric circuit

[15] F. Uppenborn, *History of the Transformer* (London: Spon, 1889), pp. 6–10. For a biography of Page that discusses his many electrical inventions, see Robert C. Post, *Physics, Patents and Politics: A Biography of Charles Grafton Page* (New York: Watson, 1976).

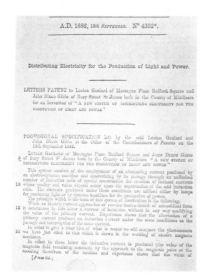

A.D. 1882, 13th *September*. N° 4362*.

Distributing Electricity for the Production of Light and Power.

LETTERS PATENT to Lucian Gaulard of Montague Place Bedford Square and
John Dixon Gibbs of Bury Street St James both in the County of Middlesex
for an Invention of "A NEW SYSTEM OF DISTRIBUTING ELECTRICITY FOR THE
PRODUCTION OF LIGHT AND POWER."

PROVISIONAL SPECIFICATION left by the said Lucian Gaulard and
John Dixon Gibbs at the Office of the Commissioners of Patents on the
13th September 1882.

LUCIAN GAULARD of Montague Place Bedford Square and JOHN DIXON GIBBS
of Bury Street St James both in the County of Middlesex "A NEW SYSTEM OF
DISTRIBUTING ELECTRICITY FOR THE PRODUCTION OF LIGHT AND POWER."

This system consists of the employment of an alternating current produced by
an electro-dynamic machine and determining by its passage through an unlimited
number of induction coils of special construction the creation of induced currents
whose quality and value depend solely upon the construction of the said induction
coils. The currents produced under these conditions are utilised either by lamps
for producing light or by dynamo machines for the production of power.
The principle which is the basis of this system of distribution is the following.
When an electric current approaches or recedes from a circuit of solenoidal form
it determines in this latter a current of induction without in any way modifying
the value of the primary current. Experience shows that the alternation of a
primary current produces an induction current under the same conditions as the
passage and interruption of the same current.
In order to give a clear idea of what is meant we will compare the phenomenon
we have just cited to that which is shown in the working of electro-magneto
machines.
In effect in these latter the inductive current is produced (the value of the
magnetic field remaining constant) by the approach to the magnetic poles or the
receding therefrom of the bobbins and experience shows that the value of

[*Price 6d.*]

*Figure IV.3. Gaulard and Gibbs's patent
for a transformer distribution system, 1882.*

but were related to the primaries through an electromagnetic field, individual lamps could be extinguished. The electrical circuit remained unbroken; only a magnetic-field relationship was interrupted. Among the several inventors who employed this plan was the Russian-born inventor Paul Jablochkoff (1849–1894), who displayed his lamps (Jablochkoff candles) at the Paris International Exhibition of 1878. To supply his lamps, which had vertical side-by-side carbon rods separated by the insulator kaolin, he used the then-famous alternating-current generator developed by the Belgian inventor and engineer, Zenobe Theophile Gramme (1826–1901). Among Jablochkoff's demonstrations in 1878 were those at the exhibition hall at the Champ de Mars, Magazins du Louvre, on l'Avenue de l'Opera. Jablochkoff experimented with varying the intensity of the lamp by varying the turns and diameter (resistance) of the wire of the induction coils. Thus he employed the transformer principle.[16] Like similar plans, however, employing a transformer with a secondary of higher resistance than the primary raised the voltage rather than lowered it. In essence, Jablochkoff used the transformer as a mode of connecting lamps to a circuit, not as a means of transmitting electricity economically at high voltage and distributing it at low voltage in a safe and usable form.

Up until 1880, the history of the transformer—as distinguished from the Rühmkorff coil—involved the connection of primaries in series, the raising of the secondary voltage in relation to the primary, and the supply of arc lamps. Prior to that time, no inventor had anticipated by demonstration the particular combination made by Gaulard: the connection of primaries in series in a high-voltage, alternating-current, extended circuit; the lowering of the secondary voltage; the supply of incandescent and arc lamps; and the supply of loads of different magnitude from different transformers.[17]

Assisted by his business partner Gibbs, Gaulard used existing technology and knowledge to solve the critical problems of transformer development. He believed that a practical transformer system would correct the reverse salient of high-cost transmission. Furthermore, in defining the problems of transformer development, he took into account the sections in the Electric Lighting Act of 1882 that governed the transmission and distribution of electricity. In responding to the provisions of the legislation, he also made possible the use of a single transformer to supply various electric lamps requiring different voltages and current. The available record does not make clear, however, whether his primary objective was to fulfill the legislative requirements, as he interpreted them, or to design a transformer suitable for various loads.

Section 18 of the law stated that "the undertakers [suppliers of electricity] shall not be entitled to prescribe any special form of lamp or burner to be used by any company or person."[18] A parliamentary committee had publicly deliberated the provisions of the legislation in the spring of 1882 and the act was dated 18 August 1882, so Gaulard and Gibbs had had an oppor-

[16] *Engineering* 23 (1877): 366; 26 (1878): 63 ff., 125–27, 321–22; and Uppenborn, *History of the Transformer*, pp. 13–16.
[17] Uppenborn, *History of the Transformer*, pp. 23–29.
[18] Electric Lighting Act of 1882, 45 & 46 Vict., ch. 56, §18.

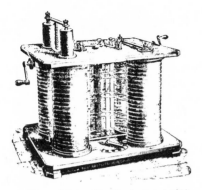

Figure IV.4. The Gaulard and Gibbs transformer. From Fleming, The Alternating Current Transformer, *2: 74.*

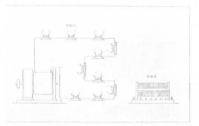

Figure IV.5. Figures from Gaulard and Gibbs's patent no. 4,362 (13 September 1882). Figure 1 shows transformer primaries connected in parallel.

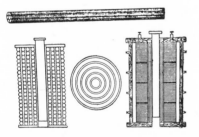

Figure IV.6. Gaulard and Gibbs's first induction coil, 1882. From Fleming, The Alternating Current Transformer, *2: 72.*

tunity to consider its provisions before leaving their patent specifications at the Office of the Commissioners of Patents on 13 September 1882.

Evidence that Gaulard and Gibbs designed their system with section 18 in mind is found in contemporary descriptions. Shortly before the system went on display at the Westminster Aquarium in London, *Engineering*, Britain's leading technical periodical, published an article written by the inventors about their system. The article began with a reminder that the Electric Lighting Act expressly proscribed the use of any particular lamp and that available incandescent lamps varied in specification from 48 to 91 volts. The reader was also reminded that "a divergence of ten percent above or below this standard is injurious to the life of the lamp, or greatly reduces its lighting power."[19] As the article continued, Gaulard and Gibbs's system allowed the supplier to use current at high tension to the point of consumption and then made the user "independent of the producer," enabling "him to apply the current he receives to any purpose he may please, such as arc lighting, incandescent lighting, the generation of power, or of heat."[20] A consumer who was "independent of the producer," therefore, could not be forced to use the lights or voltages or other specific components that a supplier, or undertaker, manufactured, sold, or had other interests in.

Gaulard and Gibbs's design cannot be fully appreciated without taking into account this response to legislation and the designers' anticipation of variations in load. The failure of some critics of the Gaulard and Gibbs system to take into consideration its response to the 1882 act explains why the system was "received with very considerable doubt, not to say contempt."[21] To an even greater extent it explains the inability of some historians to appreciate Gaulard and Gibbs's invention.[22] The several methods that Gaulard and Gibbs provided to allow consumers to vary the characteristics of supply involved a number of complications in design that appear irrational unless the designers' intent is known.

A description of the system that Gaulard and Gibbs displayed at the Westminster Aquarium exhibition in 1883 illustrates the relationship between legislation and design. The transformer introduced in 1883 consisted of four sets of four vertically stacked bobbins. The bobbins, wound with wire, constituted the secondary of the transformer. The primary consisted of an insulated copper conductor wound on a hollow paper tube. The open core (*E*) of the transformer was a soft iron bar fitted inside the tube on which the primary was wound. (See Figs. IV.4–IV.6.)

The adaptability of the transformer resulted from the fact that the bobbins could be connected to one another in different ways, the stacks of four could be combined variously, and the iron core of the primary could be cranked in and out of the stack of bobbins. Each of the sixteen bobbins had two spring terminals resting on a vertical commutator that could be

[19] "Gaulard and Gibbs' System of Electrical Distribution," *Engineering* 35 (1883): 205.
[20] Ibid.
[21] Ibid., p. 458.
[22] A. A. Halacsy and G. H. von Fuchs ("Transformer Invented Seventy-five Years Ago," *Transactions of the American Institute of Electrical Engineers: Power Apparatus and Systems* 80 [1961]: 121) make only brief reference to Gaulard and Gibbs, remarking upon the anomalies of their transformer.

turned by a crank mounted in the center on top of the transformer. By turning the commutator, the positive end of each bobbin could be connected to the negative end of the adjacent one in the stack of four; in another position of the crank, all positive ends were joined, as were all negative ends; and in a third position, the bobbins were grouped in two series of two each. Not limited to these permutations, the transformer also provided terminals on the upper board for the lower and upper ends of each stack of bobbins and, by means of short pieces of wire, the terminals could be connected in such a way as to group the stacks in series, in parallel, or "in sets of two each."

Gaulard and Gibbs designed this transformer to be connected in the circuit of an alternating-current generator. The number of transformers in a circuit varied: they had only two transformers at the aquarium installation, but their descriptive literature mentioned as many as eleven.[23] They also decided—and this turned out to be critical—to connect the primaries of the transformers in series, as was customary at that time. Critics of Gaulard and Gibbs were of the opinion that the wiring of transformers in series was an example of their ignorance of transformer principles. Within a few years, others would connect transformer primaries in parallel, an arrangement that would prove advantageous and become standard.

At least one contemporary authority misunderstood the Gaulard and Gibbs transformer, taking it to have a 1:1 ratio of transformation (the reduction of voltage was believed to result solely from the subdivision that occurred with multiple transformers in series).[24] In contrast, the descriptions, undoubtedly supplied to technical journals by Gaulard and Gibbs themselves, explicitly state that each consumer using a transformer "would have it in his power to generate currents at a potential of 45, 60, 91, or other number of volts as he chose."[25]

Another design feature of the Gaulard and Gibbs transformer that was singled out for criticism was the open core.[26] But Gaulard and Gibbs had good reason to use an open rather than a closed magnetic core. The closed core, or closed magnetic circuit, would have prevented loss of magnetism (indeed, Gaulard and Gibbs's successors in transformer design resorted to it), but Gaulard and Gibbs chose the open core because it could be cranked in and out of the coils. In this way they intended to provide the means for varying the secondary output, presumably in response to any minor variations resulting from load changes; in fact, in 1885 Gaulard wrote that he would dispense with the movable core and use a closed core if the customer load was invariable.[27]

[23] *Engineering* 35 (1883): 205.

[24] Uppenborn, *History of the Transformer*, pp. 35, 41.

[25] *Engineering* 35 (1883): 205. "If the secondary coil be made in several parts, each with independent terminals, these parts may be variously combined either in parallel arc, or compound parallel arc, or series, according to the conditions under which the second, or locally generated current, is to be employed, and thus it is possible to generate by means of one secondary generator, currents of high or low potential" (ibid., p. 206).

[26] Uppenborn,, *History of the Transformer*, pp. 34–35, 46–49; *Telegraphic Journal and Electrical Review* 17 (1885): 114–16 and 465–66.

[27] Letter from Gaulard to the journal, *Telegraphic Journal and Electrical Review* 17 (1885): 508.

When the state of the technology and circumstantial factors are considered, Gaulard and Gibbs's connecting of the transformer primaries in series also seems reasonable. Unlike Edison, who designed an entire system of incandescent electric lighting, Gaulard and Gibbs designed a transformer for combination with available components to form a system. For example, they used a Siemens alternating-current generator in the Westminster Aquarium installation. Originally designed for arc lights connected in series, the Siemens machine was a constant-current machine. This type of generator maintained a constant current through the entire circuit by means of an automatic regulator that varied the generator output and electromotive force according to the load. By varying the circuit, or terminal, voltage, the circuit current could be held constant despite variations in output. Therefore, it seems probable that Gaulard and Gibbs decided to take advantage of the constant-current characteristic of the Siemens machine by wiring the primaries of their transformers in series, as arc lights were wired. They expected that the constant-current flow through the circuit would result in the appropriate electromotive force across the secondaries. They also believed that it would be possible to obtain a variable secondary current, which was necessary for an incandescent-lamp load wired in parallel.[28] They would have been wiser to design and develop a generator especially for their system, as Edison had, but they probably were not able to obtain financing as readily as had the Wizard of Menlo Park.

Before 1885, however, a major flaw in the system had been discovered: their expectation that a constant primary current would result in a constant secondary voltage had proved erroneous. If the load in the secondary varied, as, for example, when incandescent lamps were switched out, the remaining lamps became brighter. To remedy this, Gaulard and Gibbs introduced in the secondary circuit an automatic regulator designed to vary the secondary current in accordance with the load and to maintain a constant secondary voltage. Whatever the success of the device, it added another complication to an already extremely complicated transformer design.[29]

Gaulard and Gibbs's efforts to rectify the inadequacies of their system included switching from an open to a closed magnetic core, connecting the primaries in parallel rather than in series, and varying the winding of the primaries. Nevertheless, they encountered serious problems. Use of the closed magnetic core raised the efficiency of the transformer but eliminated the advantage of the movable, rod-shaped, open-ended core. In late 1885 or early 1886 Gaulard and Gibbs also tried a combination of primaries connected in parallel and a closed core. They had been told that this combination would provide good voltage regulation, but the resulting system did not function satisfactorily.[30] Its components probably required more redesigning than they could afford or were capable of doing. In the end, it seems, voltage regulation was the major reverse salient in their system.

---

[28] *The Engineer* (London) 60 (1885): 430.

[29] Ibid.

[30] P. Strange, "Some Tests on a Gaulard and Gibbs Transformer," *Papers Presented at the Third Weekend Meeting on the History of Electrical Engineering, Manchester, England, 11–13 July 1975* (London: Institution of Electrical Engineers, n.d.), p. 4/5.

They resorted to extremely complex devices to regulate the secondary voltage under changing load conditions, but they were unable to solve the problem, and it was other inventors and other engineers who found the answer. Outstanding among those who succeeded were three men associated with one of Europe's leading electrical manufacturers, Ganz & Company of Budapest, Hungary (then a kingdom in the Austro-Hungarian Empire).

Before considering the critical revisions made by the Hungarians, however, it is important to note that Gaulard and Gibbs laid the foundations for a new system of lighting during their search for a solution to the problem of transmission and distribution in the d.c. system. In effect, they responded to a reverse salient in incandescent lighting in general and were not intent on saving the existing d.c. system, as were Edison and other inventors and engineers. The general reverse salient that Gaulard and Gibbs observed in evolving electric lighting was clearly and succinctly described by them. In March 1883, in an article in *Engineering*, they described their system as a solution to the problem "of the further industrial development of absolute distribution, that is to say, a system of distribution limited neither by the distance of the central factory from the point of consumption, nor by the number of consumers to be supplied."[31] They termed the problem "distribution"; today, because of the distance envisaged, we would term the problem "transmission." (One of the first proposals made by Gaulard and Gibbs was to illuminate the Suez Canal using 40,000 volts for transmission.)

To describe Gaulard and Gibbs's work as the invention and development of a new system is valid for several reasons. First of all, contemporaries referred to alternating current with transformers as a new system, calling the competition between d.c. lighting and a.c. lighting "the battle of the currents," or "the battle of the systems." In addition, Gaulard and Gibbs and their successors in the a.c. field had to introduce different components from those used in the d.c. system, with the major exception of the incandescent lamp. Other reasons for categorizing d.c. and a.c. as distinct systems include the development of different theoretical structures to analyze and explain them and the establishment of separate engineering-school courses to teach them. More important for an understanding of technological change is the fact that the history of the invention and development of Gaulard and Gibbs's system is a case of the dynamic movement from one system to another as a result of a flaw in the first being corrected by problem identification and an invention that became the essential core of a new system.

Because many inventors perceived the reverse salient in the direct-current system and because some saw weaknesses in the Gaulard and Gibbs system and introduced improvements, there was intense litigation over the priority of invention and patents.[32] The basic patent was a provisional specification entitled "A New System of Distributing Electricity for the Production of Light and Power" and dated 13 September 1882. Gaulard and Gibbs took out related patents in 1884 and 1885. On 11 December

---

[31] "Gaulard and Gibbs System of Electrical Distribution," *Engineering* 35 (1883): 205.

[32] P. Strange, "Transformer Patent Litigation, I: 1880–1890," *Papers Presented at the First Weekend Meeting on the History of Electrical Engineering, Birmingham, England, 13–15 July 1973* (London: Institution of Electrical Engineers, n.d.), pp. 8/1–8/8.

1886 Sebastian Ziani de Ferranti, acting for Sir Coutts Lindsay and Company of London (see pp. 97–98 below) petitioned for the revocation of the original Gaulard and Gibbs patent, alleging that it lacked novelty and utility. In March 1887 Gaulard and Gibbs countered with an action claiming infringement of their patent by Ferranti and Sir Coutts Lindsay and Company.[33] The petitioners also claimed that Gaulard and Gibbs's complete specifications did not conform to the provisional specification. In April 1888 the court allowed Gaulard and Gibbs to amend their specifications. Then, hearing the petition, the judge declared the patent void because of the insufficiency of the complete and amended specifications. Gaulard and Gibbs appealed the case to Britain's Court of Appeal, which held that the invention was useful and not anticipated, but that the patent was void because a discrepancy remained between the provisional and complete specifications and because the amendment extended the scope of the invention. Gaulard and Gibbs then presented an appeal to the House of Lords, where three members heard their case. The Lord Chancellor did not see the novelty the Court of Appeal had perceived in the combination proposed by Gaulard and Gibbs; Lord Farrer Herschell ruled that the state of public knowledge was such that the alleged invention "was a mere adaptation requiring no invention and possessed of no novel features"; and Lord Morris decided that by withdrawing through amendment their claim to a specially constructed and novel secondary generator (transformer), Gaulard and Gibbs had left Hamlet out of the play. In July 1890 the appeal was dismissed with costs.[34]

In their petition for revocation, Ferranti and his associates cited as prior publication thirteen British patents, Faraday's *Experimental Research in Electricity*, portions of three works by "Noad," two articles in the *Telegraphic Journal and Electrical Review* of 15 April 1879, and Count du Moncel's description of the Jablochkoff system in *Exposé des applications d'electricité* (1878). In citing prior use, they were less specific, claiming that the alleged invention was commonly used in telegraphy; that electricians used the alleged invention (induction coils carrying alternating current) to supply light when they found it convenient and appropriate; and that James Edward Henry Gordon, in May 1880 and subsequently, used the invention to supply light at his home in Dorking and at various places in London.[35] The reference to use of the transformer system by the well-known inventor and engineer Gordon was the most specific claim, but his installations were not discussed in the lords' ruling.

The two prior patents that were considered most important by the lords were Jablochkoff's (British Patent no. 1,996 of 1887) and Edwards and Normandy's (British Patent no. 4,611 of 1878). The lords found, however, that Jablochkoff's patent did not specify that his transformers would lower, or step down, the electromotive force (voltage); he simply did not exclude that possibility. Edwards and Normandy's patent greatly impressed Lord Herschell because it described a transformer that would lower the tension

---

[33] Great Britain, *Reports of Patent, Design, and Trade Mark Cases*, vol. 4, no. 7 (30 April 1887), pp. 189–91.

[34] Ibid., vol. 7, no. 40 (5 November 1890), pp. 382–87 and 388.

[35] Ibid., p. 371.

(voltage) and raise the quantity (amperage) of electricity. Lord Herschell recognized, however, that Edwards and Normandy contemplated using not a high-tension circuit for extended transmission but a relatively low one for distribution in a single building from a source of energy in that building. Moreover, these inventors did not have access to practical incandescent lamps.

This review of the legal process serves as a reminder that patents and legal decisions concerning them are not the same as inventions and informed judgments about them. The lords took an ambivalent position, for example, on the novelty of combinations. As Lord Herschell said none too clearly,

The truth is that the Patentee was simply applying to a greater distance and working through a greater length of wire that which had been conceived by other people before him, and if he used an alternating current of high tension, which they did not, it was simply that he applied to the purpose which he wanted to accomplish the ordinary well-known means of producing that result, namely, using for long distances an alternating current of high tension.[36]

Furthermore, Lord Herschell minimized the significance of the use of incandescent lamps by Gaulard and Gibbs and not by others by commenting, "It was only just previously to the time of the Appellant's alleged invention that these incandescent lights had come into use, and it was only just about that time therefore that people had been concerned to find a means of combining the carrying of a current to a considerable distance which could only be done at high tension, with at the same time utilizing it wherever it was wanted at low tension."[37] Because Lord Herschell's philosophy of invention excluded the definition of problems and the combination of available technology to effect their solution, one wonders what he did consider to be an invention.

In making their decision, the lords did not weigh heavily the demonstrations of transmission and distribution made by Gaulard and Gibbs in 1883 and 1884, demonstrations that were not matched in complexity, extent, and success by any other inventors of alternating-current transmission systems before 1885. The first of Gaulard and Gibbs's public demonstrations took place at the Westminster Aquarium exhibition in April 1883. In November 1883 a system was placed into operation for the Metropolitan Railway Company using generators at the Chapel Street works and a mix of arc and incandescent lamps at underground stations at Notting Hill Gate, Gower Street, King's Cross, Aldgate, and Edgeware Road. The system extended for fifteen miles and "seems to have worked well according to reports in the technical press."[38]

Lacking money to purchase new equipment, Gaulard and Gibbs moved equipment from the Metropolitan Railway Company to Turin, Italy, for an international exhibition and competition for the best system of long-distance electrical transmission. With great water-power sites in the Alps, the Italians were sensitive to the inadequacies of the Edison and other

[36] Ibid., p. 386.
[37] Ibid., p. 387.
[38] Strange, "Some Tests on a Gaulard and Gibbs Transformer," p. 4/1.

direct-current, low-voltage systems in transmitting energy from site to urban load centers. In September 1884 Gaulard and Gibbs placed their transformers on a fifty-mile circuit that lighted the exhibition buildings, the Turin railway station, and stations at Veneria Reale and at Lanzo, a small village in the Savoy Alps. For the Turin installation, Gaulard and Gibbs were awarded the grand prize of 10,000 francs.[39] Permanent installations followed in 1885 and 1886 at Tours, France; Rome, Italy; and Aschersleben, Germany. Gaulard and Gibbs's best-known installation was begun at Grosvenor Gallery, London, in March 1885.[40]

The adjudicating lords not only did not consider successful demonstrations and installations to be evidence of an *Entwicklungsfähig* ("development-feasible") invention but they failed to ponder the particular problems that Gaulard and Gibbs attempted to solve with their system. Among the conditions Gaulard and Gibbs took into account, and the lords ignored, were the Electric Lighting Act of 1882 and the varied characteristics of existing lamps. They intended to use the public streets for transmission and distribution from central stations, and therefore they correctly anticipated that their system would have to conform to the provisions of the 1882 legislation. None of the inventors whose patents were cited as anticipating Gaulard and Gibbs's had solved that particular contextual problem.

The litigation was extensive, the controversy in technical journals heated, and the negative court decisions were like trial verdicts for Gaulard and Gibbs. Broken by these experiences, Gaulard died insane on 26 November 1888 at the Sainte-Anne Hospital in Paris. By then other inventors and engineers, who acknowledged the priority of Gaulard and Gibbs's demonstrations but stressed their own improvements, were rapidly introducing the transformer system. The editor of the British journal *Telegraphic Journal and Electrical Review* wrote that the cares and disappointments of the long and bitterly contested patent litigation had aggravated "anterior tendencies" and hastened Gaulard's final breakdown.[41] Several months before his untimely death, Gaulard had appeared at the Elysée asking the concierge to conduct him to the president of France, for whom, he said, he had an urgent message. The message was, "I am God and God does not wait."[42] One French historian of technology describes Gaulard as a suffering inventor whose ingenuity cost him not only his money but reason and life as well.[43]

Gaulard's fate was tragic. His experience also suggests that it is futile for an inventor, a historian of invention, or even the courts to attempt to prove who invented a machine, device, or process such as the steam engine, the telephone, or the transformer. James Watt, for instance, did not invent the

[39] *Telegraphic Journal and Electrical Review* 15 (1884): 363–64. For descriptions of installations, see J. A. Fleming, *The Alternating Current Transformer in Theory and Practice*, 2 vols. (New York: Van Nostrand, 1892), 2: 75–77; and *Engineering* 38 (1884): 529–30.
[40] Strange, "Transformer Patent Litigation," p. 8/2.
[41] *Telegraphic Journal and Electrical Review*, 23 (1888): 620.
[42] *Electrical Engineer* (London) 1 (1888): 147.
[43] Pierre Rousseau, *L'Histoire des techniques et des inventiones* (Paris: Fayard, 1958), p. 477. Rousseau describes Gaulard as "ruiné abandonné, malade et sans famille, il lut être entfermé a l'hospice de alienés, et c'est de justesse que l'on sauva ses restes de la fosse commune" (ibid.).

steam engine; he invented a separately condensing steam engine. Alexander Graham Bell did not invent the telephone; he put together a telephone system that embodied a particular application of the principles of variable resistance and induction. And Edison did not invent the incandescent lamp; his achievement was an incandescent lamp with a high-resistance filament. Similarly, Lucien Gaulard did not invent the transformer; he built a transmission and distribution system that incorporated transformers with primaries connected in series in a high-voltage, alternating-current circuit, and secondaries capable of supplying various combinations of relatively low voltage and high current to incandescent and arc lamps. Even such qualifications simplify the nature of these inventions, however. In fact, the inventions were particular solutions to particular critical problems defined within the context of place, time, and other circumstances. The inventors improved on their early inventions, and other inventors also adapted and modified them to suit other places, times, and circumstances.

Gaulard and Gibbs merit a prominent place in history because they invented, developed, and demonstrated a system that was, to use the apt German expression, *Entwicklungsfähig*. Once demonstrated, their invention stimulated a stream of improvements. Their system was a combination of known principles and devices, but it was a unique combination that functioned to solve a problem and correct a reverse salient. Nevertheless, when this *Entwicklungsfähig* invention was introduced, its potential was quickly sensed; inventors and engineers were aware of the flaw in the d.c. system and eagerly set about solving the problem. Thus it proved to be difficult to give Gaulard and Gibbs credit for their contribution in a court of law.

As noted, three inventor-engineers associated with Ganz & Company, the prominent Hungarian electrical manufacturing firm, substantially improved the design of the Gaulard and Gibbs transformer. According to Gaulard, the Hungarian engineer Otto Titus Bláthy (1860–1939) approached him about his transformer system at the Turin exhibition. Bláthy is said to have appealed to him not to license the system in Austria until Bláthy could convince his firm, Ganz & Company of Budapest, of its value.[44] Bláthy acknowledged that he and another Ganz engineer, Charles Zipernowski (1853–1942), did see Gaulard at Turin, but that all they learned from him was how a transformer should not be constructed.[45] Ganz & Company engineers then bought a Gaulard and Gibbs transformer and in the winter of 1884/85 made two major improvements in it. Bláthy decided that a closed core could be used to provide a more effective magnetic field than the open core. Max Déri (1854–1938), also of Ganz & Company, and Zipernowski recognized ineffective voltage regulation in the secondary circuits as a reverse salient in the Gaulard and Gibbs system and connected the primaries in parallel. By May 1885 Déri, Bláthy, and Zipernowski had developed their system and demonstrated it at the Hungarian National Exhibition in Budapest. Their transformer is considered by some to have

[44] Letter from Gaulard to the journal, *Telegraphic Journal and Electrical Review* 17 (1885): 508.
[45] Letter from Bláthy to the journal, ibid., pp. 465–66.

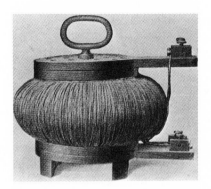

Figure IV.7. *The Déri, Bláthy, and Zipernowski transformer, 1885. From Fleming,* The Alternating Current Transformer, 2: 87.

been "the first commercial, practical transformer."[46] Fleming describes it as a major improvement upon that of Gaulard and Gibbs.[47] (See Fig. IV.7.)

The invention and development of the closed-core, parallel-connected transformer by Bláthy, Déri, and Zipernowski was not fortuitous—they and Ganz & Company had built up a momentum in the alternating-current field that carried them into transformer development generally, and parallel connections specifically. In 1878 Ganz & Company assigned part of its extensive engineering works to the manufacture of electric lighting apparatus in Austria-Hungary and by 1883 had made over fifty installations. It offered an entire system consisting of both arc and incandescent lamps, generators, and other accessories. The fact that it adopted alternating current exclusively was notable in the era of the Edison direct-current system.[48] Two of the engineers who worked on transformer design had prior experience with alternating current. Zipernowski, founder of the electrical division at the Ganz factory, designed the company's patented alternating-current system. Bláthy, technical director of the electrical division after 1883, also held many a.c. patents. Although Déri was most active in the sales division, he too, contributed to the technical development of the transformer.

In designing various alternating-current installations, Ganz engineers used parallel circuits and constant-voltage generators.[49] This experience undoubtedly influenced Zipernowski, Déri, and Bláthy to try working the transformer in parallel, for such an arrangement called for a constant-voltage generator. Since transformers with primaries wired in parallel and supplied from a constant-voltage generator were self-regulating, Ganz engineers could dispense with the complex regulator Gaulard and Gibbs had placed in the secondary circuit.

To understand why Ganz & Company succeeded where Gaulard and Gibbs failed it is important to note that Ganz & Company had customarily designed and manufactured an entire system. Because of this, it was not inclined to develop a complicated transformer that permitted adaptation to other inventors' and manufacturers' components. Thus, Zipernowski and his colleagues did not resort to such ingenious expedients as the open, removable core and sets of bobbins with connection permutations. The Ganz approach was more analogous to the systems approach of Edison. But another major explanation for the ingenious and complicated Gaulard and Gibbs design and the ingenious and relatively simple Ganz & Company design is that the Electric Lighting Act of 1882 did not apply in the Austro-Hungarian Empire. It should also be noted that by 1885, in the United Kingdom and elsewhere in the industrial West, lamp voltages were being standardized, and it was no longer likely that a fixed voltage from the supplier would dictate the capacity of a particular manufacturer's lamps.

The Ganz & Company system was widely adopted. By 1890 nearly seventy

[46] Halacsy and von Fuchs, "Transformer Invented Seventy-five Years Ago," p. 121.

[47] Fleming, *The Alternating Current Transformer*, p. 86. A balanced view attributed to Gaulard and Gibbs the introduction of a transformer system that could be developed in a practical form (*Betriebsfähig*), and to the Hungarians the actual introduction of that form. L. Schuler, *Geschichte des Transformators*, quoted in Dettmar, *Starkstromtechnik*, p. 84.

[48] *Engineering*, 35 (1883): 80–81 and 550.

[49] Ibid., pp. 550–51.

central stations of various sizes had been placed in operation using the company's alternating-current generators, transformers, and controls. These stations supplied 100,000 incandescent lamps and 1,000 arc lamps.[50] Both Ganz & Company and Gaulard and Gibbs displayed their systems at the Inventions Exhibition in the spring of 1885 at South Kensington in London. The displays drew the attention of many electricians and engineers (among them, the American George Westinghouse) to alternating-current distribution.

Others soon joined the Ganz & Company engineers in the ranks of the improvers of Gaulard and Gibbs's system. In London a train of events was set into motion that resulted in a modification to the system Gaulard and Gibbs installed in London's Grosvenor Gallery. The owners of the gallery had side-stepped the restrictions of the Electric Lighting Act of 1882 by stringing their wires along rooftops, thereby avoiding the opening of the streets. They then wanted an economical means of transmitting across the rooftops.

The Grosvenor central station had unorthodox origins. When the twenty-sixth earl of Crawford, James Ludovic Lindsay, served as British commissioner at the Paris Electrical Exhibition of 1882, he found the displays of the new incandescent light—especially that of Thomas A. Edison—impressive. He then persuaded his uncle, Sir Coutts Lindsay, proprietor of the Grosvenor Art Gallery on New Bond Street, to install a portable generator, the better to light the works of Burne-Jones, Whistler, and other contemporary artists. Electric light would not smoke the paintings as did gas and burning oils and waxes.

New art and technology soon converged. Sir Coutts Lindsay and Company was founded to supply lighting not only to the gallery but to neighboring businesses whose customers were impressed by the lighting of the art gallery and its "greenery yallery" works of art.[51] To meet the growing demand, the company turned to Gaulard and Gibbs, who installed their transformers in series in a 2,500-volt circuit with Siemens single-phase alternators. As the load increased, however, the system proved virtually unworkable.[52]

Thus, in 1886 Sir Coutts Lindsay and Company called in the young and bright engineer Sebastian Ziani de Ferranti (1864–1930), the manufacturer of meters used by the company. Born in Liverpool to an Italian father who had once been guitarist to Leopold I of Belgium and a mother who was the daughter of a portrait painter, Ferranti was a striking person both physically and intellectually.[53] Compared in his youth to Edison, he subsequently had a long and active career as an inventor, engineer, and leading British electrical manufacturer. He had no advanced scientific or engineering training before beginning work in 1881 for Siemens Brothers, the

[50] Fleming, *The Alternating Current Transformer*, p. 121.

[51] A. P. Trotter, "Early Days of the Electrical Industry and Other Reminiscences" (typescript at the library of the Institution of Electrical Engineers, London), p. 586.

[52] Parsons, *Early Days of the Power Station Industry*, p. 22.

[53] On Ferranti's life see W. L. Randell, *S. Z. de Ferranti and His Influence upon Electrical Development* (London: Longmans, Green, 1948); Arthur Ridding, *S. Z. de Ferranti, Pioneer of Electrical Power* (London: Science Museum, 1964); and Gertrude Ziani de Ferranti and Richard Ince, *The Life and Letters of Sebastian Z. de Ferranti* (London: Williams & Norgate, 1934).

British branch of the German enterprise, but at the Siemens works at Charlton, near Woolwich, his enthusiasm for the new field of electrical engineering grew when he found himself surrounded by dynamos, arc lights, and other apparatus. He soon impressed his associates with his mechanical genius.

Noting his genius and striking personality, Ferranti's lawyer advised him in 1883 to establish himself as a private manufacturer of his alternators and other equipment. Ferranti responded by marrying the lawyer's daughter and gaining his father-in-law as his financial backer. As a bright young engineer and the respected inventor of an alternator, Ferranti thus drew the attention of Sir Coutts Lindsay when numerous technical difficulties arose at the Grosvenor Gallery station.

Appointed chief engineer of Sir Coutts Lindsay and Company in 1886, Ferranti followed the precedent of Ganz & Company and changed over to the parallel connection of the Gaulard and Gibbs transformer primaries. In 1886, on behalf of his new employer, he also petitioned for revocation of the Gaulard and Gibbs patents. Gaulard and Gibbs had charged fees for the use of their patented transformer and in turn threatened Ferranti and Sir Coutts Lindsay and Company with infringement proceedings for the redesigned transformer. Ferranti also prudently replaced the Siemens alternators with machines of his own design. In short, he redesigned the station's components systematically (see Fig. IV.8). The Grosvenor Gallery station then expanded impressively until in 1888 there were five machines, five circuits, and 34,000 lamps lighting a large district of London. Its peak load in 1888 was 600 kw.[54] Grosvenor Gallery was called the only central station in England worthy of the name and the "one ever green oasis standing boldly out in the midst of a dismal waste."[55] (See Fig. IV.9.)

In the United States, still another inventor and another company improved on the Gaulard and Gibbs system at about the same time as the Déri, Bláthy, and Zipernowski modifications were being developed in Hungary and the Ferranti system was being built in England. The American, William Stanley (1858–1916), the Hungarians, and Ferranti all knew about the work of Gaulard and Gibbs; indeed, they had had intimate contact with it, despite their insistence that they had learned little from it. Patent tactics suggested that they make this disclaimer. Déri, Bláthy, and Zipernowski went so far as to say that they learned only how not to build a transformer; Ferranti claimed that the Gaulard and Gibbs invention was not patentable; and William Stanley observed years later that although the work of Gaulard was remarkable in showing a comprehensive knowledge of the laws of the induction coil and reactive effects, "it was hopelessly ill-fitted to the conditions imposed by a general lighting system."[56] As noted earlier, Bláthy

---

[54] Clipping from *Engineering*, 2 August 1912; and London Electric Supply Corp., Ltd., *Annual Report*, 1888.

[55] See *Electrical Engineer* (London) 1 (1888): 348; quotation from ibid. 2 (1888): 386.

[56] Thomas C. Martin's unpublished manuscript "Biography of William Stanley" was kindly loaned to me by Mr. Samuel Sass, librarian of the William Stanley Library, General Electric Co., Pittsfield, Mass. (hereafter cited as Martin, "Stanley Biography"). Laboratory and personal notebooks spanning Stanley's career as a professional inventor (c. 1882–1910), as well as company records and personal materials, are part of the William Stanley Papers on file at the William Stanley Library. Karen Belmore of the University of Pennsylvania is currently writing a biography of William Stanley the inventor.

Figure IV.8. Ferranti's 1885 notebook sketches for transformers installed at the Grosvenor Gallery central station, London. Courtesy of the Ferranti Ltd. Archives, Hollinwood, England.

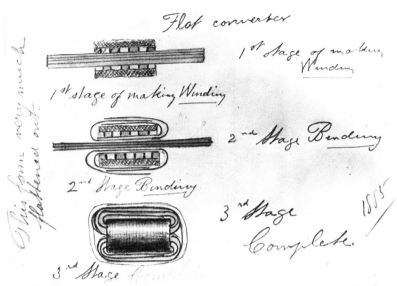

Figure IV.9. Overhead distribution system (1887) from the Grosvenor Gallery central station.

Figure IV.10. *William Stanley. From* Creating the Electrical Age, *ed. Nilo Lindgren, a special issue of* EPRI Journal *4 (1979): 37.*

discussed the Gaulard and Gibbs system with Gaulard at the Turin exhibition, and Ferranti worked with the system at Grosvenor Gallery. Stanley, too, had a Gaulard and Gibbs transformer at his disposal as he developed his own design. The information in Gaulard and Gibbs's patents and in the technical periodicals was available to all.

The Gaulard and Gibbs device came into Stanley's hands through George Westinghouse, who by 1884 owned two prosperous manufacturing firms, the Westinghouse Air Brake Company and the Union Switch & Signal Company (manufacturer of railway equipment). Westinghouse's center of operations was steel-producing Pittsburgh. His brother, H. H. Westinghouse, who made steam engines to drive electric generators, introduced William Stanley to George, who, after deciding to enter the field, contracted with Stanley to devote his time to further development of an electric lighting system. Westinghouse also purchased the incandescent-lamp and generator patents that Stanley had assigned earlier to the Swan Incandescent Electric Light Company of Boston, Massachusetts. Several of the Union Switch & Signal Company engineers were assigned to electric lighting development as well.[57]

Stanley, a native of Brooklyn, New York, was the son of a prominent lawyer. The family later moved to Englewood, New Jersey, where Stanley grew up. Encouraged by his father to study law, Stanley entered Yale University, but left after several months. In a note to his father he explained, "Have [had] enough of this—am going to New York."[58] Being inclined toward mechanical and scientific activities, Stanley first found employment with a New York manufacturer of telegraphic apparatus. His father soon became resigned to his predilections, however, and funded his entering a partnership in the electroplating business. About 1881 Stanley abandoned electroplating to join the well-known inventor of the machine gun, Hiram Maxim (1840–1916), who at the time was engaged in the invention, development, and manufacture of electric arc and incandescent lamps for the U.S. Electric Lighting Company in New York City. This company was Edison's major competitor for the incandescent lighting market until 1885, but it had only about one-fifth as many customers as the Edison enterprise. One of the U.S. Electric Lighting Company's problems was its failure to develop a coherent system. Maxim concentrated on the incandescent lamp, leaving the development of the d.c. generator to others.[59] After Maxim left the company and returned to Europe in 1881 or 1882, Edward Weston (1850–1936) became chief engineer for the company, and Stanley worked for a few months under another of the pioneer inventors and engineers of the industry.[60] In 1882 Stanley went to work for another manufacturer of incandescent lamps, the Swan Electric Company in Boston, and several of his inventions were patented by the company. After a year, however, he returned to Englewood, New Jersey, and established his

[57] Harry Douglas, *William Stanley: A Short Biography* (New York: H. S. Robinson, 1903), pp. 59, 64–66; and Passer, *Electrical Manufacturers,* p. 131. See also the biographical booklet by Laurence A. Hawkins, *William Stanley, 1858–1916* (New York: Newcomen Society, 1951).

[58] Douglas, *Stanley,* p. 33.

[59] Passer, *Electrical Manufacturers,* p. 148.

[60] D. O. Woodbury, *A Measure of Greatness: A Short Biography of Edward Weston* (New York: McGraw-Hill, 1948).

Figure IV.11. *Entry from William Stanley's notebook, 18 September 1883, showing work with alternating-current generator (G) and translating devices (T) for transmitting induced currents to secondary circuits. Courtesy of J. Church and the Stanley Library, Pittsfield, Mass.*

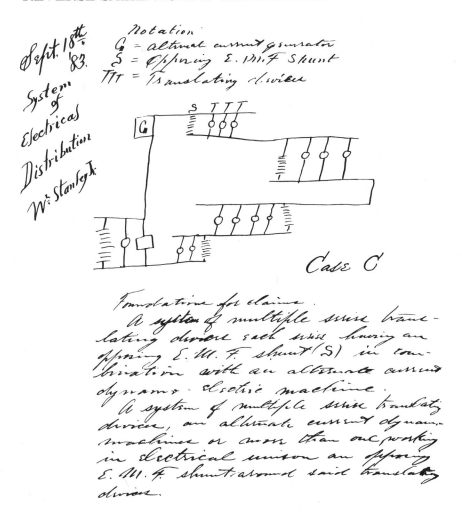

own laboratory to experiment with electrochemistry, incandescent lamps, and storage batteries (see Fig. IV.11). From there, in 1884, he was enticed to George Westinghouse's company. Stanley's having worked with most of the leaders in the field before he was thirty indicates how small and intimate the pioneer community was.

Stanley did not lose his status as a professional inventor by associating with Westinghouse, but Westinghouse's support of his work was contractual. The agreement the two men signed on 5 March 1884 specified that Stanley would assign all future inventions funded by Westinghouse to Westinghouse on condition that Westinghouse would manufacture and sell them. Stanley was to receive one-tenth of the profits and an annual salary of $5,000. Responsibility for the development of Westinghouse's electric lighting project thus fell to Stanley. In 1885 he established a factory for manufacturing incandescent lamps of his design, and he designed direct-current generators to function systematically with his lamps. He also took out ten patents on the incandescent-lamp system, plus one that was taken out jointly.

In the summer of 1885 George Westinghouse ordered several Gaulard and Gibbs transformers and secured an option on the American patent rights, which he later exercised, but there are conflicting accounts as to who at Westinghouse promoted the transformer system. Relying on the reminiscences and correspondence of one of Westinghouse's employees and protégés, Guido Pantaleoni, Harold Passer, the historian of American electrical manufacturers, concludes that George Westinghouse alone pushed through the project: "Opposition by ALL the electric part of the Westinghouse organization was such that it was only Mr. George Westinghouse's personal will that put it through."[61] At Westinghouse's request, Pantaleoni had secured the option on the Gaulard and Gibbs patent rights after discussing the system with Gaulard at Turin, had consulted Werner Siemens about it, and had also sought the advice of Ganz & Company.

Stanley himself minimizes the role played by George Westinghouse in promoting the system, maintaining that Westinghouse was advised against the alternating-current system by a "trusted expert." According to Stanley, Westinghouse's contribution was to make capital available for experimentation, but he did so on terms that were not favorable to Stanley. Only when Westinghouse saw Stanley's system in operation did he decide to enter the a.c. field actively.[62] According to Thomas Commerford Martin, biographer of Stanley and Edison, Stanley had the greatest difficulty bringing Westinghouse to a clear understanding of the transformer system.[63] Furthermore, Martin concludes, "For some reason, Westinghouse, who seems to have regarded Stanley as rather unstable and visionary, full of chimerical projects, and who was probably guided by inside influences not too friendly toward the impetuous, self-centered young genius, declined flatly to furnish Stanley with any money to build and try out his new alternating current system."[64]

The "trusted expert" and "inside influence" mentioned by Stanley and Martin was probably Pantaleoni. According to Henry M. Byllesby, a prominent pioneer in the development of electric light and power who assisted Edison in the design of the Pearl Street station and who joined Westinghouse at about the same time as Stanley, "Pantaleoni himself when I first joined the Westinghouse interests was scornfully pessimistic regarding Stanley and all his work, and the possibilities of anything worthwhile resulting from the alternating current. They were wholly antithetical characters, and the moment they came in contact trouble arose, which was destructive of any result."[65] In sum, Westinghouse probably did push the development of a transformer system, but not Stanley's approach to it.

When Byllesby became general manager of the Westinghouse electrical enterprise, he tried to help his friend Stanley, whom he saw as a temperamental inventor who did his best work away from the routine and discipline of industry. He supported the decision of Stanley and Westinghouse

---

[61] "Reminiscences of G. Pantaleoni, April 1939," in Westinghouse Electric Company files and quoted in Passer, *Electrical Manufacturers*, p. 132.

[62] William Stanley, "Alternating Current Development in America" (Paper read before a meeting of the American Institute of Electrical Engineers, 15 February 1912), *Franklin Institute Journal* 168 (1912): 568–73.

[63] Stanley memorandum of 1915, quoted in Martin, "Stanley Biography," chap. 7, p. 2.

[64] Ibid.

[65] Ibid., p. 5.

to separate—at least physically. On 24 December 1885 Westinghouse and Stanley signed a new contract. According to its terms, Stanley would leave Pittsburgh to reside in Great Barrington, Massachusetts, in the Berkshires he had known during vacations as a child. Furthermore, Westinghouse would purchase back for $12,500, $25,000 worth of securities in a new Westinghouse company in which Stanley held one-tenth of the shares, the total value of which was $100,000. The remaining shares (valued at $75,000) were to be delivered fully paid to Stanley. Stanley was to use the $12,500 to operate a laboratory at Great Barrington. He would receive a salary of $4,000 annually, or $1,000 less than he was then receiving. Westinghouse also agreed to contribute $200 a month toward general laboratory expenses. In return, Stanley was to submit his inventions to the company for patenting and acceptance. If they were not taken up, Stanley could dispose of them as he pleased.[66]

Stanley's move to his own laboratory in Great Barrington, Massachusetts, mirrored Thomas Edison's quest for greater autonomy. The freedom to conceptualize a system made up of one's own invented components won out over the restrictions of fitting one's inventions into the conceptual framework of another. Moreover, this harmonious coordination of components eliminated reverse salients. It is interesting to note that Stanley, who had been depressed and in poor health in Pittsburgh, was rejuvenated soon after his arrival in Great Barrington. Early in 1886 he acquired an abandoned mill, restored an old steam engine, set up a Siemens alternator loaned to him by the Westinghouse company, hired one assistant, and painstakingly began designing and developing new transformers. His transformers and lamps were soon installed in houses and businesses in Great Barrington, and on 6 March 1886 the system began to operate on a regular basis. His alternating-current transformer, central-station system for public service, "the very first in America beyond all dispute," used a 25-h.p. steam engine, transformers supplying either twenty-five or fifty incandescent lamps, and wires strung on insulators attached to the "grand old elms" that lined the village streets.[67] The length of the transmission circuit from central-station laboratory to village center was about 4,000 feet. Connected to it were thirteen stores, two doctors' offices, one barbershop, the telephone exchange, and the post office. Initially, Stanley stepped up the generator output from 500 to 3,000 volts, transmitted the current to the center of the village, and then distributed it through the streets after stepping it down to 500 volts.

One question remains to be answered: What did Stanley invent and develop? He had access to the patents and technical literature concerning the Gaulard and Gibbs and Ganz & Company systems. Following the well-

[66] Ibid., p. 7. Passer lists laboratory support as $600 a month and implies that the company in which Stanley held one-tenth of the shares was the Westinghouse Electric Company formed on 8 January 1886. This, however, contradicts Passer's statement on the same page that Westinghouse received all the stock in the Westinghouse Electric Company except 2,000 shares (one-tenth), which were sold to seven business associates. Passer, *Electrical Manufacturers*, p. 136.

[67] Martin, "Stanley Biography,"; quotation in chap. 7, p. 8. A decade later Franklin L. Pope, a pioneer in the field of electrical engineering, was electrocuted when he accidentally made contact with wires leading to the Stanley system transformer in his basement. Woodbury, *Measure of Greatness*, p. 143.

Figure IV.12.   *The Westinghouse transformer.*

publicized Ganz display at the Hungarian National Exhibition in Budapest in May 1885, Déri, Bláthy and Zipernowski displayed their system at the Inventions Exhibition in London in July 1885.[68] Then, in the fall of 1885, Reginald Belfield, an English electrician who had been in charge of a demonstration of the Gaulard and Gibbs system at the Inventions Exhibition in London, was brought to Pittsburgh by Westinghouse and entered his employ at about the time that the Gaulard and Gibbs transformers arrived. Belfield brought with him substantial knowledge of transformer design and operation. Despite the prior literature and exhibitions, however, Stanley faced the extremely demanding challenge of actually building transformers and a system within which they would function. On the theoretical level, Stanley believed that one of his particularly important contributions was recognizing the effectiveness of using counter electromotive force in the transformer to obtain a high degree of self-regulation.

Again, it was Byllesby who sensed Stanley's worth and recognized his achievements. In February or March 1886 he visited the installation at Great Barrington; others in Pittsburgh would pay Stanley no heed. Byllesby reported to Westinghouse and others that the transformers were properly proportioned, well regulated, and that he was "profoundly impressed." Then, after making a thorough examination himself, Westinghouse agreed that the alternating-current system had "arrived."[69] Persuaded that the system was marketable, Westinghouse tested the Stanley system more thoroughly in Pittsburgh in the summer of 1886 using an alternator designed by Stanley to replace the Siemens model, which regulated voltage poorly. Transmission ran from the plant of the Union Switch & Signal Company to East Liberty, a distance of three miles. Satisfied by the performance of this pilot system, Westinghouse began commercial production and shipped his company's first commercial plant to Buffalo, New York, where a local utility placed it in service. Orders for twenty-five alternating-current plants followed within a few months.[70]

George Westinghouse and his engineers contributed substantially to the development of a transformer system by designing components for economical and quantity production. To facilitate manufacture, Westinghouse patented the design given the transformer by Stanley (see Fig. IV.12).[71] A noted Westinghouse electrical engineer and patent lawyer, Franklin L. Pope, perceived the essence of the Westinghouse contribution when he wrote in 1887, "I saw that there were great possibilities in the system [Gaulard and Gibbs's] although it was not worked out practically at all—not worked out as we would work it out in this country by any means—and I was so well satisfied . . . of the possible results which we could get

[68] Halacsy and von Fuchs, "Transformer Invented Seventy-five Years Ago"; the authors cite the extensive contemporary literature describing the Ganz system.

[69] Martin, "Stanley Biography," chap. 7, p. 13.

[70] Passer, *Electrical Manufacturers*, p. 138. In his account of Gaulard and Gibbs, Stanley, and Westinghouse, Passer mistakenly presents the Gaulard and Gibbs system as one without voltage-transformation characteristics. He may have assumed this because the Gaulard and Gibbs transformer bought by Westinghouse was wired for one to one, primary to secondary. Ibid., p. 135.

[71] Charles F. Scott, "Tesla's Contribution to Electric Power," *Electrical Engineering* 62 (1943): 352.

when we put our young men on it in this country."[72] At about the same time, a writer in an English technical journal lamented, "It is not a very complimentary reflection for European electricians and capitalists that although all ideas and experimental work needed have come from Europe ... it should be reserved for an American firm to take up the system and make it the commercial and practical success which the Westinghouse Company is now doing."[73] Westinghouse's manufacture and marketing of the transformer system was impressive: by September 1887 the company had in service or under construction central stations with a total capacity of one hundred thirty-four thousand 16-c.p. incandescent lamps; by 1890 there were three hundred Westinghouse central stations with a total capacity of half a million 16-c.p. lamps.[74]

[72] *Telegraphic Journal and Electrical Review* 20 (1887): 349.
[73] Ibid., 21 (1887): 446.
[74] Passer, *Electrical Manufacturers*, p. 149; Fleming, *The Alternating Current Transformer*, 2: 167.

CHAPTER V

Conflict and Resolution

B Y THE end of the 1880s, Edison's direct-current system faced substantial competition from the more recently developed alternating-current system. The Westinghouse Electric Company had introduced the transformer in America; Ganz & Company had many installations in Europe; and there were small manufacturers of a.c. equipment with a few installations in Britain. During the closing years of the eighties, technical journals and even the popular press kept readers informed of "the battle of the currents," or "the battle of the systems." Professional societies held debates concerning the merits of each system, and engineers and station managers filled the technical journals with articles proclaiming the technical and economic advantages of one system as compared to the other.[1] The debate continued until the 1890s, with low-voltage direct current competing against single-phase alternating current for the incandescent-lighting market.

The single-phase alternating-current system originated as a solution to a critical (transmission) problem in the direct-current system. Engineers and managers had greatly improved the d.c. system, but they remained stymied by the cost of transmission beyond a radius of several miles. Alternating current provided relatively low-cost transmission, but the single-phase alternating-current system also had a serious drawback; it lacked a practical motor. So handicapped, the two systems competed intensely, especially during the years 1887–92. The "battle" ended in the 1890s, but not with the destruction of one side by the other. Instead, there was a notable resolution of the controversy on both the technical and the organizational level. Resourceful invention and development were necessary to bring it about. Before analyzing the resolution, however, we will follow the course of the debate, for it represents the kind of conflict between or among technological systems that has occurred throughout the history of technological change.

---

[1] For an extended discussion of the various advantages and disadvantages of the two currents, see the series of papers printed in *Electrical Engineer* 7 (1888): 166–68, 220–24. See also Frank L. Pope, "The Westinghouse Alternating System of Electric Lighting," *Electrician and Electrical Engineer* 6 (1887): 332–42, for the well-stated opinions of an advocate for alternating current who died accidentally in 1895—a victim of that current.

In some areas the conflict was based primarily on technical and economic factors. Because the needs of heavily populated cities differed from those of lightly populated towns and city suburbs, the advocates of the competing systems could point to clear technical and economic advantages for each. In city centers, heavy population concentration diminished the need for large supply areas to obtain a diverse and reasonably well-balanced load. In the lightly populated areas, economies of transmission gave the alternating-current salesman the advantage. It was in intermediate situations that the technical and economic advantages of the two systems were not clear-cut and that competition was most intense.[2]

In the areas where technical and economic considerations did not lead to straightforward, rational decisions, persons skilled in other modes of argument entered the fray. Their participation and their influence upon decisions about technology are significant reminders that technological change cannot be understood if viewed only from the perspective of the designing engineer and the calculating cost accountant. Special note should be taken of the way in which political power, particularly that exercised through law, was brought to bear during "the battle of the systems." The debate raged most furiously in the United States, where the use of electric lighting was most widespread. The leading protagonists were the Westinghouse and Edison Electric Light companies, but other manufacturers, including the firm of Thomson-Houston were involved. Neither George Westinghouse nor Thomas Edison was above entering the fracas, even though by the late 1880s Edison was spending much of his time attending to matters other than the electrical companies he had founded. Reverberations of the struggle were heard abroad, but we will concentrate here on the American scene.[3]

The Edison interests first resorted to unorthodox political tactics. For this, they had a broader range of political contacts on which to call than did the competition. They and other direct-current advocates realized that their failure to solve the transmission problem called for a nontechnical compensatory response. Before the final exchange in "the battle of the currents," an attempt was made to create public indignation against an electrical system that supposedly endangered life. Essentially the episode concerned one private enterprise's endeavor, through political power and legislation, to outlaw the technological advantage of another. It was neither the first nor the last instance of such maneuvering in the history of technology.

In 1888 the Edison interests were not unhappy to see an unknown, self-

[2] The technical periodicals have carried historical essays on the struggle; see especially the one by L. B. Stillwell in the fiftieth anniversary issue of *Electrical Engineering* 53 (1934), "Alternating Current versus Direct Current." Harold C. Passer, *The Electrical Manufacturers, 1875–1900* (Cambridge, Mass.: Harvard University Press, 1953), is informative on the manufacturers' role.

[3] There is secondary literature about the struggle in England and Europe. See, for instance, R. A. S. Hennessey, *The Electric Revolution* (Newcastle upon Tyne: Oriel, 1971), which has a chapter on "the battle of the systems" in Britain; R. H. Parsons, *The Early Days of the Power Station Industry* (Cambridge: At the University Press, 1940), which also has a chapter so titled; and Georg von Siemens, *History of the House of Siemens*, trans. A. F. Rodger, 2 vols. (Frieburg/Munich: Alber, 1957), which treats the conflict and its resolution in a chapter entitled "The Alternating Current Revolution."

styled electrician who had once sold Edison electric pens appear with a scheme. Harold Brown first achieved public attention—perhaps notoriety would be a better word—when in July 1888 at Columbia College's School of Mines he electrocuted a dog described by him as vicious. The animal was not simply electrocuted, however; it was first subjected to direct currents of varying voltages and then dispatched by alternating current. It is obvious that Brown was trying to persuade his audience that direct current was relatively safe, while alternating current was the deadly current. Indeed, Brown showed his hand not long after this demonstration when he told an audience that included New York lawmakers, "It is certain that yesterday's work [the electrocution of a second dog] will get a law passed by the legislature in the fall, limiting the voltage of alternating currents to 300 volts."[4]

A confluence of circumstances complicated the grisly affair. Because the New York State legislature was seeking a more humane mode of execution than hanging, Brown seized the opportunity to combine the effort to outlaw high voltages in the streets with the move to use alternating current for electrocutions in the prisons. He expected the association of death with alternating current to facilitate legislation. With the approval of Thomas Edison and the help of his laboratory staff, including Arthur Kennelly, later professor of electrical engineering at the Massachusetts Institute of Technology and at Harvard University, Brown executed calves and horses, animals that were larger than humans, he pointed out. After the New York State legislature heard testimony about electrocution from witnesses, including Thomas Edison, it voted in favor of the method. Despite the determined opposition of Westinghouse, Brown managed to acquire a Westinghouse generator and placed it in New York's Auburn State Prison. In August 1890, with Brown present, the first person was legally electrocuted. Then a publicity campaign was launched asking, Do you want the executioner's current in your home and running through your streets?

Brown and Edison proceeded with the scheme to outlaw the current, but met with failure in New York State and elsewhere. In one case before the Virginia legislature, they both urged that alternating current be limited to 200 volts. The legislation did not pass, however, because a member of that body learned that the Virginians were being used in a competitive struggle between two "Yankee" corporations.

Despite conflicting opinions as to the efficacy of electricity (in its first test) as a means of execution, New York State retained the system. Expressing the belief that the use of a commercial generator might have been unfair "to a particular electrical firm" (Westinghouse), the official report

[4] Harold P. Brown to Arthur E. Kennelly, 4 August 1888, Edison Archives, Edison National Historic Site, West Orange, N.J. For a detailed account based on original sources, see Thomas P. Hughes, "Harold P. Brown and the Executioner's Current: An Incident in the AC-DC Controversy," *Harvard Business History Review* 32 (1958): 143–65. Some inventors and engineers, such as Elihu Thomson, were seeking means of ensuring the safety of alternating currents rather than outlawing them. See Elihu Thomson to Thomson-Houston Electric Co., 20 December 1889, Elihu Thomson Papers (LB 4/89–1/90 867–878), Library of the American Philosophical Society, Philadelphia, Pa. I am indebted to Bernard Carlson of the University of Pennsylvania for calling my attention to this letter; Carlson is writing a biography of Elihu Thomson.

of the execution did recommend that a generator be specially constructed for future use in such cases. An unofficial recommendation was that Brown's name be associated with electrical execution in the same way that Dr. Ramon Guillotin's had become associated with his device. Brown's name soon disappeared from the nation's newspapers and periodicals, however, and there is good reason to doubt that his attempt to associate alternating current with death had a noticeable effect upon the outcome of "the battle of the currents." Statistics show a steady and rapid increase in the number of central stations using the Westinghouse system of alternating current during the period when Brown was most active, 1888–90.

Free of legal snares, the alternating-current system gathered momentum as manufacturing and installation increased, but around 1890 it, too, experienced reverse salients, and a new generation of inventors and engineers responded by defining them as critical problems. Comparable to the transmission problem that had frustrated the direct-current inventors, the most serious problem for the alternating-current system was the need for a practical motor. Another problem stemmed from the use of higher voltages in a.c. practice than in d.c. and the resulting need to insulate and ground circuits to guard against injury and death from electric shocks. While "the battle of the currents" was being fought in the nontechnical arena, the new generation of engineers and inventors worked to solve the motor and safety problems.

There was controversy over priority of invention in the case of the alternating-current motor as there had been about priority of invention of the a.c. transformer. Because the motor controversy, like the transformer debate, is an important instance of near simultaneity of invention, we need to examine the circumstances that stimulated a number of inventors to identify the motor as a reverse salient and define it as a set of critical problems for which they proposed solutions.

Inventive activity centered on the alternating-current motor in part because alternating current was well publicized and available for experimentation. Its characteristics could be observed first hand. Gaulard and Gibbs's system began to attract attention in 1883, and with each successive year, alternators and transformers became more widely known, tested, and used. Galileo Ferraris, one of the inventors associated with the a.c. motor, became interested in alternating-current phenomena, for instance, when he acted as a technical evaluator of the Gaulard and Gibbs and Ganz & Company transformers.[5] Displays at other international expositions, such as those used by Gaulard and Gibbs in London, which attracted George Westinghouse's attention, also stimulated cogitation and experimentation related to a.c. needs and opportunities. In addition, manufacturers loaned or gave alternating equipment to engineering schools so that faculty and students might use it in research and thus discover its possible applications. Engineers and managers of utilities using the new current became familiar with its strengths and weaknesses first hand. This is a reminder that new tech-

[5] Ludwig Guttman, "On the Rotary Magnetic Field and Multiphase Alternating Current Distribution," *Electrical World* 21 (1893): 276, quoted on page 4 of G. Silva, *Galileo Ferraris: The Rotating Magnetic Field and the Asynchronous Motor*, a translated excerpt from the original Italian text published in *L'Elettrotecnica* 34 (1947) (hereafter cited as Silva, *Ferraris*).

nology must be used in order to be improved, or to put it differently, postinnovation invention and development are common. The originators cannot envisage all of the opportunities and needs of diverse-use environments.

With increased competition from a.c. systems, the direct-current utilities appreciated the advantage they held with their practical motor and decided to publicize it even more. In response, the competing a.c. forces sensed more sharply than ever the desirability of correcting the salient weakness of their system. Notions of load management and attendant economies became less vague as the 1880s ended, and this clarification of operating economies clearly defined the need for a motor to fill out the load curve.[6] Utilities managers and engineers began looking beyond early problems of reliability and the market for luxury lighting to a more evenly distributed and twenty-four-hour load.

Not only did the utilities define the need for an a.c. motor, but the manufacturers with investments in a.c. lighting systems—for instance, Westinghouse—also encouraged inventors. In the 1880s the independent inventor was still an important figure; he had not yet been displaced by the research-prone academics, industrial scientists, and development engineers the manufacturers would eventually hire. So these independents responded, sensing correctly that successful invention would be rewarded because an economic need had been perceived. In the late eighties, the regular announcement of patents in numerous technical and engineering periodicals such as the *Scientific American* kept the acutely sensitive inventors aware of opportunities.

A number of makers and users responded to the need with new inventions, but many inventors and scientists simply analyzed their old discoveries and inventions from the perspective of the newly perceived need. The records of patent litigation in the case of a.c. motors include numerous references to ideas and apparatus that had obviously been rehabilitated after a reverse salient gave them technical and economic value. The original ideas and apparatus had often resulted from something's being possible rather than needed. The perception of a reverse salient transformed the playing and curiosity into critical-problem solving. A long list of antecedent ideas pertaining to the a.c. motor appeared as inventors, patent lawyers, and individuals and organizations with economic interests in the a.c. system began their purposeful attempts to establish priority.

Because the factors that stimulated interest in the development of an a.c. motor were numerous, the number of inventors who claimed priority and the variations in historical attribution were numerous as well. Patents and patent litigation shape memories so strongly that disinterested accounts based on them cannot be assumed. For instance, patent holders may recall the stimulating influences of scientists but not those of other, competing patent holders. There is also a nationalistic bias to accounts of the invention and introduction of the a.c. motor system. Italian histories stress the discoveries of Galileo Ferraris (1847–1897); American accounts often focus

---

[6] The term "load factor," which precisely identified load-management problems, was introduced by R. E. B. Crompton, in 1891; see Parsons, *Early Days of the Power Station Industry*, p. 208. For a discussion of this influential concept, see p. 218 below.

on Nikola Tesla (1856–1943) and the Westinghouse engineers with whom he worked; Yugoslavian historians recall Westinghouse's activities, but as a backdrop to Tesla's heroic contributions; and the Germans quietly but confidently note that Michael Osipowitch von Dolivo-Dobrowolsky (1862–1919) and the German General Electric Company (AEG) introduced the motor in the form that was used most widely by 1900. Swedish historians remember Jonas Wenström (1855–1893), and the Swiss stress the work of Charles Eugene Lancelot Brown (1863–1924) of Brown, Boveri & Company. The French call attention to Marcel Deprez (1843–1918), who contributed concepts to the a.c. field, as he did to so many other early electrical developments. Obviously the circumstances that focused attention on the need for an a.c. motor were transnational.

To unravel the tangled skein of priority claims, let us establish a framework of analysis. The categories to be analyzed will be defined by questions. For example: Who were the scientists or inventors who first articulated the general ideas underlying the invention and publicized them through lecture or publication? Who were the inventors who first constructed working models? Who first applied successfully for a patent? What persons or companies initially introduced the invention on the market? And what is the precise nature of the invention the priority of which is in question? The answers to these questions must be offered with qualifications and reservations.

Taking the last question first, the invention of interest is the alternating-current, polyphase motor. To establish a rotating magnetic field in its stator windings, the polyphase motor uses several out-of-phase, or polyphase, currents. By 1900 the three-phase system was common, but the two-phase system also had been used. Polyphase motors were of several designs; the most common distinction was between the induction, or asynchronous, motor and the synchronous motor. The induction motor found wide use, varying in size from the motor in a dentist's drill to the propulsion unit of a battleship, and today is used throughout the world. "Together with the petrol engine, it has become the most indispensable machine that man has made."[7]

The rotor of the induction motor has current induced in its windings, or conductors, by the rotating magnetic field of the stator. The reaction between the induced rotor current and the stator's magnetic field develops the torque of the motor. The synchronous motor has a separate source of direct current for the rotor; as in the case of the induction motor, the rotating magnetic field of the stator of the synchronous motor is established by polyphase currents. The synchronous speed of a polyphase motor is the speed of the rotating magnetic field in the stator. This speed varies directly with the frequency of the applied voltage and, inversely, with the number of poles. The magnetic revolving field runs at the same speed, pole for pole, as the alternating-current generator supplying it. As observed, the induction motor's rotor runs slightly behind synchronous speed; the synchronous motor's rotor runs at synchronous speed.[8]

[7] Von Siemens, *House of Siemens*, 1: 140.

[8] For a clear and expertly simplified description of the alternating-current motor, see U.S., Bureau of Naval Personnel, *Basic Electricity* (New York: Dover, 1970), pp. 317–37.

Figure V.1. *Nikola Tesla. Courtesy of the National Museum of American History, Smithsonian Institution, Washington, D.C.*

The most widely acclaimed of the inventors of the alternating-current motor was Nikola Tesla. Tesla is better known than the others not only because of the success of his invention but also because his native country, Yugoslavia, has rigorously cultivated his memory; because he was associated with a leading American manufacturer; because he was greatly honored by his contemporaries; and because he was a colorful, dramatic personality who attracted considerable attention in newspapers and periodicals and about whom a number of books for the general audience have been written.[9]

Nikola Tesla was born on the northernmost coast of Dalmatia in the village of Smiljan, Lika, then a part of the Austro-Hungarian Empire and now a village in Yugoslavia. His father, a Serbian, was a Greek Orthodox priest; his mother was "of a distinguished Serbian family."[10] Because his mother is remembered as being descended from a long line of inventors, Tesla was probably brought up in that tradition. He studied engineering at the Graz Polytechnic in Austria and completed his education at Prague University at the age of 23, at the time when arc lighting was attracting much attention and Edison had embarked on his well-publicized quest for incandescent lighting. Tesla, who was not reticent in autobiographical matters, recalled on the occasion of receiving the Edison Medal in 1917 that at the polytechnic school, one of the finest in Europe, he rose each morning at three o'clock and worked until eleven in the evening for an entire year without a day's exception. His professors awarded his work the highest distinctions. Later, he demonstrated his remarkable stamina by outlasting his employer, Edison, during a nine-month period when Tesla worked from half-past ten each morning until five the next. Tesla recalled that Edison said, "I never saw such a thing, you take the cake."[11]

[9] Bokšan Slavko, *Nikola Tesla und sein Werk* (Leipzig/Vienna/New York: Deutscher Verlag für Jugend und Volk, 1934); Thomas C. Martin, *The Inventions, Researches, and Writings of Nikola Tesla* (New York: Electrical Engineer, 1894); John J. O'Neill, *Prodigal Genius: The Life of Nikola Tesla* (New York: Washburn, 1944); A. J. Beckhard, *Nikola Tesla, Electrical Genius* (London: Dobson, 1961); Inez Hunt and Wanetta Draper, *Lightning in His Hand: The Life Story of Nikola Tesla* (Denver: Sage Books, 1964); H. B. Walters, *Nikola Tesla, Giant of Electricity* (New York: Crowell, 1961); C. J. Bethenod, *Tesla* (Belgrade: Société . . . de l'Institut Nikola Tesla, 1938); and Margaret Cheney, *Tesla: Man Out of Time* (Englewood Cliffs, N.J.: Prentice-Hall, 1981). In 1956, the centenary of Tesla's birth, an international meeting in Belgrade was the occasion for publication of Tesla's works, as well as essays about him, by the Nikola Tesla Museum of Belgrade. I am indebted to Professor Carl Chambers of the University of Pennsylvania, an invited participant in the meeting, for calling my attention to the meeting and its publications. Among these are V. Popovič, R. Horvat, and N. Nikolič, comps., *Nikola Tesla: Lectures, Patents, and Articles* (Belgrade: Nikola Tesla Museum, 1956); V. Popovič, comp., *Tribute to Nikola Tesla: Presented in Articles, Letters, Documents* (Belgrade: Nikola Tesla Museum, 1956), a volume that includes an extensive bibliography of books and articles; and Nikola Tesla Museum, *Centenary of the Birth of Nikola Tesla, 1856–1956* (Belgrade: Nikola Tesla Museum, 1959). The literature in technical journals is extensive. Much of this is noted in John Ratzlaff and Leland Anderson, comps., *Dr. Nikola Tesla: Bibliography* (Palo Alto, Calif.: Ragusan, 1979).

[10] A. P. M. Fleming, "Nikola Tesla," *IEE Journal* 91 (1944), reprinted in Popovič, *Tribute to Tesla*, p. A215.

[11] Tesla's "Address of Acceptance" upon receiving the Edison Medal was reprinted in Popovič, *Tribute to Tesla*, p. A103; it was originally published in *Electrical Review and Western Electrician* 70 (1917) (hereafter cited as "Address of Acceptance"). Information on Tesla's career is drawn from this address and from Tesla's "Some Personal Recollections," *Scientific American*, June 1915, reprinted in Popovič, Horvat, and Nikolič, *Nikola Tesla*, pp. A195–99 (hereafter cited as "Personal Recollections").

Tesla later insisted it was in 1877 that he first defined the critical problem which resulted in his invention of the polyphase system. A student at Graz at that time, he witnessed a professor operating a Gramme generator as a d.c. motor. The sparking from the generator's brushes and commutator was intense, and it was apparent to all that these components would soon burn out. This demonstration motivated Tesla to search for a design that would eliminate the brushes and commutator. (Other inventors who remembered being stimulated by classroom lectures or demonstrations to make their major inventive contributions were Charles Hall, who isolated aluminum, and Rudolf Diesel, father of the diesel engine.) It is not clear whether Tesla then wanted to develop a generator or a motor without a commutator, or both. Five years later, in 1882, after picturing the problem and possible solutions in his mind many times, the flash of insight came, Tesla recalled, as he recited Goethe while walking in the Budapest City Park with a friend.

Tesla, who memorized "entire books" and could select "from memory word by word," was reminded of the following passage:[12]

Sie rückt und weicht, der Tag ist überlebt,
Dort eilt sie hin und fördert neues Leben
O dass kein Flügel mich vom Boden hebt,
Ihr nach und immer nach zu streben!

Ach! zu des Geistes Flügeln wird so leicht
Kein körperlicher Flügel sich gesellen.

At that moment in the park, Tesla's imagination indeed soared: "In an instant I saw it all, and I drew with a stick on the sand the diagrams which were illustrated in my fundamental patents of May, 1888."[13] Inasmuch as five fundamental, related patents were issued to Tesla in May 1888, the detail of his insight must have been remarkable.

Five years passed from the moment of Tesla's brainstorm and his application for the patents. Tesla maintains that the vision of the machines without commutators remained vivid and compelling in his mind during that time. In essence, he was searching for an entirely new system, for his motor would require a special polyphase generator. In Budapest, where he worked as an electrician for the Austro-Hungarian state telephone system, his determination to construct the motor persisted; it also survived a move in 1882 to Paris, where he worked for the French Edison Electric Light Company, a manufacturer of direct-current apparatus. While in the employ of French Edison, he helped install central stations, one of them in Strasbourg, where he worked for two years. In Strasbourg, according to his reminiscences, he constructed a model of his motor, but was unable to find anyone to fund its further development. Then, in 1884, he sailed for America and in New York found employment in Edison's Machine

---

[12] "Personal Recollections," p. A198. The passage is from Goethe's *Faust*, pt. 1, in *Goethes sämtliche Werke*, 36 vols., introduction by Karl Goedeke (Stuttgart: Gotta, 1867–82), 10:44–45 (lines 719–22 and 737–38). Loosely translated it reads: "The sun sinks; the day is done./The heavenly orb hastens to nurture life elsewhere./Alas, no wings lift me from earth/To strive always to follow!/... Oh, that spiritual wings soaring so easily/Had companions to lift me bodily from earth." I am indebted to Dr. Horst Daemmrich for locating this passage.
[13] "Personal Recollections," p. A198.

Works; there he worked tirelessly, but for less than a year. Tesla reports that he was discouraged by Edison's opposition to alternating current, but Edison's opposition may have been indifference, for in 1884 there was little competition from alternating current. Francis Jehl recalled years later that Tesla did tell others at the Edison Machine Works of his enthusiasm for alternating-current motors without commutators.[14] Tesla then left Edison and with several financial backers established the Tesla Electric Light & Manufacturing Company of Rahway, New Jersey. Among the patent applications he filed for his company in 1885 was his design for a commutator that would prevent sparking; he also patented an arc-light generator, a regulator for a generator, and an arc lamp. This evidence that Tesla's commitment to his concept was sustained is of particular interest because some of his competitors doubted that he had the idea for an a.c. system with an induction motor five years before filing his patent applications.[15]

Tesla's activities after 1884 fit the general pattern established by other professional inventors. He set up a company bearing his name, one that was based on his existing patents and his ideas for future inventions. He concentrated on inventions and developments for which there was ready funding and a likely market. This may explain why he postponed work on the problem that was of greatest interest to him. The financial backers of the Tesla Electric Light & Manufacturing Company had little reason to invest in an alternating-current system in 1885, for at that time its use was experimental and its pioneers—especially Gaulard and Gibbs—were yet experiencing numerous difficulties. In his concentration on the invention and development of major improvements in existing systems and on the solution of critical problems in evolving systems, Tesla followed the road taken by contemporary professionals like Charles Brush, Elmer Sperry, William Stanley, Elihu Thomson, Frank Sprague, and S. Z. de Ferranti. He, as did they, took the risk of investing in a breakthrough invention. Many relatively unknown professionals have been more conservative, concentrating instead on incremental improvements.

Tesla's opportunity came in 1887, although detailed information to authoritatively document this critical year is lamentably lacking. Because Tesla was an experienced inventor, daily diaries, laboratory notebooks, and records of patent applications must have been made, but to date, the scholarly monograph that would authenticate his creative activities has not been written.[16] Therefore, Tesla's own sparse recollections, his patents, and a major technical paper covering the critical period from the fall of 1887 to the spring of 1888 constitute the information available to the historian.

Tesla puts it succinctly: "Early in 1885 people approached me with a

[14] Gordon Friedlander, "Tesla, Eccentric Genius," *IEEE Spectrum*, June 1972, p. 26. Friedlander relies on Francis Jehl, *Menlo Park Reminiscences*, 3 vols. (Dearborn, Mich.: Edison Institute, 1937–41).

[15] Silva, *Ferraris*, pp. 19–20.

[16] For an example of the help that diaries, notebooks, and other contemporary records give in establishing the history of inventive activity, see Thomas P. Hughes, *Elmer Sperry, Inventor and Engineer* (Baltimore: The Johns Hopkins Press, 1971). The Tesla Museum in Belgrade has Tesla's papers, some of which have been microfilmed and placed on deposit in the Library of Congress in Washington, D.C. I examined the microfilm and found little of interest on the 1880s.

proposition to develop an arc light system and to form a company under my name. I signed the contract, and a year and a half later I was free and in a position to devote myself to the practical development of my major discovery. I found financial supporters, and in April, 1887, a company was organized for the purpose, and what has followed since is well known."[17] In another account, he dated the formation of the second company as April 1886 and referred to the construction of a laboratory.[18] His first two patents on the alternating-current motor were filed on 12 October 1887. Part interest was assigned to Charles F. Peck of Englewood, New Jersey, who probably was one of the financial backers of whom Tesla wrote.[19] Three more patents pertaining to the system were filed in November and issued on 1 May 1888, and two were filed in December and also issued on 1 May.[20]

The patents issued on 1 May 1888 described a system for the conversion, transmission, and utilization of energy. Essentially the system involved a generator for converting mechanical to electrical energy and a motor to convert the electrical energy once again to mechanical power. According to Tesla's patents, the system would use high voltage, or tension, for transmission, and a motor would provide uniformity of speed regardless of load. Tesla depicted the rotating magnetic field as "a progressive shifting of the magnetism or of the 'lines of force.' "[21] According to expert analysis of the patents, they covered both the polyphase synchronous and the asynchronous, or induction, motors.[22]

Like the ideas of many other inventors, Tesla's concepts were geometrically symmetrical and logically ordered. His idealized images stimulated his creative endeavors. If the images did not coincide with reality, he endeavored to reorganize the real so that it would approximate his ideal. This approach required invention. For example, in the paper he presented on 16 May 1888 to the American Institute of Electrical Engineers, he described how generators and motors were and how they ought to be. The polyphase system embodied in the patents issued to him only two weeks earlier was "the ought to be." As he wrote in his paper, in direct-current systems, alternating currents were induced in the generator only to be changed into

[17] "Personal Recollections," p. A199.

[18] "Address of Acceptance," p. A106.

[19] "Electro Magnetic Motor": filed 12 October 1887; issued 1 May 1888 (Patent no. 381,968). "Electrical Transmission of Power": filed 12 October 1887; issued 1 May 1888 (Patent no. 382,280).

[20] "Electro Magnetic Motor" (Patent no. 381,969), "Electro Magnetic Motor" (Patent no. 382,279),and "Electrical Transmission of Power" (Patent no. 382,281) all were filed 30 November 1887 and issued 1 May 1888. "Method of Converting and Distributing Electric Currents" (Patent no. 382,282) and "System of Electrical Distribution" (Patent no. 381,970) were filed 23 December 1887 and issued 1 May 1888. The October and November applications have been judged the essential ones. See Silva, Ferraris, p. 14. The December patents refer to a rotating field independent of motors and to transformers of a kind no one adopted. Remarks of G. Revessi, ibid., p. 15. The seven applications made by Tesla (those listed in n. 19 and those listed here) were derived by him from four applications, three of which were divided into two applications each. I am indebted to David Rhees of the University of Pennsylvania for this information.

[21] Patent no. 382,280.

[22] G. Revessi, "Galileo Ferraris e Nicola Tesla: L'Invenzione del motore asincrono," Memorie della R. Accademia di Scienze Lettere ad Arti in Padova 53 (1936–37), quoted in Silva, Ferraris, pp. 14–15.

Figure V.2.   *Tesla's patent (no. 381,968)*
on the electric transmission of power,
1 May 1888.

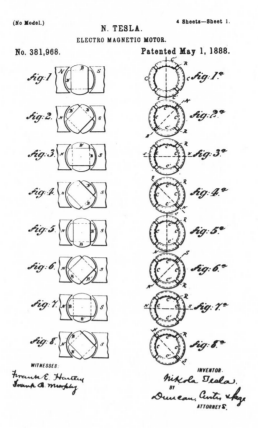

direct currents by the troublesome commutator. After distribution to motors, these direct currents were again changed by a commutator into reversing currents in the motor. This, to Tesla, was illogical. Why not use alternating currents throughout the system, and thereby dispense with the commutator—"a complicated device and, it may be justly said, the source of most of the troubles experienced in the operation of the machines"?[23]

Having dispensed with the commutator, Tesla then faced the problem of producing a rotation of the magnetic poles, an effect produced in d.c. motors by the commutator. He proceeded in his paper—and in his patents—to describe the rotating magnetic field produced by out-of-phase, alternating currents (see Fig. V.2). Using a series of diagrams to illustrate his points, he showed that a generator armature revolving within a magnetic field induced within its segments a magnetic field the polarity of which rotated around the segments. Then, Tesla told his audience—and the patent examiner—the rotating field caused the same effects as a rotating, or whirling, magnet. One such effect, long observed by electricians, was that an iron or steel disc placed in the vicinity of a rotating permanent magnet whirled sympathetically; the same thing happened when a metallic disc was brought into the field of action of the rotating electromagnet. In the latter case a magnetic field was induced in the disc by the rotating electromagnet

[23] Nikola Tesla, "A new System of Alternate Current Motors and Transformers," a paper read before the AIEE on 16 May 1888 and published in *AIEE Transactions,* 5 (1888): 308–24.

and the consequent reaction between the swirling magnetism of the electromagnet and the magnetized disc rotated the disc.

To transmit energy, Tesla ran conductors from slip rings connected to the segments of the generator's armature (rotating member) to (nonsparking) slip rings connected to the segments of the stator (stationary member) of a motor. The stator of the motor, then, had within its segments the same revolving field as the generator armature. The motor armature (rotor) then experienced the induced current and magnetism. Because of the slight lag in phase, a torque was developed. The lag or slippage gave the motor an asynchronous character; the rotor's speed was less than that of the revolving magnetic field of the stator. Tesla also described a synchronous motor that used direct current to produce the magnetism in the field.[24]

Two fundamental ideas were stressed by Tesla in his description of a polyphase motor system—the elimination of the commutator and the use of a rotating magnetic field. However, Galileo Ferraris may have conceived of the rotating magnetic field before Tesla did. The controversy among supporters of the Serbian Tesla and the Italian Ferraris has not ended and is not likely to be resolved.[25] Tesla's testimony that he had the idea of the revolving field in 1882 is not supported by disinterested witnesses or by contemporary publications, and Ferraris's claim that he discovered the rotating magnetic field (in 1885) was not substantiated by lecture or publication until March 1888,[26] months after Tesla filed for his patents.

The circumstances surrounding Ferraris's discovery are interesting as a case of converging scientific and technological influences. As is the case with so many inventions, Ferraris, a professor of physics at the University of Turin, discovered the rotating magnetic field by analogy. Furthermore, his discovery came from a convergence in his mind of technological and scientific information. Extrapolating from his experiments with alternating-current transformers, he reasoned that if two simple, harmonic, equal optical motions with a phase difference of one-quarter produced circular motion, then two similarly out-of-phase electric currents and their accompanying magnetic fields would produce a rotating magnetic field. These thoughts were probably stimulated as well by his awareness of Maxwell's electromagnetic theory.[27]

Ferraris did build an operating motor, but he did not become interested in patenting his invention until the Westinghouse Company persuaded him to allow it to take out an American patent so that the discovery could be made useful. Ferraris's compensation from Westinghouse was a gift of $1,000. Establishing patent priority here is complicated because other in-

[24] Patent no. 382,281 describes one motor element magnetized by a.c. current, the other "element" drawn by d.c. current.

[25] See Silva, *Ferraris*, p. 4; and the extensive publications about Tesla cited in n. 9 above.

[26] The 1885 models (dated by Ferraris) are in the *Istituto Elettrotecnico Nazionale* in Turin. Silva, *Ferraris*, p. 3. The March lecture was published; see Galileo Ferraris, "Rotazioni elettrodinamiche prodotte per mezzo di correnti alternate," *Atti dell'Accademia delle Scienze di Torino* 23 (1887–88): 360 ff.

[27] See the remarks made by R. Arnò, a pupil of Ferraris's, in "Commemorazione solenne di Galileo Ferraris . . . ," *L'Elettrotecnica* 9 (1922): 517, quoted in Silva, *Ferraris*, p. 2; and Sylvanus P. Thompson, "Galileo Ferraris," *The Electrician* 38 (1897): 497, also quoted in Silva, *Ferraris*, p. 2.

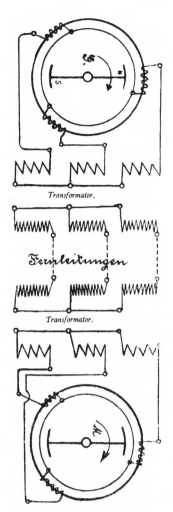

Figure V.3.   *Three-phase (polyphase) transmission system of Friedrich Haselwander: G = generator; M = motor. From* Offizielle Zeitung . . . Frankfurt am Main 1891, *p. 11.*

ventors constructed motors and took out patents about the same time that Tesla applied for his patents and Ferraris published his discovery. As suggested earlier, the simultaneity resulted in part from the fact that inventors at that time were experimenting with, or were informed about, the advances being made in the alternating-current system. They were familiar with the most recent improvements in transformers and a.c. generators, and they saw further implications in the new phenomena observed. They were like explorers in a new land who witness previously unobserved events and discover untapped resources. In the case of inventors, the new events and resources arose from the inventions and developments of other pioneers on the leading edge of technology and science. These pioneers in electrical technology were not lonely inventors laboring in intellectual isolation.

In January 1887 Friedrich August Haselwander, an engineer in Offenburg, Baden, saw the possibility of substituting slip rings for the three-section commutator of a d.c. Thomson-Houston armature he was repairing. He realized that on both motor and generator, slip rings would allow the use of alternating current rather than direct current. Subsequently he applied for a patent on both two-phase and three-phase systems with transformers capable of high-voltage transmission. (See Fig. V.3.) In August 1887 he started constructing a generator with three-phase windings, and in October 1887 he put it to work supplying lights. The manufacturing company of W. Lahmeyer & Co. of Frankfort on the Main took out a license on Haselwander's patent, but it was subsequently declared invalid when challenged by another manufacturer, AEG (German General Electric). Another pioneer in electrical technology in 1887 was the American C. S. Bradley, who applied for a patent on a polyphase generator and a synchronous motor on 8 March 1887. Subsequently, however, the courts ruled that Tesla had priority over these two inventors because of the completeness and fullness of his patent.[28]

The names of Tesla, Ferraris, Haselwander, and Bradley do not exhaust the list of inventors and scientists who made priority claims or contributions to the invention of the polyphase motor and the related generator and transmission system.[29] In retrospect, however, it appears that Ferraris substantiated his claim for priority in demonstrating the idea underlying the invention of a rotating-magnetic-field system; that Tesla made the first successful patent application; that Ferraris first publicly announced (by lecture) the invention or discovery; and that Haselwander and Bradley built the first full-scale polyphase generators. Nevertheless, the development of an economical polyphase system comprised of both motor and generator was not carried out alone by any of these inventors. Electrical manufacturers would take the lead in that endeavor.

By 1900 the larger manufacturers had settled on a three-phase system with an induction motor. The manufacturer most responsible for the de-

---

[28] Georg Dettmar, *Die Entwicklung der Starkstromtechnik in Deutschland* (Berlin: ETZ Verlag, 1940), pp. 67–69; Siemens, *House of Siemens,* 1: 122; and Anton Schwaiger, "Geschichte des Drehstroms," *Technikgeschichte* 28 (1939): 53–54.

[29] See Silva, *Ferraris,* pp. 10 and 25, and Schwaiger, "Geschichte des Drehstroms," for the claims of others, including Elihu Thomson, W. Baily, and M. Deprez. Sylvanus P. Thompson, *Polyphase Electric Currents and Alternating Current Motors* (London: E. & F.N. Spon, 1895) also is helpful. Silva's *Ferraris* includes a bibliography related to the priority question.

velopment of this system in Europe (and probably in the rest of the world) was the German company Allgemeine Elektrizitäts-Gesellschaft, of Berlin, and the engineer of this company who took the lead in developing the system's motor was Michael Dolivo-Dobrowolsky (1862–1919). Learning of the discoveries and inventions of Tesla and Ferraris through publications, Dolivo-Dobrowolsky began experimenting with polyphase machines. He doubted Ferraris's statement that the polyphase motor could obtain at best only 50 percent efficiency, and his experiments proved that he was right. He also questioned Tesla and Westinghouse's choice of a two-phase system; thoroughly investigating different phase systems with an experimental motor, he decided on the three-phase arrangement, which he gave the name now used in Germany, *Drehstrom* ("turning, or rotating, current"). In addition, he patented the simplest and most practical induction, or asynchronous, motor with a short-circuited rotor, a motor that has been widely used and copied. It is interesting to note, however, that in his later years he developed a direct-current high-voltage transmission system.[30]

In the United States, the Westinghouse Company invested in the development of the Tesla polyphase motor and transmission system. This is not surprising, considering that the Westinghouse Company–Tesla relationship was much like that between Westinghouse and William Stanley. Both show how manufacturers related to independent inventors before engineering-and-development departments and research-and-development laboratories were organized.

Westinghouse first sent H. M. Byllesby to examine Tesla's motor and patents after Tesla read his paper to the AIEE in May 1888. Advised by his staff that the Ferraris patent would have little value, George Westinghouse obtained an option to purchase the patents by payment of $20,000 in cash, $50,000 in notes, and a royalty of $2.50 for each motor horsepower manufactured. Minimum royalties were to be $5,000 in the first year, $10,000 the second, then $15,000 each succeeding year. Westinghouse purchased the patents in July 1888. When Westinghouse was in dire financial straits in the 1890s, however, Tesla agreed to accept a cash settlement of $216,000 in lieu of royalties. Bankers believed that this would place less strain on the company.[31] Years later, Tesla did not consider his compensation generous enough. Trying in 1920 to interest E. M. Herr, president of Westinghouse, in his wireless power-transmission system, Tesla wrote that he was ready to sell the system, provided "your company is willing to come to

---

[30] "Dolivo-Dobrowolsky," in *Männer der Technik*, ed. Conrad Matschoss (Berlin: VDI, 1925); and Dettmar, *Starkstromtechnik*, pp. 70–71. It is not surprising that in 1957 a German professional association, the Verband Deutscher Elektrotechniker (VDE) categorized Dolivo-Dobrowolsky with Ferraris, Tesla, Bradley, and Haselwander as an inventor of the three-phase system. The VDE committee, which was especially commissioned to investigate and establish priority in the invention of the system, also included the Swede Jonas Wenström and the Swiss Charles E. L. Brown among its inventors. Wenström's work is discussed in Åke Vrethem, *Jonas Wenström and the Three-Phase System*, Stockholm Papers in History and Philosophy of Technology (Stockholm: Royal Institute of Technology, 1980).

[31] Letter of G. Westinghouse, quoted in Passer, *Electrical Manufacturers*, p. 278. This account of Westinghouse's relationship with Tesla is based on Passer (ibid., pp. 278 ff.), who had access to Westinghouse Co. files; and on Cheney, *Tesla*, pp. 40 and 48–49. Cheney gives the original purchase price as about $60,000, of which $5,000 was to be paid in cash and 150 shares of stock plus the $2.50 royalty.

an understanding with me on terms decidedly more generous than those under which they acquired my system of power transmission thirty years ago."[32] Tesla believed that the success of Westinghouse Electric was due to his invention of the polyphase system.

As part of the agreement in 1888, Tesla worked in Pittsburgh for a year with Westinghouse engineers to develop a marketable motor and system. His relations with the Westinghouse staff were not entirely harmonious, however, and like Stanley, he soon returned to his independent inventive endeavors. Tesla later recalled that he encountered much opposition in Pittsburgh; only Albert Schmidt and Charles F. Scott had faith in the new system and assisted him. Tesla believed, however, that George Westinghouse's confidence in him never wavered.[33]

Other companies in the United States and Germany, including the Stanley Electric Manufacturing Company and the Thomson-Houston Company, contributed to the development of the polyphase system.[34] It is notable that the British stood aside. As one British historian writes, "The polyphase system and the induction motor were developed in the United States and on the Continent." English engineers and manufacturers "took no part in this."[35]

The introduction of the polyphase motor and system in the early 1890s and the earlier introduction of the transformer, which was soon adapted to polyphase function, permitted the alternating-current system to match the capacity of the older, direct-current system in supplying both power and light. Furthermore, the polyphase system corrected the fundamental flaw, or reverse salient, of direct current—its uneconomical transmission. "The battle of the systems" was not yet over, however. Existing direct-current utilities in densely populated areas continued to expand to meet load increases. Their unamortized investment in direct current was so large that it discouraged replacement with a polyphase system. If the utilities supplemented the existing direct-current system with an additional and unconnected polyphase system, they would lose the advantages of scale of a single system. Furthermore, General Electric, Siemens, AEG, and the smaller British equipment manufacturing companies remained partially committed to direct current through capital investments in patents, equipment, and operating facilities as well as in experience and expertise. Because of their experience and expertise, Edison, other engineers and managers, and skilled laborers were biased toward the older system. Technical change was frustrated by the institutionalization of direct current during the previous decade. Manufacturers, utilities, and consumers waited cautiously for the lines of future development to be clearly delineated.

Because "the battle of the systems" had become far more complicated than a technical problem awaiting a simple technical solution, it ended

[32] Tesla to E. M. Herr, 19 October 1920, Tesla Collection, Microfilm, Title 7229, Reel 6, Library of Congress, Washington, D.C.

[33] Tesla to G. E. Tripp, Chairman of the Westinghouse Electrical & Manufacturing Co., 15 April 1922, ibid.

[34] Philip L. Alger and Robert E. Arnold, "The History of Induction Motors in America," *IEEE Proceedings* 64 (1976): 1381–83.

[35] I. C. R. Byatt, *The British Electrical Industry, 1875–1914* (Oxford: Clarendon Press, 1979), pp. 69 and 144.

without the dramatic vanquishing of one system by the other, or a revolutionary transition from one paradigm to another. The conflict was resolved by synthesis, by a combination of coupling and merging. The coupling took place on the technical level; the merging, on the institutional level. The battle ended with a relatively gentle transition whereby the older system slowly gave way over decades, first as the new system supplemented it, then as the new replaced the worn-out and obsolete. On the institutional level, vested interests were gradually transferred into a stake in the new system. The history of the resolution of "the battle of the currents" helps explain why engineers and managers argue that their adjustments to change are more reasonable than those made in the realm of political and international relations.

Technical resolution of the controversy was greatly facilitated by couplers, or transducers. The most obviously contributory of these was the rotary converter. Histories of electrical technology rarely give the device the place of prominence it deserves. Because the rotary converter was the solution to a critical problem defined by a number of inventors and engineers who identified a reverse salient, it, too, is a case of simultaneity of invention. The reverse salient was econotechnical—the need to sustain vested interests in the old system while making the transition to the new.

In the United States, invention of the rotary converter is attributed to Charles S. Bradley. Bradley had worked for Edison and invented a polyphase generator before establishing a factory at Yonkers, New York, early in the 1890s to manufacture the converter he invented in 1888. General Electric soon acquired the Bradley patent and facilities.[36] The Westinghouse Electric Company developed rotary converters as well.[37] In 1891 Wilhelm Lahmeyer (1859–1907), inventor of various direct-current machines, introduced a rotary converter with a single armature for changing direct current first into polyphase and then the reverse. Haselwander, whose polyphase system Lahmeyer's company acquired, also patented a rotary converter. Zipernowski and Déri of Ganz & Company took out a German patent. Other names also are recorded in the history of the rotary converter.[38]

Single-armature rotary converters were used for other and varied transformations, including frequency and phase conversion. Motor-generators also were used for conversion; the motor, driven by one kind of current, drove a generator that produced another kind. The highly important advantage of this capacity for conversion was that various old systems, not only a.c. and d.c., could be coupled, or synthesized, with new ones. For example, in an effort to save the investment of utilities in single-phase alternating-current equipment, Westinghouse first introduced a two-phase motor, believing that it could be accommodated to the single-phase system most easily. Subsequently the two-phase motor was abandoned in favor of

[36] Passer, *Electrical Manufacturers,* pp. 300–301.

[37] Benjamin G. Lamme, an outstanding Westinghouse inventor and engineer, has recalled the early decades of the company and his contributions to it in *Benjamin Garver Lamme, Electrical Engineer: An Autobiography* (New York: Putnam, 1926); chap. 6 is about induction motors, rotary converters, and a.c. generators (1885–1900).

[38] Dettmar, *Starkstromtechnik,* pp. 74–76.

Figure V.4. *Westinghouse exhibit, Chicago, 1893. From Edward D. Adams,* Niagara Power: History of the Niagara Falls Power Company, *2 vols. (Niagara Falls, N.Y.: Niagara Falls Power Co., 1927), 2: 192.*

the three-phase arrangement, but, again, a coupler (in this case a phase-converter) saved the old system during the period of transition.

The variety of couplers on the market by the early 1890s suggested the need for a universal electric supply system, and even though it is rarely acknowledged, the definition and introduction of that system proved to be one of the most influential innovations in the history of electric light and power. The concept of a universal system completed the transition from the era of electric light to the era of electric light and power. Westinghouse displayed the system at the Chicago exposition of 1893, and the design (seen in Fig. V.4) can be attributed to the engineering staff of that company.[39] The essence of the concept was a unified system embracing, or coupling, generators (supply) and loads of varying characteristics (demand). The system was capable of supplying incandescent lamps, arc lights, direct-current motors for both stationary and traction purposes, single-phase alternating-current motors, polyphase motors, and energy for thermo-electrical and electrochemical uses from a common transmission line or ring fed by centralized, large-scale generators. (See Figs. V.5–V.8.) Its attractiveness to the ordering and systematizing mind of the engineer and manager can best be conveyed by describing the unsystematized and dis-

[39] This account of the universal system is based on the "Scott Essays," a collection of short essays written by Charles F. Scott in 1938. The collection consists of eight items, each of which is about ten pages long. Scott, who was a Westinghouse engineer from 1888 to 1911, focuses on the introduction of the universal system at Niagara Falls, but he also makes some general observations on the history of electric light and power prior to 1895. I am grateful to Robert Belfield for providing me with a copy of these papers, which he located in the archives of the Niagara Mohawk Power Co. of New York; these archives are now being processed at the George Arents Research Library at Syracuse University (hereafter cited as Niagara Archives).

Figure V.5. *The Westinghouse concept of a universal supply system. From* Cassiers Magazine *8 (1895): 358.*

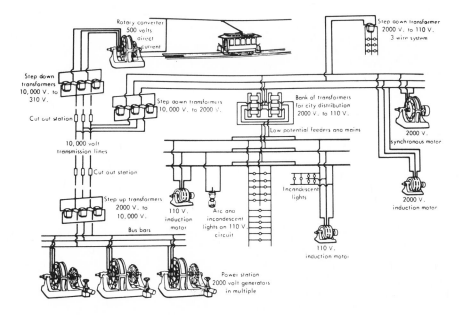

Figure V.6. *Comparison of electric power systems. From typewritten memorandum by Charles F. Scott, "What Was the Niagara Central Station Plan?" (5 April 1938). Courtesy of the Niagara Archives, George Arents Research Library, Syracuse University, Syracuse, N.Y.*

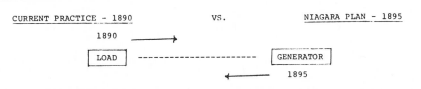

CURRENT PRACTICE - 1890   VS.   NIAGARA PLAN - 1895

1890 ——————▶

LOAD ----------------------- GENERATOR

◀—————— 1895

Current Practice began with LOAD-lamps or motors, and adopted different kinds of current for different lamps and motors.

Niagara Plan began with GENERATION. A single kind of current for all purposes.

Different circuits carried the appropriate kinds of current to the several kinds of loads; some economically limited to short distances.

A universal transmission system suited to very great distances supplies current to substations to be transformed as desired for different types of load.

Each kind of current requires special type of generator. Each generator limited to its own kind of service.

Generators may be operated in parallel as a unit.

Most generators inherently limited in size. In some cases similar generators did not run in parallel, but each supplied independent circuits.

A generator may be made of very great size, usually limited by the capacity of the power source which drives it.

Product of individual inventors, intent on a particular service.

Product of a broad scientific study of wide scope.

Product of the '80's, when cut-and-try methods prevailed.

Product of the '90's, based on prior empirical experience and on the rapidly developing scientific knowledge and engineering methods of design and operation.

Product of cut-and-try search for a way to operate certain lamps or motors.

Product of comprehensive research and coordination of old and new in a single Comprehensive System.

Figure V.7. *Nonuniversal electric supply system of 1890. From typewritten memorandum by Charles F. Scott, "What Was the Niagara Central Station Plan?" (5 April 1938). Courtesy of the Niagara Archives, George Arents Research Library, Syracuse University, Syracuse, N.Y.*

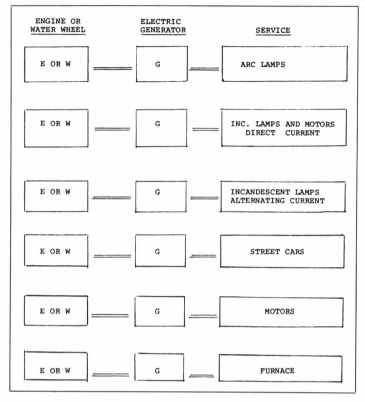

Figure V.8. *The Niagara (universal) power plan. From typewritten memorandum by Charles F. Scott, "What Was the Niagara Central Station Plan?" (5 April 1938). Courtesy of the Niagara Archives, George Arents Research Library, Syracuse University, Syracuse, N.Y.*

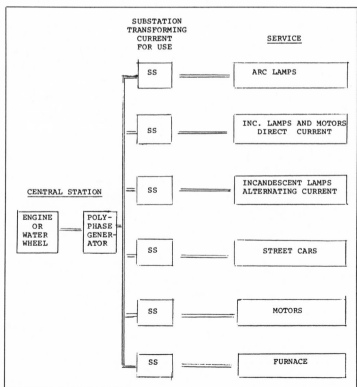

orderly state of electric supply at the time of the introduction of the universal system (about 1893). The Buffalo General Electric Company, a utility supplying part of the city from a central station at 40 Court Street, had in that station seven steam engines of several makes and fifteen generators of seven different types supplying five different kinds of customers, or loads, on separate circuits. Fragmented supply meant that when the economically critical load factor was low, economies of scale could not be realized.[40]

The universal system shown at the Chicago exposition used the new polyphase current for transmission and couplers, including transformers, to connect established modes of distribution serving different loads. In this way a utility would not have to scrap its existing distribution circuits. At the exhibition a 1,000-h.p. generator drove a two-phase Tesla induction motor. The motor in the display was intended to represent the prime mover of the universal system. The motor powered a two-phase generator, which in turn sent current to step-up transformers for transmission and to step-down transformers at distribution points. At these points a motor drove an arc-light generator, a rotary converter supplied direct current for streetcar motors, incandescent lamps were supplied, and various motors were driven. Inventors and engineers who saw or read about the display quickly grasped the significance of the universal system.

One of the creators of the universal electric supply system drew an analogy between it and water-supply systems, which combined springs, private wells, and cisterns to supply various types of consumers; telephone systems, which unified the formerly isolated systems of independent companies; and large railway systems, which integrated smaller ones that had used short roads with different track gauges. (In more recent times, computer networks would be considered analogous.) Utilities soon embraced the compelling economic and technical logic of the universal system (see Fig. V.9). By 1900, technical periodicals were publishing accounts of utility systems throughout the world that had used the universal system to make the transition to the era of electric light and power.[41] Even earlier, leading manufacturers had committed themselves as well to the new resolution of "the battle of the systems."

It is important to observe, however, that the struggle between the two systems did not come to an end with the development of a technical solution—couplers joining the old and new; it came with the introduction of institutional inventions that in turn resolved the conflict on the organizational level (a conflict so clearly revealed in the episode involving Harold Brown and the electrocutioner's current). In the United States the swing of Edison General Electric to the new way signaled the resolution of the systems conflict on the institutional level. The reasons for the shift were complex. For one thing, Thomas Edison, a bitter foe of the new current who had supported efforts to discredit and outlaw it, relinquished control of the Edison enterprises in the late 1880s. In the absence of his monu-

[40] Charles F. Scott, "What Was the Niagara Central Station Plan?" (5 April 1938), Niagara Archives.

[41] W. L. R. Emmet, "Results Accomplished in Distribution of Light and Power by Alternating Currents," *Electrical World* 27 (1896): 570–72, 593–96.

Figure V.9.   *Resolution and transition in "the battle of the systems": Percentage of direct current* (Gleichstrom), *mixed* (Gemischte), *single-phase alternating current* (Wechselstrom), *and polyphase current* (Drehstrom) *used in Germany, 1880–1935. From Rudolf von Miller's article in* Technikgeschichte *25 (1936): 112. Courtesy of the Division of Electricity, National Museum of American History, Smithsonian Institution, Washington, D.C., and Rudolf von Miller.*

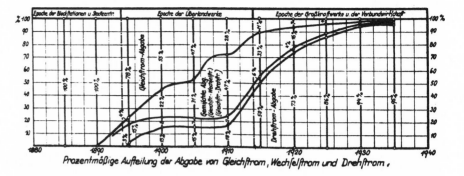

mental prestige and conservative influence, Edison General Electric's dogmatic opposition to alternating current waned. Edison's withdrawal also cleared the way for a merger with the Thomson-Houston Company, a rising manufacturer of electrical machinery that carried a line of both direct- and alternating-current equipment. After the merger in 1892, the new company, General Electric, introduced equipment for polyphase systems. Another organizational response among manufacturers came in 1896 in the form of a patent-exchange agreement between General Electric and Westinghouse. The engineers in each company, and especially the engineers in the utilities who bought generators, motors, and transformers from them, had found the proscription of obvious improvements aggravating and they welcomed this facilitation of rational technical exchange.

The development and introduction of polyphase power also affected relations among German manufacturers, but in a different fashion. In 1887 Siemens & Halske and the German Edison Company (later AEG) made an agreement based on the assumption that AEG would engage primarily in the financing and construction of large generating stations and related systems and that Siemens & Halske would manufacture the equipment for those stations. The agreement no longer suited AEG, however, after it found that German cities and towns often preferred to establish their own municipally owned utilities and after it developed a polyphase power system designed primarily by Dolivo-Dobrowolsky. Foreseeing a substantial business in the manufacture of power-transmission equipment, AEG wanted to be free of the agreement in order to establish its own facilities. In 1894, two years after Edison General Electric joined with Thomson-Houston, a merger that facilitated the move into the manufacture of alternating-current equipment, AEG and Siemens canceled their agreement and AEG gained full freedom (with the exception of cables, which were left to Siemens & Halske) to develop its manufacturing program.[42]

In both the United States and Germany the stock of a number of electric utilities was owned by the large manufacturers, and the utilities depended upon the manufacturers for the design and development of equipment. Thus, the utilities also sought institutional arrangements that would facilitate the adoption of the manufacturers' technical solution to the systems conflict. For instance, the history of utilities in Chicago reveals that the

[42] Von Siemens, *House of Siemens,* 1: 148; and Conrad Matschoss, "Die geschichtliche Entwicklung der Allgemeinen Elektricitäts-Gesellschaft in den ersten 25 Jahren ihres Bestehens," *Beiträge zur Geschichte der Technik und Industrie* 1 (1909): 60.

organizational response, or invention, there was merger. Early direct-current stations with franchises covering small central-city districts merged. Polyphase transmission allowed for large-area supply, so the alternating-current utilities supplying the suburbs merged into new organizations with the intention of gaining the entire urban market. The series of mergers that occurred in America and Germany, and to a lesser extent in Britain, are documented in numerous utility histories.[43]

Agreement about technical standards also contributed to the resolution of "the battle of the systems" and the establishment of the polyphase, or universal, system. During the period from 1887 to 1892, when the struggle was intense, utilities and manufacturers chose different frequencies. Depending upon the particular character of supply and load, different frequencies had distinct advantages. Therefore, a general agreement about frequency did not come through the establishment of one frequency's obvious technical superiority over the others; rather, a spirit of flexibility and compromise among the various utility interests, and especially among the manufacturers, was primarily responsible for the agreement. Other important factors were precedent and successful practice—e.g., Edison's introduction of 110 and 220 volts for power distribution in the United States.

[43] Utility histories vary greatly in quality. Outstanding among them are Forrest McDonald, *Let There Be Light: The Utility Industry in Wisconsin, 1881–1955* (Madison, Wis.: The American History Research Center, 1957), a study by a leading American historian which treats Wisconsin utilities and provides an overview of utility history that is applicable to the entire country; Raymond C. Miller, *Kilowatts at Work: A History of the Detroit Edison Company (Detroit, Mich.: Wayne State University Press, 1957) and The Force of Energy: A Business History of the Detroit Edison Company* (East Lansing: Michigan State University Press, 1971), two scholarly volumes about two utility magnates who greatly influenced the industry, Alex Dow and Walker L. Cisler; and Nicholas B. Wainwright, *History of the Philadelphia Electric Company, 1881–1961* (Philadelphia: Philadelphia Electric Co., 1961), an account of one of America's largest and oldest utilities. An overview of utility development is given in *Creating the Electric Age,* ed. Nilo Lindgren, a special issue of *EPRI Journal* 4 (1979). Histories of other utilities include Charles M. Coleman, *P.G. and E. of California: The Centennial Story of Pacific Gas and Electric Company, 1852–1952* (New York: McGraw-Hill, 1952); Glenn Weaver, *The Hartford Electric Light Company* (Hartford, Conn., 1969); Thomson King, *Consolidated of Baltimore, 1816–1950: A History of Consolidated Gas Electric Light and Power Company of Baltimore* (Baltimore, 1950); Jack Riley, *Carolina Power & Light Company, 1908–1958* (Raleigh, N.C., 1958), Wade H. Wright, *History of the Georgia Power Company, 1855–1956* (Atlanta, 1957); George Bush, *Future Builders: The Story of Michigan's Consumers Power Company* (New York: McGraw-Hill, 1973); Edward G. Nelson, *KPL in Kansas: A History of the Kansas Power and Light Company* (Lawrence, Kans.: Center for Research in Business, University of Kansas, 1964); John Dierdorff, *How Edison's Lamp Helped Light the West* (Portland, Oreg., 1971); Thomas C. Wright, *Otter Tail Power Company from Its Origin through 1954* (Fergus Falls, Minn., 1955); Howard R. Fussell, *A History of Gulf States Utilities Company, 1912–1947* (Houston: Texas Gulf Coast Historical Assoc., 1967); Herbert W. Meyer, *Builders of Northern States Power Company* (Minneapolis, Minn.: Northern States Power Co., 1972); Indianapolis Power & Light Co., *Electrifying Indianapolis* (Indianapolis, Ind., 1960).

Georg Boll, *Entstehung und Entwicklung des Verbundbetriebs in der deutschen Elektrizitätswirtschaft bis zum europäischen Verbund* (Frankfort on the Main: VWEW, 1969), includes in his list of sources many histories of German utilities. The history of electric supply in the state of Bavaria is given in Verband Bayerischer Elektrizitätswerke E. V., *Elektrizität in Bayern, 1919–1969* (Munich: VBE, 1969). For British utilities, see Leslie Hannah, *Electricity Before Nationalisation* (Baltimore: The Johns Hopkins University Press, 1979), with its citation of numerous utility histories; and Parson, *Early Days of the Power Station Industry,* which provides accounts of British utilities up to about 1900. Other references to utilities discussed throughout the book in hand are provided in the footnotes; see, especially, Chapters VII, VIII, and IX.

These historical circumstances and events ought to be remembered because later generations often assume that standards necessarily represent the clear technical superiority of one system over another.

During the early years of alternating and polyphase currents, 133⅓, 125, 83⅓, 66⅔, 60, 50, 40, 30, and 25 cycles were used.[44] Designers and engineers chose the frequency that was optimum for the particular set of characteristics created by the coupling of incandescent lamps, transformers, arc lighting, induction motors, synchronous converters, or other apparatus. Because design of this equipment was changing rapidly, however, an already complex situation became more complicated. In America, the engineering staff of the Westinghouse Company played a leading role in dispelling the disorder; they attempted to rationalize production by standardizing frequency. Freedom to decide on a standard frequency was constrained by the fact that a large number of Westinghouse-designed central stations were supplying incandescent lamps with 60 cycles, but the die was not cast. The Westinghouse designers also took into account the spread of the slower-rpm, directly coupled (rather than belt-driven) generators that had first been introduced in Europe. The slower generator, or alternator, was less complex when designed for lower frequencies such as 60 cycles because fewer field poles were needed for them than for higher frequencies such as 133⅓ cycles. Yet other designers advocated a high frequency because it reduced incandescent-lamp flickering. The development of a synchronous converter that operated well at 60 cycles encouraged supporters of this frequency because utilities could couple existing single-phase systems, direct-current systems, and the polyphase extensions and transmission networks. This ability to couple the old with the new lessened support for 40 cycles, a frequency that had been advocated by some because of its suitability for motors and transmission. Design of unprecedentedly large equipment for a Niagara Falls installation also shaped the ultimate decision. George Forbes, chief consulting engineer for the Niagara site, proposed 16⅔ cycles, but Westinghouse engineers then wanted 33½ cycles. The relatively low frequencies were well suited for power transmission. The midpoint of this particular difference was 25 cycles. By about 1900, Westinghouse, the other manufacturers, and the utilities were settling for two standards: 25 cycles for transmission and for large motors, and 60 cycles for the more general-purpose systems. Introduction of the high-speed turbine as prime mover accelerated the trend toward 60 cycles because generators with fewer poles could be used than with the slow-moving, reciprocating steam engines. A Westinghouse engineer who took part in the technical activity surrounding the frequency question took pains to point out that the struggle to set standards was not a competition between manufacturing interests, as was the case with many issues during "the battle of the systems"; rather, it was an effort by technical persons to find a means to reinforce an all-embracing, general system of supply. The battle was indeed over.[45]

In Germany the decision on a standard frequency may have been less

---

[44] Benjamin G. Lamme, "The Technical Story of the Frequencies," *IEEE Transactions* 37 (1918): 60.

[45] Ibid., p. 82.

difficult to reach because AEG, spurred by Emil Rathenau, encouraged the use of the slower-speed, directly coupled generators from which high frequencies, such as $133\frac{1}{3}$ cycles, were difficult to obtain and because synchronous converters, for which low frequencies, such as 25 cycles, were most suitable, had not been widely adopted. The outcome in Germany was a standard of 50 cycles.

The situation in Britain remained disorderly. The tendency before World War I was toward variation, not standardization. London led the trend; by 1914, at least ten different frequencies and a bewildering assortment of voltages were in use in that city. Reasons for the variety included the absence of oligopoly in the electrical manufacturing industry and the prestige and influence of the consulting engineers.[46] One British manufacturer complained that his competitors, confident that individualism, not business opportunism, characterized professional competence, carried individualism in the design of equipment to an extreme. Having a highly particularized design seemed to show that the manufacturer and his designers were "engaged in a superior branch of applied science, not in an industry, still less in a trade."[47] R. E. B. Crompton, a small manufacturer and leading consulting engineer, lamented that the consulting engineers with their design fads; the municipal governments that employed them and took pride in having their own ideas about how "switchboards and other plant . . . should be done"; and the small manufacturers, who were content to have a small order for individual components, retarded not only the move for standardization but the overall growth of the British electrical supply industry.[48] An exception to this rule was the system of electric supply on the northeastern coast of England, where Charles H. Merz, consulting engineer, was the primary influence on the design of a regional system (see pp. 443–60 below). Merz standardized 40 cycles in the region centered around Newcastle upon Tyne. His reasoning was logical and systematic, for he took into account the characteristics of the region. His rationality proved expensive, however, when after World War I his system had to be integrated at great cost into an all-British one in which 50 cycles was the standard frequency.[49]

About the same time that U.S. and German utilities and manufacturers were reorganizing to facilitate the introduction of the universal electric power system and the adoption of standards, another major advantage of the polyphase system was demonstrated. A display of power transmission at the International Electrical Exhibition in Frankfort on the Main in 1891 showed not only the potential of using distant water-power sites to supply electricity to heavily populated industrial areas but also the suitability of polyphase systems for long-distance power transmission. The Frankfort exhibition highlighted the possibilities of what came to be known as "point-

[46] See Chapter IX for a full discussion of British parochialism.

[47] Adam G. Whyte, *Forty Years of Electrical Progress: The Story of G.E.C.* (London: Benn, 1930), p. 20.

[48] Crompton's testimony, in *Official Report: Committee on Electrical Legislation 1901, IEEE Reports and Resolutions*, pp. 36, 43, 47; see also Alderman Pearson's testimony before the committee, ibid., pp. 83–84, 89.

[49] Charles H. Merz, "Autobiography," chapter entitled "Cork," pp. 7–9, Merz & McLellan Co. Archives, Amberley, Killingworth, near Newcastle upon Tyne, England.

Figure V.10. *Honorary chairmen of the Frankfort exhibition of 1891: Adalbert von Waltenhofen and Thomas Edison* (top); *Marcel Deprez* (center); *Silvanus P. Thompson* (bottom left); *and Werner von Siemens* (bottom right). *Von Waltenhofen was professor of electrical technology at the* Technische Hochschule *in Vienna. From* Offizielle Zeitung . . . Frankfurt am Main 1891, *p. 828.*

to-point" transmission, as distinguished from transmission by way of networks or rings. The Lauffen-to-Frankfort transmission line, 175 km. in length, is an excellent example of point-to-point transmission; the regional power systems of the 1920s are examples of network or ring transmission (see Chapter XII below).[50]

The Frankfort exhibition was a skirmish in "the battle of the systems." In 1890 the city of Frankfort had to decide on a current for its municipal electrical works. Cologne's choice of alternating current shortly before made the selection more difficult, for, until then, no large German city had opted for alternating current. The electrical engineers who were brought in as consultants could not save the municipal authorities of Frankfort from a decision by making a clear-cut recommendation in the face of viable alternatives. The issue assumed the proportions of a battle for the city of Frank-

[50] In a lecture given at the Frankfort exhibition, Sylvanus P. Thompson judged the power transmission display a major scientific event of the century. "Das neue Gebiet der Wechselströme," *Offizielle Zeitung der Internationalen Ausstellung Frankfurt am Main 1891*, p. 788.

*Figure V.11.   Charles E. L. Brown.
Courtesy of Brown, Boveri & Co., Baden,
Switzerland.*

fort as it engulfed engineers, manufacturers, and well-informed burghers.[51] Leopold Sonnemann, a banker, politician, and founder of the *Frankfurter Zeitung*, then suggested that the controversy be resolved by holding an electrotechnical exhibition in which the advocates of alternating current could demonstrate the most recent advances in their systems.

A younger generation of engineers and designers contributed substantially to this exhibition. Prominent among them were Oskar von Miller, Charles Eugene Lancelot Brown, and Michael Dolivo-Dobrowolsky. A persistent vision of harnessing the water power of the Alpine region of his Bavarian homeland stimulated von Miller; Brown brought to the exhibition the mechanical genius of the Swiss, with whom his father, an English engineer, had settled and with whom he himself worked; and Dolivo-Dobrowolsky was intensely involved in developing polyphase machinery. Behind Brown was the Swiss mechanical-engineering manufacturing firm of Maschinenfabrik Oerlikon, and supporting Dolivo-Dobrowolsky was the Berlin electrical manufacturer AEG.

In 1882 von Miller had helped organize an electrical exhibition in his birthplace, Munich. On that occasion, he showed his determination to make power transmission a reality by arranging for the leading French engineer, Marcel Deprez (1843–1918), to design a power transmission system extending 57 kilometers from the village of Miesbach, at the foot of the Bavarian Alps, to the exhibition in Munich's Glass Palace. In the palace itself an electric motor pumped a waterfall to symbolize the transmission of water power. Deprez used direct current for transmission, as he had in other demonstrations, and ordinary telegraphy wire for the high-voltage transmission line.[52] (See Figs. V.12 and V.13.)

Von Miller looked upon international technical exhibitions as an effective means of introducing new technology to a wide engineering and business audience. He wanted a dramatic new technology to be the centerpiece of each exhibition. At Munich and Frankfort he chose power transmission; when consulted about the planning of the Chicago Exhibition of 1893, he recommended as the highlight of the exhibition the electrification of a long-distance railroad running out of Chicago.

In 1890 von Miller, who headed his own consulting engineering firm in Munich and had been named technical director of the proposed Frankfort exhibition, called a planning session to discuss the long-distance transmission. Present at the meeting were representatives of Maschinenfabrik Oerlikon, whose director, P. E. Huber-Werdmüller, was an enthusiastic supporter of the project; AEG, whose director, Emil Rathenau, was a graduate of the Zurich *Eidgenössischen Polytechnikum* (technical higher school), as was Huber-Werdmüller; and a cement works at Lauffen, on the Neckar River, where von Miller had recently completed a hydroelectric installation. The two manufacturers agreed to develop and supply polyphase equipment,

---

[51] Von Siemens, *House of Siemens*, 1: 120.

[52] On the Munich electrical exhibition of 1882 and Deprez and von Miller's power transmission display, see *Offizieller Bericht über die im Königlichen Glaspalaste zu München 1882 . . . Internationale Elektrizitäts-Ausstellung*, ed. W. von Beetz, O. von Miller, and E. Pfeiffer (Munich: Autotypie Verlag, 1882).

Figure V.12. Artificial Munich waterfall driven by power from Miesbach, 57 kilometers away. From Offizieller Bericht . . . München 1882, p. 105.

Figure V.13. Direct-current motor driving the pump for waterfall at the Munich exhibition of 1882 (Miesbach-to-Munich power transmission system designed by M. Deprez). From Offizieller Bericht . . . München 1882, p. 104.

and the cement works agreed to allow its surplus power to be transmitted over the 175-kilometer line to Frankfort.[53]

In January 1891 Charles E. L. Brown, who was technical director of the electrical machinery section of Maschinenfabrik Oerlikon, provided the technical commission of the Frankfort exhibition with evidence that a high-voltage power transmission would prove successful. At the Oerlikon plant he set up a laboratory test unit to prove that the use of oil-insulated transformers stepped up voltage to 40,000 volts, that his transmission line insulators could function with voltage of such extraordinary intensity, and that stepped-down transformers reduced the voltage to distribution level (see Fig. V.15). The possibility of insulating the transformers and transmission line had been in serious doubt. The experiment was unusual in that it incorporated both a step-up transformer with a low-voltage generator and oil-bath insulation, an idea that had been tried earlier with transmission cable. Despite the success of the laboratory test, however, there was still doubt about the efficiency of a long-distance transmission.[54]

The full-scale Lauffen-to-Frankfort transmission system for the exhibition incorporated an international array of machinery (see Figs. V.14–V.17). At Lauffen, the cement factory turned over to the planners a 300-h.p. water turbine; AEG supplied a switchboard and two step-up transformers; and Maschinenfabrik Oerlikon furnished the third transformer, insulated in an oil bath, and a three-phase generator. Financial support for the erection of the 175-kilometer transmission line and right of way for the line was provided by the imperial and local governments. At Frankfort, Maschinenfabrik Oerlikon supplied one transformer and AEG provided two. AEG also built a large advertising sign with 1,000 incandescent lamps and a waterfall powered by a pump with a new AEG polyphase motor. Power from Lauffen lit the lamps and drove the waterfall.[55] (See Figs. V.18, V.24, and V.25.)

The transmission system was not operative until 24 August 1891, shortly before the exhibition ended. On that date the brilliant lighting display was illuminated and the waterfall began to flow. Leading electricians viewed the installation and made critical analyses of its performance. The transmission efficiency surprised critics, for they had expected 50 percent losses. Transmitting 190 h.p. at about 25,000 volts, the efficiency measured from

[53] *Beschreibung und Darstellung Elektrischer Werke welche nach den Projekten und Unter Leitung des Technischen Bureaus Oskar von Miller* (Munich, 1898–99). This publication by von Miller's consulting engineering firm was kindly loaned to me by Oskar von Miller's son and successor as head of the firm, Rudolf von Miller. I have drawn on the chapter on the Frankfort exhibition of 1891. Also informative on the exhibition is Walther von Miller, *Oskar von Miller Nach Eigenen Aufzeichnungen, Reden und Briefen* (Munich: Bruckmann, 1932), pp. 55 ff.

[54] B. A. Behrend's "The Debt of Electrical Engineering to C. E. L. Brown" (reprint from *Electrical World and Engineer*, 16 November 1901–1 March 1902) was kindly supplied by F. Thomas of Brown, Boveri & Co., Ltd. (BBC), Baden, Switzerland.

[55] "Die Lauffener Kraftübertragung," *Offizielle Zeitung . . . Frankfurt am Main 1891*, p. 825; Karl E. Müller, "50 Jahre Drehstrom-Kraftübertragung," *Oerlikon Bulletin*, no. 231 (May-June 1941), pp. 1437–43, and no. 232 (July-August 1941), pp. 1445–52; William Boveri, "75 Jahre Wechselstrom-Kraftübertragung: Lauffen/Neckar-Frankfurt/Main," *BBC-Nachrichten,* September-October 1966, pp. 523–25; Hans Happoldt and Karl Merz, "Die Entwicklung der Generatoren seit der Energieübertragung Lauffen-Frankfurt," ibid., pp. 526–31; Karl Schlosser, "Marksteine des Transformatorenbaues," ibid., pp. 534–39; and Heinz Mors, Karlheinz Herzig, and Joachim Ufermann, "Hochspannungs-Freileitungen," ibid., pp. 563–64.

Figure V.15.   *Model designed by Charles E. L. Brown for testing alternating-current transmission system (January 1891): Generator (center); two oil-insulated transformers (left); and motor (right), which drove the generator. Courtesy of Brown, Boveri & Co., Baden, Switzerland.*

Figure V.14.   *Oil-filled, three-phase transformer used in the Lauffen-to-Frankfort transmission system. From* Offizielle Zeitung . . . Frankfurt am Main 1891, *p. 1045.*

Figure V.16.   *Three-phase motor designed by Dolivo-Dobrowolsky (the type used in the Lauffen-to-Frankfort transmission system). From* Offizielle Zeitung . . . Frankfurt am Main 1891, *p. 971.*

Figure V.17.   *Three-phase generator (designed by Charles E. L. Brown of the Maschinenfabrik Oerlikon) driven by water turbines at Lauffen. From* Offizielle Zeitung . . . Frankfurt am Main 1891, *p. 599.*

the Lauffen turbine to the low-voltage side of the transformers in Frankfort was 74.5 percent. At 78 h.p. the efficiency was 68.5 percent. Not only did this display promise practical application for long-distance or point-to-point transmission (from a mountain waterfall, for example, to an industrial city on the lowland plain), but the transmission system, working with three-phase current, also contributed greatly to the establishment of this system as standard instead of the two-phase system that was being tried by West-inghouse in the United States and by other manufacturers abroad. Oil emerged as a practical insulator. Only the working frequency of 40 cycles was not adopted. Engineers and designers throughout the world looked upon the Frankfort exhibition as justification for further transmission ventures, including a Niagara Falls project.

Beginning in 1889, decision making over a four-year period in connection with a Niagara project reflected the complexity, the confrontations, and the ultimate resolution of "the battle of the systems." The central

Figure V.18. *Power transmission, Lauffen to Frankfort: Waterfall and incandescent lamps at Frankfort powered by natural waterfall 175 kilometers away. From* Offizielle Zeitung . . . Frankfurt am Main 1891, *p. 828.*

Figure V.19. *Logo of the official periodical of the Frankfort exhibition of 1891. From* Offizielle Zeitung . . . Frankfurt am Main 1891, *p. 1.*

Figure V.21.   *Electrical engineers and managers at the exhibition. Courtesy of Rudolf von Miller.*

Figure V.20.   *Celebrating the opening of the exhibition: "Out of darkness into light." From* Offizielle Zeitung . . . Frankfurt am Main 1891, *p. 101.*

Figure V.22.   *Advertisement for electrical manufacturing company owned by Albert Einstein's uncle. From* Offizielle Zeitung . . . Frankfurt am Main 1891, *p. 166.*

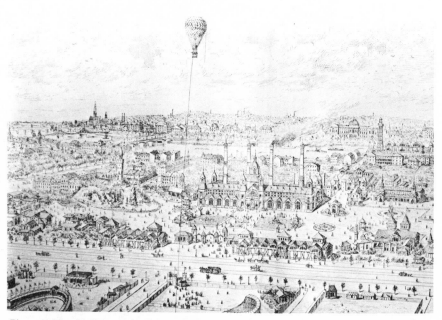

Figure V.23.   *Bird's-eye view of the exhibition. Note hill with waterfall (left center) powered by electricity transmitted from Lauffen. From* Offizielle Zeitung . . . Frankfurt am Main 1891, p. 636.

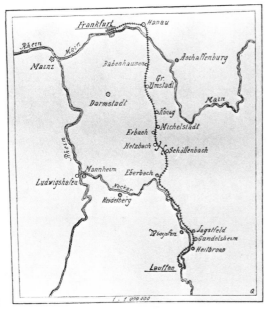

Figure V.24. *Scenes along the route of the Lauffen-to-Frankfort transmission line, 1891.* From Offizielle Zeitung . . . Frankfurt am Main 1891, *p. 30.*

Figure V.25. *Route of the Lauffen-to-Frankfurt transmission line.* From Offizielle Zeitung . . . Frankfurt am Main 1891, p. 32.

characters remained the same in several instances, for Thomas Edison and George Westinghouse were involved, the first as a consultant and the second as a manufacturer, and Charles Brown of Switzerland was consulted for his expertise. The plan to harness the power of Niagara Falls had a long history, but it took well-defined organizational form and acquired momentum when financiers, including J. P. Morgan, formed the Cataract Construction Company in 1889 to develop a coherent scheme for large-scale utilization instead of numerous small water-wheel mills. At the start, the company considered using water turbines to discharge the water into a central tunnel that would empty into the Niagara River below the falls, thereby providing a drop, or head, below the level of the surface canals that diverted water to the turbines from the Niagara River above the falls. The company was also interested in transmitting power to Buffalo, a manufacturing city twenty miles distant, and considered mechanical as well as

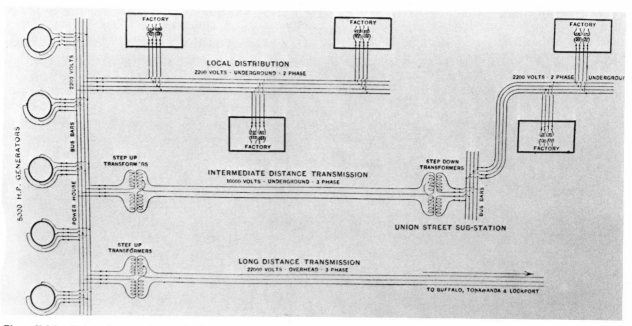

Figure V.26. *Universal system of power distribution and transmission, Niagara Falls Power Co. From AIEE,* The Niagara Falls Electrical Handbook *(1904), p. 78.*

Figure V.27. *The Niagara (universal) system: Diagram of circuits. From Adams,* Niagara Power, 2: 250.

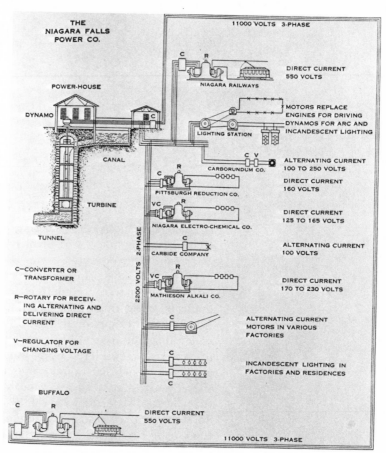

pneumatic linkages to accomplish this. Within a year, the company was leaning toward the development of one or two electric power stations that would use the discharge tunnel, but that, in contrast to the numerous intake canals and turbines envisaged earlier, would need only one surface canal. The success of the Frankfort transmission in 1891 helped persuade the engineers to support electric transmission to Buffalo. Plans from the Westinghouse Company for a two-phase system, together with the advocacy of a polyphase system by Professor George Forbes of England, chief electrical consultant, carried the day for polyphase generation and transmission.

The Westinghouse plan was a refinement of the Chicago idea (see Figs. V.26 and V.27). It involved long-distance, high-voltage transmission to Buffalo. The Westinghouse engineers who engaged in the invention and development of the universal system at Niagara included Charles F. Scott, who by inventing a coupler in the form of a transformer connection made possible the two-phase to three-phase conversion; Oliver B. Shallenberger, who developed an improved induction meter; Benjamin Lamme, who improved the magnetic circuitry of the system's generators and motors; Albert Schmid, who contributed mechanical improvements; and a group of engineers headed by Lewis B. Stillwell, chief electrical engineer at Westinghouse.[56] Indicative of the change that was taking place in the mode of invention and development is the need to identify groups rather than individuals as the sources of invention.

The Niagara power station went into service in August 1895. Westinghouse built the first two generators (two-phase), the switchgear, and auxiliary power-house equipment; General Electric constructed the transformers, the transmission line to Buffalo (three-phase), and the substation there. The transformer connection invented by Charles Scott permitted the coupling of the two- and three-phase currents. Transmission began in 1896, but consumption at the rapidly expanding industrial complex using the universal system at Niagara exceeded the power transmission to Buffalo.[57]

[56] Robert Belfield, "The Niagara System: The Evolution of an Electric Power Complex at Niagara Falls, 1883–1896," *IEEE Proceedings* 64 (1976): 1349. Belfield's Ph.D. dissertation, "The Niagara Frontier: The Evolution of Electric Power Systems in New York and Ontario, 1880–1935" (University of Pennsylvania, 1980), is an in-depth study of the Niagara system.

[57] See Belfield, "The Niagara System" and "The Niagara Frontier." See also Harold Sharlin, "The First Niagara Falls Power Project," *Business History Review* 35 (1961): 59–74; and Passer, *Electrical Manufacturers,* pp. 282–95.

CHAPTER VI

Technological Momentum

FRANKFORT in 1891, Chicago in 1893, and Niagara in 1895 were substance and symbol. Demonstrating the technical practicality of the polyphase universal electric supply system, they stimulated enthusiasm for the new technology. Moreover, it was not by chance that entrepreneurs chose to display the new technology dramatically at such world exhibitions. Nor is it surprising that the installation at the awe-inspiring site of Niagara Falls attracted unusual attention.

During the 1890s the polyphase system gathered momentum. Because it encompassed the old contenders in "the battle of the systems" (direct current versus single-phase alternating current), the new system gained widespread support from men and institutions in the rapidly growing field of electrical engineering, science, and industry. A supportive context, or culture, formed rapidly. Men and institutions developed characteristics that suited them to the characteristics of the technology. And the systematic interaction of men, ideas, and institutions, both technical and nontechnical, led to the development of a supersystem—a sociotechnical one—with mass movement and direction. An apt metaphor for this movement is "momentum." So reinforced by a cultural context, and interacting in a systematic way with the elements of that context, the universal system, like high-momentum matter, tended in time to resist changes in the direction of development. Development proceeded along lines that could be extrapolated. The universal system gathered a conservative momentum. Its growth generally was steady, and change became a diversification of function.

The momentum initially came mostly from an aggregate of manufacturers who invested heavily in resources, labor, and manufacturing plants in order to produce the machinery, devices, and apparatus required by the new system; later, educational institutions taught the science and practice of the new technology; then research institutions were founded to solve its critical problems; and all the while a growing number of engineers, skilled laborers, appliers of science, managers, and other persons invested their experience and competence in the new polyphase universal system.

The polyphase universal system gained momentum in part because of the commitment made by a new generation of inventors and engineers. The first two decades of electric light and power supply witnessed several

140

Figure VI.1. *Elihu Thomson, c. 1885.*
Courtesy of the Elihu Thomson Papers,
American Philosophical Society,
Philadelphia, Pa.

waves of pioneers: first the direct-current men led by Edison; then the single-phase alternating-current group, including Lucien Gaulard and S. Z. de Ferranti; and finally the developers of polyphase current, among them Nikola Tesla and Michael Dolivo-Dobrowolsky. Edison did not make the transition from direct current to other forms. In fact, his stubborn resistance intensified "the battle of the systems." Moreover, Edison decided not to exercise his option on the a.c. system of Ganz & Company, a system his company could have used to compete effectively against Westinghouse. After 1890 he entered into inventive enterprises that were only remotely related to electric light and power. Gaulard was unable to modify his stance successfully, and he saw others greatly improve upon his system before his untimely death. Tesla, after his breakthrough into polyphase power, embarked on discoveries in even more esoteric fields, such as wireless power transmission and communication. A notable exception to these pioneers' inflexibility was the dogged determination of Elihu Thomson to maintain a position of leadership throughout the rapid fluctuations that hit electric light and power. Thomson (1853–1937), who was born in England but grew up in Philadelphia, Pennsylvania, spanned the electric power spectrum from his early concentration on arc lights, through direct-current incandescent lighting, into alternating and polyphase systems. Thomson not only made the necessary transitions but he moved from the status of independent inventor to that of scientific consultant. After his company, Thomson-Houston, merged with Edison General Electric to form General Electric in 1892, he assumed the position of senior industrial scientist and adviser in research and development for the new firm.

Thomson was one of a small group of mostly young men of various nationalities who helped bring about the shift to polyphase, the event "which ushered in a new economic epoch."[1] These young inventors and electrical engineers differed from most of their predecessors in that their scientific or engineering training included mathematics, an extremely important competence for those wishing to analyze polyphase systems. Philadelphia's Central High School, where Thomson taught, offered advanced technical and scientific courses that were not altogether different from the courses given in the *Technischen Hochschulen* ("technical higher schools") in Central Europe. Tesla trained at the *Technischen Hochschule* in Graz and at the University of Zurich. Michael Dolivo-Dobrowolsky attended a *Technische Hochschule* in Germany (Darmstadt), as was common for Russian-born aspirants to careers in science and engineering.

Following close behind Tesla and Dolivo-Dobrowolsky was a group of electrical inventors and engineers who not only attended institutions of higher learning but also had the opportunity to take courses in electrical engineering. So trained, these men took part in the rapid professionalization of the field of electrical engineering, and expanding electrical engineering education and professionalization greatly reinforced the com-

[1] Georg von Siemens, *History of the House of Siemens,* trans. A. F. Rodger, 2 vols. (Freiburg/Munich: Alber, 1957), 1: 114. Biographies of the leading electrical engineers of the early generations are identified in Thomas J. Higgins, "A Biographical Bibliography of Electrical Engineers and Electrophysicists," *Technology and Culture* 2 (1961): 28–32, 146–65. See also idem, "A Classified Bibliography of Publications on the History and Development of Electrical Engineering and Electrophysics," *Bulletin of Bibliography* 20, nos. 3–7 (1950–52).

mitment to, and momentum of, the system of power and light that had been ushered in by the pioneers during the 1890s. Courses in the *Technischen Hochschulen* and in colleges of engineering were organized on the basis of the knowledge and experiences of the pioneers, and technical journals and meetings spread information and attitudes, thereby consolidating the predominant trends in practice.

The training and education of the electrical technicians, engineers, and scientists who filled the increasing number of positions offered by electrical utilities and manufacturers occurred on many levels and in many ways during the first two decades of electric light and power. In Germany, England, and the United States, the sources of this training included apprenticeship programs, technical and trade schools, and secondary schools with technical programs. Germany's rapid industrialization in the late nineteenth century, including the impressive rise of its electrical industry, has been attributed in part to the German system of secondary, trade, and technical education outside of institutions of higher learning.[2] Especially in Germany, electrical manufacturers established sizable apprenticeship programs to train skilled workmen, some of whom became foremen and plant superintendents. After the manufacturers began designing and producing highly complex polyphase generators, motors, and transformers, they also offered practical training for graduates of engineering schools, many of whom were thought by manufacturers to be overexposed to irrelevant theory. Cooperative programs requiring participation in the classroom and in the workplace spread throughout the United States after 1900.

Such varied modes of training notwithstanding, attention here will be limited to several leading degree- or diploma-granting three- or four-year engineering colleges in the United States and England and to the leading *Technischen Hochschulen* in Germany. Those selected were among the first to offer courses in electrical engineering, to organize sets of courses constituting programs in electrical engineering, or to establish departments of electrical engineering. Special attention will be paid to the way engineering institutions formed a bridge between physics and the practice of engineering. Concentration, therefore, will be on the early engineering professors as contributors to the science of electrical engineering. The term "applied science" will be avoided because it suggests a one-way relationship from fundamental science to practical application; instead, "science of electrical engineering" will be used as a broader reference to organized, quantified, and generalized information, or experience derived from practice.

During the transition from direct to polyphase current, the number of institutions of higher learning granting engineering degrees or diplomas continued to grow. The number of courses, programs, and departments in electrical engineering increased even more impressively. During the formative years of the electrical industry, the existence of a substantial number of engineering institutions influenced the growth of the industry, and the rise of the industry shaped electrical-engineering institutions. One result of this interaction was the scientific character of electrical engineering. In contrast, civil and mechanical engineering developed before engi-

---

[2] William H. Dawson, *The Evolution of Modern Germany* (New York: Scribner's, 1914). See especially his discussion of the technical schools in Saxony (pp. 99, 135).

neering education was considered a *sine qua non* for the professional, and their cast was less scientific.

When the Pearl Street station opened in New York in 1882, about seventy U.S. institutions offered a professional education in engineering.[3] Many of these had been founded with funds provided by land grants made under the Morrill Act of 1862, according to which each state that endowed agricultural and mechanical colleges would be given 30,000 acres of public land for each congressional district. None had courses in electrical engineering, but a number included lectures on electricity in their physics courses. By 1899, the number of these institutions was eighty-nine, and 98 percent of them offered a four-year course in engineering. Forty-nine had courses in electrical engineering; sixty-seven taught civil engineering; and sixty-one offered mechanical engineering. It should be noted, however, that students could also prepare for work in the electrical industry by studying mechanical engineering. In the United States in 1899 there were 2,397 students enrolled in electrical engineering programs and there were 370 graduates of these programs. For mechanical engineering the figures were 3,293 students and 480 graduates; for civil engineering, 2,667 students and 419 graduates.[4]

In Germany after 1860 several technical schools were given the status of institutions of higher learning. They were characterized as universities that would prepare students for industrial careers. These institutions offered students a course of study that was more advanced than the curricula of U.S. engineering schools. The institutions that received the *Technische Hochschule (TH)* constitution were Karlsruhe (in 1865), Munich (1868), Aachen (1870), Dresden (1871), Brunswick (1877), Berlin-Charlottenburg (1879), and Hannover (1880).[5] Their organization and curricula were substantially influenced by Franz Grashof (1826–1893), a leading engineering professor who called for a grounding of the engineer in the fundamentals of science.[6] Franz Reuleaux (1829–1905), another of Germany's pioneering engineering educators, also stimulated the development of higher technical education with an emphasis on the sciences by pointing out the superiority of America's industrial displays at the Philadelphia exhibition of 1876.[7] The Germans looked upon education as a means of borrowing technology from other nations and advancing beyond it; they also recognized that electrical engineering was particularly well-suited to a scientific emphasis.

In England, concern among manufacturers and engineers about the state

---

[3] This number was derived from statistics giving the number as eighty-nine in 1899 and identifying it as an increase of 21 percent over that for 1878. I am indebted to Robert Belfield for calling my attention to statistics on engineering education. His sources were I. O. Baker, "Engineering Education in the United States at the End of the Century," *Science*, 2 November 1900, pp. 666–74; and Charles R. Mann, *A Study of Engineering Education* (New York: Joint Committee on Engineering Education of the National Engineering Societies, 1918), esp. pp. 18–24.

[4] Baker, "Engineering Education," p. 668.

[5] Wilhelm Treue, "Die Geschichte des Technischen Unterrichts," in *Festschrift zur 125-Jahrfeier der Technischen Hochschule Hannover, 1831–1956* (Hannover, 1956), p. 60.

[6] F. Grashof, "Über die Organisation an polytechnischen Schulen zugrunde legende Prinzipien," *Zeitschrift Verein Deutscher Ingenieure*, 1864, p. 591.

[7] Friedrich Klemm, *A History of Western Technology*, trans. D. W. Singer (Cambridge, Mass.: M.I.T. Press, 1959), pp. 326, 339.

of technical education was widespread during the second half of the nineteenth century, especially after 1880. The convening of several parliamentary committees to consider the state of technical education in Britain revealed this concern.[8] The impressive industrial competition among foreign exhibitors at London's Great Exhibition in 1851 was a major stimulus of activity aimed at improving Britain's situation.

General concern about the erosion of Britain's industrial leadership in the late nineteenth century led to a number of efforts to improve the country's system of technical and engineering education. In 1878 the historic livery companies, or guilds, established the City and Guilds of London Institute for the Advancement of Technical Education. The institute used subsidies and its considerable influence to promote technical education throughout the country. In London the institute-supported Finsbury Technical College opened in 1883, and in 1884 the institute contributed to the opening of the Central Institution in South Kensington.[9] Finsbury College would train those desiring an intermediate post, including that of foreman, in industry and those preparing for the advanced courses offered at engineering colleges such as University College and King's College of the University of London. In 1907, after several reorganizations, the Central Institution achieved university status as part of the Imperial College of Science and Technology of the University of London.[10] Other centers of higher technical study in Britain in the 1880s included Manchester, Birmingham, Liverpool, Leeds, Bradford, and Glasgow, as well as the engineering facilities and courses at Cambridge University.

These U.S., British, and German institutions of higher technical education responded to the rise of the electric supply industry in various ways, and here the trends among them can be detected by considering the responses of several: the Massachusetts Institute of Technology and Cornell University in the United States, the Central Institution in London, and the *Technischen Hochschulen* at Darmstadt and Berlin in Germany. In each case several questions will be raised: When was electrical engineering introduced into the institution's curriculum? When was a separate department of electrical engineering established? What constituted the educational and practical background of the electrical engineering faculty, and in what research and consulting activities were faculty members engaged? How did the curricula reflect the transition to, and thereby the momentum of, the polyphase, universal system and long-distance transmission?

In Germany, Werner von Siemens, an influential figure in the engineering and science community, stimulated the establishment of professorships of electrical engineering. He realized that the need for young electricians grounded in science would exceed the number forthcoming

[8] See, for example, Great Britain, Parliament, House of Commons, *Sessional Papers: Second Report of the Royal Commissioners on Technical Instruction*, 3 vols. (London: HMSO, 1884), 1: xxix–xxxi.

[9] *Centenary of the Imperial College of Science and Technology: A Short History of the Imperial College* (London, 1945).

[10] William Wickenden, *A Comparative Study of Engineering Education in the United States and Europe,* Society for the Promotion of Engineering Education, Bulletin no. 16 (Lancaster, Pa.: The Lancaster Press, 1929), p. 38; *Centenary of the Imperial College of Science and Technology*, p. 16.

Figure VI.2. Charles Cross. Courtesy of the M.I.T. Museum and Historical Collections, Cambridge, Mass.

from related fields or from informal education. Speaking in December 1881 before Germany's Elektrotechnische Verein (Society of Electrical Engineering), von Siemens called for the establishment of a chair of electrical engineering in all *TH*s in order to familiarize young men with the theory and practice of electrical engineering.[11] So moved, the *Technische Hochschule* in Darmstadt appointed Erasmus Kittler (1852–1929) the first professor of electrical engineering in Germany, and he, in turn, established the first department of electrical engineering. He came to the Darmstadt *TH* in 1882 after serving as an instructor in physics at the Munich *TH*. There were only 137 students enrolled at the languishing Darmstadt *TH* in 1882, but within several years Kittler's electrical engineering program was internationally renowned and had turned out a large number of Germany's leading electrical engineers, among them Dolivo-Dobrowolsky.[12]

In 1882, at the Berlin-Charlottenburg *TH*, Adolf Slaby (1849–1913) inaugurated a two-hour course in electrical machinery. In 1883, also at the Berlin *TH*, Werner von Siemens established the first chair of machine construction and electrical engineering (a position held by Slaby). When the Berlin *TH* moved in 1884 into impressive new buildings on ground provided by Kaiser Wilhelm I, the faculty established an electrical laboratory, after which the study of electrical engineering expanded rapidly; by 1892, eighty students were engaged in laboratory work there. Electrical engineering was not, however, a separate department at Berlin as it was at Darmstadt; it was part of the mechanical engineering curriculum. Unlike other early German electrical engineering professors, Slaby had a background in mechanical engineering rather than physics.[13] From 1895 until 1905 Gisbert Kapp (1852–1922), a pioneer in the scientific analysis of electrical machinery, taught electrical engineering at Berlin. The presence of the two major manufacturers of German electrical machinery (Siemens and AEG) in Berlin probably stimulated the emphasis on machinery, or mechanical engineering, courses.[14]

The Massachusetts Institute of Technology established the first four-year course in electrical engineering in America in 1882. Its action was based on the belief that the rapidly increasing development "of the various branches of electrical engineering, and the consequent demand for persons conversant with the theory and applications of electricity,"[15] called for an academic response. Initially the course reflected the interest of its planners in the better-established field of electrical communications, but emphasis eventually shifted to light and power. Administratively, the course was part of the physics curriculum, and Professor Charles Cross (1848–1921), head of the physics department, planned and directed it. A separate department

[11] Georg Dettmar, *Die Entwicklung der Starkstromtechnik in Deutschland* (Berlin: ETZ Verlag, 1940), pp. 279–80.

[12] Von Siemens, *House of Siemens*, 1: 116.

[13] *Chronik der Königlichen Technischen Hochschule zu Berlin, 1799–1899* (Berlin, 1899), p. 187.

[14] Rolf Sonnemann, "Wissenschaftlich-technischer Fortschritt, Ingenieurausbildung und Monopolinteresse im Spiegel der Elektrotechnischen Zeitschrift, 1880–1900," *NTM-Schriftenr. Gesch., Technik, Med.* (Leipzig) 11 (1974): 1, 67. Other early developments in electrical engineering education in Germany included a professorship at the Munich *TH* held by W. von Beetz (1822–1886) and lectures given by W. Dietrich at Stuttgart beginning in 1882. Wilhelm Kohlrausch (1855–1936) established a program at the Hannover *TH* in 1884.

[15] *Electrical Review* 12 (1883): iii.

Figure VI.3. *William E. and Hertha Ayrton, husband and wife, leaders in the development of electrical engineering and science. Courtesy of the Institution of Electrical Engineers, London.*

Figure VI.4. *Silvanus P. Thompson. Courtesy of the Institution of Electrical Engineers, London.*

of electrical engineering was established in 1902, with Louis Duncan (1862–1916) as chairman. In 1907 Dugald C. Jackson (1865–1951), head of the electrical engineering department at the University of Wisconsin, moved to M.I.T. to replace Duncan. Six students had enrolled in M.I.T.'s four-year course at the outset, and the catalog of the institute suggested a modest outlay for special laboratory equipment. In the 1890s, however, enrollment in electrical engineering rose to match that in civil and mechanical engineering, and two years after Jackson's arrival, the department had 200 full-time students.[16]

In 1883 Cornell University announced a program in electrical engineering headed by William A. Anthony (1835–1908), a professor of physics. By 1885 Cornell was offering a four-year undergraduate program and a one-year graduate program in electrical engineering. In 1887 Professor E. P. Roberts received an appointment to teach electrical engineering and became Cornell's first full-time faculty member in electrical engineering. Professor Harris J. Ryan (1866–1934) succeeded Roberts in 1889. By 1905 the department was offering eleven courses and was staffed by one professor, one assistant professor, and one instructor.[17]

The Central Institution of the City and Guilds of London Institute opened in 1884 offering courses in electrical engineering taught by William E. Ayrton (1847–1908), a leading electrical engineering scientist and head of the physics department. Ayrton had toured Germany investigating its electrical-engineering facilities before establishing the curriculum at the Central Institution.[18] The courses offered within the physics department extended over three years. In 1899, by way of identifying electrical engineering as its major interest, the department changed its name to electrical engineering. The new designation reflected the philosophy expressed by the Prince of Wales at the school's opening-day ceremonies in 1884: "The altered conditions of apprenticeship, and the almost general substitution of machine for hand labour have made the teaching of science, in its application to productive industry, . . . necessary."[19]

Electrical engineering was also nurtured in London by Sylvanus P. Thompson (1851–1916), author of widely influential electrical engineering texts and principal of Finsbury Technical College for many years. John Hopkinson (1849–1898), a developer of engineering theory and a consultant to the Edison interests in England, was the first incumbent of the chair of electrical engineering at King's College (1890) and twice served as president of the Institution of Electrical Engineers. He had studied engineering at Owens College in Manchester and had continued at Trinity College, Cambridge, in mathematics after winning a mathematics fellow-

[16] "Report of the Visiting Committee Assigned to the Electrical Engineering Department 27 February 1922," in "GE Trans. 1915–1923, MIT," Elihu Thomson Papers, Library of the American Philosophical Society, Philadelphia, Pa. I am indebted to Bernard Carlson of the University of Pennsylvania for calling my attention to this and other items at the APS Archives.

[17] Waterman T. Hewett, *Cornell University: A History*, 4 vols. (New York, 1905), 2: 316–17; *Electrical World* 6 (1885): 154. Other early departments of electrical engineering in the United States were located at the University of Missouri (1886) and the University of Wisconsin (1891). Frederick E. Terman, "A Brief History of Electrical Engineering Education," *IEEE Proceedings* 64 (1976): 1401.

[18] *Centenary of The Imperial College of Science and Technology*, p. 16.

[19] Ibid., p. 37.

Figure VI.5. *John Hopkinson and students at King's College, London. From* Greig, John Hopkinson, *p. 35.*

ship. Focusing on electrical science and generator design, his publications in the early 1880s established him as a leading electrical engineering scientist. He became a Fellow of the Royal Society in 1877.

The character of the professors who staffed the new departments and taught the electrical engineering courses in England, Germany, and the United States influenced the field both in theory and in practice. A common characteristic of the founders of electrical engineering courses, judging from the above sample, was their study of physics, which helps explain the scientific bent of the courses. The prior inclusion of electrical science in physics courses suggests why physics professors associated with engineering programs established electrical engineering courses. Charles R. Cross of M.I.T., for instance, graduated from M.I.T. in 1870 and remained there to become head of the physics department in 1876. He not only created the four-year course in electrical engineering but also did private consulting work and published papers on acoustics and telephony. Cross also served as vice-president of the American Institute of Electrical Engineering (AIEE).

William A. Anthony, another of the first generation of professors to cultivate electrical engineering programs, held the chair in physics at Cornell University from 1872 to 1887 before organizing a four-year course in electrical engineering there. Earlier he taught mechanics and physics at Iowa State University and in secondary schools. He held a doctorate from the Yale Scientific School (1856), where he was first in his class, and shortly after graduation he worked as a mechanical engineer in a cotton-machinery plant. He left Cornell to do consulting work and night teaching at Cooper Union. Three years later he became president of the AIEE. Among his publications were several textbooks.

Louis Duncan, who headed M.I.T.'s electrical engineering program after Cross, also had a background in physics. An 1883 graduate of the U.S. Naval Academy, Duncan was sent by the navy after several years' service

Figure VI.6. *William A. Anthony.*
Courtesy of the Cornell University Archives,
Ithaca, N.Y.

Figure VI.7. *Louis Duncan. Courtesy of*
the M.I.T. Museum and Historical
Collections, Cambridge, Mass.

to Johns Hopkins University to do graduate work in physics and electricity under physics professor Henry A. Rowland. At Hopkins, Duncan helped establish the unit of electrical resistance for the U.S. government. After receiving his Ph.D. in 1887, he resigned his commission to teach in a Hopkins program in applied electricity that had been established in 1886. Duncan had major contracts as a consultant, including work on the Washington, D.C., underground trolley system and, in 1886, the first electrification of a portion of a main-line railroad, the Baltimore tunnel of the Baltimore & Ohio Railroad. He also played a leading role in the electrification of the Third Avenue elevated railroad in New York City (1897–1901). His appointment as chairman of the new electrical engineering department at M.I.T. from 1902 to 1907 was characterized as unfortunate because of his preoccupation with consulting work.[20] Among his consulting associates was Frank J. Sprague. Duncan was elected president of the AIEE in 1895. He wrote numerous research papers.[21]

Soon a second generation took over the American programs. Dugald C. Jackson and Harris J. Ryan of the second wave had remarkably similar careers. Both were born in Pennsylvania and both studied electrical science and technology at Cornell under William A. Anthony. Jackson took the postgraduate course in 1885–87 and Ryan worked as a research assistant to Anthony. Together, in 1887, the new graduates established an engineering firm in Lincoln, Nebraska, but after a year Ryan returned to Cornell to teach and Jackson became an engineer for the Sprague Electric Railway & Motor Company and then a district engineer in Chicago for the Edison manufacturing interests. Jackson was called to the University of Wisconsin in 1891 and established the electrical engineering department there; Ryan was head of electrical engineering at Cornell from 1895 to 1905. Both wrote basic texts: Jackson, *Alternating Currents and Alternating Machinery* (1896); Ryan, *Electrical Machinery* (1903).

In 1905 Ryan became chairman of the electrical engineering department at Stanford University; in 1907 Jackson assumed that position at M.I.T. Ryan did applied research as head of the high-voltage laboratory at Stanford, which was later named for him. While he was a professor at M.I.T., Jackson headed Jackson & Moreland, a Boston consulting engineering firm. Ryan also was active as a consultant, serving, among other clients, the Los Angeles power bureau. Jackson was president of the AIEE in 1907; Ryan, in 1923. Dugald Jackson also served as president of the Society for the Promotion of Engineering Education in 1905. During World War I, Jackson supervised the provision of power to American forces in France and Ryan headed a laboratory project on submarine detection.

In Germany, Erasmus Kittler (of the first wave) had the background in physics. He had a doctorate in physics and was a lecturer in physics at the Munich *TH* before moving to Darmstadt. At Darmstadt, however, he arranged for electrical technology and science to be offered in an independent department rather than under the jurisdiction of physics. His confidence

[20] Samuel Prescott, *When M.I.T. Was "Boston Tech," 1861–1916* (Cambridge, Mass.: M.I.T. Press, 1954).

[21] I am indebted to Robert Rosenberg of The Johns Hopkins University for information on electrical engineering education at Hopkins.

Figure VI.8. *D. C. Jackson. Courtesy of the M.I.T. Museum and Historical Collections, Cambridge, Mass.*

Figure VI.9. *Professor and Mrs. Harris J. Ryan. Courtesy of the Cornell University Archives, Ithaca, N.Y.*

Figure VI.10. *Erasmus Kittler. From Offizielle Zeitung . . . Frankfurt am Main 1891, p. 103.*

in the future of the subject may have been stimulated by the Munich electrical exhibition of 1882, which he served as a section judge. The exhibition generated the interest of the grand duke of Hessen, head of the province, who gave his support to Kittler and the subject.[22] Kittler's program of study served as a model for others in Germany. It involved eight semesters of natural science, mathematics, theoretical and applied mechanics, economics, and the science and practice of electrical engineering. He introduced a final, comprehensive examination (*Abgangsexamen,* later the *Diplomprüfung*) as well as a qualification requirement (*Prüfungsordnung*) that was to follow at least a year of practical experience.

Kittler stressed the need for practical experience to complement theory. He was a consultant on numerous projects, including twelve large central stations and many isolated plants for industry, commerce, and public facilities. He assumed responsibility for entire projects, admirers reported, from the planning stages to successful operation. Many of his students attributed their success to his emphasis on a thorough grounding in the practice, as well as the science, of engineering.[23] Kittler was also the author of a widely used three-volume text, *Handbuch der Elektrotechnik* (2nd ed., 1892).

Adolf Slaby and Gisbert Kapp, who developed the program in electronical engineering at the Berlin-Charlottenberg *TH,* differed from Kittler and their U.S. peers in that the background of both was mechanical engineering rather than physics. This emphasis on machine theory and design

[22] Dettmar, *Starkstromtechnik,* pp. 279–80.
[23] R. Werner, "Zu Kittlers 70. Geburtstag," *Elektrotechnische Zeitschrift* 43 (1922): 865–66.

Figure VI.11. *Adolf Slaby. Courtesy of the Siemens Museum, Siemens AG, Munich.*

Figure VI.12. *Gisbert Kapp. Courtesy of the Siemens Museum, Siemens AG, Munich.*

at the Berlin school resulted in part from the influence of Werner von Siemens, who as a manufacturer was interested in machine design, especially that of generators and motors. When Slaby's interest turned to electrical communications, a field in which he published a pioneering text, *Die Funkentelegraphie* (1897), and in which he took out major patents, he brought Gisbert Kapp, another electrical engineer with a mechanical background, to the Berlin *TH* faculty. Kapp had had extensive experience in England in the design of electrical machinery and electrical central stations. A leading advocate of polyphase transmission and power systems, Kapp's books and articles on electrical machinery and transmission were authoritative guides to the science of electrical engineering. Kapp also served as first secretary of the Verband Deutscher Elektrotechniker (Association of German Electrical Engineers). In 1905 he gave up his faculty position in Berlin and returned to England, where he became professor of electrical engineering at the University of Birmingham and eventually president of Britain's Institution of Electrical Engineers.[24]

England's William E. Ayrton, who taught electrical engineering at the Central Institution in London, had a notably diverse background. He passed the first B.A. examination of the University of London in 1867 with honors in mathematics. He later studied electricity under Sir William Thomson (Lord Kelvin) at Glasgow. From 1873 to 1878 he held the professorial chair in physics and telegraphy at the Imperial Engineering College in Tokyo, which at that time was the largest technical university in the world. In 1879 Ayrton returned to London to become a professor at the recently founded City and Guilds Institute. From 1881 to 1884 he was professor of applied physics at the institute's Finsbury Technical College. Next he became head of the physics department and professor of electrical engineering at the City-and-Guilds-supported Central Institution.

Ayrton, like his peers in Germany and the United States, had extensive practical expertise. Before moving to Tokyo, he worked for the Indian telegraph service. In England he served as a consultant to the government on electrical equipment for the navy and advised private companies as well. His applied research included work on transformers, power transmission, and measuring devices. Ayrton published numerous papers and a widely used text, *Practical Electricity* (1887). In 1892, he was elected president of the Institution of Electrical Engineers. Ayrton's wife, Hertha, also was a scientist and engineer of reputation. John Perry, Ayrton's departmental colleague at Central Institution, collaborated with him in practical research as well as in teaching.

For the most part, then, it was physicists who organized the early electrical engineering courses and departments, and soon graduates of these departments were teaching in them. Judging from the careers of these second-generation scientist-engineers, they placed considerable emphasis on practical experience as well as the science of electrical engineering in their academic and consulting activities. Thus they faced in practice the challenge of the transition from direct current to the polyphase, universal system.

[24] Wedding, "Adolf Slaby," ibid. 34 (1913): 429–30; B. A. Behrend, "Gisbert Kapp Dies at His Home," *Electrical World* 80 (1922): 499; D. G. Tucker, *Gisbert Kapp, 1852–1922* (Birmingham: University of Birmingham, 1973).

This transition, as has been seen, fundamentally altered practical endeavors and introduced new problems for those who were developing a science of electrical engineering. The nature of the response can be seen in the changes made in electrical engineering curricula between the early 1880s and the turn of the century.[25]

In 1884, during the direct-current era, M.I.T.'s description of its four-year course in electrical engineering read:

Thorough instruction is given in the theory of electricity. Also an extended course of lectures is devoted to the consideration of the various technical applications of electricity to land and submarine telegraphy, the telephone, electric lighting, and the electrical transmission of power. Instruction is given by lectures and laboratory exercises upon the processes of photometry, especially as applied to the measurement of electric lights. Advanced instruction in electrical measurements, including work with dynamo-electric machinery, together with a course in the electrical testing of telegraph lines is provided. The subjects of construction, specifications, and contracts also receive attention.[26]

This summary reflects the physicist's view of technology as the application of science. It also indicates an enduring interest in telegraphy, a subject that had engaged teachers of physics for years and that would metamorphose to dominate the curriculum again a half-century later when wireless and electronic communications evolved. The electrical transmission of power refers not to long-distance transmission in contrast to central-station distribution, but to the general study of electricity as a substitute for mechanical power transmission.

Fifteen years later, the annual M.I.T. catalog still contained much of the wording of the earlier description, a fact that perhaps reflects the propensity of faculty members to assign low priority to catalog copy. Several general statements in the 1899 description, however, manifested the revolution that had taken place in the field. "Utilization of power" had been added, and the subject matter of the new and extended series of practical lectures included alternating-current generators, motors, and transformers. Among the practicing engineers listed as guest lecturers were Louis Bell (on the transmission of power) and John Blood (on the design of alternating-current machinery). The most substantial addition to the description was a two-page listing of the equipment used in the laboratory of electrical engineering, most of it for alternating and polyphase light and power systems.[27] Because the first year of the M.I.T. curriculum was the

---

[25] For further comparison of the curricula in U.S. engineering schools, see Francis C. Caldwell, "A Comparative Study of the Electrical Engineering Course Given at the Different Institutions," *Proceedings of the Society for the Promotion of Engineering Education* 7 (1899): 128. Professor Caldwell of Ohio State University drew conclusions from a close study of the courses of study in eighteen leading American technical universities.

For the university and college catalogs and bulletins on which the present analysis is based, I am indebted to Mrs. Jeanne Pingree, archivist of the Imperial College of Science and Technology, London; the archivist of the Hessische Landes und Hochschulbibliothek Darmstadt; Mr. J. E. Boell, director of archives, University of Wisconsin; Mrs. Faith M. Towle, assistant reference librarian at M.I.T.; the reference librarian at Cornell University; and Mr. J. Louis Kuethe of The Johns Hopkins University library.

[26] *M.I.T. Annual Catalogue, 1884–1885*, pp. 43–44.

[27] Ibid., *1899–1900*, pp. 89–90.

TABLE VI.1. M.I.T.'s ELECTRICAL ENGINEERING COURSE OF STUDY, 1884–1885

Fourth Year	
First Term	*Second Term*
Physical Laboratory: Electrical Testing, and Construction of Instruments	Physical Research
Testing of Telegraph Lines, Dynamo Machines, etc.	Technical Applications of Electricity
Technical Applications of Electricity to Telegraph, Telephone, Electric Lighting, etc.: Lectures	Advanced Physics, Memoirs, etc.
Mechanical Engineering	Differential Equations
Mech. Engineering Laboratory	Theory of Probabilities
Applied Mechanics, Thermodynamics, Hydraulics, etc.	Method of Least Squares
	Calculus of Variations
	Mechanical Engineering
	Mech. Engineering Laboratory

Note: The student is advised to take Advanced German.

Source: *M.I.T. Annual Catalogue, 1884–1885*, p. 28.

same for all students, and because in the next two years, both in 1884–85 and 1899–1900, students concentrated on physics, mathematics, and mechanical engineering, the fourth-year courses best show the state of electrical technology and science from the perspective of the faculty. A comparison of the two catalogs reveals an impressive increase in complexity and richness within a decade and a half (see Tables VI.1 and VI.2).

In 1885 Cornell University's Sibley College of Mechanical Engineering and the Mechanical Arts offered an electrical engineering program that was nearly identical to its mechanical engineering curriculum for the freshman, sophomore, and junior years; included in this program were "drawing, mathematics, mechanics, mechanism, machine design, the elementary study of physics, and preliminary practice in the use of electrical and other instruments."[28] The work in the senior year in electrical engineering differed from that in mechanical engineering only in the spring term, when training in the testing, construction, and maintenance of electric lighting and telegraph systems was offered. Sibley College also offered a one-year program of graduate study in electrical engineering comprised of courses in electrical machinery and plants.

By 1900 William Anthony was no longer at Cornell, and the electrical engineering department had become closely associated with mechanical engineering. A student's course of study was the same during his first three years regardless of his intention to concentrate in mechanical or electrical engineering. In his fourth year he could choose special work in electrical engineering, but his degree was in mechanical engineering; a statement was made in the diploma to the effect that he had elected to do work in the electrical engineering department.[29]

The Department of Light and Power, in which the departments of mechanical engineering and electrical engineering were associated, asserted the "business" character of Cornell's program. This emphasis harmonized with Cornell's status as a land-grant institution dedicated to industrial and agricultural education. The Department of Light and Power insisted that

[28] Cornell University, *Register, 1885–86*, p. 107.
[29] Ibid., *1900–1901*, p. 332.

TABLE VI.2. M.I.T.'s ELECTRICAL ENGINEERING COURSE OF STUDY, 1899–1900

Fourth Year			
First Term		Second Term	
Technical Applications of Electricity to Telephony, Electric Lighting, Electrical Generation of Power, Railroad Signals, etc.	400, 403	Technical Applications of Electricity; Electric Motors; Alternating Current Machines	400, 416
Methods of Dynamo Testing (lectures)	402	Transmission and Distribution of Energy	417
General Electrical Testing	405	Methods of Dynamo Testing (lectures)	402
Electrical Engineering Laboratory; Testing of Dynamos, Electric Lamps, etc.	420	Principles of Dynamo Design	418
Electrical Measuring Instruments and Methods (lectures)	406	Telephone Engineering	419
Theory of Periodic Currents	404	Electrical Engineering Laboratory; Measurements of Dynamo Electric Machinery, Special Methods	420
Photometry	392	Theory of Periodic Currents	404
Steam Engineering	540	Discussion of the Precision Measurements	423
Dynamics of Machines	543	Engineering Laboratory	545
Hydraulics	471	Economics of Corporations	261
Engineering Laboratory	545	Thesis	
Strength of Materials; Friction	86		
Method of Least Squares	57		

Note: Students having the requisite preparation and ability may pursue more advanced courses in the mathematical theory of electricity and other subjects. With this end in view, competent students may take Fourier's Series and allied topics, also Energetics and Electro-Chemistry, as extra studies.

Source: *M.I.T. Annual Catalogue, 1899–1900*, p. 41.

its methods, like those of the Sibley College of Engineering as a whole, were, "as far as practicable, those of the business establishment or engineer's office, and admission and discharge will be governed as far as possible by business rules."[30] Unlike M.I.T., which required liberal-arts courses of its undergraduates in engineering, Sibley College assumed that students had completed their general academic education before entering its four-year program, and therefore it organized its entire course of study solely to meet the demands of the engineering profession. The result was a mix of mathematics, physical science, and shopwork courses during the first three years.

The electrical engineering courses the students were required to take in their fourth year in this business-oriented engineering environment dealt with the design and construction of electrical machinery, the distribution of electric light and the electrical transmission of power, as well as the construction and maintenance of electric lighting and power plants. Required courses in physics gave special attention to the needs of "the practical electrician." The electrical engineering laboratories at Cornell, like the one at M.I.T., were equipped with alternating and polyphase machinery. In the fourth year of study students were encouraged to show a capacity for "intelligent study of special or original problems" by writing a thesis representing special investigation, preferably both theoretical and experimental.[31] Professor Ryan, who, possibly in accordance with the business environment of the college, was identified as "consulting engineer," offered a two-semester course in electrical engineering to fourth-year students. Ryan also taught courses on "the finance of the production and utilization of electrical energy" and "a history of the development of electrical engi-

[30] Ibid., p. 330.
[31] Ibid., p. 345.

neering." (To offer a history of a field that was so young was remarkable.) For graduate students there was Ryan's year-long course in electrical engineering. Henry Norris, who held a degree in mechanical engineering and was listed as "electrician," and a "Mr. Hoxie" also taught electrical engineering courses; in addition, Norris taught a graduate course on electric railways.

Because of the slow growth of electric power systems during the 1890s in London by comparison with Chicago and Berlin, and because of the failure of polyphase power systems to take hold in England before 1900, it is interesting to compare the electrical engineering curriculum at London's City and Guilds Central Technical College with the programs offered at M.I.T. and Cornell. It should be recalled that the Central Institution, whose name was changed to Central Technical College in 1893, was founded by the City and Guilds of London Institute to promote technical education in London and throughout the country. Toward that end, the Central Technical College conducted ordinary as well as honors examinations in electrical engineering. Students prepared for these not only at the Central Technical College but at other technical institutions throughout Britain. Certification of the successful completion of the examinations was comparable to a diploma in engineering in the United States. Passing the examinations was also considered equivalent to the status obtained after completing the seven-year apprenticeship that traditionally qualified one for admission to a craft guild.

Ayrton, who was one of the first two professors appointed by City and Guilds Institute and who was briefly associated with Finsbury Technical College before becoming one of four professors at the Central Institution in 1884, presided over the institution's program in electrical engineering from 1884 to 1908. He organized the electrical laboratories, gave related lectures, and administered a small department (at first three, but by 1899 seven persons). He also administered the electrical engineering examinations conducted by his institution. As early as 1886, 151 candidates from various parts of Britain presented themselves for one of the two examinations in electrical engineering.[32] In 1885 the honors-grade examination included one question calling for the description of an alternating-current motor and another asking how two alternating-current generators might be run in parallel, or series ("coupled up"), an arrangement that had not yet been widely introduced in practice.[33]

In 1885 Ayrton was assisted by a laboratory associate, T. Mather, who became department head in 1910, and a workshop instructor. Together they provided a three-year program in what was then called the physical department, a program that focused on electrical engineering, but that also promised, according to department description, to organize courses in optical instrument making and problems of heat and ventilation.

In 1899 the three-year course remained essentially mathematics, chemistry, physics, mechanics, and electrical engineering laboratories and related lectures. Central Technical College emphasized laboratory work to the point that "instruction is given largely by tuition in the laboratories, the

[32] *Telegraphic Journal and Electrical Review* 19 (1886): 120, 170, 240.
[33] For a copy of the examination for the year, see ibid. 16 (1885): 511.

lectures being rather for the purpose of aiding the Students in their laboratory practice than of forming a distinct course by themselves."[34] Laboratory work in the first year consisted of quantitative experiments that were arranged to verify the laws of physics; the second-year laboratory prepared the students to carry out original, independent investigations. During the third year, students were expected to spend much of their time on original research projects.[35] Ayrton's lectures and related laboratory work included a large unit dealing with alternating and polyphase systems. His lectures also explained load curve and other aspects of the economics of electrical engineering. Languages, the social sciences, and the humanities were excluded from the curriculum.

Judging by these course and laboratory descriptions and by the quality of the faculty, the British student was receiving an education in electrical science and technology that equaled the offerings at M.I.T. and Cornell. This seems to have been the case despite the relatively slow development of electrical supply, especially electric power systems, in England. A partial explanation for this seeming paradox might be that the backwardness of Britain's electric supply industry was quantitative, not qualitative. Germany and the United States had more central-station, distribution, and transmission systems, but those that existed in England provided ample experience for the country's electrical engineering faculty and students. If true, this explanation raises serious doubts about the value of the parliamentary hearings that were held to investigate technical education as a cause of the slowing pace of British industrialization.

The program established by Kittler in Germany in 1884–85 was the same the first two years as for all Darmstadt *TH* students. Electrical science and practice were stressed in the third and fourth years. Comparison of the curricula for students entering M.I.T. and Darmstadt in the academic year 1884–85 reveals that the German students were almost a year more advanced than M.I.T. students. Mathematics courses, especially analytical and descriptive geometry and differential and integral calculus, were taken the first year at Darmstadt and the second year at M.I.T. The German students took general physics the first year and physics laboratory the second; M.I.T. followed this sequence in the second and third years. As noted, M.I.T. students concentrated on electrical engineering in their fourth year; Darmstadt students began this specialization in their third year.[36] As of 1884–85, Darmstadt students were required to take political economy in their third and fourth years; they could elect history of literature and foreign languages. This comparison with M.I.T. supports the generalization that German students entered institutions of higher learning at least one year more advanced in their studies than U.S. students and that a graduate of four years of study at a *Technischen Hochschule* had the equivalent of a master's degree in the United States.[37]

Of the curricula considered here, the development of the course of study

[34] *Central Technical College, Session 1899–1900, Programme,* p. 20.
[35] Ibid., p. 23.
[36] *Programm der Grossherzoglich Hessichen Technischen Hochschule zu Darmstadt f.d. Studienjahr 1884–85,* pp. 57–58.
[37] Wickenden, *A Comparative Study of Engineering Education,* pp. 248–49.

at Darmstadt from the early 1880s to 1900 was the most impressive. By 1900, Darmstadt's basic electrical engineering courses were taught in the first and second years rather than in the third year as formerly. As a result, the third and fourth years included more specialized courses such as power transmission systems, water-power technology, and the design and construction of central stations. Kittler limited his teaching in the third and fourth years to the general course on electrical engineering and the *Praktikum,* or "practical training," for examination.[38]

In addition to teaching, consulting, and curriculum-planning, academics engaged in research. They often designed their experiments to probe the reverse salients of the evolving power systems and to gain information and theory conducive to the definition and solution of critical problems. Since research occasionally involved graduate students, it depended to some extent on graduate education, but less so than in recent decades. Of the institutions considered here, only Cornell University listed graduate courses with those for undergraduates. In the United States before World War I, the general attitude was that practical experience, with large manufacturers such as General Electric or Westinghouse or with a utility, was superior to study beyond the bachelor's degree. Many employers viewed graduate study as a waste of time "hanging around college" and as a means of avoiding facing up to the real world.[39] The special programs initiated by the large manufacturers during the decade before World War I lessened interest in promoting graduate studies. There was little "hanging around." In electrical engineering during the period 1900–1909, M.I.T. awarded only six master's degrees and no doctoral degrees; the figures for Cornell were fourteen master's degrees and two doctorates.[40]

In Germany the graduate of a four-year *TH* program was at least a year ahead of graduates from an American university and therefore had earned the equivalent of a master's degree. Thus, the *Prüfungsordnung,* or "qualification requirement," offered by Kittler and in programs modeled on his was more advanced than the master's program in the United States. The *TH*s in Germany did not win the jealously guarded privilege of awarding doctoral degrees, however, until 1899, when Kaiser Wilhelm I persuaded the universities to accept the *Technischen Hochschulen* among the elect. In England, professional engineering societies established their prerogative in recognizing the highly qualified engineer by awarding him membership after examination. The string of letters signifying membership therefore signified experience and advanced learning.

Despite the paucity of graduate programs and research assistants in electrical engineering before World War I, the academics conducted research and published their results. They chose research problems related to evolving light and power systems. Their approach was scientific, if by "scientific" one means using and attempting to formulate general statements or laws; resorting to mathematics as an analytical tool and language; formulating hypotheses; and designing experiments in the laboratory or in nature to

[38] *Programm der Grossherzoglich Hessischen Technischen Hochschule zu Darmstadt f.d. Studienjahr 1900–1901,* pp. 64–65.

[39] Terman, "History of Electrical Engineering Education," p. 1402.

[40] Ibid., p. 1403.

test these theories. Not only was their research methodology scientific, but their academic credentials in some instances were comparable to those of research scientists. Because of this common methodology and shared educational experience, the academic electrical engineers can be called engineering scientists and their work the science of engineering. The academics often carried on their research in the universities and engineering colleges, but they also did it as consultants to industry, both in the plant and in the field. After 1900 more engineering and natural scientists found a congenial research environment in the industrial laboratories of the United States and Germany.

The academic engineering scientists chose their research projects in the context of reverse salients, not in response to an evolving pure-science paradigm. From about 1890 until World War I, reverse salients arose in the universal urban system and in point-to-point high-voltage transmission, developments that followed upon the resolution of "the battle of the systems" and the introduction of polyphase systems. Electrical engineers who established the earliest electrical engineering programs published early in their careers. They, too, focused on problems related to the main line of development and thereby contributed to the mounting momentum. They did not make breakthrough inventions like those of the independent inventors, but they worked on critical problems that, when solved, allowed further development of urban systems and point-to-point transmission. After the founding of the American Institute of Electrical Engineers in 1884, they often used its principal organ, the *AIEE Transactions,* as their medium of publication.

Louis Duncan, who taught at both The Johns Hopkins University and M.I.T., published an article in the *Transactions* in 1887 on the improvement of alternating-current motors, one of the critical problems of the decade. Two years later, he published a paper about conducting efficiency tests of a.c. apparatus in urban central stations. At this early date, Duncan stressed the technical and economic significance of the load diagram. Two more articles on a.c. systems followed in 1892 and 1894, when he directed his attention to the critical problem of polyphase motors.[41] Dugald C. Jackson, who succeeded Duncan at M.I.T., also published early and often. He, too, concentrated on critical problems associated with the development of alternating and polyphase systems. Furthermore, Jackson directed his attention to electrical engineering education.[42]

[41] Louis Duncan, "Alternating Current Electric Motors," *AIEE Transactions* 5 (1887–88): 211–35; idem, "Some Tests on the Efficiency of Alternating Current Apparatus," ibid. 7 (1889–90): 109–27; idem, "Note on Some Experiments with Alternating Currents," ibid. 9 (1892):179–91; and L. Duncan, S. H. Brown, W. P. Anderson, and S. Q. Hayes, "Experiments on Two-Phase Motors," ibid. 11 (1894): 617–38. I am indebted to Robert Belfield for his extensive and detailed search for, and categorization of, the academics who published in the *AIEE Transactions* from 1887 to 1904.

[42] D. C. Jackson and R. J. Ochsner, "Alternating Currents and Fuses," *AIEE Transactions* 11 (1894): 430–40; D. C. Jackson and S. B. Fortenbaugh, "Some Observations on a Direct-Connected 300 K.W. Monocyclic Alternator," ibid. 12 (1895): 350–57; D. C. Jackson, "The Commutated Current Wave of a Composite Wound Alternator," ibid. 15 (1899): 403–8; idem, "The Technical Education of the Electrical Engineer," ibid. 9 (1892): 476–99; and idem, "The Typical College Courses Dealing with the Professional and Theoretical Phases of Electrical Engineering," ibid. 22 (1903).

Harris J. Ryan of Cornell, another of the second wave of electrical engineering academics, wrote an article on transformers for the *Transactions* in 1889 and also reported on alternating-current apparatus. In 1899 he published in the same journal an article on the wave form of alternating currents. In 1903 he wrote about a cathode-ray alternating-current wave indicator, an important instrument for analysis of problems arising in high-voltage transmission. Ryan's 1905 article, "The Conductivity of the Atmosphere at High Voltages," also pertained to high-voltage transmission.[43]

Because Ryan taught in a major electrical engineering department and engineering college and was influential in the field, his early and persistent research pertaining to the critical problems of high-voltage transmission provide insight into the character of advanced academic research in electrical engineering before World War I.[44] While studying at Cornell under Anthony, Ryan first became interested in what would become his chief focus—the study of high-voltage phenomena. The circumstances of his early commitment involved a visit arranged by Anthony in about 1886 to the works of Frank Sprague, pioneer in electric traction and motors. Sprague's remarks about electric power transmission "started me out in life with a never-ending enthusiasm for the study of high voltage phenomena," Ryan recalled.[45] The fact that he was awarded the Edison Medal of the AIEE in 1925 "for his contributions to the science and the art of high-tension transmission of power" suggests that his self-analysis was sound.[46]

The bibliography of his numerous technical articles in various journals shows that after publishing thirty-seven articles on diverse subjects (but with a tendency toward alternating-current problems) he turned in 1903 or 1904 to high voltages and transmissions and that his attention remained focused on this subject until his last article, which appeared in 1925. Of the forty-four articles published by Ryan after 1903, all but ten are identifiable by title as dealing with high-voltage and transmission problems.[47]

Ryan's major paper on high-voltage problems was published in 1905 after being read before the AIEE on 26 February 1904.[48] "The fundamentals set forth in this paper were a distinct contribution to electrical science."[49] Charles F. Scott, a pioneer in the working out of the universal system and in transmission, said: "The paper we have just heard read is a remarkable one, and will probably be considered unique among the papers presented to our *Institute*."[50] Such attention and high praise are not sur-

---

[43] Harris J. Ryan's articles included "Transformers," ibid. 7 (1889–90): 1–29; "Some Experiments upon Alternating Current Apparatus," ibid., pp. 324–65; "The Determination of Wave Form of Alternating Currents without a Contact Maker," ibid. 16 (1899): 345–60; "The Cathode Ray Alternating-Current Wave Indicator," ibid. 22 (1903): 593 ff.; and "The Conductivity of the Atmosphere at High Voltages," ibid. 23 (1904): 101–34.

[44] On Ryan's research see W. F. Durand's memoir "Harris Joseph Ryan, 1866–1934," in National Academy of Sciences, *Biographical Memoirs*, vol. 19 (Washington, D.C., 1938), pp. 285–306 (hereafter cited as "Ryan," *Biographical Memoirs*).

[45] Ibid., p. 287.

[46] Ibid., p. 299.

[47] I have relied on the Ryan bibliography appended to Durand's memoir as the basis for my analysis. Ibid., pp. 302–6.

[48] Ryan, "High Voltages."

[49] "Ryan," *Biographical Memoirs*, 19: 294.

[50] Discussion of Ryan's 1904 paper, *AIEE Transactions* 23 (1904): 135.

prising, however, for by using a scientific method, Ryan had solved a critical problem defined by Scott in experiments for the Westinghouse Company— a problem that had cast serious doubt on the possibility of exceeding 40,000–50,000 volts in long-distance power transmission.[51]

Ryan had been attracted to the problem by Scott's 1899 article describing power losses in high-voltage transmission lines due to the corona effect.[52] He defined more precisely the variables affecting the onset of corona and the attendant loss between lines. After experimentation and analysis, he formulated an equation expressing the relation between the critical, corona-causing voltage, conductor sizes, and their separation. He also sought the value of the rupturing electric field and the distance from the surface of the conductor at which rupture occurred.

This is not the place to discuss the equation, its variables and constants, for Ryan has done that with admirable clarity in his paper, but the nature of Ryan's style of research merits attention. His method was scientific; his hypothesis can be summarized. Accepting electric-field theory, he assumed that an electric force applied to the terminal faces of a dielectric produced a "distortion of the atomic structure of the dielectric." Because high-voltage transmission lines in proximity were separated by a dielectric (atmosphere), the electric force of the high voltage distorted the electric field in the atmosphere. Corona, a visible effect described earlier by Scott (see p. 162 below),[53] and power loss finally occurred, Ryan believed, when the dielectric was ruptured by the force of the high-voltage and high-displacement current. When rupture came at the critical voltage, ordinary conduction current in the dielectric and corona brought substantial real-power loss from the lines because the resistance of the conducting zone produced a lag in the charging current, thereby giving it a power component in phase with the voltage. Ryan hypothesized that the critical, corona-causing voltage was a function of transmission-line diameter and spacing, for these determined the characteristics of the dielectric. In his preliminary statement of theory and hypothesis of capacitance reactions, Ryan cited the prior research of Charles P. Steinmetz, J. J. Thomson, and others.[54]

Ryan used the experimental observations of Scott, but it became evident that he needed, in addition, to plan and design a series of experiments to obtain the data necessary to write the formulas expressing the relation of the factors conditioning the start of atmospheric loss and corona. In the spring of 1903 Ryan set up the experimental apparatus with the help of his students. He suspended the wire, or conductor, that simulated the transmission line at the center of a metallic tube. The electric field was to be established between wire and tube. The tube also facilitated control of temperature and atmospheric conditions. Having already established the value of cathode-ray indicators for observing alternating currents, he used

---

[51] "Ryan," *Biographical Memoirs,* 19: 293. In his memoir on Ryan, Durand expressed the belief that Scott's work cast serious doubt on the use of voltages above 40,000.

[52] Charles F. Scott, "High-Voltage Power Transmission," *AIEE Transactions* 15 (1899): 531.

[53] Charles P. Steinmetz had used the term "corona" in an 1893 paper on high-voltage discharge (ibid. 10 (1893): 85). See Ryan, "High Voltages," p. 104.

[54] Ryan, "High Voltages," pp. 105–8.

an oscilloscope in this experiment.[55] He also placed outside and alongside the tube a brass rod in order to observe differences on the oscilloscope between the charging current for wire and tube and the charging current between tube and brass rod. He assumed no atmospheric discharge would occur between the tube and rod. Switches allowed him, for comparison, to alternate the wire and the rod in the high-voltage circuit. Thus he established his reference. His apparatus also included high-voltage and voltage-adjusting transformers and a phase-adjusting induction motor.

Ryan and his assistants watched as the charging current from wire to tube remained the same as that from the rod to the tube as the voltage was increased. Then, at the critical point in the voltage rise, the anticipated hump in the display of the current charging the wire occurred; no comparable event was seen in the other charging current. If the room was darkened, the corona appeared at the same instant. The design worked, and the experimenters turned to collecting data on different diameter wires and other factors affecting corona. The objective was not obscure; Ryan wanted to supply engineers and manufacturers with tables (calculated by means of his equation) that showed what diameter of line conductor and what spacing had to be used to avoid loss between conduction lines on which specified voltages were to be transmitted.[56] He was interested not only in the highest voltage then being used commercially but also in those that could be reasonably anticipated in the near future. Ryan's research and his results dealt with contemporary problems of high-voltage engineering; they also reached into the future.[57]

Academics like Ryan did research in the engineering schools and the universities, but the most active research site increasingly became the electrical manufacturing company. After 1900 the large electrical manufacturers established research-and-development laboratories, but until then the engineering departments of manufacturing companies were responsible for invention and development. The engineering departments therefore dominated research and development during the interval after the apogee of Edison and the independent inventors and before the establishment of the major industrial research laboratories. General histories of technology and science tend to ignore the important role of the engineering departments during the formative years of the electrical industry.[58]

---

[55] Ryan pioneered in America in the use of the oscilloscope as a research tool for the study of alternating-current phenomena. See "Ryan," *Biographical Memoirs*, pp. 202–3; and Ryan, "The Cathode Ray Alternating Current Wave Indicator," pp. 539 ff.

[56] Ryan, "High Voltages," p. 127.

[57] See remarks of Charles F. Scott in discussion of Ryan's "High Voltages," *AIEE Transactions* 23 (1904): 135.

[58] Harold Passer, *The Electrical Manufacturers, 1875–1900* (Cambridge, Mass.: Harvard University Press, 1953), describes the problem solving that was done by the engineering staffs of leading American manufacturers, especially General Electric and Westinghouse. The third part of his book deals with the invention and development associated with the rise of electric traction and alternating- and polyphase-current equipment (pp. 211–345). Arthur A. Bright, Jr., *The Electric-Lamp Industry: Technological Change and Economic Development from 1800 to 1947* (New York: Macmillan, 1949), tells of research, invention, and developments within the industry not only in the United States but in England and Europe as well. His study also shows that manufacturers presided over research and development after 1890. Von Siemens, *House of Siemens*, discusses light and power systems research and development at Siemens &

There are countless examples of critical problems being solved by manufacturers' engineers. Note has been taken of the development of transformers and motors at Westinghouse in the United States and AEG in Germany, and of rotary converters at both Westinghouse and General Electric. Manufacturers also focused on the problems of transmission. Conceivably the utilities might have taken responsibility for research and development. Nevertheless, after the independent inventors and along with the academics, equipment manufacturers such as Westinghouse took the initiative in solving the salient critical problems. As a result, solutions in the form of new equipment tended to be standardized in the United States and Germany, where the manufacturing companies were large and influential. In contrast, in Britain the consulting engineers and utilities specified characteristics, and a confusing variety resulted.

In the United States one example of company engineers solving problems involves Westinghouse and two of its engineers, Charles F. Scott and Ralph Mershon. Their experimentation and field work stimulated the laboratory research of Harris Ryan. Scott, who was born in Athens, Ohio, in 1864, attended Ohio University and then received a bachelor's degree from Ohio State University. In 1886 and 1887 he did advanced work in physics, mathematics, and chemistry at The Johns Hopkins University. In 1888 he began working for the Westinghouse Electric & Manufacturing Company, where be became assistant electrician in 1891, electrician in 1893, and chief electrician in 1896. Besides taking a lead in articulating the universal electric power system, he invented the Scott connection in 1894, which was used between the Niagara power plant and the transmission line to Buffalo. The Scott connection "T-connected" two transformers to change two-phase current to the three-phase arrangement and was widely used.[59] In 1911 Scott accepted the professorship of electrical engineering at the Sheffield Scientific School, Yale University, where he served as department head until his retirement in 1933. Ralph Mershon, his associate, was born in Zanesville, Ohio, and received a degree in mechanical engineering in 1890, also from Ohio State University. Joining the Westinghouse Company in 1891, he designed the transformers the company displayed at the Chicago exposition

---

Halske and includes data on other German manufacturers. Sigfrid von Weiher and Herbert Goetzeler, *The Siemens Company: Its Historical Role in the Progress of Electrical Engineering* (Berlin and Munich: Siemens, 1972), survey inventions and innovation within a company setting. There are no studies comparable to Passer's *Electrical Manufacturers* or Bright's *Electric-Lamp Industry* as far as the British manufacturers are concerned. I. C. R. Byatt, *The British Electrical Industry, 1875–1914* (Oxford: Clarendon Press, 1979), stresses economic rather than technological developments; Lord Hinton of Bankside, *Heavy Current Electricity in the United Kingdom: History and Development* (Oxford: Pergamon Press, 1979), offers an insightful but relatively short survey of Britain's electrical supply and manufacturing industries; and Leslie Hannah, *Electricity Before Nationalisation* (Baltimore: The Johns Hopkins University Press, 1979), concentrates on Britain's supply industry, stressing business, economic, and political developments rather than research and development.

Engineering departments in other manufacturing industries also underwent the transition from the era of the independent, professional inventor to that of the industrial research laboratory. See, for example, Lillian Hoddeson, "The Emergence of Basic Research in the Bell Telephone System, 1875–1915," *Technology and Culture* 22 (1981):512–44.

[59] James E. Brittain, ed., *Turning Points in American Electrical History* (New York: IEEE Press, 1976), pp. 131–34.

in 1893. He also designed rotary converters, another of the major system couplers. Both became president of the AIEE, Scott in 1902, Mershon in 1912.[60]

After observing transmission-line phenomena on Westinghouse experimental lines, Scott suspected energy loss. He described them vividly:

The wires began to give a hissing or crackling sound and in the dark began to appear luminous at a little below 20,000 volts. As the voltage was increased the sound became more and more intense, the wires vibrated and became more and more luminous, until at the higher voltages they were surrounded by a coating of soft blue light many times the diameter of the wire. Often there were bright points along the wire, probably corresponding to bits of dust or rough places resembling points on the wire.[61]

Scott and others suspected that this dramatic phenomenon was accompanied by substantial energy loss. Later, when Ryan investigated it, the effect was commonly named corona; for the time being, Scott simply identified it by its most unwanted effect—"loss between the lines."

These early observations led Scott and other Westinghouse engineers in 1894 to turn for field observations to the pioneering Telluride, Colorado, transmission line that had been placed in operation in 1891 by Westinghouse and the San Miguel Gold Mining Company. Because of the prohibitive cost of transporting coal in the mountains and the scarcity of wood at its high elevation, the mining company had desperately sought to draw electric power from a water-power site more than two miles away. The facility has been called the first large alternating-current power transmission line in America.[62] It was angle phase, however, with a synchronous, rather than an induction, motor.

In preparation for their work at Telluride, Scott and Westinghouse shipped recently developed high-voltage, high-capacity transformers to Colorado. One of these was designed for the unusually high output of 200 kw and could be stepped up to as much as 60,000 volts for transmission. Two generators and several interchangeable armatures also were available at Telluride. These provided different frequencies and wave-form outputs. Scott's primary interest was to observe losses "between lines" as the voltage level was varied under different climatic, load, frequency, wave-form, and spacing conditions.

Ultimately, the results of Scott and Mershon's observations and analysis were presented in a set of tables, a form of communication that was readily understood and appreciated by working engineers designing transmission lines. The research led Scott and Mershon to conclude that varying the frequencies and wave forms had little effect on very high voltage transmission and, surprisingly, that different weather conditions also had relatively little impact. Precipitation increased energy losses, but the increase

---

[60] The Mershon Papers are in the Mershon Archives at Ohio State University, Columbus, Ohio. For a biographical sketch and his published essays, see Edith Cockins, *Ralph D. Mershon,* 2 vols. (Columbus: Ohio State University, 1956). The Scott Papers are at the Yale University Library, New Haven, Conn.

[61] Scott, "High-Voltage Power Transmission," p. 536.

[62] Ibid., pp. 531–32. For a history of the Telluride installation, see Charles C. Britton, "An Early Electric Power Facility in Colorado," *Colorado Magazine* 49 (1972): 186–95.

was small. What did prove interesting to them was the high level of loss that occurred between lines once the critical, corona-causing voltage had been reached. The measurements led Scott to the general conclusion that the losses between transmission lines were economically acceptable below 45,000 volts, but increased rapidly when "about 50,000 volts" was exceeded.[63] They had clearly identified a reverse salient in the advancing front of high-voltage transmission; their response was to recommend limitations until the problem was solved.

Manufacturers enhanced the invention-and-development capacity of their engineering departments by hiring young men such as Scott and Mershon (graduates of the new electrical engineering programs) and by entering into merger and acquisitions agreements. Economic historians stress the economic importance of the numerous company mergers that occurred during the closing decades of the nineteenth century; they do not, however, sufficiently emphasize the technological consequences of mergers and acquisitions, particularly in the electrical industry. The companies acquired often brought to the merger the men, material, and ideas that were needed to rectify inadequacies in the acquiring company's system of production and products. The merger of Edison General Electric and Thomson-Houston in 1892 resulted primarily from the fact that the patent holdings of the companies were complementary.[64] Edison General Electric held a strong position in urban d.c. stations, d.c. power transmission, and street railways; Thomson-Houston's strength lay in arc lighting and alternating currents.[65]

The mergers were a way of acquiring the services of inventors and engineers as well as their patents. In some instances, an engineer or inventor with the acquired company became a salaried member of the engineering department of the acquiring company. In other cases, the acquiring company obtained the right to patent the future inventions of an inventor or engineer in a specified area. The inventor assumed the status of consultant vis-à-vis the company. In 1888 Charles J. Van Depoele (1846–1892), one of the pioneering inventors of electric streetcars, joined the electric railway department of Thomson-Houston after his company was bought. The company and patents of Elmer Sperry, who invented several improved electric streetcar components and electric mining machinery, were purchased by Thomson-Houston. Sperry agreed to act as a consultant to Thomson-Houston for two years at an annual salary of $5,000 to develop this equipment. In 1895 Sperry made a similar arrangement with General Electric, agreeing to turn over applications, patents, and future patents on the electric streetcar brake and to act as a consultant in developing the patented inventions.[66]

In 1893, when General Electric acquired the firm of Eickemeyer & Osterheld of Yonkers, New York, it also obtained the services of a young electrical engineer and mathematician whose career would be inextricably linked with the history of General Electric. In 1889 Charles Proteus Stein-

[63] Scott, "High-Voltage Power Transmission," pp. 543 and 556.
[64] Passer, *Electrical Manufacturers*, p. 321.
[65] Ibid., p. 325.
[66] Thomas P. Hughes, *Elmer Sperry, Inventor and Engineer* (Baltimore: The Johns Hopkins Press, 1971), pp. 71–74.

Figure VI.13. *Charles Steinmetz and Thomas Edison. Courtesy of the Samuel Insull Archives, Loyola University, Chicago, Ill.*

metz (1865–1923), a recent immigrant from Breslau, Germany, had found work as a draftsman at the Yonkers company, which manufactured hat-making machinery and electric motors. Realizing that Steinmetz could solve complex design problems because of his advanced education in mathematics and mechanical engineering in Germany and Switzerland, Rudolf Eickemeyer, also a German immigrant, assigned Steinmetz the task of developing an alternating-current motor. Working in a small laboratory attached to Eickemeyer's factory, Steinmetz derived equations for analyzing magnetic hysteresis and eddy currents, two critical causes of inefficiency in electrical machinery. He published his investigations and conclusions in 1890 and 1892 in the *Electrical Engineer*. In 1893 he presented a major paper on the application of the algebra of complex numbers to the analysis of alternating-current circuits. His application of mathematics to solving alternating-current-machinery and circuitry problems was especially instructive for engineers because he presented his results in a congenial format.[67] After joining General Electric, Steinmetz became a member of the company's calculating department; he would remain with GE until his death. In the eyes of many American engineers, he became the symbol of successful scientific engineering research and development. Long-distance power transmission was one of the research problems on which he, like Scott, Ryan, Mershon, and others, concentrated for years.

Steinmetz had completed his doctoral dissertation in mathematics at the University of Breslau, but fearing arrest because of his participation in the socialist student movement there, he had emigrated to Switzerland. In Zurich he studied mechanical engineering at the Polytechnic School, one of the world's outstanding engineering institutions. Thus he brought with him to America the prestige associated with a Central European education. It should be recalled, however, that American academics and industrial engineers also were moving beyond empirical studies and taking a mathematical and scientific approach to solving critical engineering problems. Moreover, in Europe, academic engineers and engineers employed by industry were carrying on research comparable to that being done in America. Educated in his much-emulated department at Darmstadt, Kittler's students enhanced the research-and-development capacity of Germany's electrical manufacturing companies.

By 1900 the capacity of scientifically trained engineers and of academics with backgrounds in the natural sciences to solve the technical problems of the expanding electric supply systems had been amply demonstrated and some leaders in the field resolved to develop an organizational form that was especially well suited to the problem-solving style of the research engineers and industrial scientists. The industrial research laboratory was well known in the chemical industry, especially in Germany, but leaders of the electrical industry saw the need and opportunity to adapt the form for their invention, research, and development functions.

An immediate reason for forming a research laboratory at General Electric was the realization that research engineers were preoccupied with rou-

[67] James E. Brittain, "Charles Proteus Steinmetz," in *Dictionary of American Biography*, ed. John A. Garraty (New York: Harper & Row, 1974), pp. 1035–36. See also John W. Hammond, *Charles Proteus Steinmetz: A Biography* (New York/London: The Century Co., 1924).

Figure VI.14.   *First home of the GE Research Laboratory, Schenectady, N.Y. Courtesy of the General Electric Co., Schenectady, N.Y.*

tine problem-solving (see Fig. VI.14). The stationary power and traction business was expanding rapidly, and the problems of designing improvements in existing generators, motors, transformers, and other machinery took an increasingly large share of the engineers' time. Men who had come to GE's engineering department with a history of invention or with aspirations to invent found themselves burdened by routine. Charles Steinmetz's work in the 1890s involved more troubleshooting than scientific research.[68] Edwin W. Rice, vice-president of General Electric, observed in 1902: "Although our engineers have always been liberally supplied with every facility for the development of new and original designs and improvements of existing standards, it has been deemed wise during the past year to establish a laboratory to be devoted exclusively to original research. It is hoped by this means that many profitable fields may be discovered."[69] Designs and standards were not the same as "discovery," so Rice was announcing the birth of the General Electric Research Laboratory, an institution which was generally considered a landmark establishment in the history of industrial research.

[68] George Wise, "A New Role for Professional Scientists in Industry: Industrial Research at General Electric, 1900–1916," *Technology and Culture* 21 (1980): 412. This is a perceptive account of the origins of the GE Research Laboratory and of the new definition of role for scientists that it institutionalized. Wise used sources at General Electric and elsewhere that were not available to earlier historians. Also valuable is Kendall Birr, *Pioneering in Industrial Research: The Story of the General Electric Research Laboratory* (Washington, D.C.: Public Affairs Press, 1957). Birr was one of the first professional historians to write about industrial research. Laurence A. Hawkins, an executive engineer of the GE laboratory, also contributed with his *Adventures into the Unknown* (New York: Morrow, 1950). The rise of the physics profession and its growing interest in applied research is placed in broad perspective by Daniel J. Kevles in *The Physicists* (New York: Knopf, 1978).

[69] General Electric Co., *Annual Report for 1902* (New York, 1903), p. 13, quoted in Wise, "A New Role for Professional Scientists," p. 408.

Figure VI.15. Nernst lamp. From 50 Jahre Electrotechnischer Verein, ed. Hans Görges (Berlin: ETV, 1929), p. 227.

General Electric also founded the laboratory in response to a technical challenge from Europe. After almost two decades of concentrating on inventions in power, distribution, and transmission systems, GE discovered that some types of European-invented incandescent lamps were threatening to displace the carbon-filament bulb invented years earlier by Thomas Edison. The mechanical and electrical engineers and inventors in GE's engineering department had the capacity to react to competitors' improvements in machinery, but the new incandescent lamps from Europe were the invention of highly educated chemists. The combination of a routine-burdened engineering department and the need for research chemists provoked the decision by GE leaders Edwin Rice, Charles Steinmetz, Albert G. Davis (a GE patent attorney), and Elihu Thomson (inventor and founder of Thomson-Houston) to recommend to GE's president, Charles Coffin, the establishment of a research laboratory. Coffin concurred.[70]

Another reason General Electric wanted a research laboratory was its awareness of the importance of patents in securing market advantages. This had been sharply impressed upon the company by the expiration of major carbon-filament patents. GE faced the alternative of either buying the patents of outside inventors or companies or inventing and applying for its own. The research laboratory was expected to provide the company with that option and the ability to negotiate with outside patent holders more effectively; for instance, the company could withdraw from negotiations and search for ways of "inventing around" the competing patent if the price for patent license was too high. These considerations explain why GE's patent attorney recommended the establishment of a research laboratory.

The market competition was forthcoming. In the fall of 1900 Steinmetz wrote to Edwin Rice, alerting him to the challenge from the Nernst lamp. Walther Nernst (1864–1941), a professor of electrochemistry at the University of Göttingen and a founder of the science of physical chemistry, had patented his lamp in 1899. The lamp had a filament of refractory metallic oxides (see Fig. VI.15). Operating at higher temperature than the carbon-filament lamp, the Nernst lamp was more efficient and economical. The Nernst lamp cost four to five times more than the improved carbon-filament lamp, but this higher first cost was offset by lower operating costs, especially in Europe, where electricity prices were higher. The Nernst lamps were widely used in Europe during the first decade of the twentieth century. The Allgemeine Elektrizitäts-Gesellschaft had patent rights in Germany and produced about 7,500,000 of them before 1907. The Westinghouse Company then obtained rights in the United States, thereby causing additional concern at GE.[71]

Even more provocative was the research under way at Siemens & Halske, the German electrical manufacturer. Wilhelm von Siemens (1855–1919), who succeeded his father, the founder, as head of the company, took a keen interest in finding a metallic filament to replace the carbon one. He assigned the task to Werner von Bolton (1868–1912), a young scientist who had studied at Leipzig under Wilhelm Ostwald (1853–1932), a pioneer,

[70] Wise, "A New Role for Professional Scientists," p. 414.
[71] Bright, *Electric-Lamp Industry*, pp. 172–73.

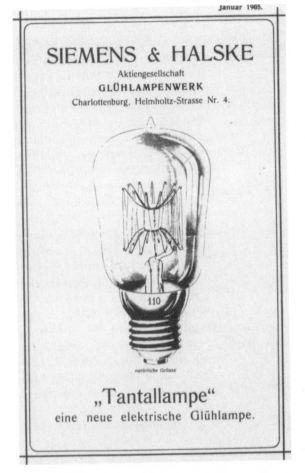

Figure VI.19. Osmium-filament lamp.
From 50 Jahre Electrotechnischer
Verein, *p. 228.*

Figure VI.20. Carl Auer von Welsbach.
Courtesy of the Siemens Museum, Siemens
AG, Munich.

like Nernst, in physical chemistry. Von Siemens provided von Bolton space in the company's laboratory in Berlin-Moabit, where he concentrated on the possibilities of tantalum. The major problem was to obtain a pure and malleable tantalum, which Bolton succeeded in doing by using an electric arc furnace. In 1902 the scientist and his assistants announced their success to Wilhelm von Siemens, and patented improvements in design and production methods followed. After 1905 the company successfully marketed the tantalum lamp. (See Figs. VI.17 and VI.18.)

Competing not only against the Nernst and tantalum lamps but also against a widely used osmium-filament lamp (Fig. VI.19) invented by the Austrian Carl Auer von Welsbach (1858–1929),[72] who had been trained by Robert Bunsen at the University of Heidelberg, the GE laboratory strove to invent and develop improved and new filaments. Challenged by inventor-scientists who had been trained in Germany in physical chemistry, a field in which Germany excelled, it is not surprising that GE brought to its laboratory research chemists who held the German doctorate. Willis R. Whitney (1868–1958), the first laboratory head, had a doctorate from the University of Leipzig, where he, like Werner von Bolton, studied under Ostwald; William D. Coolidge (1873–1975), who became the first assistant director of the laboratory, also was a physical chemist who received his Ph.D. from Leipzig; and Irving Langmuir (1881–1957), the third member of the triumvirate that dominated the GE laboratory in its early years, held a doctoral degree from the University of Göttingen, where he studied under Nernst. Whitney and Coolidge taught at M.I.T. before going to GE; Langmuir was an instructor at Stevens Institute of Technology. General Electric responded to the impact of German excellence in chemistry at least a decade before World War I and the embargo of German chemical imports made the fact dramatically clear. (See Fig. VI.22.)

Scientists at the GE laboratory soon developed new filaments. Whitney, using a heat-treatment process in an electric-resistance furnace of his own design, drove off impurities from the carbon filament and gave it a harder graphite coating. By 1904 this brought the greatest improvement in the carbon incandescent lamp since 1884. The company manufactured the new lamp under the name of GEM (General Electric Metalized). Coolidge invented and patented methods for making ductile molybdenum and tungsten. In 1910, $54,000, or one-third of the total expenditures of the GE laboratory, went to the development of the process, and substantial amounts were spent annually from 1906 to 1911. Before Coolidge's successful patent application in 1912, GE bought the tungsten-filament patents of outside companies and inventors, paying out $490,000 in 1909 alone.[73] When GE attempted to use its tungsten patent in Europe, the Siemens & Halske Company, AEG, and the Auer Gesellschaft persuaded GE that they had conflicting patents and priorities. (Auer held the patents of Carl Auer von

---

[72] In 1891 Carl Auer von Welsbach had greatly enhanced the competitiveness of gaslight with his invention of the Welsbach mantel. His company, Österreichische Gasglühlicht und Elektrizitätsgesellschaft, successfully marketed both his mantel and his osmium lamp in Germany despite the strength of AEG and Siemens. Hans-Joachim Braun, "Gas oder Elektrizität? Zur Konkurrenz zweier Beleuchtungssysteme, 1880–1914," *Technikgeschichte* 47 (1980): 11–13.

[73] Von Siemens, *House of Siemens*, 1: 290–91.

Figure VI.21. *Irving Langmuir's laboratory, 1912, Schenectady, N.Y. Courtesy of the General Electric Co., Schenectady, N.Y.*

Figure VI.22. *First permanent home of the GE Research Laboratory, Schenectady, N.Y. Courtesy of the General Electric Co., Schenectady, N.Y.*

Welsbach.) General Electric and the three German companies then agreed to share rights in the use of the Coolidge and Siemens patents covering the manufacture of tungsten-filament lamps.

The General Electric Research Laboratory not only separated invention and research from everyday production problems and minor system improvements but it offered scientists alternative roles to those of the academic and consultant. In the laboratory, persons with inventive instincts had access to facilities for research and invention without the attendant responsibilities for developing and marketing the inventions. Furthermore, Whitney engendered an atmosphere that was found congenial by Ph.D. scientists:

Willie Walker
Corrosion of iron and
a hell of a lot of other things.

Figure VI.23. *Early laboratory colloquium at the GE Research Laboratory: Willis Whitney (fourth from right); William Coolidge (far right). Courtesy of the General Electric Co., Schenectady, N.Y.*

Figure VI.24. *Staff at the GE Research Laboratory, c. 1912: Willis Whitney (third from left); Irving Langmuir (far right). Courtesy of the General Electric Co., Schenectady, N.Y.*

weekly colloquia, publication of scientific papers, and encouragement to acquire a greater fundamental understanding of technological phenomena.[74] (See Figs. VI.23 and VI.24.) Paradoxically, the professional scientists were assuming the role of the independent inventors, whose empirical and commercial style so many of them disdained.

While GE distanced its researchers from commercial activities, it looked upon their patents as commercial property. Scientists were encouraged to do work of the highest quality and to investigate relevant fundamental problems, but patent rights belonged to the company. As Whitney said categorically in 1909, "Whatever invention results from his [the laboratory staff member's] work becomes the property of the company. I believe that no other way is practicable."[75] So much for the myth that the laboratory was insulated from economic concerns.

The General Electric Research Laboratory was not the sole site for invention and research at GE, however. The company had a range of prob-

[74] Wise, "A New Role for Professional Scientists," p. 426.
[75] Willis R. Whitney, "Organization of Industrial Research," *Journal of the American Chemical Society* 32 (1910): 74, cited in Wise, "A New Role for Professional Scientists," p. 422.

lems to solve—theoretical, design, and operational. Other GE laboratories focused on problems outside the Research Laboratory's special competence. The Research Laboratory's work ranged widely, however, from manufacturing tungsten X-ray targets to fundamental studies of the structure of matter. After 1910, for instance, Langmuir worked on gas-filled incandescent lamps and electronic vacuum tubes. Others at the laboratory concentrated on early radio problems, X-ray radiography, insulation, and materials science, or metallurgy.[76] Much was left to others, however, especially in relationship to the invention, development, and design of generators, transformers, motors, and transmission systems. In Lynn, Massachusetts, before World War I, there was a research facility for Elihu Thomson and a laboratory concentrating on arc lamps. In Schenectady, New York, there were other laboratories in addition to the Research Laboratory: the General Engineering Laboratory (general development work); the Insulations Laboratory; the Illuminating Laboratory; and the Department of Consulting Engineering. General Electric also had research-and-development facilities in Erie, Pennsylvania; Harrison, New Jersey; Cleveland, Ohio; and Pittsfield, Massachusetts. By 1918 General Electric had twelve laboratories, the largest of which was the Research Laboratory, judging by the record of its expenditures.[77]

The consulting engineering department was Steinmetz's brainchild. He had declined the opportunity to head the Research Laboratory, perhaps because he and others believed him to be too individualistic for major administrative responsibilities. He may have considered himself more an engineer than a research scientist.[78] As described in 1912, the consulting engineering department should have appealed to engineer-inventors, much as the Research Laboratory was intended to attract and hold scientists with inventive talent. Steinmetz wrote that his department was to advise other GE departments about unusual phenomena, design requirements, or troubles it encountered. The department would also advise customers and work on problems of general interest to the electrical engineering profession. Steinmetz wanted each member of his department to choose his own work, and he emphasized that such a diversity of activities required an organization different from the "autocratic" system employed by traditional engineering departments.[79] Among the problems Steinmetz judged to be of general interest to the electrical engineering profession and to General Electric were high-voltage transmission, including the corona effect and instability. A loss of power in 1919 in the Chicago system of Commonwealth Edison was the type of customer problem to which he responded.[80] Steinmetz modeled the system with circuit equations in order to locate the problem.

[76] Birr, *Pioneering in Industrial Research*, pp. 41, 45, 47–48.

[77] Ibid., pp. 82–83.

[78] James E. Brittain, "C. P. Steinmetz and E. F. Alexanderson: Creative Engineering in a Corporate Setting," *IEEE Proceedings* 64 (1976): 1414.

[79] Ibid.

[80] Charles P. Steinmetz, "Some Problems of High-Voltage Transmissions," *AIEE Transactions* 31 (1912): 167–73; idem, "Instability of Electric Circuits," ibid. 32 (1913): 2005–21. For an example of how Steinmetz responded to a customer problem, see his article about a power loss in the Commonwealth Edison system in Chicago on 18 September 1919, "Power Control and Stability of Electric Generating Stations," ibid. 39 (1920): 1215–70.

The impressive development of metallic filaments in Germany notwith-standing, the head of Siemens & Halske in 1911 viewed American research achievements in the electrical industry with envy. Wilhelm von Siemens noted that as a country of thinkers and scientific researchers, Germany had risen to first rank as an industrial power, but that America's private enterprise system cultivated an uninterrupted stream of impressive research results. Von Siemens acknowledged the role in Germany of the universities, the *Technischen Hochschulen,* and the newly established Kaiser Wilhelm Gesellschaft, but feared that his country would fall behind unless other measures were taken and institutions founded.[81]

Siemens & Halske had a tradition of laboratory research, but these activities remained dispersed among the company's various works, or plants, until 1920, when a central laboratory was established. Research there, as at the General Electric Research Laboratory, was intended to be less tied to production and routine improvements than was the case in the other laboratories. At first, the central laboratory was called the Physical Chemistry Laboratory, but then the company changed the name to the Research Laboratory of the Siemens & Halske Company. Dr. Hans Gerdien, formerly a lecturer at Göttingen University, became director of the central laboratory, but because the managers of the different works preferred to maintain their specialized laboratories, he had a difficult time centralizing the basic research. To coordinate the various research centers, the company in 1919 named D. Harries to head a central office for scientific and technical research (Zentralstelle für Wissenschaftlich-Technische Forschungsarbeiten).[82]

Research laboratories provided institutional structures that supported and increased the momentum of the evolving electrical supply systems. Further reinforcement arose from the growth of professional societies with their periodicals, annual meetings, and encouragement of professional education. The professional societies, like the manufacturers and their laboratories, directed the attention of scientists and engineers to the problems of general urban systems and high-voltage transmission.

Paradoxically, considering the subsequent history of the organization, the American Institute of Electrical Engineers was established in 1884 by men primarily concerned with telegraphy. By 1900 the institute's emphasis had shifted to what the Germans aptly named *Starkstrom* ("heavy current"), in contrast to *Swachstrom* ("light current"). Heavy current involved light and power; light current was used essentially in telegraph communications before the wireless assumed importance about 1900. Of the twenty-five men who signed the call for formation of a national electrical society, twenty-one were telegraph inventors or manufacturers or were generally classified as being in the "telegraph business."[83] The first president of the AIEE was Norvin Green, president of the Western Union Telegraph Com-

---

[81] Ferdinand Trendelenburg, *Aus der Geschichte der Forschung im Hause Siemens* (Düsseldorf: VDI, 1975), p. 45.

[82] Von Siemens, *House of Siemens,* 2: 35–36; Trendelenburg, *Geschichte der Forschung,* pp. 45–48.

[83] Donald McNicol, "Telegraph Men Founders of the A.I.E.E.," *Electrical Engineering* 53 (May 1934): 675.

pany. Others among the early officers of the institute were Western Union managers or executives.[84]

The AIEE soon achieved recognition as the professional organization of U.S. electrical engineers. Electrical inventors and manufacturers held the office of president during the institute's first decade, but electrical engineers or teachers of electrical engineering dominated the office during its second decade. Of the eight presidents who served during the second decade, Edwin J. Houston (1893–1895) taught at Central High School in Philadelphia, Pennsylvania; Louis Duncan (1895–1897) taught electrical engineering at M.I.T.; Francis Crocker (1897–1898) received a Ph.D. in electrical engineering from Columbia University before teaching there; Arthur E. Kennelly (1898–1900) was appointed professor of electrical engineering at Harvard University two years after serving as AIEE president; Carl Hering (1900–1901) was a professor at the University of Pennsylvania and studied at Darmstadt under Kittler; Charles Steinmetz (1901–1902) held a professorial position at Union College in Schenectady, New York, while a GE engineer; and Charles Scott (1902–1903) took postgraduate courses at The Johns Hopkins University and later became a professor of electrical engineering at Yale after serving as AIEE president. As a group, these men bridged the academic and practical worlds.[85]

The publication of papers in the institute's *Transactions* enhanced the formation of a community of discourse centered on critical problems. A classification of AIEE papers for the years 1884–1899 shows that the subjects most frequently discussed were alternating-current practice, traction, dynamos, motors, and incandescent lighting. This concentration accords well with the focus of problem-solving found during the period of "the battle of the systems" and its resolution. Only three papers dealt explicitly with polyphase systems (as compared to fifty-two on alternating-current practice), but, as noted, the general rubric "alternating current" included single-phase and polyphase systems.[86]

Other periodicals contributed to the identification of problem areas for electrical engineers. The leading trade, as contrasted with professional society, journal was *Electrical World*. The forebear of this journal, *The Operator*, was founded in 1874 and was published for telegraphers. Reflecting the same changes as occurred in the character of the AIEE leadership, *The Operator* broadened its purview in 1883 to include electric lighting and changed its name to *The Operator and Electrical World*. James H. McGraw, publisher of a rival journal, the *American Electrician*, bought out *The Operator and Electrical World* and another journal, the *Electrical Engineer*, and by 1906 had combined them under the title *Electrical World*. The McGraw Publishing Company also owned the *Street Railway Journal*. Besides McGraw's publi-

[84] A. Michal McMahon, "Corporate Technology: The Social Origins of the American Institute of Electrical Engineers," *IEEE Proceedings* 64 (1976): 1385.

[85] "Great Names behind the Institute." *Electrical Engineering* 53 (1934): 786–825.

[86] Charles F. Scott, "The Institute's First Half Century," ibid., p. 666. I have assumed that the listing "A.I.E.E. Papers—1884–99 and 1929–33" refers to papers published in the *AIEE Transactions*. The second category on the list, "miscellaneous and theory," is too inclusive and vague to be ranked with the others. Robert Belfield, who analyzed issues of the *AIEE Transactions* dating from 1887 to 1904, found discussion of polyphase problems in many of these articles.

cations, there were numerous other trade journals in the electrical field by 1900. Other professional societies, such as the National Electric Light Association, which organized utility engineers and managers and was the organization of Edison light and power companies, also held meetings and published papers. The preferred journal for scholarly papers dealing with matters new and outstanding, however, remained the *AIEE Transactions* because of its professional prestige.[87]

In Britain, the professional society for electricians and electrical engineers had earlier origins than the AIEE. In May 1871 a group of men associated with telegraphy met in London and founded the Society of Telegraph Engineers. With the increase in interest in electric lighting, however, the society in 1880 added "and of Electricians" to its name, and in 1889 changed the name to Institution of Electrical Engineers (IEE). It was recognized as the professional society for British electrical engineers. Professor William Ayrton of the City-and-Guilds-sponsored Central Institution, South Kensington, edited the *IEE Journal*.

As in the United States, the latest news in the field was supplied in part by trade journals. Among these were *The Electrician*, founded in 1878; the *Telegraphic Journal and Electrical Review*, established in 1872; the *Electrical Engineer* (London), begun in 1882; and *Lighting* (later *The Electrical Times*), which first appeared in 1891. General publications such as *The Engineer* and *Engineering*, which were founded in 1856 and 1866 respectively and enjoyed an international reputation, also reported developments in electrical engineering.[88]

In Berlin in 1879 a group organized by Postmaster General Heinrich von Stephan and stimulated by the support of Werner von Siemens founded the Elektrotechnischer Verein zu Berlin, which, beginning in 1880, published a periodical entitled *Elektrotechnische Zeitschrift*. Another early periodical, the *Zeitschrift für angewandte Elektrizitätslehre* (1880), later renamed *Centralblatt für Elektrotechnik* (1883), merged with the *Elektrotechnische Zeitschrift* in 1889 to become the leading professional journal in the field. Among other German professional electrical engineering societies was the Elektrotechnische Gesellschaft founded in Frankfort on the Main. In 1893 the Verband Deutscher Elektrotechniker was formed through a merger of local electrical societies, thereby establishing a national society.[89]

[87] Scott, "The Institute's First Half Century," p. 667.

[88] For this account of the Institution of Electrical Engineers, I have drawn upon Percy Dunsheath, *A History of Electrical Power Engineering* (Cambridge, Mass.: M.I.T. Press, 1962), pp. 319–32. The authoritative work is Rollo Appleyard, *History of the Institution of Electrical Engineers, 1871–1931* (London: Institution of Electrical Engineers, 1939).

[89] Dettmar, *Starkstromtechnik*, pp. 287–91; and Sigwart Ruppel, ed., *Elektrotechnische Gesellschaft Frankfurt, 1881–1931: Geschichtstafeln der Elektrotechnik* (Frankfort on the Main: Elektrotechnische Gesellschaft, 1931?).

CHAPTER VII

Berlin: The Coordination of Technology and Politics

DURING the late-eighteenth, nineteenth, and early twentieth centuries, two waves of industrialization swept over the Western world. The first, commonly known as the Industrial Revolution, impressively transformed regions of England and Scotland; the second, occasionally called the Second Industrial Revolution, changed newly united Germany and the United States. The 1870s are often cited as the decade of transition from the first to the second industrialization phase. The British Industrial Revolution is usually associated with a system of production and transportation that was heavily dependent on coal, steam, and iron; the rapid industrialization of Germany and the United States involved these components, but steel, electricity, and man-made chemicals, both organic and inorganic, were layered on the earlier technologies as well. The second era of rapid industrialization is not as well understood or defined as the first, but when future historians probe its phenomena and structure, electric power—especially high-voltage transmission and the universal system of electric light and power made known to the world at Frankfort and Niagara Falls—will figure intricately and prominently in their interpretations and presentations. Just as Manchester, Birmingham, and other British cities were the centers of the first wave of industrialization, certain cities will be clearly identified as the sites of the second. Among the urban centers of the second wave will surely be Chicago and Berlin; among the cities by-passed will be London. Consideration of the electrification of the first two before 1915 and the laggard development of the third will provide some understanding of urban industrialization in the modern era.

During the four decades between the founding of the German empire and the outbreak of World War I, Berlin, the empire's capital, was a dynamic urban center. It became the leading industrial and commercial city on the Continent, and in wealth and population ranked behind only London and Paris. With regard to administration and organization, contemporaries named it the model city of Europe and praised its citizens for the remarkable burst of energy that transformed a provincial capital of Prussia into a cosmopolitan city of international eminence.[1] The municipal au-

[1] A. Shadwell, *Industrial Efficiency*, 2 vols. (London: Longmans, 1906), 1: 159–60; and Sigfrid von Weiher, *Berlins Weg zur Elektropolis: Technik-und Industriegeschichte an der Spree* (Berlin: Stapp, 1974), pp. 83–134.

thority was widely judged the ultimate in enlightened, efficient, and socially purposeful government in a progressive era that looked to its cities as standard-bearers in an urbanizing and industrializing civilization. On the Continent, and especially in Germany, a city was "idealized as a great family," organized for economic, political, and social ends and its government was seen as the means for obtaining these ends.[2] Among the means were public utilities, including transportation, water and gas, waste disposal, and lighting. By World War I, a hallmark of the modern progressive city was cultivation of a new and subtle form of light and power—electricity. Berlin was not found lacking.

Because the city government in Berlin so effectively regulated the private utility that provided the metropolis with light and power, and because the history of technological change in the city cannot be understood without taking the regulatory activities of the government into account, a brief description of the government's structure and functions is needed. In 1881 Berlin became an administrative area distinct from the Prussian province of Brandenburg, and subsequently the civil government of Berlin was organized into two major divisions, the *Magistrat* and the *Stadtverordnetenversammlung* ("common council"). The *Magistrat* consisted of an *Oberbürgermeister* ("chief mayor"), a *Bürgermeister* ("mayor"), and a *Stadtrat* ("city council") of thirty-two members, fifteen of whom were paid administrators and seventeen of whom served without pay. The 144 members of the *Stadtverordnetenversammlung* were chosen by manhood suffrage. For general decisions, the *Magistrat* and *Stadtverordnetenversammlung* met in common and sometimes named committees that were representative of the whole. These committees had jurisdiction over public services such as water supply, the lighting and cleaning of the streets, and drainage. Detailed contractual arrangements with the private company supplying Berlin with electric light and power were executed by the *Magistrat*.

Not only was Berlin progressive but it was industrial. As in Chicago, this fact deeply affected the city's electric utility. London and other European capitals had become world centers earlier, when cities were the focus of government, commerce, and religious activities. Berlin's formative years as a national capital coincided with Germany's rapid industrialization and rise to technological preeminence following the Franco-Prussian War and the country's unification. Berlin, more than Vienna, Paris, or London, was shaped by industrial imperatives and technological opportunities. By 1900 the quaint two-story houses so characteristic of Prussian Berlin had given way to business blocks that housed not only commercial enterprises but the headquarters of the industrial corporations and financial institutions (including the Reichsbank, Deutsche Bank, Diskonto Gesellschaft, and Jacob Landau) that served them. The administrative boundaries of "old Berlin," with a population of about two million, remained as they had been since 1860, encompassing only about twenty-nine square miles—six miles across and about five and a half miles from north to south. Industry, however, mushroomed in so-called "greater Berlin," radiating ten miles from city

---

[2] Frank S. Hoffman, "Municipal Activities in Germany," *Outlook* 58 (1898): 1063. I am indebted to Alan Steiner for calling my attention to the Hoffman article and to sources about Berlin's transportation systems.

center. More than 50 percent of the working population of old Berlin was employed by industry, and the percentage in greater Berlin, with a population around three million, increased as new industries and factories found ground there. Berlin manufactured wool, worked iron and steel, turned out heavy machinery (including steam engines and locomotives), produced sewing machines and bicycles, organized chemical processes, and designed and assembled heavy and light electrical equipment for Germany and the rest of the world. When burgeoning industry within the city threatened to demean the architecture and the environment, the municipal authorities constructed a great canal south of the city to connect the Spree and Havel rivers and provide a transport artery that would draw industry away from city center.

Berlin was not only an industrial city but one whose industry was notably electrical. Two of the world's leading manufacturers of electrical machinery, Siemens & Halske and Allgemeine Elektrizitäts-Gesellschaft (AEG), had central offices and factories in Berlin and its immediate surroundings. About 50 percent of the personnel of Germany's electrical industry worked there until World War II. With good reason, then, it has been called the *Elektropolis*.[3] This burgeoning industry helped maintain the vitality of Berlin's—and Germany's—economy during periods of mild recession in the three decades preceding World War I. Siemens & Halske's and AEG's presence also furthered the electrification of industry and transportation in Berlin. The histories of both companies involved at one time or another not only electrical manufacturing but also electrical supply in the city. In 1883 Siemens & Halske was party to the founding of Deutsche Edison Gesellschaft, which in time metamorphosed into AEG, which in turn, in 1887, founded the Berliner Elektricitäts-Werke (see pp. 68–77 above).

The founder of Siemens & Halske, Werner von Siemens (1816–1892), his associates, and the company stimulated an ethic of science and professionalization as well as invention and industry in Berlin.[4] In 1879 von Siemens was instrumental in the establishment of the Berlin Elektrotechnischer Verein (Electrotechnical Association), which in turn, in 1894, stimulated the founding of the Verband Deutscher Elektrotechniker (later the leading professional organization for the field in all of Germany). Von Siemens also urged the government to establish in Berlin an institute de-

[3] See von Weiher, *Berlins Weg zur Elektropolis*. On AEG in Berlin, see *75 Jahre AEG*, a booklet published by the firm in 1958, especially pp. 29–30. On the Berlin electrical industry after World War I, see Peter Czada, *Die Berliner Elektroindustrie in der Weimarer Zeit* (Berlin: Colloquium, 1969).

[4] Biographies of Werner von Siemens and histories of Siemens & Halske are numerous, due in part to the existence of the company's archives in Munich. Among the works are Werner von Siemens, *Inventor and Entrepreneur: Recollections of Werner von Siemens* (1892; reprinted ed., London: Lund Humphries, 1966); Georg von Siemens, *History of the House of Siemens*, 2 vols. (Freiburg/Munich: Alber, 1957); Jürgen Kocka, *Unternehmensverwaltung und Angestelltenschaft; am Beispiel Siemens 1847–1914: Zur Verhältnis von Kapitalismus und Bürokratie in der deutschen Industrialisierung* (Stuttgart: Klett, 1969); Sigfrid von Weiher, *Werner von Siemens: Ein Leben für Wissenschaft, Technik und Wirtschaft* (Göttingen: Musterschmidt, 1970); and Sigfrid von Weiher and Herbert Goetzeler, *The Siemens Company: Its Historical Role in the Progress of Electrical Engineering* (Berlin and Munich: Siemens, 1972). See also Jürgen Kocka, "Siemens und der Aufhaltsame Aufstieg der AEG," *Tradition* 17 (1972): 125–42; and, primarily for later developments, Czada, *Die Berliner Elektroindustrie*.

voted to basic research. In 1883, in an appeal to the Prussian government, he wrote, "Research is the firm foundation of technological progress; a country's industry has no hope of attaining an international, leading position and sustaining itself unless it is in the forefront of scientific research."[5] Siemens offered to finance the building of such an institute with funds that had been left to him in the estate of his brother, William. The German Reichstag eventually accepted a modification of Siemens's proposal and established the Physikalisch-Technische Reichsanstalt (Imperial Physical-Technical Institute), of which Hermann von Helmholtz (1821–1894), then Germany's leading physicist, became the first president in 1887. Besides these contributions to the scientific life and intellectual spirit of Berlin, Werner von Siemens was responsible for the installation of chairs in the new field of electrical engineering at German *Technischen Hochschulen* (see pp. 144–45 above).[6] These institutions, their engineers, and their scientists helped ensure that technology in general, and such advanced fields as electrical engineering specifically, would flourish in Berlin.

By 1900 Berlin's Allgemeine Elektrizitäts-Gesellschaft was a larger manufacturer of electrical machinery and apparatus than Siemens & Halske. It had 17,300 employees, while Siemens had 13,600; it had 60 million marks in share capital, while Siemens (including its foreign subsidiaries) had 54.5 million.[7] When founded in 1887, AEG was primarily a financier and operator of electrical utilities and a maker of incandescent lamps. In contrast, Siemens was a manufacturer of equipment. Within a decade, however, AEG had expanded its manufacturing to include power equipment and had introduced a line of polyphase machinery. The company continued to finance and build central power stations and during the 1890s became especially active in the construction of electric streetcar systems. It acquired the rights to the designs of Frank Sprague, the American pioneer in electric traction. The company followed a policy of financing its customers, whether they were horsecar companies converting to electricity or companies with new franchises. In cooperation with banks in Berlin and elsewhere, it also financed and constructed power plants for electric supply utilities and industrial facilities that used heavy-power equipment. In 1888, for example, with the help of several banks, AEG established Aluminum-Industrie A.G. in Neuhausen, Switzerland. By 1900 AEG had established 248 power plants with a total capacity of 210,000 h.p.[8] AEG also grew by amalgamation, the establishment of *Interessengemeinschaft* (ad hoc common-interest organizations), and cartel (price-control) arrangements. With the Union Elektrizi-

---

[5] Quoted in Sigfrid von Weiher, *Werner von Siemens: A Life in the Service of Science, Technology and Industry* (1970; Eng. trans., Göttingen: Musterschmidt, 1975), p. 73.

[6] Von Siemens, *House of Siemens*, 1:115.

[7] If, however, the 7,400 employees of Schuckert & Co. are included, then the Siemens-controlled enterprise was larger. In 1903 Siemens & Halske and Schuckert & Co. of Nürnberg established a holding company, Siemens-Schuckertwerke. The new entity operated the electric power equipment manufacturing facilities of the two companies. The shareholding, management, and profits of Siemens-Schuckertwerke were weighted slightly in favor of Siemens & Halske. Siemens & Halske held 45.05 million marks of the holding company's capital; Schuckert, 44.95 million. Von Siemens, *House of Siemens*, 1:192, 196.

[8] Conrad Matschoss, "Die geschichtliche Entwicklung der Allgemeinen Elektricitäts-Gesellschaft in den ersten 25 Jahren ihres Bestehens," *Beiträge zur Geschichte der Technik und Industrie* 1 (1909): 62.

täts-Gesellschaft (founded by Ludwig Loewe and Thomson-Houston International), AEG formed an *Interessengemeinschaft* that divided the market and profits in the electric streetcar field, where Union had a firm position. In 1903 Emil Rathenau, the founder of AEG, reached an agreement with General Electric in America to divide their world markets: AEG would continue to be preeminent in Europe; GE, in North America. AEG and GE also agreed to cooperate in the development of the Riedler-Stumpf and Curtis steam-turbine patents when steam turbines began displacing reciprocating steam engines in power stations. AEG further strengthened its position in the turbine field by exchanging stock with the Swiss electrical manufacturing firm Brown, Boveri & Company, which held rights to manufacture the Parsons turbine.[9]

AEG's influence among German utilities through stock ownership was extensive. Because the control was that of a supplier over its market, the relationship resembled vertical integration. In 1911 AEG owned some part of 114 public power plants (*Elektrizitätswerke*), and these supplied 31 percent of the connected electric load in Germany. Siemens-Schuckert owned a part of 80 power plants, and these supplied 6.3 percent of the connected load. The two manufacturers held a disproportionate interest in the relatively few large-capacity plants rather than in the large number of small ones. Power plants with a capacity of more than 10,000 kw. supplied 34.6 percent of the connected load, and the two manufacturers owned an interest in the twelve of these that supplied 26.4 percent of the load.[10] Much of the capacity of the central stations in which AEG had invested was located in Berlin or its suburbs and was operated by the AEG subsidiary, the Berliner Elektricitäts-Werke (BEW).

Considering the investment network and AEG's pivotal role in it, it is not surprising that Emil Rathenau, the company's founder, was known in Germany as the inventor of the principle of market creation through investment financing. Rathenau also founded and presided over the Berliner Elektricitäts-Werke until his death.[11] Like that of Werner von Siemens, Rathenau's name looms large in the history of the electrical industry in Germany and also in the establishment of Berlin as its *Elektropolis,* but while von Siemens came to represent invention, engineering, and industrial science, Rathenau stood for the powerful and widely influential interaction of investment capital, industrial enterprise, and highly organized marketing. His son, Walther (1867–1922), succeeded him as head of AEG in 1915 and went on to display organizational abilities on an even greater stage than had his father. During World War I, Walther directed the allocation of Germany's resources, and shortly thereafter he served as minister of reconstruction (1921) and foreign minister (1922). He was assassinated by nationalistic and anti-Semitic fanatics who opposed, among other things,

[9] Ibid., pp. 64–65.

[10] Helga Nussbaum, "Versuche zur reichsgesetzlichen Regelung der deutschen Elektrizitätswirtschaft und zu ihrer Überfuhrung in Reichseigentum, 1909 bis 1914," *Jahrbuch für Wirtschaftsgeschichte* (Berlin, 1968), pt. 3, pp. 137–38.

[11] See pp. 66–78 above for a discussion of Rathenau and the founding of AEG and BEW. Biographies of Emil Rathenau include A. Riedler, *Emil Rathenau und das Werden der Grosswirtschaft* (Berlin: Springer, 1916), and Felix Pinner, *Emil Rathenau und das elektrische Zeitalter* (Leipzig: Akademische Verlagsgesellschaft, 1918).

Figure VII.1. *Headquarters of Berliner Electricitäts-Werke and Allgemeine Elektrizitäts-Gesellschaft, Berlin, 1891. From* Offizielle Zeitung . . . Frankfurt am Main 1891, *p. 163.*

his desire to fulfill the reparations terms of the peace settlement. Walther, who was passionately committed to philosophy and the arts, probably encouraged his father to appoint the famous architect Peter Behrens (1868–1940) as art and architectural adviser to AEG. Behrens designed the famous 1909 AEG turbine factory in Berlin. He also designed various electrical appliances manufactured by the company.[12]

No account of the Berlin environment in which the Rathenaus nurtured AEG and BEW would be adequate without due emphasis on the role of the investment banks (*Kreditbanken*) there. Consideration of these and their relationship with the electrical industry focuses attention on Georg von Siemens. A cousin of Werner's, Georg rose through the ranks to become a head of the Deutsche Bank in Berlin after having worked for Siemens & Halske. The Deutsche Bank was one of the leading *Kreditbanken*, which were, in effect, a combination of commercial and investment banks, banks that have been characterized by many historians as the centers of great industrial influence—almost control—during the rapid industrialization of Germany after 1871.[13] Georg von Siemens advised and assisted in the financing of both Siemens & Halske and Rathenau's enterprises until AEG

---

[12] Books and articles about and by Walther Rathenau are numerous, but most focus on his political views, economic and social philosophy, and aesthetic interests. Hermann Brinckmeyer, *Die Rathenaus* (Munich: Wieland, 1922), does direct attention to his AEG association.

[13] Hugh Neuburger, "The Industrial Politics of the *Kreditbanken*, 1880–1914," *Business History Review* 51 (1977): 190–207, challenges the argument that the banks dominated the bank-industry relationship in Germany, a thesis advanced by Rudolf Hilferding, *Das Finanzkapital* (Vienna: Verlag der Weiner Volksbuchhandlung, 1923), and others. Neuburger sees the relationship as one among negotiating, autonomous powers. This view is also advanced in Kocka, *Unternehmensverwaltung*, p. 431.

became a strong manufacturing competitor of the Siemens firm. In 1896 Georg resigned his chairmanship of the board of AEG when his plan, which was for AEG to function primarily as a financer of utilities and power plants and Siemens as a manufacturer, was obviously no longer viable. Georg von Siemens had great respect for Rathenau, believing him to be the best financial and commercial head in the industry, but AEG's managers believed that Georg, because of his relatives, was too deeply involved in a conflict-of-interest situation. For his part, Georg thought that Rathenau was overextended as financier, manufacturer, and operator of electrical utilities. AEG made a profit from selling heavy machinery, selling the utility shares it took in payment for the equipment, and then from operating some of the utilities. Although Georg von Siemens, with his German sense of order and cooperation, was frustrated in his attempt to balance the activities of the two leading manufacturers, he was able to bring about the transformation of Siemens & Halske from a nonshare, family-owned enterprise to a stock company whose shares were owned by the family. This made it possible for banks such as the Deutsche Bank, which made loans to Siemens & Halske, to place members on the board of directors (*Aufsichstrat*) of the share company.[14]

Motivated by industrial needs and the desire to seize technological opportunities, Berlin developed an exemplary transportation system. As was the case in Chicago, Berlin's electric utility supplied the power when much of the city's transportation system was electrified. Earlier, the Prussian state became involved in the planning and funding of a transit system in order to facilitate military mobilization by railway. The location of industry, worker settlements, and middle-class suburban neighborhoods, as well as military considerations, shaped the structure of the state's transportation network. Berlin, like London and Paris, was a railway nodal point; twelve main lines converged on it. Concerned, as London had been, that additional main-line railway stations in center city would mar and congest, Berlin ringed itself with main-line stations. These were connected by a *Ringbahn* ("ring railroad"), which opened in sections beginning in 1872 and which, after 1882, fed into the *Stadtbahn* ("city railway") that extended across Berlin from east to west. Connecting with the *Stadtbahn*, streetcar lines honeycombed the city. The elevated portion of the *Stadtbahn* was carried about twenty feet above the streets, and its stations are rich in historical connotations—Friedrichstrasse, Zoologischer Garten, and Alexanderplatz. Despite both the excellence and the original excess capacity of the *Stadtbahn*, as well as the extensions of the streetcar lines, increased demand and growing congestion led in 1896 to the start of construction of an elevated and subway system across center city, mostly to the south of the *Stadtbahn*. The city demanded that the system go underground in the western section to eliminate noise and traffic in the choice residential areas in the vicinity of Charlottenburg. Earlier the city insisted that electric streetcars run off storage batteries in center city to avoid unsightly and dangerous overhead power lines. Because of technical problems with storage batteries, however, the requirement was subsequently eased. At the turn of the century the

---

[14] Von Siemens, *House of Siemens*, 1: 148–49, 154–58.

U.S. consul general in Berlin attested to the artistic beauty, the architectural charm, the fitness, and the general excellence of the Berlin systems of mass transit.[15]

As in other industrializing cities, electric light and power helped shape Berlin's architecture; deeply influenced the design of its factories, and workshops, and chemical plants; stimulated industrial growth; determined the location of the city's transportation systems; provided telephone and telegraph communication; and, when substituted for steam power, lessened noise and dirt. In short, electrification affected the way in which workers labored, management organized, and Berliners lived. Because of this, the private company that supplied most of the electric light and power for Berlin until 1915, and the municipal government that regulated electrical supply, shaped the history of Berlin.

On the eve of World War I, the Berliner Elektricitäts-Werke (BEW) ranked as one of the world's leading electric supply utilities. Engineers and managers looked to it, along with the Commonwealth Edison Company of Chicago, as pacesetters in the establishment of world standards, both technical and commercial, for the electric supply industry. The city engineers of Melbourne, Australia, included Berlin and BEW on their world tour in 1912, observing that "on the Continent one naturally visits Berlin, being electrically the most important city, not only from an electric supply point of view, but on account of the fact that the two most powerful electrical manufacturing concerns in Europe have their factories here."[16] In 1913 Georg Klingenberg, head of AEG's power-plant division, brought the details of Berlin's electric supply system to the attention of the technical world in his analysis of the utilities in the German capital, Chicago, and London.[17] Samuel Insull of Commonwealth Edison included the Berlin system in his comparisons of the technical and economic characteristics of leading utilities. In Germany, BEW was the largest of the urban utilities.

Those who compared the state of electrification in various cities found that Berlin had lit the streets of the metropolis beautifully, in part with arc lighting from BEW and in part with gaslight from the city-owned plant. The delegation of engineers from Melbourne waxed eloquent. Berlin's streets, they said, were among the finest in the world—wide, smooth, and clean. Nowhere else was there a thoroughfare more beautiful than Unter den Linden as it stretched through the heart of the city and into the *Tiergarten*. Its lighting was supplied by newly introduced flame-arc lamps. The street was so smooth that "a considerable proportion of the youth, male and female, during their leisure, disport themselves on roller skates, incurring thereby some considerable risk from the rapid automobiles that abound everywhere." The leading thoroughfares, with or without skaters, had arc lights, and other streets had excellent gaslighting.[18]

[15] Frank H. Mason, "Transportation Problems and Progress in Germany," *U.S. Consular Reports*, no. 273 (June 1903), p. 176; idem, "New Electric Underground and Elevated Railway at Berlin," *Advance Sheets of U.S. Consular Reports*, 29 March 1902, p. 7.
[16] City of Melbourne, *City Electrical Engineers' Notes on Tour Abroad* (Melbourne, Australia, 1912), p. 13.
[17] Georg Klingenberg, "Electricity Supply in Large Cities," *The Electrician* 72 (1913): 398–400.
[18] City of Melbourne, *Engineers' Notes*, p. 38.

1900

70,2 Mill. kWh

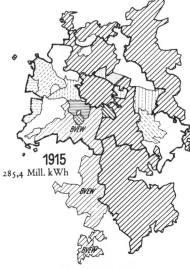

1915

285,4 Mill. kWh

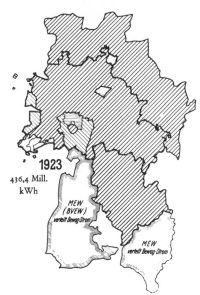

1923

436,4 Mill. kWh

Figure VII.2. Growth of supply areas in Berlin, 1900, 1915, and 1923. Area supplied by BEW = ▨. From Matschoss et al., 50 Jahre, p. 63.

As early as 1890 Berlin's utility was superior, in many particulars, to that in New York, the city where Edison had inaugurated central-station electric lighting. John Beggs, vice-president of the Edison Electric Illuminating Company of New York, paid a professional visit to Rathenau's Berlin facilities in the spring of 1890, and his reports glowed with admiration. In his estimation—and he had inspected stations in America and Europe—Berlin's central stations had "attained the greatest degree of perfection reached up to the present time." He found them, "architecturally, mechanically and electrically considered, models of neatness, efficiency, reliability, economy and permanency."[19]

The history of central-station supply in Berlin before the establishment of AEG or BEW, its fully owned subsidiary, has already been told (see pp. 66–76 above). In 1884 Deutsche Edison Gesellschaft (DEG), the Rathenau company that held Edison patents, established the Städtische Elektrizitäts-Werke (StEW) to take over the supply franchise that had been given to DEG by the city of Berlin. The new company built central stations in Berlin on Markgrafenstrasse and Mauerstrasse, opening them in 1885 and 1886 respectively. In 1887 StEW's name was changed to Berliner Electricitäts-Werke, and Rathenau founded AEG. AEG then took over the management of BEW, located the BEW offices in the same building with its own on Schlegelstrasse, and formed a common management, or executive, board (*Vorstand*) consisting of Rathenau, Oskar von Miller, and Felix Deutsch.

The tone of the annual report of the Berliner Elektricitäts-Werke was sanguine for a year and a half after this reorganization. The lighting of Unter den Linden, the broad thoroughfare from Pariser Platz to Spandauerstrasse, was greeted "[by] the citizens of Berlin and by the entire electrical industry with joy."[20] The rapid increase in demand emboldened the company's directors to install capacity beyond the load that could be immediately connected. They were not concerned that capacity would stand idle; the demonstrated "preference of the public for electric light, especially the lively nightlife of the city, the erection of numerous new buildings and the rising living standards of the city" assured them that the system's load would soon again reach the limits of capacity.[21]

BEW soon ranked as the largest of the German utilities. The first annual statistics published in the 1895 *Elektrotechnische Zeitschrift*, Germany's most authoritative journal in the field, showed the Berlin utility ranking well ahead of its closest rivals, Hamburg, Frankfort on the Main, and Munich. At that time BEW had four central stations; all supplied direct current, and they had a total generating capacity of about 9,900 kw. The municipal works in Hamburg had a capacity of about 2,400 kw. A decade later, BEW, with a capacity of 85,100 kw., continued to outpace other urban utilities. The Hamburger Elektricitätswerke followed with 25,400 kw. In 1913, on

[19] John Beggs to Board of Directors of EEI Co., New York, 8 April 1890, published in Payson Jones, *A Power History of the Consolidated Edison System, 1878–1900* (New York: Consolidated Edison Co., 1940), p. 331.

[20] Berliner Elektricitäts-Werke, *Sechster Geschäftsbericht der Actien-Gesellschaft Berliner Elektricitäts-Werke betreffend das Geschäftsjahr vom 1. Januar 1887 bis 30. Juni 1888* (Berlin: BEW, 1888), p. 1.

[21] Ibid., p. 2.

the eve of World War I, BEW was still ahead, with a capacity of 192,700 kw., compared to the Hamburg utility's 44,300 kw.[22]

Berlin, then, was an industrializing, well-ordered capital that sustained the development of modern technology, especially electrical technology. BEW used the most advanced light and power equipment and presided over the introduction of the general urban system that had been invented and developed during "the battle of the systems." AEG owned BEW and used it as a site for full-scale testing of new technology. Besides AEG engineers and managers, the experts at Siemens & Halske, by virtue of their presence in Berlin, stimulated BEW to excel. The Charlottenburg *Technische Hochschule* and the University of Berlin also enhanced the spirit of achievement. The municipal government was not, however, overawed by the technical, managerial, scientific, and financial power of its private utility. Berlin's officials drew upon the proud tradition of the Prussian civil service, demanding and receiving the same respect and authority that public officials of the state and national governments received. The strong Berlin government strove to direct and regulate its public utilities, including electricity, in order to fulfill its high standards and to satisfy the more broadly enfranchised electorate's increasing demand for public services. Always in the background, in an era of spreading municipal socialism, or municipal ownership of utilities, was the possibility that socialists and their allies in the *Stadtverordnetenversammlung* would take over the utility when its franchise expired. The socialists argued that government ownership would bring service that was more responsive to the electorate than that provided by the private utility. Until such an eventuality, however, the city was determined to tax the revenues of the utility in order to meet the increasing cost of social welfare.

Berlin before World War I, therefore, is an excellent setting for a study of the interactions of an effective, informed, and strong regulatory authority with a well-managed, well-financed, and technologically advanced private enterprise supported by various financial and banking interests. The various contracts negotiated between the city and the utility manifest not only the city's determination to share in the revenues of the private company but also the kind and extent of services the municipality believed its consumers—whether household, commercial, traction, or industrial—should have. In addition to the profit drive, the contracts reflect the utility's decisions about what was, or was to be, technically feasible. Finally, the arrangements made between the two parties manifest the technical style of a large-scale, capital-intensive, science-based enterprise. As in England, regulatory legislation could have retarded technological change by imposing conditions that reflected an outmoded state of technology or by imposing limits to growth, but in Berlin this was not the case. As in Chicago, politicians could have been so pliant as to have little effect on the utility, but, again, this was not true in Berlin. Berlin managed to coordinate technical and political power and create a working political economy.

---

[22] These statistics are taken from *Elektrotechnische Zeitschrift* 16 (1895): 223–26; 26 (1905): 24; and 34 (1913): 1450. Numbers have been rounded off to the nearest 100 kw. In the statistics for 1913, Düsseldorf-Reisholz and Gleiwitz rank ahead of Hamburg, but these are assumed to be area supply stations rather than urban utilities.

The first contractual agreement was reached in 1884 between the city and Deutsche Edison Gesellschaft and incorporated essential categories found in later agreements. Berlin required an income from the utility, and the contract specified that 10 percent of the utility's gross income and 25 percent of its annual net profit would go to the city. These percentages were calculated after a dividend of 6 percent on share capital was declared and deducted. In addition, the city would regulate rates by requiring *Magistrat* approval of prices for electricity supplied to customers. Because the city wanted the exciting new light for its streets and public buildings, the price for this service was specified. The utility was required to supply any customers who agreed to take service for three years.

The wording on compulsory purchase and the duration of the franchise caused debate and created problems. The Berlin negotiators undoubtedly recalled the confusing and generally negative impact of the compulsory-purchase clause enacted by the British Parliament in 1882. According to the Berlin-DEG agreement, the franchise was intended to extend for thirty years, but upon two years' notice after 1 October 1895 the city would be able to demand purchase at the tax value. If the city raised the taxes on the utility, the purchase price for the city then increased (a subtle control). The agreement specified as well that the utility had to set aside 2 percent of its gross income yearly until it had a depreciation and replacement fund amounting to 20 percent of the share capital. This provision guarded against a private concern's running down its plant before the time of compulsory purchase.

An interesting article in the 1884 contract provided that the city would have to pay 150 percent of the appraised value if the utility was taken over after fifteen years, and 3.33 percent more for each year less than the fifteen. For each year in excess of fifteen, the price would be reduced by 3.75 percent. The intent was to compensate the owners for too short a period in which to obtain a return on their investment. The effect was that after twenty-eight years Berlin could purchase the utility without paying a premium.[23]

What did the private company receive? In essence, the city conceded a monopoly on an inner-city area with a radius, centered on the Werdersch Market, of about half a mile. The area matched the economical distribution area of the early d.c. plant at the Pearl Street station in New York. In this area the utility could lay conductors under the street, thus also following the Pearl Street precedent. Considering the state of the technology, the area was appropriate. Technical change would soon extend the range of distribution, however, and then the company would want the franchise extended.

Before the two sides reached an agreement, the *Stadtverordnetenversammlung* engaged in heated debate. To some members, the granting of a monopoly to a private enterprise was a blow struck against progressive municipal practice. Others found it hard to believe that the city would consider cultivating an electric lighting system in private hands while it

[23] Hugo Meyer, "Municipal Ownership in Germany," *Journal of Political Economy* 14 (1906): 563; Robert C. Brooks, "Municipalization of the Berlin Electric Works," *Quarterly Journal of Economics* 30 (1916):192.

owned an operating gaslighting plant in which it had invested heavily. The persuasive argument in favor of the contract was the same as that used in England: because electric lighting was a new, unproven technology, private enterprise should take the risks, and the franchise-granting government should share the income and look forward to compulsory purchase after the utility had been tried and proven. Taxpayers' money was not—it was thought then—to be invested in technical innovations of a commercial kind. In time a large majority of the common council accepted the Berlin-DEG agreement. According to the enthusiastic endorsement of one Berlin newspaper, "Nothing characterizes the significance of a new creation better than that at the moment of birth it is recognized as a common good for all." Other cities and utilities throughout Germany patterned their relationships on the Berlin arrangement.[24]

In later years BEW negotiated contractual revisions as Berliners outside the original monopoly zone sought service. Requests for extension of service brought a response from the municipal authorities. The franchise service area was extended to provide a load for the direct-current stations on Markgrafenstrasse and Mauerstrasse and for the Spandauerstrasse station, which opened in 1889. In 1890, however, the *Magistrat* refused to allow expansion of BEW's facilities beyond the three central stations already in operation and another under construction (Schiffbauerdamm). Furthermore, the city limited expansion of BEW's total plant capacity to 28,000 h.p. This was probably done to protect the lighting market of the city-owned gasworks. The rapidly increasing demand for electricity, however, soon brought strong pressure against these constraints. Within the territory of supply, BEW made adjustments by thickening its network of distribution lines and, when the distances of transmission exceeded good d.c. practice, by building storage-battery substations that could be charged from the central-station generators during the low-load daylight hours and then be used to supplement the supply during the heavy-load evening hours. But the moratorium on the building of central stations and the limits on plant capacity frustrated the drive of BEW's managers and engineers to enhance the development of electric light and power.

The demonstration of a universal supply system at the Chicago exposition in 1893 and of the technology of high-voltage power transmission at Lauffen-Frankfort in 1891 made obsolete—from the viewpoint of BEW's engineers—the 1884 contract between the city of Berlin and BEW. In 1896, polyphase equipment was introduced at a new central station on Schiffbauerdamm (Fig. VII.3). At the central station on Markgrafenstrasse, engineers installed transformers and converters to change 3,000-volt polyphase current from Schiffbauerdamm into 220-volt direct current for distribution. The transformation of the Markgrafenstrasse station—the first central station in Germany—into a substation was indicative of the shift to the universal system. The new technology challenged old legislation, and the contradiction between technology and legislation sharpened further as

---

[24] Conrad Matschoss, "Geschichtliche Entwicklung der Berliner Elektricitäts-Werke von ihrer Begründung bis zur Übernahme durch die Stadt," *Beiträge zur Geschichte der Technik und Industrie* 7 (1916): 8; Conrad Matschoss et al., *50 Jahre Berliner Elektrizitätswerke, 1884–1934* (Berlin: VDI Verlag, n.d.), p. 12.

Figure VII.3. *First polyphase generator in Berlin, Schiffbauerdamm central station, 1896. Courtesy of Berliner Kraft-und Licht AG (formerly BEW).*

the central-station managers in Berlin sought to supply the recently electrified urban transit system with power. BEW wanted to build a large polyphase station outside city center in order to satisfy the demand not only for lighting but for stationary and traction power. The utility's goal reflected the interest of AEG, the manufacturer-owner, in introducing the new polyphase equipment its engineer Michael Dolivo-Dobrowolsky had developed. Emil Rathenau considered building a "gigantic central station" on the Spree River about ten miles from the city where real estate was cheaper, cooling water was available, and coal delivery and ash removal were easier. The station would supply Berlin by using three-phase transmission to reach substations and the center-city distribution system.[25]

BEW's negotiations with the municipality to obtain the rights to fulfill these technical possibilities were drawn out and difficult. Discussion extended from the summer of 1897 until the end of 1898. Besides the utility's determination to expand by fulfilling technical possibilities and to obtain profits from related economies, there was the underlying issue of private versus public ownership, which the Social Democrats in the city government kept alive, and the related practical question of when the city should exercise the power of purchase. BEW wanted a clear and precise understanding about the purchase option so that it could plan its expansion and

[25] "The History of a Great Electrical Company," *Electrical Review* 37 (1895): 627.

establish profitable rates. On the other hand, the majority of the *Magistrat* and *Stadtverordnetenversammlung* wanted the utility to flourish so that consumer needs would be met and the utility would be sound when taken over. The situation called for constructive negotiation rather than adversary confrontation.

The rapid increase in electric street railway load after 1896 posed a major issue for negotiation. In 1879 Werner von Siemens had displayed an electric locomotive at a Berlin trade fair, but the importance of this mode of transportation did not become clear until the 1890s. In 1894 the power load (traction and stationary) on BEW was only one-tenth of the total; in 1900 the traction load matched the combination of stationary power and light load. The city wanted the rate for traction power to be low in order to facilitate the transportation of lower-income groups, especially the increasing number of industrial workers. The utility wanted to increase the power load in order to utilize enlarged plant capacity more fully, improve the load factor, and lower unit capital costs. The two parties finally agreed that all electric-streetcar franchises granted by the city would require that the enfranchised enterprise take electricity from BEW. In return, the utility agreed not to exceed charges of 10 pfg./kwh. for the electric-streetcar power supply; this was substantially lower than the 16 pfg./kwh. that had been charged since 1894 for power current, a rate which had scarcely exceeded costs. The company took the risk that economy of scale and improved load factor would lower costs below the price. Each five years the price was to be renegotiated, but it was not to be higher than that charged by the three largest German cities that supplied current for electric traction from steam-generating plants.

On the issue of the area of BEW's supply monopoly and the duration of its tenure before compulsory purchase was possible, the 1899 contract provided a monopoly for all of Berlin and specified that the compulsory-purchase option could not be exercised before 1915. The ceilings placed on generating capacity were raised but not eliminated in 1899, again suggesting a reluctance on the part of various interests represented in the city government to see electric light and power rapidly displace gas lighting. The agreement of 1899 stated that capacity inside Berlin should be limited to 42,500 kw. and that power for the city originating outside Berlin should not exceed 37,000 kw. This provided for the high-voltage polyphase capacity BEW wanted outside the city and at the same time assured compulsory purchase of these facilities along with those in the city (see Fig. VII.4). The contract also specified that all electric generating plants and supply franchises held by AEG as well as BEW within a radius of 30 kilometers from the center of Berlin would be subject to purchase in 1915. By means of this understanding, the two parties avoided the possible dilemma of the private enterprise extending its facilities beyond the political jurisdiction of the government authority holding a purchase option—a chronic problem in London before World War I.

Increasingly burdened by expenditures for city services, Berlin took advantage of the renegotiation of the contract to raise its share of BEW's income. Berlin continued to take 10 percent of the gross income, but doubled its share of the company's profits. The 1899 contract specified that the municipality would receive 50 percent of the net profits after a 6 percent

Figure VII.4. *BEW polyphase system,*
1900: Central stations (△); motor-
generator conversion substations (○); and
the city limits of Berlin (– – –). From
Matschoss et al., 50 Jahre, p. 127.

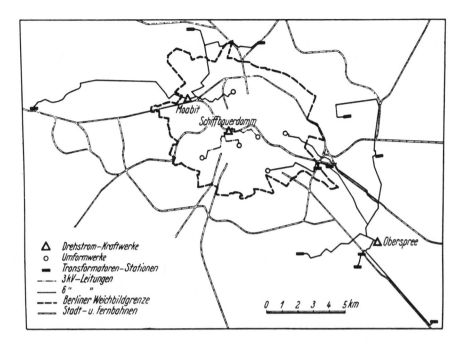

dividend on stock shares had been declared and deducted.[26] (In 1911–12 the total income of the city from BEW was 7 million marks.)

The city also specified that when profits (*Reingewinn*) exceeded 12.5 percent of the share capital, customer rates would be reduced by up to 10 percent. By 1913 (a year for which comparative rates with other major world cities are available) BEW, with city approval, charged 40 pfg./kwh. for lighting, with discounts based on the annual bill for current; 16 pfg./kwh. for night service and advertising; and 11 pfg./kwh. for power. (There was no charge for meters.)[27]

Differential rates favoring industrial consumers, both traction and stationary, kept light a luxury for private consumers but spurred industrialization. In 1914 only 6.6 percent of Berlin households were connected for electricity supply; only in 1927 did the number reach 50 percent (see Fig. VII.5).[28] As observed, the share of light in the total load dropped sharply after 1894, and the traction load far surpassed the light load at the turn of the century. The most notable change in the load mix after 1900 was the steady increase in stationary power load (see Fig. VII.6). In 1900 BEW

[26] *Saling's Börsen-Jahrbuch, 1900/1901,* p. 1090; 1901/1902, pp. 1176–77.

[27] The County of London Electric Supply Co., Ltd., *Public Inquiry Held by the Electricity Commissioners in Connection with Application for Consent for the Erection of a Power Station at Barking: October 6, 7, 8, 9, 1920: Justification of Evidence Given by Mr. Charles H. Merz,* app. 3, "Electricity Supply in Great Cities Throughout the World" (item in Merz & McLellan Co. Archives, Amberley, Killingworth, near Newcastle upon Tyne, England). The information on Berlin is for the years 1911–13; see p. 258 below for a comparison of rates for Berlin and other cities. Another source, City of Melbourne, *Engineers' Notes,* p. 13, reported the municipality's income from BEW revenues in 1910–11 as £174,000. The city's share of the utility's income combined with its share of the profits totaled 37 percent of BEW's gross income. W. Fellenberg, "Die Entwicklung der Starkstrom technik in Deutschland und in der Vereinigten Staaten von Nordamerika," *Elektrotechnische Zeitschrift* 30 (1909): 1199.

[28] Matschoss et al., *50 Jahre,* p. 56.

Figure VII.5.   *Increase in electricity consumption in Berlin: Percentage of Berlin households connected* (table left); *increase in kilowatt-hours consumed per capita* (graph right). *From Matschoss et al., 50 Jahre, p. 56.*

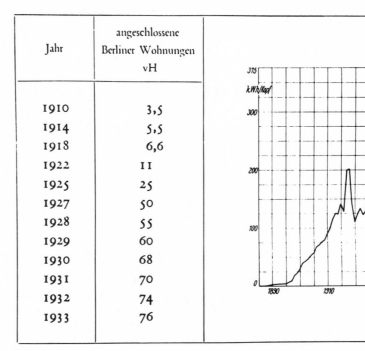

Jahr	angeschlossene Berliner Wohnungen vH
1910	3,5
1914	5,5
1918	6,6
1922	11
1925	25
1927	50
1928	55
1929	60
1930	68
1931	70
1932	74
1933	76

Figure VII.6.   *Development of various loads, BEW: Light and power* (Licht und Kraft); *electric traction* (Bahnen); *and high-voltage transmission* (Hochspannung). *From Matschoss et al., 50 Jahre, p. 89.*

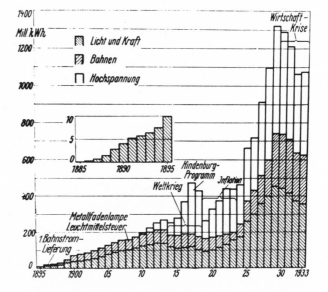

supplied 52,014,612 kwh. Connected to the Berlin network by the end of fiscal year 1899/1900 were 4,964 motors, 303,304 incandescent lamps, 12,396 arc lamps, and 676 miscellaneous apparatus. The total connected load was 36,350 kw. In addition, the streetcar load amounted to 8,000 h.p. Of the total consumption, in contrast to connected load, light amounted to 28 percent; stationary motor power, 24 percent; and streetcar, or traction power, 48 percent.[29] In 1900 the connected stationary power load surpassed

---

[29] *Saling's Börsen-Jahrbuch, 1901/1902,* p. 1176. These statistics refer to the business year 1 July 1899 to 30 June 1900.

TABLE VII.1. MOTOR LOAD OF BERLINER ELEKTRICITÄTS-WERKE IN BERLIN CLASSIFIED ACCORDING TO USE, 1914

Load	No. of Motors	Kw.
Metal Working	7,023	24,155
Elevators	4,940	28,939
Woodworking	3,816	11,464
Presses	3,749	8,997
Ventilators	3,297	1,240
Meat Packing	1,724	5,287
Sewing Machines	1,638	1,280
Washing Machines	1,026	1,947
Pumps	880	5,813
Paper Mills	698	1,831
Cloth Cutting	630	390
Leather Working	571	1,615
Grinding and Polishing	493	1,381
Dough and Butter Machines	400	1,881
Stirring, Mixing, and Grinding	395	2,154
Spinning	274	558
Dynamo Drive	188	2,743
Coffee Making	156	273
Tobacco Manufacturing	146	393
Hat Pressing	53	113
Various	4,686	14,030
Total	36,783	116,484

Source: Berliner Electricitäts-Werke, *Geschäftsbericht . . . 1914/15*, pp. 10–11.

the connected light load for the first time. (As noted, traction load about matched the combination of stationary power load and lighting load.) From that point on, BEW was primarily a power company. By 1907 the traction and stationary power loads were about equal. After 1910, when BEW began supplying industrial consumers by means of 100-h.p. motors directly from a high-voltage network rather than from the regular 220-volt system, the stationary power load surpassed the traction load and continued to grow larger. BEW promoted in factories the introduction of electric motors to replace steam engines with belt-transmission systems by renting motors at a modest cost and allowing their return or purchase after thorough testing and evaluation. The close coordination between AEG, the manufacturer of these motors, and BEW furthered the electrification and industrialization process.[30] (See Table VII.1.)

Further breakdown of the load according to type of consumer throws light on the nature of industrialization in Berlin. In 1910–11 the total output of BEW, including energy used in its own central stations, amounted to 198,031,743 kwh. The total was categorized as follows: lighting, 26 percent; power, 37 percent; traction, 32 percent; and high-voltage supply (probably stationary power), 5 percent. In contrast to the output, the connected load was 206,726 kw., 39 percent of which was in lighting, 54 percent in power and heating (mostly stationary and traction power), and 6 percent in high-voltage supply.[31] In 1911 the connected load for all German central stations was 38 percent lighting (incandescent and arc) and 60 percent

[30] Matschoss, "Berliner Elektricitäts-Werke," pp. 20–21.
[31] City of Melbourne, *Engineers' Notes*, p. 13. Klingenberg, "Electricity Supply in Large Cities," p. 398, lists BEW's 1910–11 output as 192,100,000 kwh.

Figure VII.7. BEW load curves on days of highest load (Höchstlast). From Matschoss et al., 50 Jahre, p. 71.

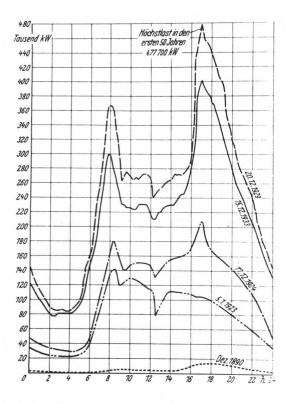

power and heating (stationary, 43 percent; traction, 15 percent; and heat, 2 percent).[32] The correlation between Berlin and the rest of Germany, urban and rural, in load mix is partially explained by the fact that while Berlin had about 8 percent of the total connected load on central stations in Germany, other large cities also had substantial shares. Like Berlin, the large cities were industrializing and providing mass transit. The traction share of the total power load, however, was much higher in Berlin than in the rest of Germany.

By World War I, BEW had created a universal supply system of the kind found in Chicago, a system that was still notably absent in London. BEW had taken the initial step toward establishing such a system in 1896 with the inauguration of polyphase supply at the Schiffbauerdamm central station. Its major move was made in 1899 (at the time of its new contract with Berlin) with the takeover of the Oberspree power station from AEG (see Fig. VII.8). This was a large station located outside Berlin. AEG had put Oberspree into service in 1897, before BEW was authorized by the city to operate power plants outside the city. With an initial capacity of 1,000 kw., the power plant had the advantages of lower-cost real estate, available cooling water, and easier coal and ash handling. Oberspree was also equipped for power transmission to western Berlin and the city's western suburbs at 6 kv., but before 1899 the plant supplied industry in the immediate vicinity, including an AEG cable-manufacturing plant (see Fig. VII.8, p. 193). After

[32] Georg Dettmar, "Die Statistik der Elektrizitätswerke in Deutschland nach dem Stande vom 1 April 1913," *Elektrotechnische Zeitschrift* 34 (1913): 1447–50.

Figure VII.8.   *The polyphase central station at Oberspree (left) and the AEG Cable Works, its primary load (right),* 1897. *From Matschoss et al.,* 50 Jahre, *p. 30.*

capacity restrictions on BEW were eliminated in 1907 by a revision of the 1899 contract, BEW increased the capacity of the Oberspree plant, installing 52,000 kw. by 1912.

As the load within the area of the city already supplied with two-wire, 110-volt current steadily increased, BEW decided (in 1899) to enlarge the capacity of its distribution system by raising the voltage to the consumer from 110 to 220. BEW was the first utility in Germany to move to the 220-volt system, and it set a precedent that spread throughout the country, much of the Continent, and to England. (The United States stayed with the 110 system.) With the change, BEW assumed the cost of converting consumers' appliances and motors to the higher voltage. The consequent lowering of distribution costs because of more economical distribution (less copper) would compensate the company. Development of a metallic incandescent-lamp filament capable of withstanding the higher voltage made the changeover possible.[33]

The metallic-filament lamp introduced in Berlin in 1906 used only one watt for each candle power of illumination and about one-third the energy of the carbon filament. The attendant reduction in the cost of lighting

[33] Matschoss, "Berliner Elektricitäts-Werke," p. 19.

Figure VII.9.   *Polyphase supply system,*
BEW, 1915: △ *= central stations;* ○ *=*
current conversion (motor-generators); ▬
= transformer stations. From Matschoss et
al., 50 Jahre, *p. 168.*

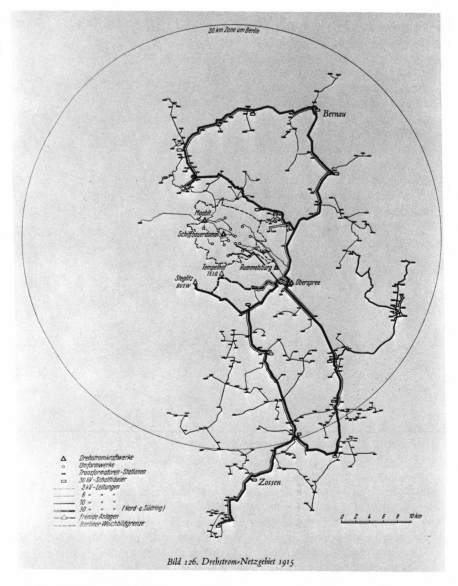

Bild 126. Drehstrom-Netzgebiet 1915

made the incandescent lamp attractive to an even larger segment of the population that had formerly used gaslight.[34] Several years earlier BEW had reached out for lower-income customers after being sharply challenged by the lower rates of small isolated stations operated by the owners of large apartment and office buildings for their own and their tenants' use. These "block stations" were able to offer lower rates because they did not use the streets for distribution, needed no revenue-sharing franchise from the city, and concentrated on high-load districts. To meet the challenge, BEW had lowered its general lighting rates in January 1904 to 40 pfg./kwh., a reduction of 25 percent. Instead of an increase in demand, however, BEW experienced a sharp downward movement in consumption during the win-

[34] Matschoss et al., *50 Jahre,* p. 38; Hans-Joachim Braun, "Gas oder Elektrizität? Zur Konkurrenz zweier Beleuchtungssysteme, 1880–1914," *Technikgeschichte* 47 (1980):13–14.

ter of 1908/9. This resulted in part from the introduction of the more efficient metallic-filament lamp, but it also followed the government's limiting shop hours to eight, the more efficient use of electricity by streetcars, and the city-owned gas plant's reduction in rates. BEW countered with further rate changes, especially those that encouraged the large power users.[35] The enticement of heavy power consumers was a response to the endemic problem of the isolated plant from which large industrial consumers supplied their own needs. As the graphs of BEW output show, the trend was again steeply upward in 1909. (See Fig. VII.5.) Isolated plants supplying individual industries continued to offer substantial competition, however; in 1913 it was estimated that isolated plants and "block stations" supplied about 40 million kwh., or the equivalent of almost one-fifth of the energy sold by BEW.[36]

BEW continued to build a generalized system by introducing steam turbines beginning in 1902 (see Fig. VII.10). The early history of turbine use by BEW shows the close relationship of AEG as manufacturer and BEW as full-scale testing ground. AEG entered the turbine field soon after early installations of Parsons turbogenerators in England and Parsons-type turbines built by Brown, Boveri & Co. of Baden/Mannheim, a subsidiary of the Swiss firm, in Elberfeld, Germany, provided strong indications of the invention's practicality and potential improvements. AEG made a false start by attempting to develop the Riedler-Stumpf design. The Riedler-Stumpf turbine was an adaptation of the basic constant-pressure design introduced by the Swedish inventor Carl G. P. de Laval in 1883. By means of large cross-sectional turbine blades and the reordering of additional steam nozzles, AEG hoped to hold the speed of rotation to practical limits. The first large turbine of this design was installed in 1902 in BEW's Moabit power station (see Figs. VII.11 and VII.12). At its maximum speed, 3,800 rpm,

[35] Matschoss et al., *50 Jahre*, pp. 37–38.
[36] Klingenberg, "Electricity Supply in Large Cities," pp. 398–99.

Figure VII.11.   *Moabit central station, 1907: Reciprocating steam engines and generators* (foreground); *turbines* (rear). *From Matschoss et al.,* 50 Jahre, *p. 36.*

Figure VII.12.   *Moabit central station, 1914. Drawing by Joseph Pennell. From Matschoss et al.,* 50 Jahre, *p. 33.*

the Riedler-Stumpf turbine produced 2,000 h.p., but AEG decided that larger models would have stability problems and be prohibitively expensive to install, so it abandoned the design.

AEG then made an agreement with the General Electric Company of America to use the Curtis design that GE had developed and patented. The manufacturers also agreed to exchange development and operational experience as they worked on the turbine. From 1904 until 1917 all BEW turbines except one were of the Curtis type. The exception arose when AEG was unable to supply within a short time a Curtis turbine when BEW's load demand suddenly increased sharply in 1905. The utility therefore placed a Brown-Boveri-built Parsons type of turbine of 5,800-kw. capacity into operation at Oberspree. More turbines were installed at Oberspree, but BEW's first all-turbine station—a station comparable to the Fisk Street station opened in 1903 in Chicago—was Rummelsburg, which began operating in November 1907 with a 13,500-kw. capacity. Further evidence that the post-Edison era had arrived was the complete conversion, in the same year, of the original Markgrafenstrasse direct-current station to a substation.[37]

By the eve of World War I—and the takeover of the utility by Berlin in 1915—BEW had acquired salient characteristics that define leading urban systems even today. Its supply area spread beyond the 5-kilometer radius of old Berlin out to the 30-kilometer radius specified in the 1899 contract. Drawing heavily on the Oberspree plant, BEW supplied about a hundred small districts, or local authorities, mostly to the east, south, and north of Berlin (see Fig. VII.13). This suburban region had a population of about 2.5 million (compared to the 2 million within old Berlin). By 1915 BEW had two other polyphase plants (Moabit and Rummelsburg) and three inner-city plants (Mauerstrasse, Spandauer, and Schiffbauerdamm). These six plants had a combined capacity of 155,000 kw. There were twenty-four substations (twelve within the city) for the conversion of polyphase current to direct current and for transformation of voltage. Transmission and distribution cable extended 7,740 kilometers. Within the city, general distribution was almost entirely 440/220-volt direct current for the lighting and small-power consumers. Also available in this area were 550-600-volt direct current for traction and 6-kv. polyphase current for heavy stationary-power users. After 1904 the level of transmission voltage was raised. Oberspree and Moabit supplied distant substations with 10-kv. transmission, and in 1911 BEW pioneered on the Continent by constructing a 30-kv. ring supply for Berlin's northern and southern suburbs. The increase in voltage necessitated the introduction of oil-cooled switches into the system.[38]

Georg Klingenberg, the engineering head of AEG's power plant design and construction division, an engineer of international reputation and author of definitive works on power-plant design and operation, described and analyzed BEW around 1913.[39] Comparing the Berlin utility with those in Chicago and London, he found that the capital cost of the power stations

[37] Matschoss et al., *50 Jahre*, pp. 133–5, 218–19.
[38] Ibid., pp. 43, 165–70.
[39] Klingenberg, "Electricity Supply in Large Cities," pp. 398–401.

Figure VII.13.   *Expansion of BEW:*
Inner boundary, *Berlin before 1920;*
outer boundary, *Berlin after 1920;*
crosshatching, *other utilities. From
Matschoss et al., 50 Jahre, p. 89.*

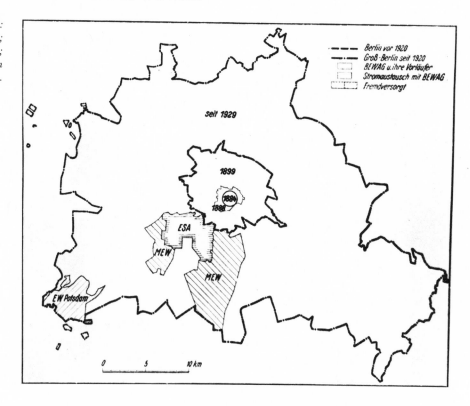

and the capital cost of the distribution system per installed kilowatt were lowest in Berlin. Berlin's per-capita consumption of current was higher than London's and lower than Chicago's, but the comparison did not take into account block and isolated stations, of which there were an unusually large number in Berlin. Klingenberg also noted that Berlin had the most evenly balanced lighting, traction, and stationary-motor loads. Berlin's load factor, the highly significant indicator, was higher than London's and lower than Chicago's. Examining the cost characteristics of the three systems, Klingenberg observed that the Berlin utility's operating costs were the lowest, and he concluded that this resulted from lower wages and lower fuel consumption per kilowatt-hour of output. Economy in the consumption of coal was counterbalanced, however, by the higher price of coal in Berlin compared to Chicago and London. The total cost per kilowatt-hour, then, was higher in Berlin than in Chicago. Not only was the price of coal higher there, but the Berlin government took a large share of BEW's revenues. (See Fig. IX.9, p. 258 below, for a detailed comparison of the three cities' utilities.)

In view of the large revenues the city derived from BEW, some contemporaries expressed surprise that Berlin would seriously consider exercising its contractual right to take over BEW in 1915. The Social Democrats in the common council advocated municipal ownership for ideological as well as economic reasons. The utility could be more easily directed toward the fulfillment of various social objectives if it was owned and operated by the city, they argued. Rates, for example, could be adjusted to favor lower-income groups. In addition, the socialists argued, profits taken by the

Figure VII.14. Georg Klingenberg, an AEG engineer noted for central-station designs. Courtesy of Berliner Kraft-und Licht AG (formerly BEW).

private owners, especially AEG, should come to the city. By 1915, those who favored the takeover believed that since the period of high financial risk had passed, it was appropriate for the utility to be government owned and managed, as were other, older utilities such as gas. They also believed that full government control of the city's electric power supply would bring increased control of the city's industrial development.

Empowered by contract to purchase the utility in October 1915 if notice of intent was given two years in advance, the city began negotiating with BEW in 1911 the question of extending the contract or purchasing the utility. The principal points at issue were rates and the relationship between BEW and AEG. The city wanted lower rates (for which there was precedent in Germany) and less AEG influence. Having lowered its rates consistently over the past thirty years, BEW argued that further reduction was impossible unless costs were substantially lowered. If, BEW insisted, the city allowed it to take advantage of advanced transmission and generation technology by constructing a large mine-mouth power plant at the Bitterfeld lignite fields almost eighty miles from Berlin, lower costs would follow.

The imperial German government, however, frustrated BEW's plans for a Bitterfeld power plant. It seized the lignite deposits to exploit them for critical wartime needs and then, in 1916, supported the construction of the Golpa-Zschornewitz power plant there (see pp. 288–89 below). Deprived of the cost-reducing technology, BEW agreed only to an extension of the existing franchise—and the existing or slightly modified rate structure for the next six years. Led by an *Oberbürgermeister* who favored municipal ownership, a *Stadtverordnetenversammlung* dominated by socialists and their allies, and a *Magistrat* that had come to agree with the common council, the city then purchased the Berliner Elektricitäts-Werke and took over management of the utility in October 1915.

Like the general contractual arrangements between the city and the utility, the specific provisions for purchase were complex and ingenious. Before the renegotiation of the contract in 1899, the city had had the right to purchase BEW at the tax value determined by two experts, one appointed by the city and the other named by the utility. They were to "appraise the properties as an interdependent whole according to commercial principles."[40] In 1899, the parties had agreed that the plant and real estate would instead be purchased according to book value. Subsequent analysis showed that the book value (original cost) of the utility's real estate was 4,195,000 marks lower than the appraised value in 1915, but that the book value of the plant was 1,700,000 marks higher than the appraised value in 1915. The book-value purchase price of plant and real estate in 1915 was, therefore, 2,495,000 marks lower than the appraised value. The city council voted to purchase BEW for 130 million marks; the exact sale price was 132,400,000 marks.[41]

Judging by its unwillingness to negotiate purposefully after the collapse of the mine-mouth power-plant proposal, AEG may not have been reluctant to sell BEW in 1915. The sale provided AEG with substantial capital for new investments; moreover, the years of BEW ownership had brought

[40] Brooks, "Berlin Electric Works," p. 192.
[41] Ibid., pp. 188, 193; Matschoss et al., *50 Jahre*, p. 223.

"quite acceptable dividends to shareholders" and had proved "a veritable gold mine to the holding company [AEG]."[42]

In America, utility managers, who often found themselves in an adversary, even hostile, relationship with government, may have been surprised to learn that leading executives from the private German company were taking jobs in the municipally owned plant. The newly named Städtische Elektrizitäts-Werke Berlin (StEW) was determined to have the best managerial and technical talent available and was willing "to pay accordingly."[43] Members of the *Vorstand* ("executive committee") were to receive a guaranteed salary of 15,000 marks as well as increment according to net income, so long as the total did not exceed 39,000 marks. Dr. Herman Passavant from BEW's *Vorstand* became a member of StEW's *Vorstand.* Gustav Wilkens, who had been a deputy member of the old *Vorstand,* held the same position in the new committee. Emar Wikander, a manager at BEW, became a member of StEW's *Vorstand* and technical director of the plant.[44] On 18 April 1915 the Berlin common council voted to purchase BEW, and on 1 October 1915 the utility passed into new ownership. Emil Rathenau died on 20 June of that year.

[42] Brooks, "Berlin Electric Works," pp. 190, 191–92.
[43] Ibid., p. 193.
[44] Matschoss et al., *50 Jahre,* pp. 44, 95.

CHAPTER VIII

Chicago: The Dominance of Technology

THE early history of many utilities is told in terms of collective enterprise and organization. In those instances, no individual serves well as the focus or center about which to structure such an account. Chicago, however, offers an exception. The history of the Commonwealth Edison Company, the utility that came to dominate electric supply in "the windy city," can be organized most effectively around the person of Samuel Insull (1859–1938). Presenting the history of electric supply in Chicago in this manner does not mean that Insull was a great man whose actions made history; it means that he was the pivotal figure during four decades of the company's history. A look at the electric supply industry in Chicago through his eyes provides an encompassing view. Furthermore, Insull was a master of public relations, and his speeches, addresses, and papers are informed sources of the company's history.

By World War I, Chicago was a creature of the industrial age, in contrast to Berlin and London, the two cities with which it will be compared. Berlin's long history as a capital city, as well as the strong Prussian traditions, set the stage for the remarkable transition the city underwent during the period of rapid industrialization in Germany. London bore the marks of modern industrial influences, but its shape and substance reflected a long preindustrial experience. Chicago, however, was pliable and open to influences, and so transportation, industry, and electric power, the primary components of industrialization, interacted and developed there relatively free of conservative, or traditional, constraints. The electric supply industry in Chicago did not face the challenges posed by the forces of parochialism in London or the strong-willed, rational government of Berlin.

Railroads stimulated the first great wave of physical, economic, and social growth in Chicago. By 1905, about 14 percent of the world's railway mileage passed through Chicago, and the city was the acknowledged railway center of the United States. The Midwest, the nation's great agricultural region, transported its goods by railroad and by Lake Michigan from Chicago to the other great cities of the United States, and from their ports to many other parts of the world. Movement within the city was facilitated by an elevated railway and by streetcars, an extensive system that was largely electrified before World War I. Many of Chicago's citizens, however, con-

Figure VIII.1. Samuel Insull, 1895.
Courtesy of Commonwealth Edison Co.,
Chicago, Ill.

sidered the system inadequate and its management riddled with corruption, and this attitude led to many reform efforts, including endeavors to establish municipal rather than private ownership of the transportation network. Thus Commonwealth Edison became deeply involved in politics as it sought to power the system.

The increase in Chicago's population at the end of the nineteenth century was impressive, even by the elevated standards of urban growth in the United States. From 1890 to 1910, the formative years of the electric light and power industry, Chicago's population increased by 599,000; the growth during these two decades was the largest in the city's history. The value of products manufactured in Chicago in the first decade of the twentieth century was second only to the value of products made in New York City. Slaughtering and meat packing was the city's leading industry, but foundries, machine shops, and other producers of iron and steel products operated on a large scale and were numerous. There were countless opportunities for a utility seeking to electrify industrial operations and to include them in a system of central-station service.

Unlike Berlin, Chicago was not known for the efficiency and honesty of its government. But venal, pliable politicians and ill-defined political institutions did not automatically frustrate the growth of public electric supply. In the first decade of this century, Chicago was subsumed under the antiquated general charter of the state of Illinois, a charter the city had accepted in 1875. Subsequently, the city grew dramatically and faced new problems to which the charter did not speak. A city council and mayor shared power, but their influence and effectiveness were hampered by the conflicting and contradictory powers still exercised by state and county authorities and by the surviving autonomy of townships that had been absorbed by the city in its growth. The power of the city to tax was constrained by state law, and city services suffered. Major exceptions were the water-supply system and the great drainage canal, which had been constructed between 1892 and 1900 to provide for sewage disposal without pollution of Lake Michigan. On the other hand, the transit system, though consolidated and electrified, was in the hands of private owners and subject to the influence of corruptible politicians. The city had erected an electric-lighting plant in 1887, but by 1900 the electric supply system also was for the most part privately owned.

In 1892, blustery, vigorous Chicago became the home of Samuel Insull, a manager-entrepreneur of incomparable vitality.[1] When Insull left the newly formed General Electric Company to assume the presidency of the Chicago Edison Company (after 1907 the Commonwealth Edison Company), he gave up the vice-presidency of a $50 million corporation to become the head of an $885,000 one. He also left the electrical manufacturing business to enter the utilities field. Insull had recommended himself for the new position after being asked by the Chicago concern for suggestions. He had realized that General Electric, the company formed from the merger of Edison General Electric, in which he had played such a

[1] This account of Insull and Commonwealth Edison is drawn in part from Thomas P. Hughes, "The Electrification of America: The System Builders," *Technology and Culture* 20 (1979): 124–61.

prominent role as head of sales and manufacturing, and Thomson-Houston, the company led by the impressive Charles Coffin, would be dominated by Thomson-Houston men.[2] Insull, who was thirty-two years old, accustomed to exercising authority, and inspired by Edison's drive to create and construct, was ready to build his own system.

Chicago Edison's acceptance of Insull's nomination is easily understood, for Insull, despite his youth, had an impressive reputation in the electric lighting field. In 1880 he had emigrated from England to America to become Edison's personal secretary. Edison's initial impression of him was not positive—Insull was young, weighed 117 pounds, and had been seasick for eight successive days. Soon, however, Edison understood why Edward H. Johnson had recommended the young man who had worked as private secretary for Edison's London representative: Insull was knowledgeable, intelligent, ambitious, and bold. Only hours after meeting Edison, Insull suggested a way to raise the additional $150,000 needed to establish companies to manufacture components for the electric lighting system being built at the Pearl Street station and soon to be erected elsewhere in America.[3]

Insull was of great value to Edison, but Edison's influence on Insull was of greater consequence. Between 1880 and 1892, Insull was the inventor's secretary and personal representative and then became manager of the Edison General Electric plant at Schenectady. These were formative years for the electric light and power industry and for Insull. He was in at the start when Edison presided over the construction and early operation of New York's Pearl Street station; he took part in the establishment of the manufacturing facilities that ultimately became General Electric; he participated in numerous conferences involving inventors, engineers, entrepreneurs, mechanics, financiers, managers, electricians, and others who shaped the early history of the electrical industry, both utility and manufacturing. In short, he studied in the Edison school, absorbing its creative, problem-solving, inclusive, systematic, innovative, and expansionist approach. He knew its graduates, who spread far and wide to take part in the growth of the industry. Furthermore, as a leading member of the school, Insull was very close to the master. Insull later said: "[Edison] grounded me in the fundamentals. . . . no one could have had a more considerate and fascinating teacher."[4] Insull never tired of characterizing Edison as the greatest man he had ever known and the one who had the strongest influence on his character.

Years later, in 1934, when Samuel Insull went on trial for using the mails to defraud in connection with his bankrupt holding company, his defense was essentially that he, like Edison, was a creative man and one greatly interested in managing productive technology; he denied that he was a predatory holding-company tycoon. His son, Samuel Insull, Jr., recalls that after the government prosecutor heard the defense, he said privately to

[2] Forrest McDonald, *Insull* (Chicago: University of Chicago Press, 1962), pp. 50–52.

[3] Burton Y. Berry, "Mr. Samuel Insull" (typescript), pp. 33–34, Samuel Insull Papers, Loyola University, Chicago, Ill. These papers were used by kind permission of Samuel Insull, Jr., and are hereafter cited as Insull Papers. I would like to thank Janet Halder for her help in sorting through the collection.

[4] Samuel Insull, "Memoirs of Samuel Insull" (typescript written in 1934–35), p. 37, Insull Papers.

the son, "Say, you fellows were legitimate businessmen."[5] Samuel Insull was acquitted, but he was exhausted; the immense system of utilities he had built up lay fractured, never to be in his hands again. History has not dealt generously with Insull; the disastrous climax of his career burned itself into the public's memory. For newspapermen, politicians, and former competitors, Insull was a Depression scapegoat; the decades of complex system-building were easier to ignore or forget—they involved difficult concepts, esoteric technology, uncommon economics, and sophisticated management.

The creation of an all-embracing system of electric light and power for Chicago was Insull's prime objective for almost two decades. In the end he reached out beyond the city and interconnected the Chicago system with suburban companies and then linked these with neighboring municipalities. The scope of the system became regional. The history of the Public Service Corporation of Illinois (a regional system) and of Insull's Middle West Utilities Company (a holding company on a national scale) are important chapters in the life of Insull and the history of electric light and power.[6] An account of the Chicago system alone, however, reveals the powerful and effective conceptual synthesis of technology and economics that guided Insull and his associates as they knit together and managed the Chicago system—considered by many as early as 1910 to be the world's greatest.

When Insull arrived, the Chicago Edison Company was but one of more than twenty small electric-lighting utilities. Within two decades, Insull and his associates had created a monopolistic, mass-producing, technologically efficient, and economically operated company for all of Chicago.[7] Insull became a spokesmen for the utilities industry, and his company was a pacesetter in both technological and management policy. (See Fig. VIII.2.) The system and the way in which it was created became models for other urban utilities. For the historian, it was both the leading edge and a representative case of the waves of development that followed.

Under Insull's direction, his company took the lead among utilities in technological innovation. The growth of the Chicago system cannot be explained without taking into account this technological component. Neither, however, can the evolution of the system be explained without considering the political situation in Chicago and the state of Illinois and the role Insull played in this particular scene. Insull did not simply manipulate politicians and political forces; the entire situation was ripe for manipulation. At about the same time, another entrepreneur of electrical supply, Charles Merz, found London to be a remarkably different problem (see pp. 249–54 below). The developments in London and Chicago direct attention not only to the two men but to their situations.

Insull, of Chicago Edison, and Merz, of the Newcastle upon Tyne Electric

[5] McDonald, *Insull*, p. 331.

[6] Historical sketches of the Middle West Utilities Co. can be found in the company's annual reports on file in the Insull Papers (see especially "Report . . . for the Fiscal Year Ending December 31, 1926"). See also "Presentation of Public Service Company of Northern Illinois in Competition for the Charles A. Coffin Prize Award April 1st 1924," Insull Papers.

[7] H. A. Seymour, "History of Commonwealth Edison Company," (typescript completed in 1935), p. 200, Commonwealth Edison Co. Archives, Chicago, Ill. I am grateful to Helen P. Thompson, librarian, for help in using the archives and the library.

Figure VIII.2.   *Prominent electrical industry executives, 1897: Samuel Insull* (second row, eighth from left); *John Kruesi, General Electric Co.* (second row, third from left); *Charles Batchelor* (second row, seventh from left); *Louis Ferguson, Chicago Edison Co.* (second row, ninth from left); *J. W. Lieb, Edison Electric Illuminating Co., N.Y.* (third row, third from left); *and T. C. Martin, editor of* Electrical World (fourth row, seventh from left). *From the author's private collection.*

Supply Company, were good friends and professional associates. Both displayed an impressive drive to build systems. Both, like Edison, were hedgehogs. Within the spectrum of entrepreneurial talents extending from engineering to management, Merz would be placed toward the engineering end, but the abilities of both men ranged from engineering to management and to politics and finance as well. Yet Merz was frustrated by political power in his efforts to bring a unified electrical system to London, while Insull used politics to create a system in Chicago that the utilities world envied.

In London during the decades before World War I, Merz and his associates found themselves confronted by the effective resistance of proponents of local government and the foes of private ownership (see pp. 253–54 below). In Chicago, Insull learned how to deal with the local politicians and eventually organized representatives of the state government in support of his system building. In addition, he held the advocates of public ownership at bay with a steady barrage of publicity. In London, the politicians were not receptive to Merz's plans; in Chicago and Illinois, politics became a part of Insull's system. Forrest McDonald, Insull's biographer, believes that Insull was for about thirty-five years the most powerful political operator on the American business scene.[8]

[8] McDonald, *Insull*, p. 82. My discussion of Insull in politics is based on McDonald's biography.

Chicago's politicians, especially those on the city council, exploited the franchise-granting power of government. An outstanding example of the ingenious way in which they obtained wealth from political power without having to steal public money is the Ogden Gas Company case. Insull learned much from observing their maneuvers. In 1895 a group of politicians that later became known as the Ogden Gas Crowd helped pass an ordinance that granted a long-term franchise to the Ogden Gas Company, an enterprise they had founded and in which they held a large block of stock. The already existing gas company, the Peoples Gas Light & Coke Company, envied the franchise holders and regretted the competition the new company could offer. So the Ogden Gas Crowd sold the franchise to the Peoples Company for over $7 million and several politicians made an individual profit of almost two-thirds of a million dollars. There were other variations on the franchise theme, as Insull, head of Chicago Edison, would learn.[9]

Insull also took note of how Charles Tyson Yerkes, Chicago's leading transportation-utilities magnate, survived, even flourished, in Chicago. By 1897 Yerkes, also a system builder, had forged a unified and electrified urban transportation system. To do this he had been forced regularly to bribe members of the city council to extend the many short-term franchises he acquired in building his transportation network. Because it was difficult to obtain long-term loans or to market long-term bonds with short-lived franchises, and because the bribes were unending, Yerkes in 1897 made a bold and rational move. He arranged for the introduction in the state legislature of a set of bills to extend all streetcar and elevated railway franchises in the state of Illinois for fifty years, require companies to pay the cities they served an initial fee and an annual percentage of gross revenues, and remove control of the transportation companies from city councils and place regulation in the hands of a state commission. Shifting responsibility to the state level might reduce corruption; it would surely extend the jurisdiction of the regulatory authority to accord with the expanding transportation networks. The bills did not pass, but the legislators did authorize city councils to issue fifty-year franchises.

Insull got a taste of the city council's franchise enterprise soon thereafter. Members of the Chicago City Council asked how much he would pay to block the granting of a fifty-year franchise to a rival company. Following the Ogden Gas Crowd pattern, the council voted a franchise to the Commonwealth Electric Company, which had been established to hold the franchise. Insull did not yield, however, so members of the council and their friends went ahead with plans to transform the dummy company into an operating competitor. Then they learned why Insull had stood his ground: he held exclusive rights from every American manufacturer to buy electrical equipment for Chicago. So daunted, the creators and owners of Commonwealth Electric sold the valuable fifty-year franchise and the company to Chicago Edison for $50,000.[10]

Insull and like-minded utilities men were to succeed where Yerkes had failed; they would bring about the transfer of regulatory powers to the state. Insull first mustered the support of leading utilities executives through

[9] Ibid., p. 83.
[10] Ibid., pp. 88–89.

the National Electric Light Association (NELA), a trade organization. In his NELA presidential address in 1898, he called for state regulation of rates and service; he also advocated that municipalities be empowered to purchase, at cost minus depreciation, the plant of any utility that failed to render satisfactory service. On this occasion and throughout most of his career, Insull showed his respect for the public service responsibilities of the utilities, and he let the public know of his concern. He was a pioneer in utility public relations. Although the NELA membership was not ready for Insull's policy in 1898 and relegated the proposal to a committee, within a decade it realized that state regulation would both facilitate expansion and counter the rising tide of municipal ownership. A new NELA Public Policy Committee submitted a strong report favoring state regulation and the membership endorsed it. In 1907, when the municipal-bond market contracted severely and municipal ownership therefore became less at- tractive, its proponents joined the call for state regulation as a reform. The National Civic Federation (in which Insull played an important role as member of the executive committee made up of business and labor leaders and public representatives) also decided that state regulation of public utilities was an important step in its effort to bring about urban reform throughout the nation. Its report of 1907, which coincided with Insull's and NELA's views, provided the basis for most regulatory laws, including those that governed the pioneering regulatory commissions established in Wisconsin, Massachusetts, and New York in 1907. Ironically, because of complexities in Illinois law, especially in the field of transportation utilities, Insull's state did not establish a state regulatory commission until 1914.[11]

When the effectiveness of London's local authorities in resisting the tran- scendence of their authority is recalled, it seems remarkable that Insull and others encountered relatively ineffectual resistance in Chicago as regulatory jurisdiction was moved up to the state level. There was, however, a delayed reaction. Almost simultaneously with the establishment of the Illinois Public Utility Commission in 1914, a movement called "Home Rule" was launched in Chicago to return jurisdiction to the municipal government. The strength of this movement was reinforced in the next few years by a complex alliance of Chicago politicians who were acutely aware of their loss of influence; by critics of profitmaking from defense expenditures, especially by public utilities; and by advocates of government ownership. The movement took on a midwestern character as Progressive leaders like Robert La Follette of Wisconsin, William Borah of Idaho, Burton Wheeler of Montana, and George Norris of Nebraska turned against war profiteers and, by 1919, especially the public utilities. (As in England, advocates of local and central government ownership combined against the expanding, privately owned utility system.)

During World War I, Insull headed the State Council of Defense in Illinois and was especially active in mobilizing public opinion in the war effort. After the war, with utilities, not the United States, under attack, he used this public relations–propaganda organization as a model for the Illinois Public Utility Information Committee. Other utilities and NELA followed his lead, and speeches and publications poured forth. "By the

---

[11] Ibid., pp. 113–14, 117, 121, 177–78.

mid-twenties, hating utilities was, in Chicago, as rare as hating mother and the flag."[12] The threat of "Home Rule" had passed.

Relegating political authority to the subservient position deemed appropriate by American entrepreneurs, Insull proceeded with the fulfillment of his primary objectives, which were in the technical and economic realms. Within two years of his arrival in Chicago, Insull demonstrated the boldness of his intentions by raising funds to build the Harrison Street central station during a year of financial panic. This generating station became known for its advanced engineering and rapid growth during its first decade of operation. Given his thorough exposure to the technology of electric light and power while he was with Edison and his companies, Insull took far more than an abstract managerial interest in the way the Harrison Street station was equipped. Opening in 1894 with 400-kw. direct-current generators totaling 2,400 kw. and driven in pairs by highly efficient 1,250-h.p. condensing, reciprocating steam engines, the station expanded within a decade to a total capacity of 16,200 kw. generated by 3,500-kw units. Located on navigable water, it had easy access to coal and cooling water. Moreover, as its capacity increased, a larger and more diverse market evolved. On the business level, Chicago Edison acquired twenty Chicago utilities and their market areas. The culmination of the expansion was its formal merger with Commonwealth Electric to form Commonwealth Edison in 1907. (The two companies had been jointly managed for some years.)[13]

In presiding over the growth of the Chicago system, Insull demonstrated how the new technology could be coupled to the old in the universal supply system. As the small and inefficient companies were absorbed, their generating plants were used for a time. After 1896 and the introduction of polyphase current, however, obsolete generating stations were transformed into substations. Insull believed that after Sebastian Z. de Ferranti, he was the first to transform inefficient stations into substations. (The very prefix "sub-" indicated the emerging hierarchical system.) Ferranti, the forward-looking British engineer and entrepreneur, had converted the small Grosvenor Gallery station into a transformer substation for the mammoth Deptford generating plant across the Thames.[14] Like Ferranti, Insull and his associates not only utilized polyphase current, transformers, and high-voltage distribution lines but also employed the newly invented rotary converter. (See pp. 121–22 above.)

The rotary converter proved to be an effective tool in the hands of the Chicago electrical engineers and managers who were intent upon growth. A frustration for them had been the combined heavy investment in downtown direct-current stations of the Edison type and isolated suburban alternating-current stations. Knowing that economies of scale would result whether one measured generator capacity or area of district served, they introduced the converter to couple the a.c. and d.c. systems at their interface, thus integrating the old and the new and building a universal system.

At the Chicago exposition in 1893 a rotary converter was exhibited by

---

[12] Ibid., p. 185.

[13] "Commonwealth Edison Company" (typescript prepared c. 1934 for Samuel Insull prior to his federal trial), pp. 2–3, Insull Papers.

[14] Arthur Ridding, *S. Z. de Ferranti, Pioneer of Electric Power* (London: HMSO, 1964).

the Westinghouse Corporation. Insull and his chief engineer, Louis Ferguson, believed that they were the first to order the new invention (in May 1896) but acknowledged that Boston Edison had used it first. In the case of Chicago Edison, Ferguson and Insull first shut down an inefficient d.c. station on Wabash Avenue, installed a reversed rotary converter at the Harrison Street station (direct current from the generators was changed into alternating current), and sent the high-voltage alternating current over the transmission line three miles to the former generating station, now converted into a substation, where a rotary converter changed alternating current to direct current for use by consumers in the district.[15] This was a variation on the usual pattern of conversion.

By 1900 Insull, who usually depended on General Electric for electrical equipment, had the rotary converter, the frequency changer, and, of course, the transformer. Schooled by Edison, Insull never lost his sensitivity to technological developments; his economic and business acumen allowed him to grasp quickly how to exploit them. At his side were Louis Ferguson, his chief engineer, who was recognized as a leader by his peers, and Frank Sargent, a consulting engineer whose reputation as a power-plant designer grew rapidly as he became primary consultant to Chicago Edison (see Fig. VIII.3).[16]

Insull not only spread his distribution system but he increased the scale of the generating units within the system. As in the case of his coupling of the old and new technologies, he was motivated by a highly developed economic sense and an aptitude for identifying the technology of the future. One historic technological achievement of which he was immensely proud was the Fisk Street station, which opened in 1903 (see Figs. VIII.4 and VIII.5). His controversial decision to build the station followed from the realization that large generating units were more efficient than small ones. Also influencing him was the realization that reciprocating steam engines with more than a 5,000-h.p. capacity (the limit of the unit recently installed in the Harrison Street station) would be impractical because of their size.[17] Adhering to his objective of amassing production in large units, he turned to the steam turbine, which was winning praise in Europe for its relatively small size, simplicity of mechanism, and moderate first cost. In this decision, as in many others, Insull demonstrated his awareness of European technological developments and his inclination to transfer and adapt technology.[18]

Having seen steam turbines in a central station in Germany, Insull sent Ferguson and Sargent to Europe early in the summer of 1901 to inspect

[15] Seymour, "History of Commonwealth Edison," pp. 308–9. In a paper presented at the Organization of American Historians' Seventy-fifth Annual Meeting, Philadelphia, Pa., 1 April 1982, Harold Platt discussed Insull's supply-side marketing techniques, stressing Insull's appreciation of the economic importance and use of the Wright Demand Meter and the rotary converter. See Harold L. Platt, "Sam Insull and the Electrification of Chicago, 1880–1925."

[16] On Sargent see Sargent & Lundy, *The Sargent & Lundy Story: Seventy Years of Engineering Service* (Chicago: Sargent & Lundy 1961).

[17] Seymour, "History of Commonwealth Edison," p. 319.

[18] When traveling in Europe as a young man, Samuel Insull, Jr., investigated technological developments there at his father's request. He later confirmed that his father had paid close attention to European technological developments (interview with Samuel Insull, Jr., Chicago, 14 August 1975).

210 NETWORKS OF POWER

Figure VIII.3. Offices of the consulting
engineering firm Sargent & Lundy.
Frederick Sargent, consultant to Samuel
Insull (left); Ayres Lundy (seated at
desk). Courtesy of Sargent & Luncy
Engineers, Chicago, Ill.

Figure VIII.4. Fisk Street station,
Chicago, under construction, 1903. From
Sargent & Lundy, The Sargent &
Lundy Story, p. 15.

Figure VIII.5. Fisk Street station. From
Sargent & Lundy, The Sargent &
Lundy Story, p. 22.

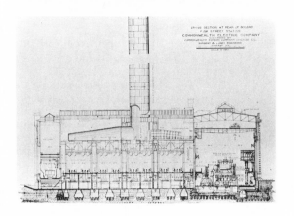

turbine installations. In England, Charles Parsons had introduced the steam turbine as a generator drive in 1884. There were small steam-turbine installations there, and in 1900 a major central-station installation was completed in Elberfeld, Germany.[19] Development of the steam turbine brought a marked decrease in weight per horsepower as compared to the reciprocating engine. If Insull had been content to try a 1,000-kw. unit, there would have been little controversy about the decision; the turbines at Elberfeld were 1,000-kw. units, and the Hartford Electric Company in America had ordered a 1,500-kw. turbogenerator from Westinghouse in 1900.[20] Committed to size and efficiency, however, Insull wanted 5,000-kw. units for the Fisk Street station.

In Europe, Ferguson's and Sargent's enthusiasm for turbines increased. Successful installations of water turbines also encouraged them. They saw the Milan Edison Company's water-power station at Paderno, with its seven 2,000-h.p. turbines; the company's chief engineer informed them that the turbines had been so successful that he had ordered two steam turbines rated at 4,000 h.p. each. Greatly stimulated, Ferguson and Sargent then visited the works of Brown, Boveri & Company in Baden, Switzerland, where the larger Milan Edison turbines would be built. There they also learned of an order from Frankfort, Germany, for a 4,000-h.p. turbogenerator; they were convinced that Europe "was tending very definitely to the turbogenerator unit."[21]

Back in the States, Charles Coffin, president of General Electric, was interested in supplying Chicago Edison with a small, 1,000-kw. turbogenerator, but he would offer none of the customary guarantees for larger units, certainly not the 5,000-kw. unit Insull had in mind. Insull persisted, having decided to go ahead and "bore with a big auger" rather than live "from hand to mouth" by adding a small turbine to a reciprocating-engine station.[22] The compromise reached was that General Electric would take the manufacturing risk and Chicago Edison would assume the expense of the installation whether it succeeded or failed. (For example, the building would have to be remodeled extensively if the turbine proved unsatisfactory.) Insull placed his order for the 5,000-kw. turbine on 28 December 1901 and the unit went into service on 2 October 1903 (see Fig. VIII.6). Three more 5,000-kw. units followed by 1905, but these were then replaced by 12,000-kw. units. By 1911 there were ten 12,000-kw. units at the Fisk Street station.[23]

Insull and other Edison officials celebrated their pioneering decision,

[19] Arnold T. Gross, "Zeittafel zur Entwicklung der Elektrizitätsversorgung," *Technikgeschichte* 25 (1936): 131; R. H. Parsons, *The Early Days of the Power Station Industry* (Cambridge: At the University Press, 1940), pp. 170 ff. On the introduction of the steam turbine in central stations, see also Kurt Mauel, "Die Bedeutung der Dampfturbine für die Entwicklung der elektrischen Energieerzeugung," *Technikgeschichte* 42 (1975): 229–42.

[20] The Hartford Electric Light Co. installed a single 1,500-kw. steam turbine in its existing Pearl Street plant in 1901. This Westinghouse-Parsons turbine was the first built in America for a utility. See Glenn Weaver, *The Hartford Electric Light Company* (Hartford, Conn., 1969), p. 87.

[21] Commonwealth Edison Co., *Edison Round Table*, November 1928, p. 2.

[22] Ibid., p. 11.

[23] Commonwealth Edison Co., *Principal Generating Stations and Transmission System* (Chicago: CEC, 1911), pp. 3, 16.

Figure VIII.6. *Fisk Street station: First 5,000-kw. turbine unit* (foreground), *the largest steam turbine in existence in 1903. From Sargent & Lundy,* The Sargent & Lundy Story, *p. 14.*

but Insull later admitted that the first steam turbines were not economical to operate, perhaps not as economical as the reciprocating engines. "I think it was the fourth turbine," he recalled, "that was a really efficient turbine."[24] However, the low first cost and labor savings of the unit offset operating problems. The installation also helped introduce the turbine to America; within a year after the Fisk Street station began operating, General Electric and Westinghouse received orders for a total of 540,000 kw. of turbo-generator capacity. Ferguson considered Insull's introduction of the large turbine daring, despite the European precedent. "European conditions," he remarked, "individualized [the] design of each machine, the careful coddling of new devices and the infinite pains and complication which is of the nature of European engineering gave an utterly different environment from American conditions."[25]

A look at Commonwealth Edison about 1910 shows how the Harrison Street, Fisk Street, and other components had been shaped into a universal system (see Fig. I.1, p. 3). In addition to the central generating stations, there were sixty-seven substations in the Chicago system; some were converted, obsolete generating stations, but others had been specially built as substations. In 1910 Fisk Street remained the largest of the generating stations, but just across a slip of the Chicago River stood the Quarry Street station, which began operating in 1908 and by 1910 had six 14,000-kw. turbogenerators. Frequencies and voltages were manipulated with ingenuity. All of the Fisk Street units generated three-phase alternating current at 9,000 volts and 25 cycles, but three 5,000-kw.-capacity transformers raised the output of three of the generators to 20,000 volts. The higher-voltage current was for transmission to the more distant substations. In the

[24] *Edison Round Table,* November 1928, p. 12.
[25] Ibid., p. 3.

Figure VIII.7. *Generator capacity,*
Commonwealth Edison Co., 1903–11.
Courtesy of the Samuel Insull Archives,
Loyola University, Chicago, Ill.

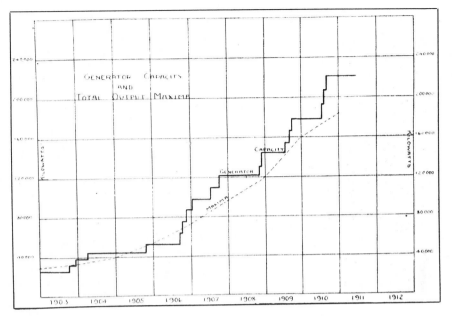

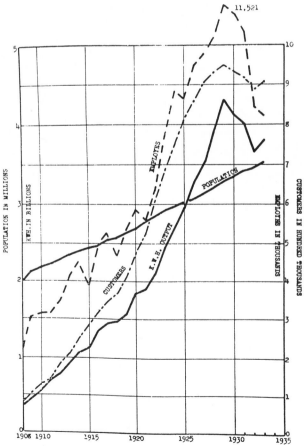

Figure VIII.8. *General statistics, Commonwealth Edison Co. Courtesy*
of the Samuel Insull Archives, Loyola University, Chicago, Ill.

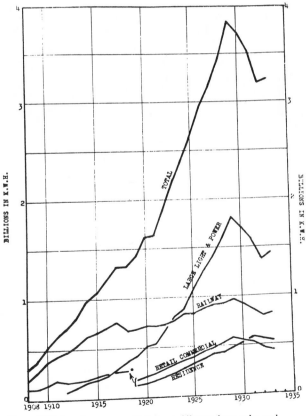

Figure VIII.9. *Commonwealth Edison's kilowatt-hour sales and*
changing loads. Courtesy of the Samuel Insull Archives, Loyola
University, Chicago, Ill.

Quarry Street station four of the generators worked at 9,000 volts and 25 cycles, which facilitated the interchange of current with Fisk, but two other generators produced current at 12,000 volts and 60 cycles. The Quarry Street station also used frequency changers to convert 25-cycle current for the 60-cycle load. The system's substations received current by direct line from the Fisk, Quarry, and Harrison Street stations.

The combination of technologies made possible the supply of incandescent-light load, stationary-motor load, and streetcar/elevated railroad load. Transformer substations reduced the 25-cycle current to usable voltages for the large motors of industrial enterprises; rotary converter substations received very high voltage, alternating, polyphase current direct from Fisk or Quarry and transformed it into direct current at the 600 volts needed by the streetcars and the elevated railway; and the 12,000-volt, 60-cycle current was transformed in substations into the low voltages used by incandescent-light consumers. Since large-capacity tie lines connected Fisk and Quarry, the stations assisted each other in handling the varying peak and low loads. The single Chicago system was impressive. Its total generating capacity was 219,600 kw.; its high-voltage transmission lines extended for 525 miles; and its low-voltage distribution system (mostly underground) spread to all parts of the growing metropolis.[26]

No description of the technology of the universal electric supply system around 1910 would be complete without consideration of the load dispatcher, for he functioned as the control center of the system. After his role was defined, the system concept was consciously and operationally articulated. The load dispatcher's office at Commonwealth Edison dated back to 1903. According to a company publication of that year, "[The] system had assumed sufficiently definite proportions to call for the organization of an operating office prepared to handle at all times the steadily increasing intricacy of detail connected with the operation of a centralized system of electrical supply." In other words, the system could be operated as a "coherent whole."[27]

A primary responsibility of the load dispatcher was to run the Fisk and Quarry Street stations in tandem, or parallel, so that each would carry a reasonable share of the varying system load. Also, when generators in one of the stations failed or had to be taken off load for repair, it was the load dispatcher's responsibility to locate the combination of generators and substations to carry the load in the interim. (There were storage batteries in some of the substations and these facilitated the meeting of peaks under usual and unusual conditions.) The load dispatcher was not only a troubleshooter but something of a historian. At his disposal were load curves, or graphic records, of the varying output from different parts of the system for each hour and each day during past years. By analyzing these, the dispatcher anticipated loads resulting from the social customs and industrial routines of Chicago's population. (The load dispatcher even knew when

[26] Commonwealth Edison Co., *Principal Generating Stations*, pp. 21–22.
[27] National Electric Light Assoc., Commonwealth Edison Section, comp., *How Commonwealth Edison Works* (Chicago: NELA, 1914), p. 62. The following section on the load dispatcher is based on the essay by R. B. Kennedy, "Operating Department: Load Dispatching of Old, and of Today," ibid., pp. 62–66.

Figure VIII.10. Generating-station operating chart, 1914. From National Electric Light Assoc., How Commonwealth Edison Works, *p. 131.*

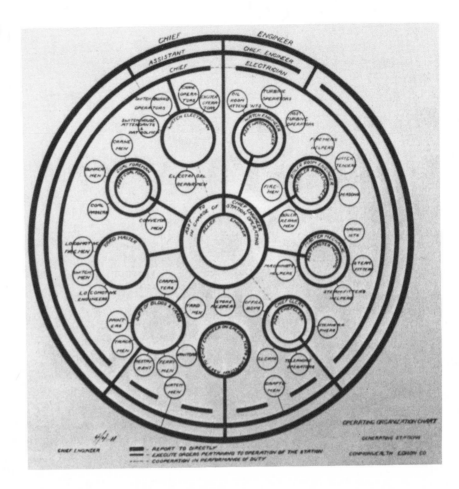

people left work early on Christmas Eve, for the industrial, stationary-motor load dropped and that of the streetcar and elevated railway rose.) He also knew that Chicago observed the Sabbath, and he planned system operations accordingly (see Fig. VIII.11).

The load dispatcher's operating board provided a graphic model of the system for information and control. It showed all generating units, the frequency changers, the transformers, and the layout of the high-voltage transmission lines and the low-voltage distribution system. Indicators showed the condition of all switches. Measuring instruments gave continuous readings of the output and load in various parts of the system. Of critical importance to the load dispatcher was the direct-connected telephone system that provided control over all generating units and switches. The complex control problems of the system involved a number of simultaneous variables, and thus it is not surprising that early analog computers—network and differential analyzers—were used several decades later to solve some of the load dispatcher's problems.

In its technology and organization, the Chicago system represented a synthesis of the ideas and activities of innumerable inventors, engineers, entrepreneurs, manufacturers, and managers from all parts of the world. Insull did not invent the Chicago system in the sense that Edison invented

Figure VIII.11. *Incidents in the routine of load dispatching, 1914. From National Electric Light Assoc.,* How Commonwealth Edison Works, *p. 63.*

Figure VIII.11. *Incidents in the routine of load dispatching, 1914. From National Electric Light Assoc.,* How Commonwealth Edison Works, *p. 63.*

the Pearl Street system. Edison acquired patents on the essential components of his system, and the organizing concept of the system was clearly his. Insull, on the other hand, was not a professional inventor or an engineer. He was, however, a systems conceptualizer comparable to Edison, though on a highly abstract level. Edison was deeply aware of the interwoven fabric of economics and technology, but was relatively naïve about the long-term economic and social factors that made up the environment within which his system functioned. Others, like Grosvenor Lowrey, advised him on such matters as the expansion of consumer and money markets. Furthermore, Edison did not articulate his technological and economic concepts sufficiently to enable a large organization to make decisions and carry out policies without his immediate supervision.

Insull, by contrast, analyzed and articulated concepts that could and did guide policies not only for the Chicago system but for other utilities as well. His conceptual synthesis involved social and market needs, financial trends, economic principles, technological innovations, engineering design, and managerial techniques. Insull discussed his concepts, policies, and experiences in addresses to utilities groups and the public.[28] The speeches dating from around 1910 are especially informative and important because they were given after he had had time to reflect on the formative years in Chicago and while he was presiding over the development of a large and complex enterprise. From the start, Insull was committed by policy to large units—generators, steam engines, transformers, and others. In order to use large, efficient units, he amassed production and distribution facilities into the

---

[28] Samuel Insull, *Central-Station Electric Service: Its Commercial Development and Economic Significance as Set Forth in the Public Addresses (1897–1914) of Samuel Insull,* ed. William E. Keily (Chicago: Privately printed, 1915); idem, *Public Utilities in Modern Life: Selected Speeches, 1914–1923,* ed. William E. Keily (Chicago: Privately printed, 1924).

large generating stations at Harrison, Fisk, and Quarry streets. He also adhered to the principle of low rates to increase consumption and of a tariff structure differentiated according to the pattern of customer demand. Insull also consistently sought to improve the diversity of load and load factor by means of carefully planned sales and load management. In order to reduce the cost of borrowing money and thereby the unit cost of output, which included interest paid on capital invested, he sought a regular return on his investments; in this way he hoped to establish a highly favorable market for his company's stocks, bonds, and debentures. His concepts were interrelated. By means of a mass-producing monopoly he could supply a diverse load broadly based on low, differentiated rates and obtain a fair return from efficient production and maximum utilization of capital borrowed at reasonable rates of interest. Insull's pursuit of his policies was facilitated by his company's superb statistics department, which kept him informed of output, performance, costs, and other variables in his business and technological system.

The way Insull's utility managed Chicago's power load deserves special attention. It is comparable to the historic managerial contributions made by railway men in the nineteenth century and is as interesting as the widely publicized managerial concepts and policies of John D. Rockefeller and Henry Ford. Insull explained his method on many occasions, but never more clearly and graphically than in a 1914 address.[29] He began by explaining diversity of demand. Referring to data on a block of apartment buildings in Chicago, he defined the critical items as the "customers' separate maxima" (92 kw.) and the "maximum at transformers" (29 kw.). The first amount was simply the sum of the individual annual maximum demands of the 193 customers in the apartment block. Because of the varied habits and activities of the customers, the maximum demand of each occurred at a different time during the year. Therefore, the maximum simultaneous demand measured at the transformers supplying the apartment block was only 29 kw. A diversity factor of 2.3 was arrived at by dividing the separate consumer maxima by the maximum at the transformers (see Fig. VIII.12).

Citing eleven different classes of consumers, Insull explained the advantages of a diversified load. If the maximum demand of each had occurred simultaneously, Commonwealth Edison's peak load from these customers would have been 26,640 kw. As a matter of fact, on the day the utility carried its maximum load in 1914 (6 January), the eleven customer categories made a simultaneous demand of only 9,770 kw. Insull's graphics showed the maximum demand made by each class of customer during the winter of 1913/14; the shaded portion of each column represented the demand made on 6 January 1914, indicating thereby the responsibility— or lack of it—of each class of customer for the utility's *bête noire*, the annual peak load. Insull enjoyed expanding on his theme, describing the various characteristics of each customer or class of customer, and explaining how these could be analyzed and combined to achieve economies of scale. For instance, the profile for ice manufacturers was optimal because their level

[29] Samuel Insull, "Centralization of Energy Supply," in *Central-Station Electric Service,* pp. 445–75.

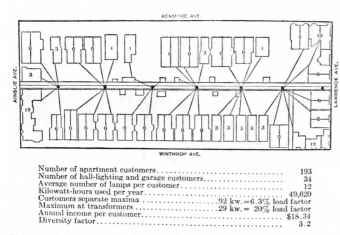

Number of apartment customers............................ 193
Number of hall-lighting and garage customers................. 34
Average number of lamps per customer....................... 12
Kilowatt-hours used per year............................. 49,620
Customers separate maxima.................92 kw.=6.3% load factor
Maximum at transformers ....29 kw.= 20% load factor
Annual income per customer............................... $18.34
Diversity factor.. 3.2

Figure VIII.12. *Diversity of demand for electric service in a block of apartment buildings, Chicago. From Insull,* Central-Station Electric Service, *p. 448.*

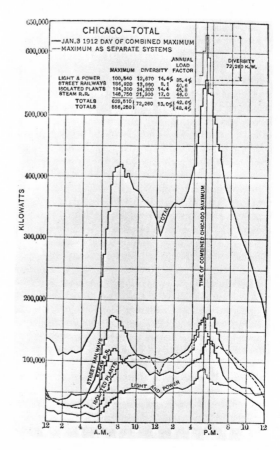

Figure VIII.13. *Chicago load on the Commonwealth Edison system, 3 January 1912: Diversity factor advantage, 72,260 kw. From Insull,* Central-Station Electric Service, *p. 267.*

of activity was low on cold January days when Commonwealth Edison's capacity was strained by the heaviest loads of the year. The brickyards and quarries also fit in well with Insull's scheme of things, for "frost interferes with their business."[30] (See Fig. VIII.14.) If it had not been for the complex metering and information-gathering techniques of Commonwealth Edison and the analytical services of the company's statistics department, Insull would not have had the indicators that allowed him to exploit diversity and manage the system's load.

Closely related to the concept of diversity was load factor. Early in his career as a central-station manager, Insull recognized the load factor as the most important operating principle in central-station economy.[31] The concept was not his invention, but he relentlessly pursued the policies that followed from it. Load factor is the ratio of the average load to the maximum load of a customer, group of customers, or the entire system during a specified period.

Insull considered load factor to be the paramount economic principle of central-station operation, but others had defined it for electrical engineers, and especially for station managers, earlier. The term may have been introduced in 1891 by Colonel R. E. Crompton, one of the pioneers of British

[30] Ibid., p. 452.
[31] Samuel Insull, "The Development of the Central Station," in *Central-Station Electric Service,* p. 27.

Figure VIII.14. *Diversity of large customers, Chicago. From Insull, Central-Station Electric Service, p. 450.*

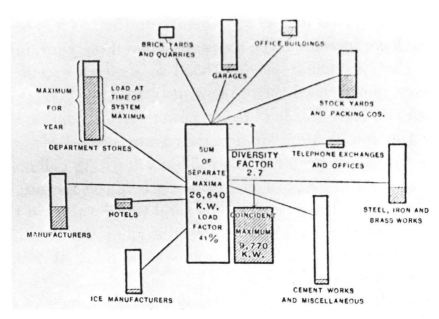

electrical engineering (see p. 58 above).[32] Even before it was defined, Edison and others who were concerned about the economics of electric supply realized that regularity of load and maximum practical utilization of generating capacity were the dominant commercial aspects of central-station supply. Long before the rise of electric supply, managers in other fields, such as railroad operation, also realized the importance of utilizing capital (plant and related facilities) fully in order to obtain a maximum return on investment. As articulated so well by electrical engineers and managers, the reason load factor was critical for electric supply is that electricity cannot readily be stored; it must be used when generated. (Storage batteries and pump-storage facilities are exceptions to the rule.) Because electricity cannot be stored, generating capacity cannot be used to produce power for storage when demand is off, say, in shoe or sewing-machine manufacture.

To raise the load factor, Insull sought customers whose load curves lacked high peaks and low valleys. He also wanted types of customers whose combined load curves were diverse and combined well. (He recognized the relationship between diversity and load factor.) To have had several classes of customers with good individual load curves but simultaneous peak demands would have resulted in a poor system load factor despite the good characteristics of the individual loads. Through load management and planned sales of machinery and appliances that would improve the load, Commonwealth Edison improved the load factor from year to year. As a result, the cost per kilowatt resulting from fixed, or investment, charges fell, which greatly pleased Insull. He identified interest on investment as the "most important factor in cost in any public service business."[33]

Insull later recalled that he became aware of the importance and complexity of the load factor when he first met Arthur Wright, manager of a

[32] Parsons, *Early Days of the Power Station Industry,* pp. 208–9.
[33] Insull, "Development of the Central Station," p. 29.

small municipal station in Brighton, England. Wright also informed him of the critical relationship between metering and load factor. "We had to go to Europe," Insull noted, "to learn something about the principles underlying the sale of the product."[34] Vacationing in England during the Christmas season of 1894, Insull first encountered the results of Wright's system of metering. In Brighton, Insull noticed that small shops burned electric light as though indifferent to the amount they consumed. Upon inquiry he found that Wright had developed an ingenious metering system that measured not only actual use but also the extent to which each customer used his installed capacity. So metered, the customer paid a charge that was fair and economical. His bill reflected his share of the central station's fixed costs that resulted from the generating capacity the central station had installed in order to carry his peak load; the bill also mirrored his share of operating costs, which varied with the amount of electricity he consumed. A customer with many seldom-used electric lamps paid more than one who had the same number of lamps installed and used them often—unless the rate he was charged reflected the installed load as well as use. In essence, Wright's metering system took into account the customer's load factor. Insull acknowledged that Boston Edison Company may have considered adopting the Wright system before Commonwealth Edison did, but he asserted that it was his company that developed and promoted the system in the United States until it became a widely used basis of charging for energy.[35]

The usefulness of the load factor and related load curves as guides to utility policy making was greatly enhanced by the introduction of the universal supply system in Chicago and elsewhere. As noted, the universal system provided for load diversity, so by inspecting load curves and analyzing the load factor, a manager like Insull could see where his utility's load curve needed to be raised and could cultivate the appropriate load, or customers (see Fig. VIII.15). One early discussion of load factor was succinct and direct:

"Load factor" is the convenient term which has been introduced to express this percentage, and a load factor of 40% is rare; 20% is more like the average results. Take for instance a load factor of 25%, with a station which pays 6% dividends. This means, that with one quarter its greatest possible output, the investment is earning the whole of its interest in addition to the running expenses and depreciation, due to the actual output. Consequently, if additional load can be added to the same plant, any excess of income from it, over the entire running expense and depreciation due to it, is so much net profit. For special services limited to the slack hours, the station can profitably make special prices. With the load curve before his eyes, showing plainly and definitely the hollows to be filled up and the low levels to be raised, the manager can best apply his local knowledge of his customers and possible customers, to devising plans for meeting their needs without increasing his plant—plans increasing his customers or their profits, and showing to his own satisfaction on his annual balance sheet.[36]

[34] Samuel Insull, "Stepping Stones of Central-Station Development through Three Decades," in Central-Station Electric Service, p. 351.
[35] On the adoption of Wright's meter system, see Insull, "Memoirs," pp. 88–89; Seymour, "History of Commonwealth Edison," pp. 278–79.
[36] Canadian Electrical News, December 1894, p. 168.

Figure VIII.15. *Load diagrams,*
Commonwealth Edison Co., 1907–8. From
Insull, Central-Station Electric Service,
p. 56.

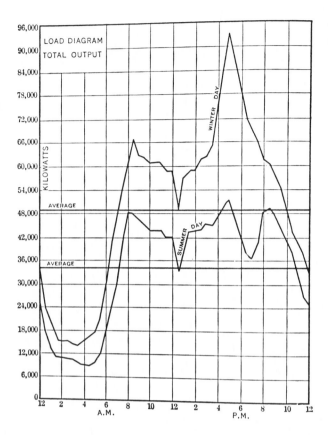

Not only could station managers plot business strategy using the load factor, but manufacturers could as well. They could anticipate the kinds of machines and appliances the utility manager would want to have available for those customers who would improve the load factor. For instance, in the 1920s, household appliances were pushed by manufacturers and utilities. Before World War I, Insull and others were cultivating the traction (streetcar, elevated railway, subway, and interurban lines) and industrial, or stationary-motor, loads. The manufacture of equipment in these areas was correspondingly brisk.

The remarkably large traction load cultivated by Commonwealth Edison reflects the management strategy adopted by Insull and his associates in 1901. By 1908, a year for which statistics are available, the company's traction load was 65 percent of its sales. For several years this percentage increased slightly and began to decline only at the outbreak of World War I as the percentage for large power users swung rapidly upward. (See Fig. VIII.8, p.213.) The disproportionately large traction load followed from Commonwealth Edison's decision to take advantage of an opportunity created by the departure from Chicago of Charles T. Yerkes, the transportation magnate, in 1901. With Yerkes no longer in the picture, streetcar and elevated railway operators were free to make choices. In the past, they had constructed and operated their own power plants because Yerkes wanted the transportation companies he controlled to buy electric generating equipment from the manufacturers he controlled. (Insull did not have exclusive rights in Chicago for traction equipment as he did for power equipment.)[37]

Drawing upon the capacity of his staff to make statistical analyses, Insull had data prepared showing that the transportation companies could buy power more inexpensively from his company than from their own isolated plants. His advantage was that Commonwealth Edison could exploit diversity of load more effectively than could isolated plants, with their concentration on one type of load. To put it differently, Insull realized that an increased traction load would improve his company's load factor and thereby reduce the unit cost of power produced. His analysis of load curves revealed that the lighting, stationary-motor, and traction loads peaked at different times during the average twenty-four-hour period and that therefore the general-purpose utility could take on a traction load without increasing its generating capacity to the same extent. The traction load would consume some of the unused stationary-power and light-generating capacity that was available during the early morning and evening rush hours. Similarly, the lighting and stationary-power customers would use some of the newly installed capacity when the traction load decreased.

Insull's company obtained its first major transportation contract in 1902 from the Lake Street Elevated Railroad and within a relatively short time had signed contracts with all elevated railway and surface streetcar companies in the city.[38] To obtain these contracts, Insull and his salesmen negotiated especially low rates with their large customers. In about 1910 Commonwealth Edison's traction rates ranged from less than a cent per kilowatt-hour down to four-tenths of a cent. These rates differed from those for relatively small commercial and residential customers, to whom a uniform rate schedule was applied. Regularly reduced, these uniform rates dropped from twenty cents per kilowatt-hour in 1892 to ten cents in 1908. (The average rate for all classes of customers dropped from twenty cents per kilowatt-hour in 1892 to two and a half cents in 1901.)[39] The company's small customers may have complained that they were subsidizing the low rates for the traction companies, but Commonwealth Edison would most likely have argued that the cost of electricity in Chicago for all customers was below that in most other large cities, and that furthermore, the precise costing of service was a bookkeeping decision.

Insull's innovative decision in 1901 to introduce 5,000-kw. turbines in the Fisk Street station and his decision to sell electricity on a large scale to the transportation companies was interrelated. His anticipation of the large-capacity power units and of lower generation costs fueled his company's drive for the large customers. At the same time, care and foresight were exercised to avoid scheduling new demand before additional capacity—especially the new turbines—had been installed and satisfactorily tested. In the case of electric utilities, sales pressure—or the lack of it—signals idle capacity or threatened overload.

[37] McDonald, *Insull,* pp. 96–97.
[38] Notes prepared on 8 May 1926 by William E. Keily for a speech delivered by B. E. Sunny at the Franklin Institute in Philadelphia on the occasion of the unveiling of a portrait of Samuel Insull, p. 6, Insull Papers. Keily edited Insull's two books of speeches, *Central-Station Electric Service* and *Public Utilities in Modern Life.*
[39] McDonald, *Insull,* p. 104.

Many businessmen considered the rapidly increasing traction load to be the important market for electricity. The original names of many electric utilities included the word "railway." After 1910 Insull saw the danger of their myopic concentration on traction, and the kilowatt-hour sales of Commonwealth Edison began to include an increasing percentage of sales to large industrial customers with whom special contracts and rates were negotiated. A disproportionate traction load violated Insull's commitment to diversity. By 1925 the company's large stationary-power and lighting loads amounted to 40 percent of the total, the traction load only 32 percent; in 1929 the figures were 47 percent for the large light and power customers and 25 percent for traction.[40]

Insull was determined to replace all of Chicago's isolated power plants, both traction and stationary, with his system. In May 1908 the *Electrical World and Engineer* reported that "although isolated plants are still numerous in Chicago, they were never so hard pressed by central station service as now. . . . The Commonwealth Edison Company has among its customers establishments formerly run by some of the largest isolated plants in the city. . . . The manufacturing districts also have been invaded and all-day motor load secured."[41] The *Electrical Review and Western Electrician* confirmed the trend, noting that in addition to the great growth in traction load there was an increase in stationary-motor load, including a number of smaller factories and shops and "a large number of great manufacturing and industrial plants."[42] Testifying before the London County Council in 1914, Insull boasted: "Now we do a big industrial business. We have six customers who buy from us 100,000,000 units a year."[43] This amounted to about one-tenth of the total power generated by Commonwealth Edison in 1913. Insull also told the London County Council that his company, in spending the equivalent of £100,000 annually to get new business, matched the amount spent toward this end by all the utilities in Great Britain and Ireland. (Opponents of municipally owned companies in Britain also criticized their poor sales effort.)

In 1909 the Commonwealth Edison Company opened an "Electric Shop" in downtown Chicago. The ground floor emphasized domestic appliances; the floor below, an "Industrial Power Room" with a wide variety of motor-driven machines, including lathes, drills, punches, and so on.[44] City engineers from Melbourne, Australia, who stopped in Chicago on their tour of electric supply businesses and technology centers in North America and Europe, found the "Electric Shop" a remarkably effective marketing scheme. They did not find that individual Chicagoans or Americans "hustled" no-

[40] Statistics from a set of tables and graphs in the Insull Papers labeled "used in federal material"; see especially pp. 7 and 7A. I have assumed that the figures labeled "railway" include all types of traction load.

[41] "The Systems and Operating Practice of the Commonwealth Edison Company of Chicago," *Electrical World and Engineer* 51 (1908): 1023.

[42] F. B. Bernhardt, "New Features of Central Station Service in Chicago," *Electrical Review and Western Electrician* 54 (1909): 975.

[43] Testimony of Samuel Insull, 23 January 1914, in London County Council, *Special Committee on London Electricity Supply, 1913–14* (privately printed paper on file in the Merz & McLellan Co. Archives, Amberly, Killingworth, near Newcastle upon Tyne, England), p. 7.

[44] "The Chicago 'Electric Shop,' " *Electrical World and Engineer* 53 (1909): 684–89.

ticeably more than individuals elsewhere in the world; what impressed them was the organized "hustling" they observed. ("Of individual hustle little was to be seen, but of efficient organisation there was a great deal.")[45] They were also impressed by Commonwealth Edison's contract agent, who was generally responsible for sales and who employed five assistants at $3,000 annually, one of whom concentrated entirely on sales to very large consumers with isolated plants. They also took note of the company's 125 salesmen, or solicitors, who did canvassing for an annual wage of $1,000 plus commissions. This sales machinery impressed the Australians.[46]

Commonwealth Edison also made good use of newspaper and periodical advertising and publicity. It published the names of the large industrial customers it incorporated into its system and ran newspaper advertisements appealing to "Mr. Private Plant Owner."[47] Beginning in January 1912, a leading trade periodical, the *Electrical Review and Western Electrician,* carried a series of articles on the introduction of stationary power in industry. Between 20 January 1912 and 7 February 1914, thirty-four major articles appeared, and half of these dealt specifically with Chicago industries, nine of which were identified as being supplied by Commonwealth Edison. Many of these articles probably originated with Commonwealth Edison, a practice that was not unusual among trade journals.[48]

By providing inexpensive and convenient power, both stationary and traction, Commonwealth Edison enhanced Chicago's attractiveness to industry. As Insull said in one of his many addresses, Chicago's greatness depended on outstanding leaders and material advantages, and among the latter "the possibility of procuring cheap electric power is most important."[49] That he made this extravagant claim for electrification as early as

[45] City of Melbourne, *City Electrical Engineers' Notes on Tour Abroad* (Melbourne, Australia, 1912), p. 26.

[46] Ibid., p. 27.

[47] "Electric Service in Chicago," *Electrical World and Engineer* 61 (1913): 1144.

[48] The following industrial customers were listed in these articles as taking power from Commonwealth Edison: Hafner Furniture, Palmer House Laundry, Empire Mattress Co., Garden City Sand Co., Bunte Bros., Roberts Sash & Door Co., Daniel Burkhartsmeier Cooperage Co., A. O. Fisher Co., Morand Bros., Imperial Leather Manufacturing Co., and E. B. Miller & Co.

Listed as Commonwealth Edison customers in "Electric Service in Chicago," *Electrical World and Engineer* 61 (1913): 1137–45, were L. Wolff Manufacturing Co. (plumbers' supplies), Crane Co. (pipes, valves, fittings), Joseph T. Ryerson & Son, Ajax Forge Co., Union Drop Forge Co., Vierling Steel Works, Federal Engraving & Colortype Co., Oscar Heineman Co. (silk manufacturers), and Hugo DuBrock & Co. (maker of shirtwaists).

In addition, "Electric Service in Chicage Suburbs," ibid., pp. 1243–54, listed the following as the company's customers: American Brake Shoe Foundry, Melrose Park; Bliss & Laughlin, Harvey; Chicago Hardware Foundry Co., North Chicago (est. in Chicago area 1902, but according to *Moody's Industrials,* 1912); Edgar Allen American Manganese Steel Co., Chicago Heights; Franklin Park Foundry Co., Franklin Park; King & Andrews Co., Chicago Heights; The Buda Co., Harvey; Central Locomotive & Car Works, Chicago Heights; Cyclone Fence Co., North Chicago; Forsyth Bros. Co., North Chicago; Victor Chemical Co., Chicago Heights; United Car Co., Chicago Heights; Vulcan Louisville Smelting Co., North Chicago; Acme Brick Co., Chicago Heights; Whiteacre Fireproofing Co., Chicago Heights; and Dolese & Shepard, Hodgekins.

I am indebted to Barbara Kimmelman of the University of Pennsylvania for this compilation.

[49] "Mr. Insull on the Sale of Electric Energy in Chicago," *Electrical World and Engineer* 54 (1909): 1024.

1909 is notable, even though he made it before an audience as sympathetic as the Electric Club of Chicago. Cheap electric power attracted industry to the city in the way that natural resources drew industry to the countryside. After the exploitation of Niagara Falls, many of Insull's contemporaries naturally assumed that water power would become the focus of a cluster of industries. However, the combination of a large, concentrated market and cheap electricity in the nation's large cities had a more powerful, if less dramatic, attraction for industry.[50] This trend was ironic in that some visionaries, including Peter Kropotkin in 1899 and Ebenezer Howard in 1902, had predicted that transmission of electric power would disperse the population across the green countryside.[51]

By equipping production and distribution facilities with large, high-efficiency units, increasing the Chicago system's diversity and load factors, and adopting a metering system that reflected the principle of load factor, Insull articulated, clarified, and synthesized many technological and economic principles during his first two decades at Commonwealth Edison. Moreover, by 1914 his concepts had brought an order that embraced even more complex and abstract variables. For example, he explicated the relationship between kilowatt-hours sold per capita, a utility's income per kilowatt-hour sold, and a system's annual load factor. For purposes of comparison he chose the municipal utilities of large American cities, Berlin, and London. (Insull, who had close contacts abroad, often compared conditions in America with those in England and Germany.) According to his analysis, as the income per kilowatt-hour fell, the output per capita increased; correspondingly, as the income per kilowatt-hour decreased and the output per capita increased, the load factor increased. Of the systems Insull analyzed, the one at Niagara Falls had the highest output per capita, the lowest income per kilowatt-hour, and the best load factor; London had the poorest. Cheap hydroelectric power and a wide area of transmission explained the data on Niagara; London had dispersed, small generating stations and a relatively small industrial (or motor) load, among, from Insull's point of view, other problems. Insull took satisfaction in pointing out that despite its dependence on thermal plants, Chicago compared favorably with Niagara and San Francisco, which utilized hydroelectric power.[52] (See Fig. IX.10, p. 259.)

Insull usually had at his fingertips an impressive array of statistical data. Besides what has been noted, he could cite for Commonwealth Edison— and often for other utilities—total capitalization year by year (growth was steady until 1930); the number of the company's customers and the size of Chicago's population (the former increasing more rapidly than the latter, again until 1930); the gross operating revenue and total operating expenses of the company; its total payroll; the connected load by class of consumer

---

[50] After 1860 the availability of steam power in the urban centers of the United States drew industry away from the countryside with its water-power sites, into the urban centers located at the crossroads of transportation systems. Louis Hunter, *Waterpower: A History of Industrial Power in the United States, 1780–1930* (Charlottesville: University Press of Virginia, 1979), pp. 485–90.

[51] Lewis Mumford, *The City in History* (New York: Harcourt, Brace & World, 1961), pp. 514–24.

[52] Insull, "Centralization," pp. 461–63.

(especially a breakdown into light, stationary power, and railway); taxes; and interest charges (see Figs. VIII.8 and VIII.9, p. 213).

Insull's blend of technology, finance, management, and entrepreneurial drive was by no means exhausted in the mass production of cheap electricity for Chicago. As early as 1910 Insull applied his concepts, technology, and managerial techniques to small-town and rural electrification in Lake County, Illinois. His biographer, Forrest McDonald, believes that "the Lake County experiment was the first demonstration anywhere that systematized electric service was economically and technically possible in large areas, rural as well as urban. The news of it exploded upon the industry."[53] To fulfill his concept of an integrated, centrally supplied system, Insull acquired isolated utilities serving small towns and their vicinities, shut down their inefficient generating plants, erected substations, and supplied electricity from a few large generating plants by means of transmission lines. By 1911 he had organized the entire Lake County enterprise into the Public Service Company of Northern Illinois. By 1923 the company was supplying a territory of 6,000 square miles and 195 communities. In the process fifty-five municipal or privately owned electrical plants were shut down or dismantled and replaced by substations and 875 miles of high-tension transmission lines from four large, efficient central stations.[54]

After successfully interconnecting the central stations of a Chicago utility, and interconnecting Lake County utilities physically by high-voltage transmission lines and then organizationally by merger or acquisition, Insull continued to expand by means of the holding-company structure. By the mid-twenties, when holding companies were flourishing in the American utilities field, Insull operated one of the largest, the Middle West Utilities Company, which had subsidiaries in nineteen states. This company supplied 8 percent of the commercial total of kilowatt-hours for the United States.[55]

[53] McDonald, *Insull,* p. 139.
[54] "Presentation of Public Service Company of Northern Illinois," p. 3.
[55] U.S., Congress, Senate, *Electric-Power Industry: Control of Power Companies,* 69th Cong., 2d sess., 1927, S. Doc. 213, p. xxxvii.

CHAPTER IX

London:
The Primacy
of Politics

COMPARED to Chicago and Berlin, London was a backward metropolis in the early twentieth century. Electric power was believed to be the hallmark of progress in enlightened cities like Chicago and Berlin, and London had notably little of it. Furthermore, London's electrical supply industry was disordered and small-scale in a time of rationalization and mass production. Chicago and Berlin each had a centralized light and power system supplying the entire city from a handful of modern power stations; Greater London had sixty-five electrical utilities, seventy generating stations averaging only 5,285 kw. in capacity, forty-nine different types of supply systems, ten different frequencies, thirty-two voltage levels for transmission and twenty-four for distribution, and about seventy different methods of charging and pricing.[1] Londoners who could afford electricity toasted bread in the morning with one kind, lit their offices with another, visited associates in nearby office buildings using still another variety, and walked home along streets that were illuminated by yet another kind. This was a night-mare for the supplier of electric lamps and appliances; interesting problems arose when one moved one's home or business. London's engineers and businessmen made such remarks as "The largest city in the world offers an excellent example of what electric supply ought not to be" and "What exists in London to-day is hardly worth talking about."[2] By World War I, London's progressive citizens were waiting eagerly for the city to get in step and catch up.

Those who equated technological change with progress and wanted to stimulate remedial activity referred to the Chicago and Berlin utilities as models of technological achievement in lamenting the state of affairs in London. Closer at hand was another exemplar, the electric supply system on the northeast coast of England. Among those who testified before Par-

---

[1] These statistics are for the year 1913. See London County Council, *Report on London Electricity Supply,* prepared by Merz & McLellan (London, 1914), app. 2, pp. 37, 38.

[2] "Electrical Development in England," *Electrical World* 82 (1923):1056. See also testimony of Charles H. Merz, 6 June 1913, in London County Council, "Special Committee on London Electricity Supply, 1913–14" (privately printed paper supplied to council members for their information only), p. 8; copies in Merz & McLellan Co. Archives, Amberley, Killingworth, near Newcastle upon Tyne, England.

liament, the London County Council, and engineering society committees on the subject of electricity in London were Samuel Insull of Chicago, Georg Klingenberg of Berlin, and Charles Merz of the Newcastle upon Tyne Electric Supply Company. Merz and Insull were their respective countries' leading spokesmen for large electric power systems. They were good friends, exchanged engineering and managerial information, and lamented the situation in London. Merz used Chicago and his own district as models for London and called on Samuel Insull to explain to Londoners his enviable success in Chicago. Georg Klingenberg also was brought to London to tell the city how backward it was electrically. Klingenberg's authority came from his position as chief engineer for power systems for Allgemeine Elektrizitäts-Gesellschaft, the electrical manufacturing company that also owned and operated the Berlin utility.

On the eve of World War I, Merz, Insull, and Klingenberg all testified to London's retrograde condition. (See Fig. I.5, p. 16.) Merz also prepared a report for the London County Council suggesting ways to reform the electric supply industry in the metropolis. Speaking in London, Insull, who was mindful of English civility, tactfully avoided making any explicit comparisons; he simply showed how advanced his Chicago system was. Back in America, however, he displayed a chart that revealed London's place in the hierarchy of the electrical supply world. His graph showed that of ten leading Western cities, including Chicago and Berlin, London had the lowest output of electricity per capita, the highest utility income per kilowatt-hour, and the lowest load factor.[3] These were the antitheses of America's mass-production style. Insull, a champion of American mass production (whether of kilowatts or mechanical gadgets), was appalled to find that most London utilities managers were content to earn a high rate of return on a relatively small investment from profits that were distributed over relatively few units.

Merz never tired of describing the irrationality of electric supply in London. In 1913 he advised the London County Council that if London used electricity on the same scale as Chicago, consumption would rise threefold. He said: "We use . . . an absurdly small amount of electricity. . . . we ought to be first, not second, or third, or last; and I am afraid that at the present rate of progress there is very great danger of our not only being last, but of our remaining last."[4]

Klingenberg, in a paper read in 1913 before the Institution of Electrical Engineers in London, also compared London with Chicago and Berlin. He told his audience that use of electricity in London was only 110 kwh. per capita as compared to 310 kwh. in Chicago and 170 kwh. in Berlin.[5] He noted that London had the most-fragmented supply system of the three cities. He also reported that London's power plants worked the least eco-

[3] Insull's data are presumed to be for public supply and for the year 1913 or 1914. See Samuel Insull, "Centralization of Energy Supply," an address given in New York on 20 April 1914 and published in his *Central-Station Electric Service: Its Commercial Development and Economic Significance as Set Forth in the Public Addresses (1897–1914) of Samuel Insull*, ed. William E. Keily (Chicago: Privately printed, 1915), p. 461.

[4] London County Council, *Report* (1913–14), p. 3.

[5] Georg Klingenberg, "Electricity Supply in Large Cities," *The Electrician* 72 (1913): 398–400.

nomically. Following his comparative analysis, Klingenberg volunteered a plan to make London more like Berlin and Chicago. His paper created a stir among the British engineers; while some cited it as more evidence of the need for reform, others defended London by interpreting the statistics differently. For example, the statistics for the centralized Berlin and Chicago systems included a substantial supply of power for traction, but this was not the case in London, where traction load was supplied mostly by isolated, private plants. Others, however, argued that the fact that traction was not supplied by public power stations in London was an additional indication of the need for a centralized system. The editor of the trade journal *The Electrician* heard the various arguments and then sided with those who found London's electricity supply "the maddest sort of performance."[6]

In 1913 London was the largest city in the Western world. The population of Greater London was 7.25 million; Greater Berlin had a population of only 3 million; Chicago counted about 2 million; Paris, nearly 3 million; and New York, 5 million.[7] The City, the small financial district in the heart of London, remained the financial heart not only of the empire but of the world as well. The port of London still drew the world's largest annual tonnage.

This great city had an extremely complicated administrative structure that defied comprehension by non-Londoners and reform-minded Londoners alike. The complex structure had a long history and was reinforced by a strong respect for tradition. London's pluralistic and historic administrative structure deeply affected the growth of the electric supply industry there.

On the local government level, electricity supply in London, as elsewhere in England, was under the jurisdiction of local authorities, a complex array of administrative units. After passage of the Local Government Act of 1888, these administrative units, or local authorities, included counties, municipal boroughs, and parishes. As would be expected in a country so sensitive to history, there were numerous anomalies that defied taxonomic analysis. The situation in London, for instance, involved exceptions. The City of London, 673 acres of financial activity, was one of the local authorities of Greater London, and one with substantial autonomy and influence. The City and twenty-eight other local authorities (municipal boroughs) made up the structure of the Administrative County of London, which at the turn of the century was a part of Greater London. In 1888 a move toward centralized authority for London was made through the creation of the London County Council, whose members would be elected. The council's jurisdictional aspirations included the counties of London, Middlesex, and portions of four other counties. The power of the county council was

---

[6] Ibid. J. F. C. Snell presented the London situation more favorably in the ensuing discussion (see ibid., p. 400). The editorial appeared on p. 393 of the same issue.

[7] The population count depended upon the area being defined as the city. Greater London, for instance, included not only the County of London, with an area of 117 square miles, but also the whole of Middlesex County and parts of four other contiguous counties, or an additional 693 square miles. The radius of Greater Berlin extended ten miles from city center, but "old" Berlin was smaller and its population was about 2 million.

challenged by the local authorities, or the counties, borough councils, and the City.

The functioning of the aggregate of administrative authorities was further complicated by Parliament's powers, which included the authority to enact legislation defining electric supply undertakings (privately owned companies and government-owned enterprises) and establishing the conditions of regulation. The parliamentary legislation affecting electrical supply in the country and in London consisted of a series of enactments. Generally, each of the electric lighting (later, power) acts was passed following the deliberations of a parliamentary committee. The committees were formed when Parliament deemed that the state of the technology had changed to the extent that the introduction or modification of regulations was necessary. Usually, however, the legislators were unable to keep pace with the inventors and engineers. The first Electric Lighting Act (see pp. 58–61 above) came in 1882 after a committee headed by Lyon Playfair decided that electricity supply enterprises were of sufficient import to require statutory recognition and regulation. The act of 1882 enabled the Board of Trade, a department of the central government, to grant provisional orders for electric supply to local government authorities or to private companies. Provisional orders by-passed the expensive and time-consuming private bill in Parliament, but after issuance had to be confirmed by Parliament. This law, in the opinion of private enterprise, stifled the development of the electric supply industry because the provisional orders provided that private companies could be purchased after twenty-one years upon demand of the local authority within whose territory the company supplied electricity. The Electric Lighting Act of 1888, the provisions of which were recommended by another parliamentary committee, extended this period to forty-two years, but left electric supply in the hands of local authorities, such as the municipal boroughs of London, or with private companies whose territory was coextensive with that of the local authority. Political boundaries continued to define the range of technology.

Further parliamentary deliberations and decisions affected electric supply following the development of the universal power and light system in the 1890s. Because the British electric supply undertakings (private companies and local-authority-owned enterprises) were not introducing the new system to the extent it was being adopted abroad, a joint committee of both houses of Parliament in 1898 investigated and recommended modifications in the Electric Lighting Act of 1882 and the amending act of 1888. The committee, headed by Viscount Cross, recommended that authority be given to utilities to supply power in bulk, or wholesale, to other undertakings, whether privately or local-authority owned. It was expected that smaller enterprises located in the territory assigned to the large power undertakings would thus distribute the power to consumers. The power undertakings could also supply power—not light—directly in districts where no supply undertaking existed. Parliament then proceeded to enact a number of bills establishing the power undertakings, most of which were private companies. The bills were all opposed by local authorities, but they included clauses recommended by a House of Commons committee headed by Sir James Kitson stating that power should not be supplied by a power company to any district already being supplied by an electric supply undertaking unless that undertaking approved. Unreasonable denial was precluded and could be overruled by the Board of Trade. Power companies, however, were slow to grow under the new authority, and they argued that the reasons were the withholding of approval by local authorities for the power undertakings to supply power directly to large customers in their districts and the failure of local authorities to negotiate seriously to buy power in bulk for distribution. All of these conditions and effects applied to London, where the local authorities guarded their privileges in the face of would-be power suppliers.

In addition to administrative complexity, both local-authority and parliamentary, the state of industrial development in London must be taken into account in explaining the history of electric supply there. Statistics suggest that the symbiotic relationship between industrialization and electrification that was observed in Chicago and Berlin was not rife in London. In 1911 only 39 percent of London's kilowatt-hours were used for power; in Berlin, stationary motors and traction used 66 percent of the kilowatt-hour output supplied by BEW; and in Chicago, stationary and traction power took 81 percent of the kilowatt-hour total.[8] Insull boasted that Commonwealth Edison had six large industrial customers who bought 100,000,000 units annually; this was about one-third of the total electricity sold by the London utilities in 1911.[9]

London in the early twentieth century was a manufacturing center and therefore a potential market for electric power. It was the site of the greatest amount of economic activity in the country as measured by the number of workers employed. About one in five English and Welsh workers in manufacturing and services were located in Greater London. On the other hand, large-scale and heavy industries were situated in the Midlands and in the north. Most of London's factories were small. Despite the commitment of twentieth-century engineers and industrialists to large-scale undertakings, London maintained its historic character as a sustainer of small-scale enterprises. As late as the 1930s there were 55,000 factories and workshops in the London factory area (an area slightly larger than Greater London), but only 34 factories employed more than two thousand workers; 22,000 employed fewer than twenty-five workers, and more than 10,000 employed under ten. This smallness of scale retarded mechanization.[10] The most important industries in London were engineering works (metal trades such as watchmaking, for instance), building, clothing, printing and paper, food, and furniture. The heavier engineering enterprises were concentrated in southern and eastern London, near the Thames River; furniture and clothing manufacturers were located in Bethnal Green, Shoreditch, and Hackney; and printing works were concentrated in the southwest part of the City and in Southwark. Later in the century, with the development of road transport and suburban railways, industry would migrate to the outer ring of London.

[8] Klingenberg, "Electricity Supply in Large Cities," p. 399.
[9] Testimony of Samuel Insull, 23 January 1914, in London County Council, *Report* (1913–14), p. 7.
[10] Merz & McLellan, *Report on Electric Service in New York, London, Paris, and Berlin*, prepared for Consolidated Edison Co. of New York (London: Merz & McLellan, 1937), p. 41. This report estimated that about 20,000 factories in London were using power, mostly electricity.

In London the authorized electrical utilities carried not only a small industrial load but, compared to Berlin and Chicago, a small traction load as well: in London, traction was 12 percent; in Berlin, 31 percent; and in Chicago, 69 percent of the total kilowatt-hours sold.[11] Traction companies in London generated power in their own plants. If the public utilities had supplied the city's underground, trams, and suburban lines, London's total supply figures would have been more impressive, for the metropolis was rapidly electrifying existing transportation systems and constructing new electrified routes. The first section of the London underground, or subway, was opened in 1863. Before 1900 this line and others used steam traction. Because of poor ventilation, however, the underground was extremely uncomfortable, so with the coming of electric power, experimental electrification was tried in 1900. Beginning in 1906 the District system was extensively electrified and the Metropolitan system also introduced electricity. During the next decade large power stations were constructed at Chelsea and Neasden to supply the network, and new electrified lines were constructed. London's surface trams were owned and operated mostly by the county council. These, too, were converted to electric motors. As some observers commented, "Nowhere in the world is there a finer system of electric tramways than those to be seen in London, and controlled by the L.C.C."[12] The failure of the transportation system to draw more power from the multitude of London general-service utilities also partially explains the utilities' failure to develop into large-scale systems. The underlying question remains Why did the London transportation system fail to take power from the utilities as the Berlin and Chicago systems did?

Berlin in the early years of the twentieth century could with reason be called the electrical metropolis. Two manufacturers, Allgemeine Elektrizitäts-Gesellschaft and Siemens, had large plants for the building of power-station equipment there. In contrast, on the eve of World War I the leading British electrical manufacturers as a group produced less heavy electrical machinery than either of the two leading German firms and less than either General Electric or Westinghouse in the United States.[13] Furthermore, not one of the foremost British manufacturing companies had built a major plant for the production of heavy electrical machinery in London. The leading British producers of power equipment were subsidiaries of foreign manufacturers. British Westinghouse had its major plant in Trafford, near Manchester; British Thomson-Houston, a subsidiary of American General Electric, had a major manufacturing facility in Rugby; and Siemens in England manufactured dynamos at Stafford, in central England. The leading British-owned manufacturers of power-plant equipment before World War I were Dick, Kerr & Company and Parsons. These companies had manufacturing facilities for electrical machinery in Preston and near Newcastle upon Tyne respectively. The British-based companies were especially

---

[11] These statistics are for the period 1910–12; see Klingenberg, "Electricity Supply in Large Cities," p. 399.

[12] City of Melbourne, *City Electrical Engineers' Notes on Tour Abroad* (Melbourne, Australia, 1912), p. 52.

[13] I. C. R. Byatt, *The British Electrical Industry, 1875–1914* (Oxford: Clarendon Press, 1979), tables on pp. 150 and 166. Compared are the gross output in electrical machinery in Britain and the electrical manufacturing sales of the German and American companies.

active in the manufacture of equipment for power plants supplying a traction load. The only substantial electrical manufacturing plants located in or near London produced incandescent lamps and telegraph cable. The head of the British General Electric Company (GEC) (not associated with American General Electric) explained that manufacturing plants were situated where the relevant manufacturing skills had become almost hereditary. GEC, for instance, chose Witton, near Birmingham, for heavy machinery because of the presence of the iron and steel industries there; it situated its lamp-manufacturing plant (Osram) in Hammersmith because London had an almost limitless reservoir of skilled female workers.[14]

In London, as in Berlin, outstanding members of the academic community taught electrical engineering (see pp. 146–47 above). The academics' influence was complemented by consulting engineering firms with worldwide experience and by the Institution of Electrical Engineers. These professionals were abreast of the latest technological developments in electrical engineering even though, by international standards of size and rationality, London's utilities were considered backward. In papers and lectures the academics and practicing electrical engineers reminded Londoners of their perilous condition in an intensely competitive technological world. They also lamented England's inability to practice its skills in its own capital. Those engineers who were satisfied with things small and beautiful (some of the London stations were models of that kind) were definitely in a minority.

IEE lectures and papers called for the development of large, centralized power systems of the kind that were being built in Chicago, Berlin, New York, and elsewhere. In his presidential address to the IEE in 1901, Sylvanus Thompson said: "Larger areas of supply and fewer generating centres in an area are necessities—it is recognized that the secret of economic working is to generate on a large scale and to distribute over a large area at an appropriate high voltage."[15] The IEE collectively tried to reorganize the electric supply network in London and all of England. Holding meetings and taking testimony from experts for several months in 1901, an IEE committee looked into the backward state of the electrical supply industry. These hearings focused on legislation that affected the industry. The report adopted by the committee on 10 April 1902 identified parochialism on the part of local authorities as the major cause of the problem. In June a deputation from the IEE presented the committee's findings to Gerald Balfour, president of the Board of Trade. Speaking for the deputation, James Swinburne said: "We hold that electrical enterprises should have their limits and boundaries set by economical considerations only, and that arbitrary boundaries, mostly of medieval ecclesiastical origin, should not

[14] Adam G. Whyte, *Forty Years of Electrical Progress: The Story of G.E.C.* (London: Benn, 1930), pp. 50–51, 73. On the British electrical manufacturers, see Byatt, *British Electrical Industry,* chap. 8; and Robert Jones and Oliver Marriott, *Anatomy of a Merger* (London: Cape, 1971). Electrical manufacture in London included cables, lamps, domestic appliances, accumulators, and, later, electronic apparatus. On London industries in general, see P. G. Hall, *The Industries of London since 1861* (London: Hutchinson, 1962), pp. 21, 26, 35–36, 148, 151–52.

[15] Quoted in R. A. S. Hennessey, *The Electric Revolution* (Newcastle upon Tyne: Oriel, 1971), p. 36.

limit the distribution or the growth of electrical systems."[16] The recurring and persistent clash of values between those favoring technological order and those attached to the historical constitution of things was thus expressed succinctly.

The IEE committee wanted to lay to rest the concern that backwardness of science might be the reason why London failed to develop a centralized electric supply system. IEE worthies insisted that England still led the world in science and especially in those disciplines on which electrical invention depended. Paradoxically, the nation's great institutions of science—The Royal Society, the Royal Institution of Great Britain, the British Association for the Advancement of Science, and the Society of Arts—were all headquartered in London, a city that was unable to organize its electric supply network according to "scientific" principles.[17]

The opinion of so many contemporary experts notwithstanding, interesting developments in electric supply were taking place in London. The city provides an unusual opportunity to consider the development of small-scale technology in an era when large-scale enterprises held the interest of most observers. These small undertakings were often appropriate for the demand they met and for the legislative circumstances of the day, and London station managers and engineers took pride in their style of supply. The relatively small scale of the undertakings and of the technology they employed encouraged a sense of personal involvement and influence; the wide variations in choice of equipment and organization manifested personal expression. Furthermore, several of the London companies showed operating efficiencies and economic achievements that were comparable to those of the largest urban stations. Before considering efforts to introduce large-scale electrification in London, we should survey the small—and by some believed to be the more appropriate—utilities.

The Westminster Electric Supply Corporation, Ltd., was one of the oldest of the privately owned companies; like several others, it supplied London's wealthy West End.[18] A. B. W. Kennedy, chief engineer of the Westminster Company from 1890 to 1926, was conservative in his attitude toward technology. A consultant to the London County Council on electrical matters, he testified in 1905 against Merz's plan for a unified supply system in the metropolis. Kennedy argued, for example, that the current of the future was single-phase alternating current, not the three-phase arrangement (a prediction that was far off the mark).[19] He designed for the Westminster Company the small plants at Millbank, Eccleston Place, and Davies Street, which in 1890–91 supplied, among other districts, Belgravia and Mayfair, London's most fashionable quarters. The system is an excellent example of the use of low-voltage, direct-current supply at a time when many engineers judged alternating current to be the way of the future.[20] All three stations had storage batteries charged in parallel with the load during lower-

[16] *IEE Journal* 31 (1901–2): 1324.
[17] *The Electrician* 49 (1902): 46.
[18] The account of the Westminster Electric Supply Corp. and other supply utilities given in this section is based on R. H. Parsons, *The Early Days of the Power Station Industry* (Cambridge: At the University Press, 1940), pp. 71–135.
[19] Byatt, *British Electrical Industry*, p. 125.
[20] Parsons, *Early Days of the Power Station Industry*, p. 98.

demand, daylight hours; late in the evening, after the load had lightened, the generators were shut down and batteries supplied the system, most of which was lighting. Kennedy selected equipment from a number of manufacturers and in so doing ignored the move toward standardization of equipment and concentration of manufacture that was then taking place in the United States and Germany. His choices included Siemens, Crompton, Willans, and Parsons machinery.

Faced with a shortage of cooling water for its steam condensers, the Westminster Company ingeniously collected rain water from the roof of the Victoria Railway Station and carried it in underground mains to the supply stations. In summer, when the utility's load was relatively light, the company operated ice-making machines at its stations to improve the load curve without expanding the area of supply. In 1902 Westminster joined with the St. James' & Pall Mall Company to construct a joint generating station at Grove Road, St. Marylebone. This station transmitted with three-phase current at 6,600 volts to the older stations, whose motor generators converted the supply to direct current for distribution.

The Metropolitan Supply Company, another privately owned London supply company, operated a central station at Whitehall Court. Designed in 1888 by J. E. H. Gordon, the station attracted attention because Gordon, who had been a leading exponent of alternating current during "the battle of the systems," used direct current at Whitehall Court. He anticipated a much more regular load there as well as increased economies from the use of storage batteries charged by direct current during periods of low load and discharged during peaks. Lows and peaks were accentuated in a station that carried mostly a lighting load. Willans high-speed engines were directly coupled to 110-volt generators, a common arrangement at that time in Britain.[21]

After passage of the Electric Lighting Act of 1888, the Metropolitan Company obtained provisional orders to supply the St. Marylebone, Bloomsbury, Lincoln's Inn, and Covent Garden districts, an area with theaters, shops, fashionable residences, and educational and public institutions. Before the turn of the century and the increase in power load, such an area was highly desirable because of its large lighting load. In 1889 Metropolitan commissioned the Sardinia Street station at the southwest corner of Lincoln's Inn Fields. Shifting to alternating current for this station, the company installed American Westinghouse single-phase, 125-kw. alternators with a generating capacity of 1,000 volts.[22] The importation of machinery by English utilities, both from the United States and Germany, was increasingly common after 1890.[23] After a fire almost destroyed the Sardinia Street station in September 1889, the company reequipped it with turbines, and in 1897 converted it from alternating to direct current—a rational decision, even in 1897, for an urban utility supplying lighting to a densely populated area from a centrally located power station.

By installing Parsons's turbo-alternators (turbines and alternating-cur-

---

[21] Ibid., pp. 72–73.
[22] Ibid., p. 76.
[23] On the export and import of electrical machinery and apparatus, see Byatt, *British Electrical Industry*, pp. 167–170.

rent generators combined) in another station at Manchester Square in 1894, Metropolitan proved that London stations need not be technical laggards. Each unit had an output of 350 kw. at 1,000 volts and 100 cycles, which was more than twice the capacity of any other turbogenerator constructed up to that time.[24] Precipitating the choice of turbines was legal action taken by neighbors of the Marylebone station who complained that vibration from the originally installed reciprocating steam engines was an intolerable nuisance. The grounds for this decision differed from those usually given for the introduction of turbines: large units for a large area of supply. Turbines therefore both prolonged the life of, and made obsolete, the relatively small urban central stations.

A rare account of the experiences of engineers and workmen operating an early central station concerns the Manchester Square station after the first turbines were installed. During the first year and a half of the station's operation, these resourceful men kept the turbines running despite the loss of as many as one-third of the turbine blades. They cooled overheating bearings with water and oil—at least until the smoke drove the crew out of the engine house. They held relay valves open with corks and tied weights onto starting levers to keep turbines operating after the steam pressure had dropped below recommended levels. Despite these difficulties, the engineers seem to have approved highly of the new steam engines—at least years afterward when writing their memoirs.

Increasing demand, rising real estate costs, and complaints about smoke, dirt, and noise caused Metropolitan finally to locate a plant outside the district it would supply. In 1897 the company began constructing the Willesden station along the Grand Junction Canal and two railroad lines (six miles from the station's demand center). Again the company turned to American Westinghouse, this time for two-phase, 60-cycle alternators from which current would be transmitted at 10,000 volts to the load center, where rotary converters would change the supply to direct current for distribution. The universal system had finally reached London.

Other privately owned London companies at the turn of the century were the Charing Cross Electric Supply Company, which supplied customers in the City and in the Strand; the City of London Electric Lighting Company; and the County of London Electric Lighting Company. To meet demand, the Charing Cross Company in 1900 constructed a station four miles from the City and transmitted with three-phase alternators, at 11,000 volts, to substations for conversion to direct current. Because of the shortage of cooling water, the station used sixteen circular steel-cooling towers. The City of London Company's Bankside station became one of London's largest. Among other customers, it supplied the City's printing establishments, which needed direct-current distribution; the company persuaded the printing plants to abandon their own isolated installations.

In 1911 in the area under the jurisdiction of the London County Council (LCC) there were thirteen privately owned and fifteen local-authority-owned electric supply utilities. The number of utilities and the number of municipal boroughs were the same (twenty-eight). (There were sixty-five utilities in the Greater London area.) In the LCC area, private companies had

---

[24] Parsons, *Early Days of the Power Station Industry*, p. 79.

Figure IX.1. *Sebastian Ziani de Ferranti at the age of twenty-two, shortly before his appointment as engineer of the London Electric Supply Corp., Ltd. Courtesy of the Ferranti Ltd. Archives, Hollinwood, England.*

invested £14 million, compared to local authorities' investment of £6.25 million.[25] The leading local-authority-owned supply companies were those for St. Marylebone, Stepney, St. Pancras, Hammersmith, and Poplar. Generally, the local-authority utilities had to supply the industrial and poorer housing areas of east and south London. The vestry of St. Pancras was the first local authority in London to supply electricity. Its original intent in 1883 in obtaining a provisional order was to keep private companies out of its district, but in 1890 the early policy of " 'masterly inactivity' could no longer be maintained, in view of the intentions of the Board of Trade that a supply of electricity should be available to all who needed it."[26] Seeking the best advice, the vestry turned to Professor Henry Robinson, Dr. John Hopkinson, and Sir William Preece. The resulting direct-current station with storage batteries near Regent's Park was a model of its kind. Water for its condensers was cooled on top of the central-station building, and the steam clouds rising from it became a well-known part of the cityscape. The vestry's second power station, on King's Road, also was interesting; it operated in conjunction with a refuse destructor. According to availability, either hot gases from the burning refuse or coal could be used to fuel the station.

There were other refuse-burning, electric power stations in London. The Ealing Local Board opened a central station in 1894 which supplied seventy arc lamps and the equivalent of eight thousand 8-c.p. incandescent lamps. The refuse plant raised sufficient heat to supply an estimated fifty horsepower; the remainder came from burning coal. Not content with this ingenious use of waste matter, however, the designers of the station used converted sewage as condenser cooling water. In 1896 the local authority at Shoreditch carried out an even more ambitious scheme. Having paid a contractor nearly £2,000 annually to remove ashbin refuse, the station's designers installed twelve "cells," each of which had the capacity to burn 8–12 tons of refuse per day as fuel. Considerable controversy—especially over regularity of supply—arose concerning the value of refuse as fuel, but a spokesman for the Shoreditch station stated in 1897 that enough heat was raised from the refuse to drive the 250-kw. plant.[27]

The small-area station, both privately and local-authority-owned, prevailed in London before World War I. Parliamentary legislation, the small scale of the city's industrial enterprises, the absence of electrical manufacturers, and concern for the preservation of local government, among other factors, sustained this style of electrification. There were, however, two considerable attempts, led by two outstanding British engineers, to swim against the tide and bring large-scale, centralized supply to London. Both efforts followed modifications in existing legislation and the development of political attitudes that seemed to favor a change of style. The first endeavor, headed by the engineer Sebastian Z. de Ferranti, was stimulated by the amending Electric Lighting Act of 1888, and the second, led by Charles H. Merz, was intended to take advantage of the recommendation of the Viscount Cross Committee that Parliament authorize power com-

---

[25] City of Melbourne, *Engineers' Notes*, p. 11.

[26] Parsons, *Early Days of the Power Station Industry*, p. 124.

[27] Ibid., p. 135.

panies. It is worth noting that Ferranti's endeavor coincided with the development of alternating current and that Merz's endeavor followed the wide acceptance of the universal system.

In conjunction with the easing of the recession of the mid-1880s and the rapid rise of the central-station industry in America, the Electric Lighting Act of 1888 dramatically increased interest in electric power in Britain, especially in London. Faced by a number of questions concerning the regulation of future supply in London and with requests from fifteen private companies and local authorities for authorization to supply London, the Board of Trade, which exercised parliamentary authority over electric utilities, resorted to a committee headed by F. A. Marindin for guidance in issuing provisional orders. Among the questions the committee considered were whether utilities should be allowed to compete in the same local-authority district and to what extent a local authority could use parliamentary powers to exclude a private company from supplying electricity within its jurisdiction. The ongoing struggle between proponents of the direct-current and alternating-current systems brought additional complexity to the debate. Among the applicants for provisional orders was the London Electric Supply Corporation, which wanted to build a power station of unprecedented size across the Thames at Deptford. Other utilities wanted to establish relatively small area supply systems.[28]

After hearing the testimony of experts and representatives of the local-authority and private interests from 3 April to 1 May 1889, the Marindin Committee recommended that local authorities be given priority in obtaining provisional orders, but that if they did not intend to apply, they should not be allowed to block the granting of provisional orders to private companies. The committee also urged the Board of Trade to grant provisional orders to two suppliers within the same local-authority district if one of the suppliers used alternating current and the other used direct. (The committee, influenced by "the battle of the systems," believed that alternating current had not been proven and that therefore customers should be given the alternative of the more reliable, direct current.) Apparently the committee did not consider the possibility that with further development alternating current would become as reliable as direct current and then would be discriminated against because of the competition provision. Nor did it provide well for the a.c. system's greatest strength: large-area supply using high-voltage transmission. Furthermore, the committee did not anticipate the increase in the city's power load and the resulting need for large power stations to supply numerous districts. The Marindin Committee perpetuated the pattern of small-scale electric supply first defined in the Electric Lighting Act of 1882.

Despite the committee's failure to support alternating current and large-area supply, however, the London Electric Supply Company went ahead with its bold scheme for a central station on the Thames in Deptford, a southeast London borough. It chose S. Z. de Ferranti to head the project. Ferranti's reputation as an inventor and engineer and his success in reorganizing the Grosvenor Gallery system notwithstanding, it is difficult to understand why he and his financial backers took such a great leap beyond the state of existing technology in their Deptford project. Ferranti's alternators, or a.c. generators, at the Grosvenor Gallery station were 700-h.p. units transmitting at 2,500 volts; what he proposed for the new central station were 10,000-h.p. alternators transmitting at 10,000 volts. Not until the first decade of the next century did engineers design, and financial backers risk, 10,000-h.p. units, and these were then driven by turbine—a steam engine that was more adaptable to large-scale transmission than the reciprocating engines envisaged by Ferranti.[29]

In proposing this remarkably large increase in scale, Ferranti not only defied traditional engineering conservatism in the development and use of new inventions but he introduced a system that had not been fully proven by widespread use under various conditions. The installations of Ganz & Company and Westinghouse in the United States were then relatively small in scale, involving alternators rated in hundreds, not thousands, of kilowatts. Therefore, mistakes in these systems would not be so costly. Furthermore, Ganz and Westinghouse intended to get involved in the serial production of machinery, the problems of which could be solved by modest and frequent modifications, or model changes. Even though Ferranti had established a manufacturing enterprise in 1887 at Charterhouse Square in London to make dynamos, transformers, switches, and other apparatus, the Deptford station could not be justified as a "tryout" for equipment he intended to improve upon and sell in quantity, for few financiers and engineers in the world were ready to scale up so dramatically.[30] What Ferranti was bringing into use was a system that had not been successfully developed and for which there was little foreseeable opportunity for development.

Why was the excessively bold project embarked upon? First, it should be noted that Ferranti did envisage the future, at least in part. He wanted his giant power plant to be located on the Thames, where land could be purchased cheaply, coal transported conveniently and inexpensively, waste removed expeditiously, and cooling water used easily. From this site electricity could be transmitted across the river to the heavily settled districts of London and there distributed by transformers. Therefore, Ferranti saw beyond the local direct-current stations that were soon to mushroom in London. On the other hand, despite the absence of a practical motor, he used single-phase alternating current. But the future lay with motors (power) and lighting, not just with lighting. Others considered the lack of a practical motor for the a.c. system to be a critical problem and were energetically trying either to solve it or to avoid using alternating current altogether (see p. 110 above).

Another explanation for Ferranti's project was the entrepreneurial spirit of his financial backers. Ferranti's youth and brilliance undoubtedly captured their imagination. Sir Coutts Lindsay and Lord Wantage (Robert James Lindsay, a brother of Coutts Lindsay) were the heaviest subscribers

[29] *A Description of the Inception and Growth of the London Electric Supply Corporation Limited and Its Deptford Generating Station, 1889–1912* (n.p., n.d.), a 20-page pamphlet loaned to the author by Mr. J. Hood of London.

[30] W. L. Randell, *S. Z. de Ferranti and His Influence upon Electrical Development* (London: Longmans, Green, 1948), p. 3.

to the undertaking. These two were members of the Lindsay family headed by their nephew the earl of Crawford. Lord Wantage, with shares valued at £220,000, was the largest shareholder. His relationship to Ferranti was similar to that of Wall Street financiers to Edison. Wantage represented that segment of the British aristocracy which had a history of financing technological change. (Edward Johnson, for instance, appealed to it when he brought Edison's system to London.) Distinguished of bearing, Eton-educated, a member of the Scots Fusilier Guards, husband of the heiress to a princely fortune, soldier, and patron of worthy causes, Wantage had financial daring to match the daring of Ferranti's engineering. R. H. Parsons believed that Lord Wantage's financial courage was comparable to the physical courage that had earned him the Victoria Cross in the Crimean War.[31]

Account should also be taken of the bullish spirit that had resurfaced among British investors by 1887. A number of entrepreneurs who had pulled out of electrical investments in 1882 and 1883 were now optimistic that Parliament would soon pass a bill more favorable to private ventures than the act of 1882 had been.[32] There was talk of a mighty Britain shrugging off her lethargy and closing the gap between herself and the enterprising Yankees in the development of electrical supply.[33] Soon the newspapers were comparing the young Ferranti with the young Edison; perhaps Sir Coutts Lindsay and Lord Wantage had been struck earlier by the same analogy.

On 26 August 1887 enthusiastic financiers registered the London Electric Supply Company (LESC) with an authorized capital of £1 million.[34] The company's twenty-three-year-old chief engineer designed the Deptford station house with stark, rectangular, 150-foot chimneys. The building was to be 210 feet long and 195 feet wide. There were to be two engine rooms within the structure: one containing two 1,250-h.p. Corliss engines driving 5,000-volt Ferranti alternators; the other housing four mammoth 10,000-h.p. engines, each directly coupled to a 10,000-volt Ferranti alternator. The station's ultimate capacity was to be 120,000 h.p.

Like Edison before him at the Pearl Street station, Ferranti specified or designed most of the major components of his system and planned the layout of his plant. A British electrical journal dubbed him a Michelangelo, and *Punch* created a dialogue between John Bull and Edison in which Ferranti was a champion of Britain's electrical progress.

> Edison
> . . . you still have some go,
> You haven't yet mastered the big dynamo.

[31] Parsons, *Early Days of the Power Station Industry*, p. 23.

[32] Clipping from *Money*, 7 November 1888, "Deptford Clipping Book," Science Museum, London, used with the kind permission of Dame M. K. Weston, then keeper of the Electrical Engineering Collection at the museum.

[33] Clipping from *Electrical Engineer* (London), 26 October 1888, "Deptford Clipping Book," Science Museum, London.

[34] *Deptford*, a short historical pamphlet reprinted from *Notes and Records of the Royal Society of London* 19 (1964): 2; and *Money*, 7 November 1888.

Figure IX.2. *S. Z. de Ferranti, the modern-day Colossus, astride the London he intended to light. Cartoon in* Electrical Plant, *May 1889.*

John Bull
No, that's what I fear, my own knowledge is scanty,
And I can't decide between you and Ferranti;
But, if we are licked by Berlin I must try
To stir up the slugs of the "London Supply."[35]

Construction of the Deptford station began in 1888, and Ferranti hoped to be supplying London from the station's two 1,250-h.p. sets by 1889. These would carry the load until the 10,000-h.p. units were in place. In October 1888 *The Electrician* proudly referred to the Deptford station as an answer to "Yankee boasts," and *Electrical Engineer* said it would "cause our American cousins to reconsider the verdict of continued backwardness of England in the matter of central stations for electric lighting."[36]

By 1889, however, problems had developed, among them the deliberations of the Marindin Committee, which pertained especially to Ferranti's new company. Almost a dozen companies had applied for the right to light London districts under the provisions of the Electric Lighting Act of 1888, and LESC sought a lion's share of the metropolitan area. As would be the case in 1905, when another bold scheme for London was proposed by a private company, the local authorities of London now opposed LESC's request for a large supply area. The Strand and St. Marylebone authorities,

---

[35] *Punch,* 5 October 1889, quoted in Gertrude Ziani de Ferranti and Richard Ince, *The Life and Letters of Sebastian Z. de Ferranti* (London: Williams & Norgate, 1934), p. 64.
[36] Clippings from *The Electrician,* 26 October 1888, and *Electrical Engineer* (London), 26 October 1888, "Deptford Clipping Book," Science Museum, London.

the St. Margaret and St. John (Westminster) vestries, and the St. Pancras vestry all objected before the Marindin Committee. Each stated the intention to establish its own system and claimed priority over the private company.[37] Unfortunately, from the point of view of the Deptford plan, the Marindin Committee in May 1889 decided to reduce the territory sought by LESC and to allow competition between a.c. and d.c. systems within the company's London districts. Even R. E. H. Crompton, of the competing Metropolitan Supply Company, later called unfair to LESC the Marindin Committee's district apportionment. In justice to the committee, it should be noted that telling criticisms of the new a.c. system were raised during the hearings. These included the prediction that a.c. transmission lines would interfere with telegraph and telephone service and the warning that the concentration of energy in one station would introduce the risk of a single accident's darkening much of London. The d.c. stations, transmitting over a limited area, spread the risk of blackout over a number of stations. Deptford—a great experimental venture—would have to compete with the small d.c. stations that had almost a decade of operating experience.

LESC's directors adjusted to the Marindin Committee's report by limiting Ferranti to two instead of four 10,000-h.p. generator sets and by deciding to build elsewhere another 20,000-h.p. station. By the summer of 1889 the two 1,250-h.p. steam engines had been erected and the alternators were near completion. The 10,000-h.p. steam engines were being built by Messrs. Hick, Hargreaves & Company, Ltd., which had the heaviest steel ingots in British history cast for forging into engine crankshafts. At Deptford, Ferranti used immense machine tools to finish the station's alternator components. (See Figs. IX.3 and IX.4.)

In September 1889 Thomas A. Edison visited the Deptford station, which was gaily bedecked for his arrival. The *London Daily News* reporter who accompanied him wrote that Edison looked enervated by his cold, but that his expression brightened when he saw the 1,250-h.p. units. Edison was also said to have listened with interest to the description of the 10,000-h.p. units, despite his firm advocacy of d.c. stations and his partisan role in "the battle of the currents."[38]

When a reporter asked Edison, "Now, Mr. Edison . . . how do you think we are getting along in this country?" he replied, "You may be slow to begin but I must say that when you do go ahead, you may even beat us." Edison pointed out the danger to life of 10,000-volt transmission, as compared to his own low-voltage system, which did not endanger even a child. Yet when summing up his impressions of Deptford for the reporter, he exclaimed in his way, "Oh, it will go!"[39]

But it did not go, at least not as Ferranti had envisaged it would. The station transmitted no electricity in 1889—a disappointment to investors, who had been led to expect transmission in 1888. Ferranti's enthusiasm did not lag, however. As his wife wrote, "The principal thing I remember

---

[37] *The Electrician* 23 (1889): 148.

[38] *London Daily News*, 26 September 1889, quoted in Ferranti and Ince, *Life of Ferranti*, pp. 61–63.

[39] *Engineering News & Post*, 26 September 1889, "Deptford Clipping Book," Science Museum, London.

Figure IX.3. Ferranti's sketch of the Deptford station boiler and economizer, dated 1887. Courtesy of the Ferranti Ltd. Archives, Hollinwood, England.

Figure IX.4. A Ferranti alternator under construction, Deptford station, 1889. Courtesy of the Ferranti Ltd. Archives, Hollinwood, England.

during those first months of married life was Deptford, and again Deptford. We talked Deptford and dreamed Deptford."[40]

When, for example, Ferranti found that the cable he had purchased to transmit 10,000 volts to London would not do, he designed his own. The Ferranti "mains" excited much comment at the time and are recognized today as a landmark in the development of paper-insulated cable. Ferranti maintained the outer conductor of his cable at earth potential high-voltage and by means of a dramatic demonstration proved the cable safe under these circumstances (thereby also relieving the anxiety of public authori-

[40] Ferranti and Ince, *Life of Ferranti*, p. 95.

Figure IX.5. *Deptford central station.*
From London Illustrated News, *26
October 1889. Courtesy of the Ferranti Ltd.
Archives, Hollinwood, England.*

Figure IX.5. *Deptford central station.
From* London Illustrated News, *26
October 1889. Courtesy of the Ferranti Ltd.
Archives, Hollinwood, England.*

ties). While one of Ferranti's assistants held a cold chisel, another drove it through the live 10,000-volt cable. The chisel short-circuited the main, and a fuse cut the supply. Years later, on 3 March 1910, Ferranti wrote with justifiable pride to his son, Vincent: "Although these mains were made in a great hurry and with somewhat faulty appliances, and with a great want of knowledge, notwithstanding that, at least half the quantity of mains originally laid down are still supplying London on 10,000 volts today, the remaining portion of these same mains being still in use but working at a lower pressure."[41]

In October 1890 the Deptford station transmitted current to London, and hopes ran high. On 15 November 1890, however, the first of a series of mishaps occurred. The old Grosvenor Gallery station, now converted to a substation with a bank of transformers to step down the voltage, was destroyed by fire. The fire began when an arc formed as a station attendant removed a plug switch carrying high voltage. The attendant extended the arc by pulling the plug back instead of replacing it, and the arc ran up the woodwork, setting fire to the ceiling. The blaze destroyed the substation in twenty minutes and cost LESC about £15,000 in uninsured apparatus. Ferranti blamed the attendant. In an interview he argued that the man could have reinserted the plug switch, pulled the switch nearby to cut off the current to the plug switch, or pulled the main safety switch to cut off the supply from Deptford. With hindsight, an engineer today might judge the human components of the system overloaded.

The company made repairs quickly and by December had restored service from Deptford via the Grosvenor substation, but on 3 December 1890 the system again failed when one repaired transformer at Deptford burned out and transferred its load to the already fully loaded remaining transformers. The bank of transformers burned out in succession. All Deptford-

[41] Ferranti Private Letters, vol. 4, letter no. 477b. I am indebted to C. J. Sommers, company archivist for Ferranti Ltd., Hollinwood, England, for calling my attention to this letter.

Figure IX.6.   *A London Electric Supply Corp., Ltd., leaflet, c. 1893. From Ridding, S. Z. de Ferranti, p. 17.*

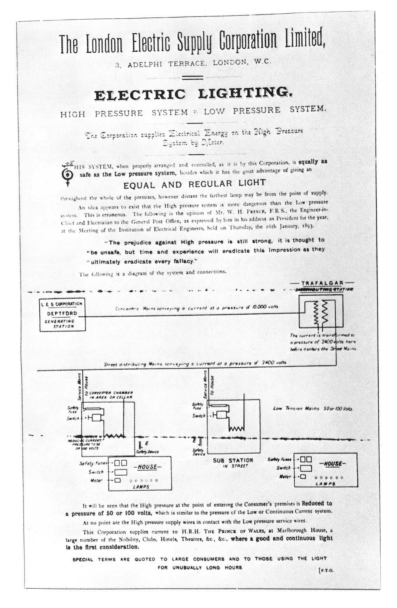

supplied lights blacked out. After the second disaster, the company discontinued service for three months to carry out major modifications in the system. Rewinding of the two 1,250-h.p. units would raise their output voltage from 5,000 to 10,000, thereby making it possible to feed the output into the 10,000-volt mains without using a step-up transformer. In the meantime, Ferranti planned to use two 200-kw. alternators brought over from the old Grosvenor station. The small output offered no problem, since three-quarters of the Deptford station's customers had been lost during the shutdown. The company went ahead with the construction of additional substations at Trafalgar Square, Blackfriars, and Deptford, and by August 1891, 30,000 lamps were connected. When Ferranti learned that the company would not complete the giant alternators, however, he resigned. After his departure, failure of system components continued to plague the Deptford station. In November, the board chairman remarked, "The whole thing came to a collapse; the dynamos, mains, and everything went wrong. . . ."[42] A much-changed Deptford survived, but a reevaluation of stock in 1898 revealed that the London Electric Supply Corporation had lost at least £400,000.

Two leading British electrical engineers were brought in to assess the situation. John Hopkinson and John Ambrose Fleming concluded that because of its high-voltage, long-distance transmission, the Deptford station was constitutionally afflicted and was surviving "on very thin ice." The consulting engineers hoped that the ice could be thickened. The only total solution would be to build new, small d.c. stations near the area of consumption. In an era when "the battle of the currents" was being waged by the advocates of alternating and direct current, this was an especially humiliating blow for the a.c. champion, Ferranti. The consultants made many technical suggestions to salvage the best of what they judged a bad engineering situation. They recommended reducing the transmission voltage to 5,000 if the company continued a.c. transmission, but would give no assurances of the system's performance.[43]

Historians' and engineers' assessments of Ferranti have been more sympathetic. R. H. Parsons, in *The Early Days of the Power Station Industry*, says that at Deptford, Ferranti's prescience with regard to electrical development was "little short of marvelous; his engineering courage in attempting to realize his vision . . . no less wonderful." W. L. Randell, in *S. Z. de Ferranti*, dismisses the system's failures by stating that in that magnificent and daring enterprise Ferranti achieved what engineers of standing considered impossible. In Oxford's five-volume *History of Technology* the "collapse" of the Deptford station is ignored and Ferranti's experiment is characterized as the forerunner of all large central stations. Percy Dunsheath, in *A History of Electrical Power Engineering*, writes of the great reputation Ferranti earned from his scheme. On the other hand, Lord Hinton of Bankside notes "that if you feel that the great engineer ought to be not merely an innovative genius but should also give good value to his financial sponsors then Ferranti cannot stand beside men like George Stephenson or Telford or Watt."[44] At least Ferranti is associated with the great Smilesian engineers.[45]

Ferranti's reflections almost half a century later provide an interesting defense of his plan and execution of it. After noting a number of mechanical and electrical problems solved in the course of construction, testing, and early operation, he lamented the company's failure to install the 10,000-

[42] Parsons, *Early Days of the Power Station Industry*, p. 39.

[43] J. A. Fleming and John Hopkinson, "Report of December 1891"; copy in the Electrical Engineering Collection, Science Museum, London.

[44] Parsons, *Early Days of the Power Station Industry*, p. 28; Randell, *Ferranti*, pp. 9–19; Percy Dunsheath, *A History of Electrical Power Engineering* (Cambridge, Mass.: M.I.T. Press, 1962), p. 167; C. Singer, E. J. Holmyard, A. R. Hall, and T. I. Williams, eds., *A History of Technology*, 5 vols. (New York: Oxford University Press, 1958), 5: 199–200; Lord Hinton of Bankside, *Heavy Current Electricity in the United Kingdom: History and Development* (Oxford: Pergamon Press, 1979), p. 30.

[45] On the character of the Smilesian engineers, see Thomas P. Hughes, ed., *Selections from Lives of the Engineers* (Cambridge, Mass.: M.I.T. Press, 1966), pp. 1–29.

h.p. units, which, he states, were built and never erected. He found it highly significant that reciprocating steam engines for the New York subway constructed ten years later were also 10,000 h.p., had two low-pressure cylinders of the same diameter (88 in.) as his, and the same piston speed (about 750 ft./min.). "[It] seems a curious coincidence," he wrote, "that ten years later on the best engineering practice was still considered to be what we had planned and built for Deptford." The failure of the Deptford station was, he recalled, "settled before it started." He referred to the decision made by the Marindin Committee to greatly reduce the presumed area of supply, and "worse than this," he continued, "we had competition from local stations arranged for in every part of our area." These competitors took actual and potential customers as Deptford experienced its breakdowns. With a monopoly, the company could have held on and modification would have been possible, Ferranti concluded.[46]

Ferranti's Deptford station was begun in 1888, about the time Tesla and the other inventors of the polyphase system were taking out their patents; it was abandoned in its grand dimensions in 1891, the year that polyphase power transmission was successfully demonstrated at Frankfort on the Main. During the years when it was being built a large number of single-phase alternating-current central stations were constructed, but by the mid-nineties engineers and station managers had begun to comprehend the advantages of the universal, polyphase system for supplying light, stationary power, and traction. In short, Ferranti set out on his bold venture in a time of transition and pursued a course that briefly appeared to be the mainstream but took him out of the rising tide of technological change.

Other London utilities also expanded greatly during the period of transition and then failed to commit themselves to the universal, polyphase system. There was a flurry of construction of new ventures after passage of the Electric Lighting Act of 1888. London seems to have been the European city best lighted by electricity in 1891, but it opted for direct current and single-phase alternating current. During the Frankfort exhibition the organizers conducted a survey among the leading electrical manufacturers in attendance to ascertain the location of major central stations in Europe and discovered that London had more incandescent lights lit from central stations than Berlin, Paris, or any of the other metropolises. According to the 1891 compilation, London had 473,000 incandescent lamps rated at 16 c.p., Berlin had 75,000, and Paris 67,600.[47] Of the London lamps, 307,000 were connected to single-phase alternating current. Even though the compilers stressed the incompleteness of the survey, and even though there are obvious errors in it, London's lead was significant. After this burst of light, however, London—and England—virtually ignored the polyphase system during the 1890s, formative decade of the new system. Natives could once again at the turn of the century lament, as they had in the eighties, Britain's electric lighting—and now power—lag.[48]

By 1900 Parliament was being told that electrical technology pushed to

---

[46] Clipping from *The Engineer*, 16 June 1944, p. 469.

[47] *Offizielle Zeitung der Internationalen Ausstellung Frankfurt am Main 1891*, pp. 634–39. There are apparent omissions and errors in this list.

[48] Byatt, *British Electrical Industry*, p. 69.

TABLE IX.1. LEGISLATION RELATING TO ELECTRIC SERVICE IN THE LONDON AREA

Title	General Scope of Act
General Acts	
Electric Lighting Act, 1882	Conditions of franchises: early terms of purchase (21 years)
Electric Lighting Act, 1888	Extended period of franchises and improved terms of purchase
Electric Lighting (Clauses) Act, 1899	Standard clauses for special orders and private acts
Electric Lighting Act, 1909	Prohibited unauthorized competition with statutory utilities
Electricity (Supply) Act, 1919	Established electricity commissioners and provided for establishment of joint electricity authorities
Electricity (Supply) Act, 1922	Conferred further powers on electricity commissioners and joint electricity authorities
Electricity (Supply) Act, 1926	Established Central Electricity Board and the Grid
Electricity (Supply) Act, 1928	Minor amendments to the 1919 act
Electricity (Supply) Act, 1933	Minor amendments to the 1919 and 1922 acts
Electricity (Supply) Act, 1935	CEB authorized to sell energy to railroads
Electricity Supply (Meters) Act, 1936	Minor importance
London Acts	
London Electric Lighting Areas Act, 1904	Adjustment of service areas of utilities
London Electric Supply Act, 1908	Purchasing powers of local authorities transferred to London County Council
London Electric Supply Act, 1910	Constituted London County Council purchasing authority for utilities to which 1908 act did not apply
London Electricity (Nos. 1 and 2) Acts, 1925	Repealed purchase provisions of earlier acts. Imposed sliding scale
London and Home Counties Joint Electricity Authority Act, 1927	Conferred additional powers on Joint Electricity Authority

Source: Merz & McLellan, *Report on Electric Service in New York, London, Paris, and Berlin*, p. 49.

bursting the seams of local-authority confines. Private capital sought authority to establish power companies outside London in order to supply electricity in bulk over a large area from large-scale power plants housing the newly introduced steam turbine as prime mover. Some engineers considered London to be suitable ground for an embracing, universal system. There was also discussion of taking advantage of Viscount Cross's recommendations for parliamentary power bills and establishing a power company for London. Once again, however, the political power of England's local authorities was manifest. In London, the London County Council and the local authorities conferred about the future of large-area supply and adopted a resolution that expressed well the realities of the situation there:

While reserving intact the rights of the local authorities in regard to electric lighting and energy, it is desirable that the London County Council should be in a position to undertake if it so determine, or if so requested by the local authorities, to supply electrical energy for the convenience of any districts desirous of being supplied, provided the Council does not become competitive with the local authorities.[49]

So laced with "ifs" and "provideds," the resolution should have been fair

[49] London County Council, *Report* (1913–14), app. 2, p. 34.

warning for would-be entrepreneurs who envisioned a centralized London system.

When Charles Merz surveyed the London scene in 1904 from the perspective of Newcastle upon Tyne, his sense of system, technological scale, and economic efficiency must have been offended. London motivated him to attempt to reorder the situation in accordance with his achievements in systematizing large-scale supply at Newcastle. Because Merz is among the leading figures in the history of British supply, and because he is encountered not only in connection with his bold London scheme but as a pioneer in the establishment of the British Grid, his background is of interest.

Charles Hesterman Merz (1874–1940) was born into an influential and affluent family (see Fig. XIV.29, p. 449). His mother, a Quaker, was a sister of the Newcastle shipbuilder John Wigham Richardson, and his father was a man of remarkable culture. John Theodore Merz (1840–1922) is known to scholars as the author of *A History of European Thought in the Nineteenth Century* (4 vols., 1896–1914). Among his peers he was known as a chemical manufacturer on Tyneside (his company was eventually absorbed by Imperial Chemical Industries). Finally, he was a founder and chairman of the Newcastle upon Tyne Electric Supply Company. Young Merz therefore had access to persons of wealth and political influence. As a consulting engineer presiding over large-scale projects with political and financial dimensions, he must have found associations of this kind helpful.

Charles Merz attended Bootham School at York Armstrong College in Newcastle, but left without taking a degree to pursue more worldly matters as an apprentice at Pandon Dene, the first central station of his father's Newcastle upon Tyne Electric Supply Company (NESCO), and later at British Thomson-Houston (BTH). He then joined BTH and became managing engineer of a utility it owned at Croydon. These early experiences help explain Merz's subsequent invention of improvements in power-plant equipment. In 1899 Merz set up a consulting engineering firm, the name of which was changed to Merz & McLellan after William McLellan joined him in 1902. From the beginning Merz and the firm manifested a bias toward—and aptitude for—the creation of large-area electric power systems.

Prior to his introduction of parliamentary legislation in 1905, which was intended to revolutionize electrical supply in London, Merz gathered with a close-knit group of Northeast Coast leaders. This meeting and those that followed in 1905 and 1906 in connection with the London project provide an example of the nature of technological entrepreneurship. Merz was a system builder, a hedgehog like Thomas Edison and Samuel Insull; he had less inventive genius than Edison, less managerial experience than Insull, but he manifested sufficient instincts for invention, talent for management, and exceptional sensitivity to the world of politics and finance to perform as a superb originator and to preside over the fulfillment of bold technological projects. (See pp. 449–60 below for his experiences as a consulting engineer for NESCO, and pp. 350–57 for his role in the building of the British Grid.)

On 28 September 1904 Merz met in Newcastle upon Tyne with men of experience and influence in large industry, railroads, finance, law, parliamentary politics, and electrical supply. Their collective perspective was needed for the encompassing project that was to be discussed. Present were G. B. Richardson, a director of the shipbuilding company Wigham Richardson & Company, James Falconer of the Mersey Railway Company; and Robert Spence Watson, a Quaker who married the sister of Merz's mother and who was a member of a firm of solicitors, a leader of the Liberal party in the Tyneside region, and a champion of educational and social reform. Also participating was Merz's father, who had become chairman of the electrical supply company NESCO and of the chemical concern Sir C. Tennant & Company of Carnoustie, as well as a director of British Thomson-Houston. Not present, but wishing to be kept informed, were Sir Andrew Noble, chairman of Armstrong Whitworth & Company, a heavy engineering concern famous for armor and guns; and, from outside the district, Samuel Insull, John Annan of Annan, Kirby, Dexter & Soc., and A. S. Northcote of London. At the meeting, Merz outlined the London electric supply problem, Spence Watson discussed the difficulties in obtaining parliamentary powers, and the group cheered itself on by referring to the successes of NESCO in supplying power to industry and the railways. The group decided to form the Electric Power Development Company in an effort to solve London's problem in a similar manner.[50]

Charles Merz, who had been mulling over the London problem for several years and who had established a branch of his consulting engineering firm on Victoria Street in Westminster, London, spearheaded the Newcastle-born project. His enthusiasm was fired by the analogy that electric supply in the Newcastle region provided for the London project. The Newcastle upon Tyne Electric Supply Company, which his father chaired and he advised as consulting engineer, entered a period of dramatic growth in 1900 when the company began to focus on power rather than lighting (see pp. 447–48 below). Persuading heavy industries in the Tyneside area to take power from its system rather than generate from their own power plants, NESCO dramatically reduced its production costs per kilowatt-hour and, correspondingly, the price it charged for electricity. In so doing, the company's management followed the strategy of mass production adopted by Insull and other progressive power companies in the United States and Germany. Merz also promoted steam turbines, one of which he had recently installed at the Neptune Bank station, as the key to large-area supply from units of unprecedented size.

By November 1904 Merz had formulated his basic argument. As director of the London project, he pointed out that London was a manufacturing center with over 30,000 factories and workshops and 600,000 employees. (He did not stress the average small size of these plants.) The startling paradox, to Merz, was the fact that the metropolis used only one-fifth as much electric power per capita as Berlin and one-tenth as much as his own Newcastle district. His major explanation for this paradox was the smallness of scale of the utilities in London, which meant large capital and operating costs—three times more in both instances than large-scale utilities experienced elsewhere. Drawing again on an analogy with the NESCO system,

[50] Notes of meeting held in Newcastle upon Tyne, 28 September 1904, Merz & McLellan Papers (Folio 70 /M 53.5/55), Tyne and Wear County Council Archives, Newcastle upon Tyne, England (hereafter cited as Merz & McLellan Papers).

from which 80 miles of suburban railroad took power, Merz projected a system for London that would feed railroads, tramways, and the underground.[51]

In a strategic move to gain approval for their power company, Merz and his associates decided to present a power bill to Parliament. They named their company the Administrative County of London and District Electric Power Company because power bills were generally taken up in alphabetical order. The tactics and strategy used in the effort to gain passage of the bill demonstrated young Merz's sensitivity to the problem of communicating the essence of technology and economics to the nontechnically trained, but they also revealed, in the long run, an overconfidence in the rational technical argument. During the course of Merz's sustained endeavor Lloyd George, who was then president of the Board of Trade, took the pains to say to the then relatively inexperienced Merz, when he, "with the usual enthusiasm of an Engineer," spoke about the technical merits of his London scheme, "My dear young friend, this is not a question of engineering, it is a question of politics."[52]

From a technical and economic point of view, the campaign for Merz's bill was well mounted. Because there was no census of industrial production available, Merz had his associate T. H. Minshall, who headed the consulting firm's London office, spend several weeks at the British Museum going through three hundred Ordnance Survey maps of London in order to locate the factory sites. Then Minshall checked with the owners to find out what kind of power they used and if cheap electricity would be a boon to them. The information Minshall gathered about London was then compared with statistics for other industrial cities. Included in the data were estimates of the electrical requirement of London, coal consumption in Greater London, the anticipated savings from adoption of electricity, the total amount of horsepower installed in London, the percentage of power supplied by existing utilities, and various comparative costs of existing power systems and of power under the Merz project (see Fig. IX.7). By the time the parliamentary hearing on Merz's bill began on 16 March 1905, Merz had had his data printed and bound in a handsome leather book and copies of it had been distributed to committee members. "The whole scale and style of the preparatory data was something new in the Parliamentary Committee Rooms." Merz recalled that "such tables became the fashion afterwards but up to then I do not think anything had ever been more completely or clearly stated and shown."[53]

According to Merz and his associates, their power company would operate a power station with six turbo-alternators capable of supplying a

[51] Charles H. Merz to The Electric Power Development Co., Ltd., November 1904, Merz & McLellan Papers.

[52] Charles H. Merz, "Autobiography," May 1936, p. 19; copy in Merz & McLellan Co. Archives. The account of the London project given here draws upon the chapter entitled "London," and the page numbers cited hereafter refer to this chapter. John Rowland's *Progress in Power: The Contribution of Charles Merz and His Associates to Sixty Years of Electrical Development, 1899–1959* (London: Newman Neame, 1960) closely follows Merz's autobiography; on London see pp. 38–50. I am indebted to Ms. Kathleen Bramley, former librarian at Merz & McLellan, for calling my attention to this privately published work.

[53] Merz, "Autobiography," p. 12.

	Population.	Output of Electricity in Board of Trade Units for all purposes.	Units output per head of population.
Boston	614,522	211,000,000	343
New York	3,437,202	971,465,995	282
Chicago	1,932,315	383,000,000	198
London (Industrial Area)	3,485,368	78,005,848	22
London (Area scheduled in London County Council Electric Supply Bill) ...	6,048,379	253,153,548	42

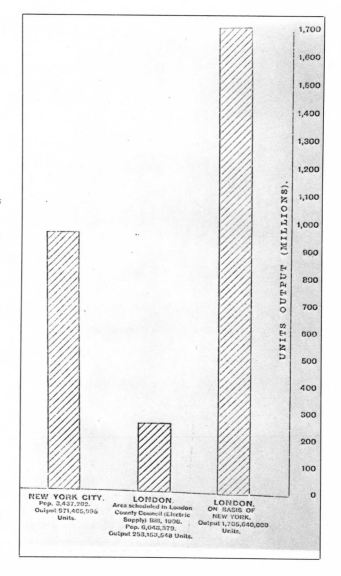

Figure IX.7. Table and graph prepared by Merz & McLellan for Charles Merz's testimony before a parliamentary committee in support of a London electric supply bill. From Merz & McLellan, Administrative County of London and District Power Company, 1906. *Courtesy of the Tyne and Wear County Council Archives, Newcastle upon Tyne, England.*

maximum load of 20,000 kw. (Insull had introduced 5,000-kw. units in Chicago in 1903, and the Metropolitan Electric Supply Company had installed a 3,000-kw. unit for London supply in 1904.) From this 20,000-kw. station, transmission lines would deliver power directly from the power company to industries in east London and would supply power in bulk, or wholesale, to existing utilities, especially in west London. After conversion to direct current or other forms, this bulk current would be distributed by the existing utilities, through their own systems, mostly for lighting. By leaving much of the lighting supply to the small utilities, Merz hoped to win their support for his plan. The new power company would be especially geared to supply London's growing traction load. Merz predicted that by taking advantage of economies of scale and diversity of load, his company could sell in bulk to the smaller London utilities for a lower price than

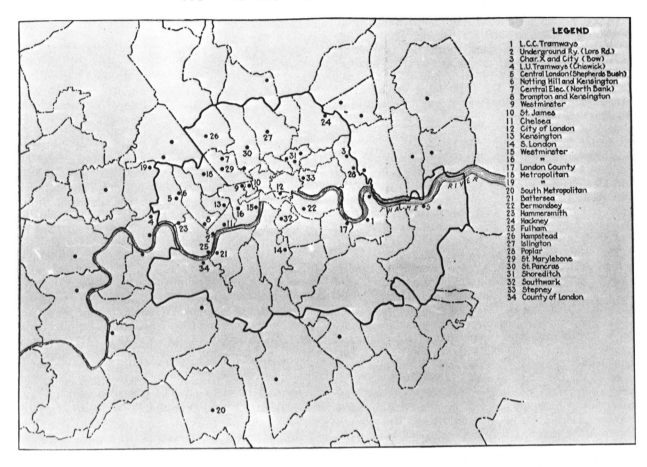

LEGEND

1 L.C.C. Tramways
2 Underground Ry. (Lots Rd.)
3 Char. X and City (Bow)
4 L.U. Tramways (Chiswick)
5 Central London (Shepherds Bush)
6 Notting Hill and Kensington
7 Central Elec. (North Bank)
8 Brompton and Kensington
9 Westminster
10 St. James
11 Chelsea
12 City of London
13 Kensington
14 S. London
15 Westminster   "
16   "
17 London County
18 Metropolitan
19   "
20 South Metropolitan
21 Battersea
22 Bermondsey
23 Hammersmith
24 Hackney
25 Fulham
26 Hampstead
27 Islington
28 Poplar
29 St. Marylebone
30 St. Pancras
31 Shoreditch
32 Southwark
33 Stepney
34 County of London

Figure IX.8. *Location of London electric light and power stations, c. 1923. From* Electrical World *82 (1923): 1057.*

what it would cost them to generate that power, even in new, expanded plants.

Merz and Minshall won support for their bill in the form of a petition signed by 250 London manufacturing firms. In addition, more influential figures in the field of finance and industry became supporters of the project and invested in shares in the syndicate, among them, Walter Cunliffe, later governor of the Bank of England; through him, Merz hoped to bring in more City-financed interests. Because Merz also had the support of Baring Brothers, he believed that he "never had any criticism of the financial stability of the promotion."[54] It is interesting to note, however, the absence of leading representatives of Britain's electrical manufacturing companies, interests that would have figured prominently in a similar American or German scheme.

Arraigned against the bill on the opening day of the hearings were over one hundred local authorities from London and elsewhere in England as represented by thirty-five counsel. In the size, diversity, and influence of Merz's opposition lies a key to understanding the history of electrical supply in London. Moreover, the bill brought concert between parties who were normally opposed. The London County Council (LCC) firmly opposed the bill. The LCC and local authorities, which were engaged in a long-term

[54] Ibid., p. 8.

struggle over the governance of London, united to oppose a private company that, it was said, would both prevent the LCC from establishing a London power company and subvert existing local-authority- and privately owned undertakings. The Progressive party, which dominated the LCC, opposed Merz's bill as a challenge to the LCC's potential role in electric supply, and some of the more radical Progressives challenged the bill because of their advocacy of municipal socialism. Issues involving London often involved national political parties, and in the case of electrical supply, this was to be expected because of the role Parliament and the Board of Trade played in questions regarding the regulation of electrical supply. The Conservative party was in power in Parliament in 1905. Although normally it would be expected to favor a private enterprise scheme, in this case it felt that the new power company might overwhelm existing private utilities in London and also the political authority of local government, which Conservatives championed in London. Even private London gas companies joined the opposition.

During the committee hearings and in the face of such opposition, Merz modified his bill. Agreements were reached assuring the future of private companies, especially in London's West End. The Metropolitan Supply Company, for example, was promised a large supply district there, and a special arrangement was made with the County of London Electric Lighting Company. With some private companies agreeing to the modifications, the Board of Trade lent the bill its support. The bill then passed the Lords' committee, the second reading in the House of Commons, and, after a protracted hearing, the vote of the select committee. Delay by the select committee, however, doomed the bill because there was not time to have the third reading in the House of Commons before the Conservative government of Arthur Balfour collapsed. Merz viewed the delays in committee as a subtle tactic intended to defeat the bill.[55]

Introduced again in 1906 after the Liberals came to power in Parliament, the Administrative County of London bill competed with an LCC proposal and with a power bill promoted by the Hackney and Shoreditch boroughs, authorities that wanted to build and interconnect stations. Acting on the advice of the Board of Trade, the newly convened parliamentary committee rejected them all. David Lloyd George, chancellor of the Exchequer in the new government, was of the opinion that any scheme for unified and centralized supply would continue to meet considerable opposition from existing interests. With regard to the Merz bill, he believed that any scheme placing electricity supply in the unfettered control of a private company would continue to arouse such opposition as to exclude it from the realm of practical politics. The LCC proposal was, in his view, too tentative, too protective of the vested interests of the existing utilities and therefore destined to fail in the effort to bring about the efficiency and economy of a truly centralized supply system. Because of the scheme's speculative nature, he also had doubts about a public authority's right to carry out such a bold plan with taxpayers', or ratepayers', money. Lloyd George leaned toward a cooperative venture in which a private company would carry out the more speculative steps and the LCC would handle the nonspeculative

[55] Ibid., pp. 17–18.

details. After such a project was proven and private investors had reaped a return, the LCC would be allowed to acquire the entire system.[56] In 1907 an LCC bill along the lines of the compromise envisaged by Lloyd George and following the recommendations of a select parliamentary Committee seemed likely to pass the House of Commons, given its large Liberal majority, but the victory of the conservative Municipal Reform party over the Progressives in the LCC election in March of that year led to the abandonment of the proposal. According to the *Times*, the LCC election of 1907 turned on the electricity issue. The Municipal Reform party then modified the LCC bill and called for the cooperation of the LCC and Merz's Administrative County of London Company. Despite Lloyd George's support, however, the parliamentary committee representing a Liberal majority rejected the compromise.[57]

The politics of technology and the controversy over private versus public ownership of electrical utilities continued after 1900 with the rise of the Fabian Society. Led by Sidney and Beatrice Webb, G. Bernard Shaw, and other progressives, the Fabians viewed municipal enterprise, and especially municipal ownership of public utilities, as an essential part of a well-organized society. The challenge of municipal ownership—sometimes called municipal socialism—to traditional private enterprise and the persuasive tracts of the Fabians help explain why, in the words of H. H. Ballin, "what should have been a comparison of the economics of small-scale and large-scale supply, developed into a dispute on the merits of private versus public enterprise, in which the technical aspect was only of secondary importance."[58]

After 1907 the utilities of London improved their technical performance but made little progress toward establishing a unified, centralized supply system before the outbreak of World War I. The technical options open to them included the interconnection of load and supply among utilities, the transformation of obsolete generating stations into substations through the conversion of bulk power, cooperation among utilities in the construction and operation of large power plants, and the standardization of distribution and transmission currents. Managerial conservatism and resistance to institutional innovation, however, usually frustrated the adoption of these options. Parliamentary law forbade the institutional amalgamation of utilities; and engineers and managers of the existing utilities showed little enthusiasm for the formation of joint committees to preside over technical interconnection—probably because they would lose their individual authority. Even though several utilities cooperated in building and operating power stations, the private companies opposed a London-wide

[56] Memorandum by David Lloyd George, 13 June 1906, in the files of the Electricity Council, London (COAAB. 37/83-no. 56) and brought to my attention by Dr. Leslie Hannah.

[57] H. H. Ballin, *The Organisation of Electricity Supply in Great Britain* (London: Electrical Press, 1946), pp. 81–83; Leslie Hannah, *Electricity Before Nationalisation* (Baltimore: The Johns Hopkins University Press, 1979), p. 47.

[58] Ballin, *Electricity Supply in Great Britain*, p. 22. On the relationship between private and public enterprise in France, see Pierre Lanthier, "The Relationship between State and Private Electric Industry, France, 1880–1920," in *Recht und Entwicklung der Grossunternehmen im 19. und frühen 20. Jahrhundert*, ed. Norbert Horn and Jürgen Kocka (Göttingen: Vandenhoeck & Ruprecht, 1979), pp. 590–603.

endeavor presided over by the LCC. The municipal companies and the LCC continued to be wary of a London power company. In view of the investment that had been made in innumerable voltage, frequency, and other current variations, standardization seemed improbable. Through parliamentary acts, however, the LCC did acquire the right to purchase London companies in 1931, a right that had formerly been held by the local authorities.[59]

In 1914 Merz & McLellan again made a proposal, this time at the request of the London County Council. A committee of the council had taken testimony from Insull, Ferranti, and others, and Merz & McLellan submitted a detailed report on the London supply situation. They suggested that a company formed by the existing private London utility companies would be the most promising institutional arrangement by which to coordinate a London-wide supply network, provided the LCC was supportive. An LCC committee rejected this arrangement, however, and adopted, instead, a plan for a mixed undertaking that combined local-authority ownership and private-company management. References were made to similar endeavors in Germany. The London County Council as a whole, however, refused to send a bill to Parliament incorporating the plan. The opposition of the London local authorities was substantial.

The war that began in 1914 would greatly influence the development of electric supply in London and elsewhere in Britain. Therefore, the year 1914 is an appropriate point from which to assess a quarter-century of electric supply history in London and in the country at large. It should be noted that a discussion of the state of London's electric supply industry is to an extent a discussion of the British supply industry. London's population was so large and its economic activities were so extensive that they constituted a substantial share of the entire nation's population and economic endeavors. In the decade before World War I, the population of Greater London amounted to about 15 percent of that of the United Kingdom, and the electricity sold in the Greater London area amounted to between one-quarter and one-third of that sold in the nation as a whole.[60]

During the second half of the nineteenth century, the British had expressed growing concern about the state of the country's industrial development. After the opening of the era of incandescent lighting in 1882, this concern was exacerbated by an awareness that the country, especially London, was not doing well with this new and exciting technology. A century later, they would have spoken of a lag in high technology. Testimony of contemporary witnesses and the meager statistics that are available do leave little doubt that Britain fell behind the United States in electric lighting during the period between passage of the two electric lighting acts (see pp. 62–64 above), but London, at least, experienced an increase in lighting installations and central station construction once the amending clauses were passed in 1888. The explanations offered for the earlier lag were

[59] Ballin, *Electricity Supply in Great Britain*, pp. 84–89.

[60] These are approximations based on statistics in Hannah, *Before Nationalisation*, pp. 427–28; Byatt, *British Electrical Industry*, p. 98; the Merz statistics presented to a parliamentary committee in 1906 (see p. 252 above); and Klingenberg, "Electricity Supply in Large Cities," pp. 398–99.

varied, ranging from the compulsory-purchase clauses of the 1882 act to the availability of gas lighting at lower cost. The recovery was generally attributed to the extension of the period before compulsory purchase could be effected and the resulting availability of finance—especially for London's private utility companies. The argument based on the comparative price of gaslight and electric light seems, then, to have lost its validity; at least the inventors believed it had.

As this chapter has shown, London did not make the transition from electric lighting to electric power after 1890. Therefore, the question was no longer the dimness of London's electric lights but the feebleness of the electric power supplied from her central stations. By the turn of the century, there was again a chorus of lament over London's failure to keep abreast of developments in the advancing technology, a technology that was increasingly seen as the technology of the new century. By 1900, statistical information and professional judgments confirmed the lag. The Institution of Electrical Engineers called attention to the backward state of affairs in 1902, and Merz tried to make the condition of London alarmingly clear in 1905. There were qualifications to be made—such as the generation of traction load in isolated plants in London instead of by utilities as in Chicago and Berlin—but this was not stressed in the comparative statements and figures on the output of central stations.

In 1913 Klingenberg provided a comparison of the Chicago, Berlin, and London systems that was Germanic in detail but British in emphasis on the backwardness of London's electrical technology. Klingenberg's audience of British engineers hardly questioned his presentation, for it reinforced what most of them had been saying. The exceptions were the managers and engineers of the small and often well-run London stations. Having announced the most obvious of the statistics—consumption of current in kilowatt-hours per capita—Klingenberg proceeded to the more esoteric figures (see Fig. IX.9). Chicago and Berlin each had capacity concentrated in six large central stations; London, in sixty-four stations. The installation cost per kilowatt was higher in London than in the other two cities. In Chicago the traction load predominated; in Berlin the power, lighting, and traction loads were more nearly equal; and in London, as the audience knew, the central stations' supply was used mostly for lighting. As seen in Fig. IX.10, London had the poorest load factor, as Samuel Insull had often stressed. Despite the higher cost of coal in Berlin, the total operating costs per kilowatt-hour sold were lower in Berlin than in London, and the ratio of profit to real value was highest in Berlin and lowest in London.[61] The British had reason to be concerned.

The testimony of contemporaries leads to the question, Why were London's utilities judged backward by such progressive engineers as Klingenberg, Merz, and the members of the IEE? To them, a progressive urban utility was, as defined by Insull, a mass-producing monopoly that supplied by means of a universal system a diverse load over a wide area at broadly based low rates and obtained a fair return resulting from efficient production and maximum utilization of capital borrowed at reasonable interest.

---

[61] Klingenberg, "Electricity Supply in Large Cities," p. 399.

Figure IX.9. Comparison of electric supply from central stations in Berlin, Chicago, and London. From Klingenberg's article in The Electrician *72 (1915).*

Table I.—*General Data.*

Item.	Berlin, 1911-12.	Chicago, 1911.	London, 1910-11.
Company	Berliner Elektricitätswerke	Commonwealth Edison Co.	Authorities and companies in and around London.
Population, approx..	2,600,000	2,200,000	6,500,000
Power stations :—			
Number	6	6	64
Installed capacity	137,000 kw.	221,700 kw.	298,400 kw.
Average size	23,000 ,,	37,000 ,,	4,670 ,,

Table II.—*Capital.*

Item.	Berlin, 1911-12.		Chicago, 1911.		London,1910-11	
Assets	160,426,000s.		280,321,000s.		547,060,000s.	
Sinking fund, &c. ...	30,380,000s.		14,148,000s.		111,800,000s.	
Real value	130,046,000s.		266,173,000s.		435,260,000s.	
Cost per kilowatt installed						
Power station........	355·93s.	38 %	477·10s.	40 %	662·00s.	45 %
Distribution system, including meters..	593·69s.	62	723·26s.	60	797·00s.	55
	949·62s.	100	1,200·36s.	100	1,459·00s.	100

Table III.—*Working Results.*

Item.	Berlin, 1911-12.	Chicago, 1911.	London, 1910-11.
Capacity—			
Total peak load	94,600 kw.	199,300 kw.	185,500 kw.
Peak load per power station	15,800 ,,	31,700 ,,	2,900 ,,
Kw.-hours generated	274,000,000	684,000,000	405,000,000
Kw.-hours purchased	—	32,000,000	—
Kw.-hours sold	216,300,000	640,000,000	319,243,000
Comprising—			
Lighting	24 per cent.	19 per cent.	61 per cent.
Power	45 ,,	12 ,,	27 ,,
Traction	31 ,,	69 ,,	12 ,,
*Factors—*Load factor (kw.-hrs. generated)	33·1 ,,	41 ,,	24·9 ,,
Efficiency of transmission system— $\frac{\text{kw.-hours sold}}{\text{kw.-hours generated}}$	0·79	0·894	0·788
Reserve factor	1·450	1·11	1·61
Utility factor (total)........	0·18	0·33	0·122
Coal—			
Price per ton	17·41s.	App. 8s.	App. 12·8s.
Consumption per kw-hr sold	1·38 kg.	,, 1·61 kg.	,, 2·37 kg.
Overall efficiency of plant	9·7%	,, 7·6%	,, 5%
Working costs—			
Revenue from kw.-hrs. sold	34,334,000s.	57,845,000s.	62,400,000s.
Expenses	17,636,000s.	28,881,000s.	27,820,000s.
Profit absolute	16,698,000s.	28,964,000s.	34,580,000s.
Percentage of real value....	12·83%	10·87%	7·85%
Operating costs per kw.-hr. sold-	centi-shillings	centi-shillings	centi-shillings
Selling price	15·856	9·044	19·530
Expenses—			
Fuel...........................	2·393	1·146*	3·051
Oil, stores, &c.	0·039	0·045	0·262
Wages	0·510	0·614	1·047
Repairs, maintenance	0·937	0·830	1·482
Rent, taxes, insurance, &c.	0·348	0·605	1·482
General expenses	0·840	0·836	1·396
Current purchased	...	0·167	...
Municipal participation	3·067	0·271	...
Total expenses	8·134	4·514	8·720
Profit (gross)	7·722	4·530	10·810

* The conversion to kw.-hours sold includes the number of kw.-hours purchased, therefore the actual value is approximately 4-5 per cent. higher.

Figure IX.10.   *Utilities of the Western world: Relation of income to output and load factor. From Insull,* Central-Station Electric Service, *p. 461.*

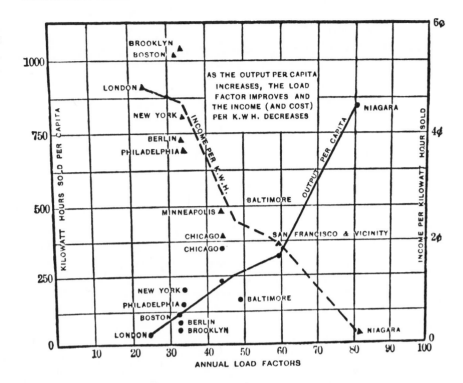

London's sixty-four utilities did not fill the bill of particulars. The reasons were diverse and complex.

A partial explanation is suggested by a comparison of the context in which the three cities' utilities developed. Insull boasted of his large industrial power consumers, and in Berlin, BEW supplied the heavy industries that were flourishing in the old city and in the newer suburbs. There was even a special category in Berlin's load statistics for the large industrial power users. London's industrial base was relatively small-scale. More specifically, the utilities in Berlin and Chicago cooperated with large electrical manufacturers that had the research-and-development strength to develop capital-intensive machinery that would not only raise consumer demand but improve the load factor. The London utilities had to work with relatively small British electrical manufacturers or import equipment from the large manufacturers in Germany and the United States. Furthermore, the Berlin utility was owned and managed by a large manufacturer and therefore represented a symbiotic, vertical coordination of supply and demand. All three cities were electrifying their transit systems, but while Commonwealth Edison and BEW integrated traction into their diversified and universal electric power systems, London's traction load remained small, only 12 percent of the utilities' total load.

Although contemporaries often looked to the condition of technological and scientific education and the health of science in general for an explanation of the state of technology, the situations in Berlin, Chicago, and London were so much alike that this rationale is of little help in explaining the differences in the development of the three utility systems. Major technical and scientific institutions were located in Berlin and London, and

Chicago's electrical manufacturers could employ engineers and scientists who had been trained in leading engineering colleges throughout the United States.

Another contemporary explanation might be called the "American view." The editor of a leading American trade journal, *Electrical World*, opined as late as 1923, after a visit to London, that electricity was "almost, but not quite, as unknown" in Great Britain—and especially London—as the Christian's God was in ancient Athens. Referring to an altar St. Paul had found there on Mars Hill and its inscription, "to the Unknown God," the editor wrote that, similarly, the British showed signs of only dimly comprehending the god electricity. There was some hope, he added, that the streets of London would not always be lit by gas lamps. To drive home his point, he further observed that the estimated output of electricity from generating stations in the United Kingdom was approximately 5 billion kwh., or less than the total generated in the metropolitan district of New York. (The United Kingdom, he hastened to add, had 45 million people; New York City, 7 million.) The American blamed government ownership for "as we have reason to know in the United States," he said, "that is a severe handicap."[62] The American editor's diagnosis of Britain's "sickness" as a case of repressive government ownership was biased by his own ideology. Perhaps he, and Americans generally, were not aware of comparisons of the performance of local-authority and privately owned utilities in London and Britain; such comparisons showed that one was not clearly superior to the other when measured by criteria including costs and return on investment.[63]

The political, or legislative, explanation for the overall backwardness of London's utilities is persuasive. The two great technological failures in London—Deptford and the Merz power scheme—foundered on the rocks of legislative obstruction. Ferranti took too giant a step and committed himself to a kind of current which had little potential for development, but his testimony that the Electric Lighting Act of 1888 and the Marindin Committee's interpretation of it imposed crippling handicaps must be taken seriously in view of supportive evidence. There is no doubt that Merz's plan was frustrated by legislation. Without legislative support, he simply could not go ahead. At the same time, the Chicago and Berlin utilities were acquiring franchises that were coextensive with technological limits.

The question remains, however, What frustrating forces were behind the legislation? What interests, values, and traditions did the legislation manifest—or better, not represent? The question can best be approached indirectly by reference to Berlin (see Fig. IX.10). Chicago does not provide as instructive a comparison, because Chicago's politicians were pliable. Technology and economics dominated politics in Chicago, where political power was a component in an embracing technological-economic system. The Chicago system was American and "go-ahead." It was coupled to the future, not rooted in the past. In Berlin, on the other hand, political and technological power were not as thoroughly integrated as they were in

---

[62] W. H. Onken, Jr., "Electrical Development in England," *Electrical World* 82 (1923): 1055.

[63] Hannah, *Electricity Before Nationalisation*, pp. 50–51; Ballin, *Electricity Supply in Great Britain*, p. 27.

Chicago, and they could—and often did—conflict. Unlike the situation in London, however, the confrontations culminated in negotiated coordination and cooperation rather than in stalemate. In Berlin, there were individuals who built bridges between the world of technology and that of politics. The presence of bankers and industrialists on the advisory board of BEW was a manifestation of the coherence and coordination of finance, industry, and utility in Berlin. Such a progressive combination of coordinated forces was needed to overcome vested and historical interests in London, but such a combination did not exist. Instead, the proponents of local government authority, of municipal socialism, and of private enterprise confronted one another in a pluralistic debate that from the point of view of the forces for technological change produced a stalemate. When the IEE delegation presented its resolution to Gerald Balfour, president of the Board of Trade, in 1902, he acknowledged that Britain's supply industry lagged behind that of the United States and Germany, but he expressed doubt that this was altogether due to legislative regulation. He insisted that there were in Britain "a large number of strongly established interests which produced a very strongly developed instinct of conservatism."[64] He recognized the primacy of politics—in this instance, the power of conservative political interests.

[64] *IEE Journal* 31 (1901–2): 1326.

CHAPTER X

California
White Coal

THE confluence of steam-turbine technology and the universal supply system pushed the development of urban power systems fueled by coal, and in the symbiotic interaction of water-turbine technology and the universal system lies a major explanation for the rapid utilization of white coal. Because the efficiency of a turbine increases with capacity, the steam-turbine power plants were located in or near cities where demand was heavy; for the same reason, water-turbine power plants situated at the water-power sites had to reach across many miles with transmission lines in order to supply a large load. Long-distance transmission and the universal system provided the market needed for large, efficient hydroelectric plants. This chapter deals with the integration of these plants, long-distance point-to-point transmission, and urban distribution systems in California.

Point-to-point transmission, as distinguished from transmission networks, or grids, is defined as the long, overland transmission of high voltage uninterrupted by large-capacity distribution and switching stations along the way. Transmission networks, or grids, also use high voltage, but they consist of a number of relatively short transmission sections tied together at major nexus, or load centers, from which distribution lines spread. As will be seen, point-to-point transmission is appropriate where a great distance separates the power sources (water or coal) and the urban and industrial load. At the turn of the century, more densely populated regions, such as the urban centers of Germany, tended toward the use of transmission networks. In California, point-to-point transmission reached its greatest heights—or distances.

The early history of transmission involves many so-called firsts throughout the world. There were major transmission demonstrations in Italy and France, as well as in Germany and the United States. Before the 1880s, these demonstrations involved direct current. Canada, Japan, Mexico, Chile, India, Brazil, Italy, Tasmania, South Africa, Spain, Sweden, France, Switzerland, the United States, and Germany all had transmission systems of 70,000 volts or more before World War I. Of the fifty-five systems then using 70,000 or more volts, forty-nine transmitted from large hydroelectric power plants; the remaining six transmitted from coal-fired plants, several from mine-mouth plants. The transmission distance was over a hundred miles in half of the installations.[1] More of these high-voltage transmission systems were in the region of California (eight) than in any other region of the world, and all eight California systems transmitted hydraulic power.

The highly developed state of water-turbine technology at the time high-voltage transmission became feasible greatly facilitated the use of hydro-electricity. Water turbines had been around for a long time. Essential turbine principles were incorporated in the horizontal water wheels shown in drawings by sixteenth-century European engineers. The history of the turbine principle in other parts of the world may be even older.[2] Turbines, which are generally classified as water wheels, differ from traditional water wheels in that they depend for power more on the momentum of the water as it enters the turbine housing than on the fall due to gravity of the mass of water at the site of the wheel. In effect, the blades of the turbine wheel must be forced aside (rotated) in order for the high-momentum water to exit, or escape, from the turbine housing. The importance of the difference for the engineers designing, and the capitalists financing, hydroelectric projects is that turbine wheels of much smaller size can be used to obtain the same power as a much larger water wheel of the traditional kind. Because buildings, penstocks, flumes, dams, and other civil engineering works make up the largest share of the cost of a hydroelectric installation, reduction in size from the traditional water-wheel type is of great importance. Furthermore, turbine efficiency had been notably improved by 1890 through experiment and analysis that drew on scientific knowledge.

Turbines, especially those of the Francis type, which had been developed in the United States but were based on European experience and hydraulic science, were well-suited both for the high-head sites of California and the lower-head sites of the eastern seaboard of the United States. The California engineers also utilized an impulse, or tangential, water wheel developed by Lester Pelton and others in California and in other mountainous regions of the West.[3] The impulse wheel was driven by an extremely high velocity jet of water playing on buckets at the periphery of a high-strength iron or steel wheel. These water wheels depended not on the mass of water falling the distance of the diameter of the wheel, but on the velocity and mass of the impacting jet of water. Therefore, these wheels, like turbine wheels, were much smaller in diameter than the conventional water wheel.[4]

[1] U.S., Department of Commerce, Bureau of the Census, *Central Electric Light and Power Stations and Street and Electric Railways, with Summary of the Electrical Industries, 1912* (Washington, D.C.: GPO, 1915), foldout facing p. 132.

[2] F. M. Feldhaus, *Die Technik Der Vorzeit, Der Geschichtlichen Zeit und Der Naturvölker* (Leipzig and Berlin: Engelmann, 1914), p. 1304.

[3] Charles M. Coleman, *P.G. and E. of California: The Centennial Story of Pacific Gas and Electric Company, 1852–1952* (New York: McGraw-Hill, 1952), pp. 112–15.

[4] On the history of the turbine, see Louis C. Hunter, *Waterpower: A History of Industrial Power in the United States, 1780–1930* (Charlottesville: University Press of Virginia, 1979), pp. 292–342; Edwin T. Layton, Jr., "Scientific Technology, 1845–1900: The Hydraulic Turbine and Origins of American Industrial Research," *Technology and Culture* 20 (1979):64–89; Hans-Joachim Braun, "Technische Neuerungen um die Mitte des 19. Jahrhunderts: Das Beispiel der Wasserturbinen," *Technikgeschichte* 46 (1979): 285–305; and Terry Reynolds, *Stronger Than a Hundred Men: A History of the Vertical Water Wheel* (Baltimore: The Johns Hopkins University Press, 1983).

The development of electric power transmission inaugurated a new phase in the history of hydraulic power. Water-power sites were exploited only by local crafts, mining, and industry so long as power transmission was limited to mechanical means, primarily belts and gears. As a result, water-power resources in mountainous areas throughout the world were only partially developed. After the introduction of direct-current generators and motors during the 1870s, however, engineers and entrepreneurs saw the possibilities of long-distance electric power transmission. Among those who led in power-transmission experiments and demonstrations was Marcel Deprez, the French engineer who tried direct-current high-voltage transmission in many demonstrations. Oskar von Miller, as noted earlier, invited Deprez to install a power-transmission demonstration on the occasion of the Munich electrical exhibition in 1882 (see p. 131 above). Two years later, at the Turin exhibition, Gaulard and Gibbs demonstrated transmission of electricity over long distances for use in lighting (see p. 94 above). In 1891 von Miller, whose interest in power transmission was greatly stimulated by the huge water-power potential of the Bavarian Alps not far from his native Munich, also helped organize the world's first substantial demonstration of polyphase-current transmission over a long distance (see p. 131 above). About a decade later von Miller began the investigations and planning that would culminate in the construction of a large Alpine hydroelectric station at Walchensee, in the Bavarian Alps, from which power was transmitted in a grid throughout Bavaria (see pp. 337–50 below).

Early demonstrations of power transmission were made in Germany, Britain, and the United States, but the history of high-voltage long-distance power transmission from water-power sites does not lend itself to comparison, for developments in the three countries were not comparable. Neither Britain nor Germany had a major (ranked on a world scale) point-to-point hydraulic power transmission system by 1914.[5] The major reason for the less impressive development of the technology in Germany was the density of that country's settlement and industrial development. The major causes in Britain were density of settlement and the lack of major mountain ranges. Germany had major water-power sites in the Alps spread along her southern and southeastern borders, but load centers were within tens of miles rather than hundreds. Britain had much less potential water power, and her industrial areas were within a relatively few miles of her high hills and low mountains, with the exception of the Highlands of Scotland. Both countries had hydroelectric stations before World War I—Germany's were large-capacity installations—but the combination of large capacity and transmission distances of a hundred miles or more was a California phenomenon.

Because of its dramatic beauty and the immense amount of water power it harnessed, some historians have concluded that the electric power project at Niagara Falls, not the California systems, was the turning point in the history of long-distance power transmission in the United States. The engineers and entrepreneurs who presided over its construction expected Buffalo, New York, a heavily industrialized city almost twenty miles from the power site at the falls, to be the load center. Instead, an industrial center

---

[5] Bureau of the Census, *Central Electric Light and Power Stations*, foldout facing p. 132.

sprang up at the falls. Most of the 15,000 h.p. that was made available in the initial installation of 1895 was put to use in the metallurgical and electrochemical plants that located there to use the cheap electricity. A transmission line to Buffalo did not open until 1896, and had a capacity of only 1,000 h.p. Not until 1910 did a major long-distance transmission line extend from Niagara Falls, and this was the work of the Hydro-Electric Power Commission of Ontario, a government-owned system supplying Toronto and the province of Ontario. In 1914 the Toronto Power Company constructed the second major transmission line from a power plant at the falls.[6]

Within the United States, only the southeastern region extending from Virginia to Alabama witnessed development of hydroelectric power transmission comparable to that of the California region. There the water power available in the Appalachian Mountains, the relatively thinly populated uplands, and the concentration of industry a hundred or more miles distant on the coastal plain all favored long-distance transmission. By 1914 North Carolina had two major transmission systems in operation, one extending 96 miles and the other 210 miles; Virginia–West Virginia's system covered 75 miles; Georgia had a 170-mile system; and Alabama's and Tennessee's systems extended 150 and 140 miles respectively. Distribution systems in urban centers and regional grids or networks would connect to these southeastern transmission lines in a way that was similar to the California connections. The only major difference between the hydroelectric plants in the two regions was the classification of those in California as high-head plants, ranging from 1,000 to 2,000 feet, and those in the Southeast as medium-head plants, measuring in the hundreds of feet.[7]

It was in California, however, in the 1890s, that an era of long-distance hydroelectric power transmission began. "The first faint glimmerings of a new dawn," the *Electrical World* observed in 1912, "were just visible and the new dawn was in the West." "California," the journal had claimed earlier, "is the birthplace of real long-distance power transmission on this continent." Subsequently the region became "the great center of developments, the world's laboratory of brilliant and successful experiments in high-tension energy transmission."[8] The history of the episode is a striking example of human beings using technology to mediate between themselves and nature, a relationship as old as human history.

The history of hydroelectric development and power transmission in California is an outstanding example of the way in which technology can compensate for natural inadequacies. Nature had left the California region

[6] For a comparative study of the use of Niagara power in Canada and the United States, see Robert Belfield, "The Niagara Frontier: The Evolution of Electric Power Systems in New York and Ontario, 1880–1935" (Ph.D. diss., University of Pennsylvania, 1980). The contemporary and historical literature on the Niagara project is extensive. Belfield cites many of these sources.

[7] "The Great Southern Transmission Network," *Electrical World* 63 (1914): 1201, reprinted in *Turning Points in American Electrical History*, ed. James G. Brittain (New York: IEEE Press, 1976), p. 163.

[8] "The Larger Aspect of Central Station Work," *Electrical World* 59 (1912):1141; "The World's Longest Power Transmissions," ibid. 39 (1902): 188.

short of energy resources.[9] Forest covered the lower elevations of the Sierras, but the rapidly growing population on the coast could not afford to transport this wood for fuel. The *Electrical World* summarized the plight of the region, with its expanding population and limited energy supply:

> The unwooded plains and bare foothills [on the coast] smoothed as with a Titan's trowel, give small supply above ground and save for a modest supply of petroleum near Los Angeles there is little below. Until the beginning of power transmission by electricity the main reliance was Australian coal and the cost of power was almost prohibitive.[10]

Considering the San Francisco area, another contemporary observed:

> Until of late power for industrial uses has been a commodity as expensive as it is indispensable. Except in mountainous localities, wood is not to be had for fuel, while the supply of California coal has thus far been really insignificant when compared with its total consumption.[11]

On the positive side, California did have natural endowments that were transformable by technology into energy resources.[12] The history of electric power in California is a notable reminder that changing technology can make nature a resource, a factor of production. The coming of long-distance power transmission made the remote lakes, streams, and rivers of the Sierra Nevada power sources first for relatively nearby mining towns, then for neighboring farm communities of the great central valley of California, and finally for the far more heavily settled coastal cities. Until the advent of power transmission, the Sierras were perceived by coastal inhabitants as obstacles to the wagon trains and railroads, as the location of gold and silver mines, and as the site of winter snows, which, during the spring thaw, brought water for irrigation but floods as well.[13]

Before describing the way in which technology moved energy a hundred miles or more, however, it is pertinent to consider the natural endowments in California that engineers and others exploited to accomplish this power transmission. Natural geography invariably figures prominently in accounts of hydroelectric projects. The moisture-laden winds passing eastward from the Pacific first encounter the Coast Range running along the ocean and ranging in height from about 2,000 to 8,000 feet. Twenty to forty miles in width, these hills and mountains receive some rain as they deflect the moist air upward, but not enough to maintain regularly flowing streams. Nor is water accumulated as deep snowfall. Therefore, the Coast Range has been of little consequence to the energy-seeking coastal communities. In contrast, in the eastern reaches of the state, rivers and streams flow from the sloping face of the Sierra Nevada. This north-south range extends about 430 miles

[9] Hamilton Wright, "Long-Distance High-Tension Transmission of Power in California," *Scientific American* 88 (1903): 373; H. G. Butler, "Pools of Power," *Survey Graphic* 51 (1924): 605; Albert W. Atwood, "Testing Big Business in the West," *Saturday Evening Post,* 22 December 1923, p. 16.

[10] "A Notable Transmission System," *Electrical World and Engineer* 45 (1905): 375.

[11] George P. Low, "The World's Longest Electric Power Transmission," *Journal of Electricity, Power and Gas* 11 (1901): 145.

[12] I have concentrated on California in this section on the history of power transmission. Similar developments occurred in other mountain states of the American West.

[13] "Rivers of the West," *Electrical West* 129 (1962): 344–45.

with elevations ranging up to 14,495 feet. The northern half normally has an annual rainfall of more than 400 inches; the southern portion usually has 10–20 inches, with the exception of the southernmost tip, which reaches into the low-rainfall area below Fresno and above Los Angeles. Natural basins and deep, river-worn canyons punctuate the Sierras and offer power and flood-control sites for dams, making possible year-round utilization of the abundant winter precipitation. The Yuba, the Feather, and the Stanislaus are among the rivers that provide sites for dams.

The Coast Range and the Sierras rim, and virtually close in at both ends, an extensive valley. From north to south it extends about 430 miles from latitude 40° to latitude 35°. In width it averages about 40 miles. The resulting 17,200-square-mile area is drained by the Sacramento and San Joaquin rivers into the San Francisco Bay. The cities in the valley, which include Sacramento, Fresno, San Jose, Stockton, and, at the bay's opening to the sea, Oakland, Alameda, San Francisco, and Berkeley, waited for irrigation powered by electric pumps to make possible the intensive cultivation of fruits, vegetables, and other crops. Similarly, these communities awaited the arrival of energy to relieve the constraints on their growth. They would spring to life dramatically as the transmission lines reached them, a situation not unlike that which occurred when canals and railroads first spread in the United States.

As was the case throughout the world before the era of electric power transmission, the water power in the mountains benefited mostly the local inhabitants, who were few in number because of the difficult terrain. Among those few were usually miners, who took advantage of the natural proximity of minerals and water. The water was used to wash ore free from the earth, and the rushing streams powered the wheels that drove the bellows and other machinery used by the processors of the ore and metal. Californians in the 1850s and sixties constructed waterworks to facilitate mining and metallurgical operations. In Nevada, Placer, Eldorado, Amador, Calaveras, Stanislaus, and Tuolumne counties, mining or water companies built storage reservoirs and canals to carry the water to mining sites to wash away the sand and gravel hills in which the gold lay buried. By the early 1860s, over 5,000 miles of artificial water courses had been built. As mining techniques changed and mining activity waned, some of these facilities were modified for use in irrigating the nearby valleys, and even later some of the engineering works were taken over for hydroelectric development.[14]

The history of hydraulic development, the abundance of rainfall, and the high elevations with rushing streams in eastern California, together with the coastal communities aspiring to grow, the extensive valley lands awaiting irrigation, and the nascent coastal industry needing cheap energy, formed a reverse salient, defined a critical problem, and suggested a solution. The situation called for engineers, financiers, manufacturers, construction crews, and entrepreneurs to provide technology, to bring new power where it was disposable. The history of the organization of hydroelectric power and point-to-point transmission in California is extremely complex and can only be highlighted here; this approach will concentrate on representative acts of long-distance power transmission from the Sierra

[14] Coleman, *P.G. and E. of California*, pp. 92–94.

Nevada to the bustling San Francisco Bay area and later to sprawling Los Angeles and its environs.

Local financiers with eastern sources of capital, as well as local representatives of eastern electrical manufacturers, were involved, but during the first two decades of power transmission in California, most of the electrical equipment and technical advice came from eastern manufacturers and consultants. A list of turbines, generators, and transformers used in major hydroelectric stations includes the names of the Stanley Electrical Company of Massachusetts, the Westinghouse Company, and the General Electric Company. The exceptions to the eastern bias were the Allis-Chalmers Manufacturing Company of Wisconsin, supplier of generators, transformers, and turbines; the Ohio Brass Company, maker of critically important strain insulators for the transmission line; and the Pelton Company of California and I. P. Morris of Philadelphia, Pennsylvania, producers of water wheels and turbines.[15] The modification of equipment to suit California conditions, the supervision of construction, and the overall design of systems, however, increasingly became the work of California engineers, some of whom had been educated at Stanford, the University of California at Berkeley, and other California institutions.[16] Among these Californians there was "no respect for constituted authority . . . [on] hydraulic, mechanical and electrical matters. If an impossible dam has to be erected . . . they build it . . . if an altogether unheard-of bit of tunnel has to be made to connect with a quite impracticable flume . . . they bore the tunnel and build the flume . . . if three or four stations must be operated together in defiance of all precedents, in go the switches and the plants operate."[17]

Several of the early power plants and a short-distance point-to-point transmission line were concentrated in the Sierras to the north and west of Sacramento, the capital of the state. The Folsom power plant on the American River, which transmitted 20 miles to the city, was conceived of and constructed under circumstances that in many ways were representative of the early development of hydroelectric power in California (see Fig. X.1). The history of the small plant has almost been forgotten, but when it began transmitting electricity to Sacramento in the summer of 1895, contemporaries celebrated the occasion by illuminating the capital with thousands of incandescent lamps, firing hundreds of guns in salute, and making boastful addresses. Local newspapers and journals proclaimed it the largest power plant in the world as well as the longest point-to-point

[15] From a survey of the *Journal of Electricity,* Dr. Arthur Norberg and his staff at the Center for the History of Science and Technology, University of California at Berkeley, concluded that through 1925 there was no major manufacturer of dynamos, generators, or large motors in the West; instead, there were many sales representatives from the East. There were, however, western manufacturers of power-line switches (Pacific Electric Manufacturing Co. of Napa, K-P-F Electric Co. of San Francisco, and Agutter-Griswold of Seattle, Wash.), water wheels, and turbines.

[16] A number of instructors of electrical engineering in California colleges and universities contributed to the growth of the utility systems through their teaching and also by working for the utilities and manufacturers or as consulting engineers. Utilities and manufacturers also contributed to research projects and facilities at leading departments of electrical engineering, such as those at Stanford and the Throop Institute of Technology (later the California Institute of Technology). See the section on high-voltage laboratories, pp. 377–80 below.

[17] "The Transmission Systems of the Great West," *Electrical World* 59 (1912): 1142.

Figure X.1. *Folsom power plant, 1958.*
Courtesy of the Pacific Gas & Electric Co.,
San Francisco, Calif.

transmission of power. The engraved invitation for a "Grand Electrical Carnival" referred to "the greatest operative electrical plant on the American continent."[18]

In several aspects, Folsom, California, was like Lowell, Massachusetts, before the development of Niagara Falls, the best-known water-power site in America. In 1850 Horatio Gates Livermore had come from New England to Eldorado County, California, where gold was first discovered, with an image of water-powered Lowell in his mind. He envisioned a similar industrial city at Folsom, a stop on the Sacramento Valley Railroad, and he pictured logging, mining, and other industries along the nearby American River. In 1850 Livermore intended to build a Lowell, but his initial steps culminated quite differently, in 1895, because of electricity.

In 1867 Livermore constructed a dam on the American River as part of his grand design for the economic development of Folsom. His plans also called for a canal on each side of the river extending about two miles from the dam to the town of Folsom: one, a power canal, would provide a head of about 80 feet above the river for powering water wheels; the other would carry logs and irrigate the potentially rich farmland in the region. Earlier, Livermore had invested in a system of water-carrying ditches to supply the placer mines with water to wash the gold-bearing soil. The elder Livermore, like so many other enterprising Californians, ingeniously utilized water as a multifaceted resource.

---

[18] Coleman, *P.G. & E. of California*, pp. 116–17. The account of Folsom is based on ibid., pp. 116–27; and Archie Rice, "Folsom Power Plant," *Journal of Electricity, Power and Gas*, as reprinted in *Pacific Gas and Electric Company: Historical and Descriptive* (n.p., n.d.), pp. 1–21.

Figure X.2.   *Horatio P. Livermore.*
Courtesy of the Pacific Gas & Electric Co.,
San Francisco, Calif.

Various troubles delayed completion of the power canal and the dam until 1893, however. (Convict labor had been used in the construction.) By then, Horatio P. Livermore, a son of Horatio Gates Livermore, had read reports of the power-transmission experiment at Frankfort on the Main and of the successful completion in 1888 of Frank Sprague's pioneering streetcar system in Richmond, Virginia. So stimulated, H. P. Livermore decided to build a hydroelectric plant at the termination of the power canal in Folsom, transmit with alternating current over the 20 miles to Sacramento, and, to ensure the system a load, or market, acquire a streetcar franchise in Sacramento. Lowell, Massachusetts, was no longer the working model for the system. For financing, the younger Livermore turned first to San Francisco, where local capitalists, including the head of the Edison Electric Light & Power Company, appealed to the Electrical Securities Company of Boston. Electrical Securities underwrote the Folsom utility's bonds, drawing upon the advice of Charles Coffin, president of the newly formed General Electric. Financed by Electric Securities, Livermore ordered his generators, transformers, and other equipment from Coffin's General Electric. The Folsom plant first transmitted 3,000 kw. from four GE alternators at 11,000 volts by means of 60-cycle three-phase current. The transmission line was carried atop 40-foot cedar poles set up in two lines along each side of a country road. At Sacramento, converters supplied direct current for the streetcar load, and lower-voltage current went to the utility's other customers. Industry did not need to move to the water-power site; the power had traveled over small wires to industry.

The Livermores (another of Horatio Gates's sons, Charles, also had become involved) did not anticipate, however, the region's summer droughts, which caused the American River to run low. In 1897 and 1898, despite the construction of a supplementary, 750-kw. hydroelectric plant at Folsom, the increasing industrial and street-lighting load in Sacramento could not be fully met because of drought conditions. Nor did an auxiliary steam plant under construction in Sacramento promise to provide a solution to the problem. Therefore, in 1899 the Livermores and their associates in the Sacramento Power & Light Company contracted to take power from the Yuba Electric Power Company's Colgate hydroelectric power plant 61 miles away.

Newer and larger than Folsom, the Colgate power plant was to become much better known because its owners and engineers would reach beyond Sacramento in the great transmission leap to Oakland, on San Francisco Bay. The construction of the Colgate plant was begun in 1895 by two California entrepreneurs, Eugene J. de Sabla, Jr., and John Martin. It would become the nucleus of a system-building endeavor that culminated in the Pacific Gas & Electric Company, one of the world's largest utilities. Both de Sabla and Martin came to the field without technical experience. De Sabla in 1892 headed an import-export house established by his father in San Francisco, and Martin dealt in pig-iron and cast-iron pipe.[19] Also involved was Romulus Riggs Colgate, a financier, an enthusiastic advocate

[19] The account of de Sabla and Martin is based on Coleman, *P.G. & E. of California,* pp. 128–43; and Archie Rice, "Colgate and Yuba Power Plants," *Journal of Electricity, Power and Gas,* as reprinted in *Pacific Gas and Electric Company,* pp. 75–97.

Figure X.3. *John Martin, 1904.*
Courtesy of the Pacific Gas & Electric Co.,
San Francisco, Calif.

Figure X.4. *Eugene de Sabla, Jr., 1910.*
Courtesy of the Pacific Gas & Electric Co.,
San Francisco, Calif.

of hydroelectric power, and the wealthy grandson of the founder of the Colgate Soap & Perfume Company.

William Stanley, the developer of Westinghouse's alternating-current system and, by 1895, head of the Stanley Electrical Manufacturing Company, a supplier of polyphase power equipment, encouraged and advised Martin and his associates on the Colgate and other hydroelectric projects. Stanley assigned Martin the agency for Stanley equipment in California after interviewing him at the Stanley plant in Pittsfield, Massachusetts. Later Stanley dispatched one of his engineers to California to instruct Martin in electrical theory and practice and in the installation of Stanley equipment. Stanley became a familiar name in California power transmission, and Stanley-designed equipment continued to be used in the region after the Stanley Electrical Manufacturing Company was absorbed by the General Electric Company.

A common financial interest in the gold mines of the Yuba River region to the north of Sacramento brought de Sabla and Martin together. In search of a power source for their mines that would be more economical than steam, they turned to hydroelectric power. They may have been spurred on by the successful development of the nearby Folsom plant. Before building the Colgate plant, they constructed a 6,100-kw. power plant on the South Yuba River, about four miles from Nevada City, California, a center of gold-mining activity. De Sabla raised the capital and found the customers for the Nevada City plant, and Martin, the self-taught hydraulic and electrical engineer, assisted by Stanley, installed 133-cycle two-phase Stanley generators and Pelton wheels in the plant and erected a 5,500-volt transmission line to Nevada City and Grass Valley. Work began on the Nevada plant (Fig. X.5) on 5 July 1895, and seven months later the system was in operation. As de Sabla succinctly summed up the entrepreneurial history of the Nevada power plant,

I owned some mining interests at Nevada City and I started by trying to get electric power for use in the mines. The problem was to take water by ditch from the South Yuba and, by a gradient less than the river's, carry it down to some point where a sufficient fall could be secured to operate electric generators. Land and water rights were acquired, plans made for the undertaking, a site selected for a dam, and, some three miles further down stream, a spot chosen for a power house at a point on the river about 15,000 ft. above the level of the ocean.[20]

With the success of the Nevada City plant, de Sabla became confident there was a market for more power in nearby Sacramento, and even in San Francisco. He raised more money for hydroelectric projects from R. R. Colgate, who later helped finance the power plant that was named for him. Martin then bought power rights from water companies that owned systems of ditches, flumes, and canals that had been built for irrigation and for supplying water to mines. One of the most promising properties for conversion to hydroelectric power was located on the north fork of the Yuba. There the partners erected the Yuba powerhouse, taking advantage of the head of water provided by an irrigation canal situated alongside a

[20] Archie Rice, "Story of the Nevada Power Plant," *Journal of Electricity, Power and Gas,* as reprinted in *Pacific Gas and Electric Company,* p. 59.

mountain. By locating the power plant below in Brown Valley, they obtained a fall of 295 feet. Completed in 1898, the Yuba power plant had a capacity of 1,080 kw. An 18-mile transmission line carrying 16,000 volts ran from it to mines and power customers in Marysville.

The success of the Nevada and Yuba power plants kindled the enthusiasm, provided the experience, and engendered the commitment that led Martin, de Sabla, and Colgate to cooperate in the construction of the Colgate plant (see Fig. X.6). With a contract to supply 2,200 kw. to the Livermores' water- and power-short Folsom plant in Sacramento, Martin, de Sabla, and Colgate in January 1899 formed the Yuba Power Company, a million-dollar corporation. The company's immediate objective was to construct, on the Yuba River eight miles above the 1898 Yuba plant, the Colgate plant, which would supply both the local load and Sacramento, over 60 miles away.

The Colgate plant, which was named for the man who helped finance it, had many of the technical features that came to characterize the high-head, California hydroelectric plants. Extensive civil-engineering construction was needed. Water diverted from the highly elevated river at the dam site flowed in a wooden flume (7 feet in width and 5 feet in depth) that snaked along the course of the river. The fall of the river was steep, but the flume lowered the elevation of the diverted water gradually, the gradient being a gentle 12⅔ feet to the mile. By the time the flume reached the power plant, which was situated at river level 8 miles downstream from the dam, the fall from the flume to the power plant was 702 feet, or more than four times the height of Niagara (see Fig. X.7).

Figure X.6. *Colgate power house and penstock (rear). Courtesy of the Pacific Gas & Electric Co., San Francisco, Calif.*

Figure X.7. *Flume line to Colgate power house. Courtesy of the Pacific Gas & Electric Co., San Francisco, Calif.*

Originally two, and later five, 30-inch pipes, or penstocks, carried water down to the power plant where nozzle-ends played streams of water, the thickness of a man's arm, upon the steel buckets of "man high" impulse wheels. The force of the water inspired awe among contemporary observers. Visitors were warned that a slip into the icy, surging waters of the flume meant sure death; they also heard that a man could hammer upon the jet of water issuing from the nozzles as if using a sledge upon an anvil. The Colgate power plant began operating in 1899 carrying the local mine and industrial load and supplying Sacramento. Because of the variations in existing motors, the plant had both three-phase 60-cycle and two-phase

133-cycle Stanley generators. By 1901 Colgate's capacity was 14,000 h.p., 3,000 of which could be delivered to Oakland.[21]

In 1901 Martin, de Sabla, and Colgate made a historic decision that resulted in the construction of the world's longest transmission line from Colgate to Oakland, on San Francisco Bay, a distance of 140 miles. The eastern bond firms E. H. Rollins & Sons and N. W. Harris & Company helped facilitate the project. Seeking refinancing, the Oakland Transit Company, which operated 126 miles of street railway, negotiated with the Rollins and Harris companies. The financiers decided that the transit company needed low-cost power to improve its economic health. Engineers of the bond houses examined the feasibility of transmission from Colgate. Upon receipt of a favorable report, the financiers suggested in June 1900 the formation of the Bay Counties Power Company to consolidate the companies operating the Nevada, Yuba, and Colgate power plants. De Sabla became president and general manager, R. R. Colgate a member of the board of directors, and Martin helped carry out the transmission project, drawing upon his Stanley connections.[22]

After consulting in Pittsfield with William Stanley and two of his associates, John F. Kelly and C. C. Chesney, both prominent engineers, Martin decided on 60,000-volt transmission, double the voltage recommended by General Electric and Westinghouse. Two parallel lines, one copper and one aluminum, carried on cedar poles 25 feet apart, reached from Colgate to the substation at Oakland. The duplication seemed necessary to insure the reliability of the pioneer project. San Francisco's *Journal of Electricity* heralded the delivery of power on 27 April from the Sierras to the Bay area as "the greatest electrical engineering triumph ever accomplished."[23]

Not only the transmission line, with its unprecedented length and remarkably high voltages, but also an impressive water crossing excited that comment. Because it was necessary to cross San Francisco Bay in order to reach Oakland, the engineers had to decide between a subterranean and an overhead crossing. They chose suspension at a narrow gap, the Carquinez Strait, where the water reached a depth of 120 feet and ran in heavy and dangerous currents near the northern extremity of San Francisco Bay. The strait was 2,750 feet wide, with a bluff rising to 162 feet to the north and a hill rising to about 400 feet to the south. The north side presented a special problem because the land was marshy behind the bluff, making anchorage for the cables difficult. The towers could not withstand the tons of lateral pull from the four suspended cables, so they had to be anchored in rock or concrete behind and below the main towers. Two principal towers were erected, a 225-foot steel structure on the north and a 64-foot one on the 400-foot elevation of the hill. Behind the north tower, at a distance of 1,385 feet, a short tower was needed to suspend the cables over the marshy ground before they were carried down to their anchorages (see Fig. X.8). The horizontal distance between the principal towers was more than three-quarters of a mile, the clearance above the bay of the suspended cables at

[21] Rice, "Colgate and Yuba Power Plants," p. 95; Low, "World's Longest Electric Power Transmission," pp. 154, 167.
[22] Coleman, *P.G. & E. of California,* pp. 144–45.
[23] Low, "World's Longest Electric Power Transmission," p. 145.

Figure X.8. Transmission Span, Carquinez Strait, 1936. Courtesy of the Pacific Gas & Electric Co., San Francisco, Calif.

their lowest point was just over 200 feet, and the cable length from anchorage to anchorage was 6,292 feet. Each cable exerted a pull of 12 tons on its anchorage. The problem of insulating the cables at the towers and anchorages was a difficult one.[24]

After completion of its great point-to-point transmission project, the Bay Counties Power Company entered a new phase of development—interconnecting within its 14,000-square-mile supply area its several Sierra power plants and various kinds of loads in the Sierras with consumers in the farmlands along the route of the transmission line and with those at the terminus of the line in the rapidly developing Oakland region. According to the practice of the day, the company tied together the Nevada, Yuba, and Colgate plants to enable them "to be operated in unison as a single station as easily as though they were combined under a single roof."[25] Through interconnections with other independent companies, power from the Colgate, Nevada, and Yuba plants in the Sierras reached California counties to the north and south of Oakland and San Francisco. The industry and agriculture in the region served thereby included street railway, manufacture, shipping, mining, stock raising, and fruit growing.[26] Yet Bay Counties remained primarily a power-generating and -transmitting company, in contrast to a utility that is strongly based in distribution. Within a few years, however, mergers and the acquisition of distribution systems, especially in San Francisco, would change the system's character.

[24] George P. Low, "The Great Carquinez Transmission Span," *Journal of Electricity, Power and Gas* 11 (1901): 91–102.

[25] Low, "World's Longest Electric Power Transmission," p. 151; Wright, "Long-Distance Transmission," p. 373.

[26] Low, "World's Longest Electric Power Transmission," p. 164.

In 1903 de Sabla and Martin, seeing the economic advantages—especially the improvement in load factor—that would follow further interconnection and system-building, formed the California Gas & Electric Company (CG&E) to acquire and merge power companies, not only to interconnect them. CG&E merged several Sierra hydroelectric power companies with its own Bay Counties Company. One of these, the Standard Electric Company, operated the Electra hydroelectric plant in the Sierras, which transmitted power to the Bay area using 30,000 volts. California Gas & Electric also absorbed the Sacramento Electric Gas & Railway Company, which owned the pioneering Folsom plant and took power from Colgate as well. A third company that was acquired early was the Central California Electric Company, which served a district, between the Folsom and Colgate plants, with three hydroelectric plants. By 1905, further expansion had brought into the CG&E system ten hydroelectric plants with interconnecting transmission lines and two long-distance, high-voltage lines extending to the Bay area.

On the demand (or load) side, CG&E obtained, in addition to the relatively small distribution systems acquired with the Sierra power companies, several utilities in the Bay area. In 1903 Martin purchased the long-established Oakland Gas Light & Heat Company, which supplied Oakland and Berkeley, and the United Gas & Electric Company, which served communities south of San Francisco, including San Jose. Obviously the next prime objective for de Sabla and Martin as system builders was a distribution system in San Francisco. They fulfilled this goal in 1906 when they acquired the San Francisco Gas & Electric Company. After more than a decade of sharp competition in San Francisco, the San Francisco Gas & Electric Company had established itself as the dominant utility there. The prior history of competition, merger, and acquisition in San Francisco was similar to that in other urban centers such as Chicago. As in Chicago, an Edison direct-current lighting system was the nucleus of early supply in San Francisco. The Edison Light & Power Company of San Francisco, established in 1891 and funded in part by the Edison General Electric Company of New York, merged with the San Francisco Gas Light Company in 1896 to form the San Francisco Gas & Electric Company. This company acquired competing San Francisco utilities during the next decade before being acquired by CG&E. (See Fig. X.9.)

Just as New York financiers had stimulated the formation of the Bay Counties Power Company and the transmission from Colgate to Oakland, N. W. Halsey & Company, a New York investment and brokerage house, took the lead in merging the San Francisco utility and CG&E to form a new corporate entity, the Pacific Gas & Electric Company. Halsey & Company was acquainted with the development of California power, having underwritten a $3 million bond issue of the Valley Counties Power Company, a subsidiary of Bay Counties Power Company. Martin and de Sabla approached Halsey again in 1905 with the proposal that it organize the merger. After an audit and engineering study by Halsey auditors and engineers, the financial house agreed.

The creation of the new utility was an ingenious exercise of eastern financial expertise. The Pacific Gas & Electric Company (PG&E) was incorporated in October 1905 with a capital of $20 million in common stock,

Figure X.9. *PG&E mergers and consolidations. Courtesy of the Pacific Gas & Electric Co., San Francisco, Calif.*

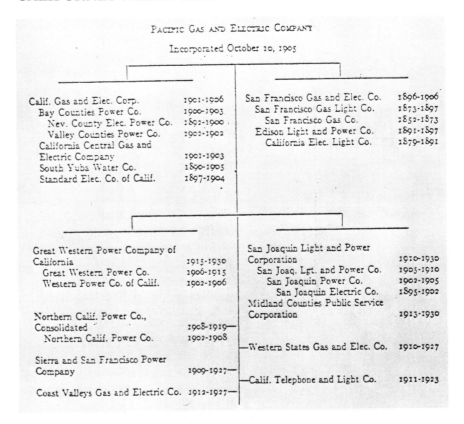

$10 million in preferred stock, $10 million in collateral trust bonds, and $4.5 million in debentures. The Pacific Gas & Electric Company acquired the almost $16 million in outstanding stock of the San Francisco Gas & Electric Company (SFG&E) by payment of PG&E bonds (worth $10 million) and cash, some of which derived from the sale of PG&E debentures. The outstanding stock of the California Gas & Electric Company (CG&E) was acquired by PG&E in exchange for PG&E preferred stock worth $10 million. The stock purchased from the San Francisco Gas & Electric Company, which retained its corporate identity, was deposited with the Union Trust Company as security for the PG&E bond issue.

At about the same time that PG&E was incorporated, Halsey incorporated the Pacific Gas & Electric Investment Company, the purpose of which was to acquire and hold the stocks, bonds, mortgages, and so on, of other corporations. The company's authorized capital was $20 million. Halsey turned over to the investment company the options to purchase the stock of CG&E and SFG&E and agreed to advance not more than $4.5 million (secured by note) to the investment company. The investment company gave all but 10 of its shares to Halsey for the formation of PG&E, assignment of rights, advance of funds, and Halsey's year of expensive investigations and negotiations. The investment company then transferred to the PG&E the rights and the advance of cash which it needed to pay to the SFG&E for its stock. In return, PG&E turned over all but 15 shares of its common stock to the investment company. As a result of this train of transactions, N. S. Halsey & Company received common stock of the investment company worth potentially $20 million because of the $20 million in PG&E stock the investment company owned. The shares of the investment company, it was understood, were divided equally among Halsey, de Sabla, Martin, Frank Drum (a San Francisco financier who participated in de Sabla utility enterprises), and the company's treasurer, in a potential $20 million return for creating property of value in excess of the value of the two formerly independent utilities.[27] Pacific Gas & Electric began business in January 1906.[28]

The San Francisco company brought steam power plants to the new Pacific Gas & Electric Company. These plants complemented the PG&E's hydroelectric power by carrying peak loads when demand rose high and when winter freezes or droughts lowered the capacity of the hydroelectric system. The steam plants also aided in regulating the whole system because of their ability to respond rapidly to load changes. Such a symbiotic relationship was not unusual at this point in the history of electrical utilities, but the use of petroleum-fired boilers for the steam plants was. A confluence of circumstances made California petroleum available at about the same time that hydroelectric power reached San Francisco. In 1907 PG&E's largest steam plant, located in San Francisco, used 546,000 barrels of California crude and generated 73 million kwh. In 1913 the company's hydroelectric capacity of over 123,000 h.p. was supplemented by oil-fired steam plants in San Francisco, Oakland, Sacramento, and San Jose with a combined capacity of more than 110,000 kw.[29]

A brief survey of the history of the company up to 1914 will show that by then the transition from a power company to a company presiding over an integrated regional system was complete (see Fig. X.10). PG&E had become the largest system on the Pacific Coast and ranked among America's five largest utilities. Operating in thirty counties of central California, the company, virtually a single operating unit of interconnected plants, supplied an area of 37,000 square miles and a population of 1.3 million (1910 census). PG&E had 36 percent of the electric and gas business in California. The number of hydroelectric plants remained about the same during the first eight years of the company's existence, but the installed hydroelectric capacity increased by 64 percent. The number of steam plants declined, but larger units brought a 160 percent increase in capacity. The consumption of oil as fuel increased. The peak demand on the system increased from 63,000 h.p. in 1906 to 160,000 h.p. in 1913.[30]

The relative increase in steam-generating capacity as compared to hydroelectric capacity is notable. The utility had begun, in 1906, with almost twice as much hydroelectric plant as steam plant and with extensive water-power rights in the Sierras. The decreasing price of California crude oil and the increasing Bay Area load partially explain the company's failure to invest in additional hydroelectric facilities. Another reason for not ac-

[27] Coleman, *P.G. & E. of California*, pp. 230–31.

[28] Ibid., pp. 227–33.

[29] Pacific Gas & Electric Co., *Eighth Annual Report of the Pacific Gas and Electric Company for the Fiscal Year Ended December 31, 1913* (San Francisco: PG&E, 1914). During the winter of 1910, the company installed its first steam turbine; four years later 85 percent of its steam capacity was the turbine type.

[30] Ibid., p. 11.

Figure X.10. *California Gas & Electric Co. (later (PG&E) distribution and transmission system, central California, 1904. Courtesy of the Pacific Gas & Electric Co., San Francisco, Calif.*

quiring additional facilities was the availability of low-cost hydroelectric power from the Great Western Power Company. Although Great Western had owned a power line from Big Bend in the Sierras to Oakland since 1909, it had lacked a distribution system in San Francisco, and so it wholesaled power to Pacific Gas & Electric for sale there.

In 1911 Great Western finally acquired a distribution system in San Francisco, and in 1912 it laid a transmission cable under the Bay from Oakland to San Francisco. Anticipating the loss of wholesale power, Pacific Gas & Electric embarked on a hydroelectric project to increase the generating capacity of the company's system by 80 percent. The Drum station, the first power plant in a related projected series of six, came into operation in 1913, adding more than 33,000 hydroelectric horsepower to the Pacific Gas & Electric system. Drum accounted both for the first substantial increase in the system's hydroelectric power since the founding of the company and for the relative rise of hydroelectric generation to the level of steam generation in 1913.[31]

The purpose of the Drum project was to utilize the watershed of the South Yuba and Bear rivers. The scope of the concept extended far beyond

---

[31] For accounts of the project see ibid., pp. 12, 14; and Rudolph W. Van Norden, "Lake Spaulding–Drum Power Development," *Journal of Electricity* 31 (1913): 525–41.

the Nevada, Yuba, and Colgate schemes of slightly more than a decade earlier, however. Rights and facilities had been acquired in the 120-square-mile watershed in 1905. Two company engineers, Frank Baum and J. H. Wise, the first a graduate of Stanford and the other of the University of California at Berkeley, foresaw the enormous potential of the watershed for power and for irrigation and in 1905 began surveying and planning. The two men were in charge when construction on the Drum power plant began in 1912, but Wise died later that year.

The comprehensive plan in 1912 called for an increase in the capacity of Lake Spaulding to 97,000 acre-feet in order to utilize more of the flow of the South Yuba River and the runoff from the watershed. Drum and five other powerhouses were to be stepped along the Bear River and along artificial waterways for a distance of about 50 miles in order to make use of the 4,600-foot difference in elevation from the reservoir to the last powerhouse. The ultimate capacity of the six plants was projected as 190,750 h.p. A 110-mile steel-tower transmission line carried the power at 100,000 volts to Cordelia, Pacific Gas & Electric's load center and switching station near the Carquinez span across the San Francisco Bay. After passing through the last of the power plants, the harnessed water was to supply an irrigation system.

Construction of point-to-point transmission lines from the distant mountains to the energy-short coast was carried out by other utilities elsewhere in California. The preeminence of California among regions of the world that were developing long-distance transmission systems prior to World War I was made clear by the U.S. Census Bureau's listing of transmission systems operating at 70,000 volts or more throughout the world in 1914. Twenty-eight of the systems were located in the United States and, as observed earlier, eight of these were in California. Four, including Pacific Gas & Electric's Drum-to-Cordelia system, terminated in the San Francisco Bay area, and four served the Los Angeles region. Michigan had three transmission systems operating at 70,000 volts or above, but no other American state had so many. Canada and Italy led the other nations with four each; England had none; and Germany had two, both of which were coal-fired plants (see Fig. X.11).[32]

The history of Pacific Gas & Electric is representative in many ways of the history of the other utility companies operating the major California transmission lines and related generation and distribution systems. These utilities—like Pacific Gas & Electric—resulted from mergers of power transmission companies and urban utilities. In northern California the Great Western Power Company competed with Pacific Gas & Electric for the Oakland and San Francisco markets. With hydroelectric facilities at Big Bend in the Sierras, Great Western also built up a distribution system in the Bay area by acquiring small companies that survived alongside the San Francisco Gas & Electric Company and thereby avoided merger with Pacific Gas & Electric. Held by an eastern power trust with headquarters in New Jersey, however, Great Western merged with its competitor in 1930. Another of the northern California long-distance transmission enterprises,

[32] Bureau of the Census, *Central Electric Light and Power Stations,* foldout facing p. 132.

the Sierra & San Francisco Company, sold all of its power to San Francisco transit operators, but Pacific Gas & Electric acquired it as well in 1927.[33]

Two of the southern California utilities, the Pacific Light & Power Company (PL&P) and the Southern Sierras Power Company, had point-to-point power transmission of greater length than the companies in the north. In 1913 Pacific Light & Power completed the 241-mile Big Creek–to–Los Angeles transmission.[34] It competed in the Los Angeles area with the Southern California Edison Company, which operated the 117-mile-plus Kern River–to–Los Angeles transmission. Pacific Light & Power supplied power from Big Creek to the Los Angeles Railway Corporation and the Pacific Electric Railway, the largest intercity railway system in the United States. By 1915 Harry E. Huntingdon, the railway and industrial financier, owned substantially all of the stock of Pacific Light & Power and all of the stock of the Los Angeles Railway Corporation. In 1917 PL&E merged with Southern California Edison.[35] The Southern Sierras Power Company, later the California Electric Power Company, was the first to transmit power from the Sierras southward to the southern California communities. The company's Bishop–to–San Bernardino line, stretching across the Mojave Desert and using 2,000 steel towers, extended 238 miles. A network of distribution lines radiated out from San Bernardino into Riverside and San Bernardino counties.

The most unorthodox of the California power transmission and distribution utilities was the Los Angeles aqueduct system with its transmission lines extending from San Francisco to Los Angeles over a distance of 383 miles at 100,000 volts. The system was unusual because an extended aqueduct was used to supply water to Los Angeles, the water was used to generate electric power, and the system was municipally owned.

The transmission lines and related irrigation systems of southern California, like those in the north, transformed a region "by Nature left with a wholly insufficient water supply and parched from lack of it" into "[one] of the garden spots of the continent."[36] The power systems in the south, along with those in the north, also moved the entire state to the forefront among electric-power-generating regions. As early as 1902 and still in 1912, California ranked second only to New York State in the capacity of water-powered prime movers in central electric stations; and by 1912 California had risen from seventh to fourth behind New York, Illinois, and Pennsylvania in steam-power-station capacity. The combination placed the state at that time second to New York in the general category of primary power capacity in central electric stations. And in the ranking of the states according to total kilowatt-hour output, California's central stations, com-

---

[33] Sketches of the history of the Pacific Coast utilities can be found in the seventy-fifth anniversary issue of *Electrical West,* 129 (1962).

[34] The Big Creek–to–Los Angeles transmission line, because it operated with the highest voltage, drew upon the largest capacity, and was the longest of the major world transmission systems, attracted considerable publicity. See, for example, A. J. Farnsworth, "The Big Creek Power Development," *Stone & Webster Journal* 11 (1912): 167–75; J. H. Anderton, "Electrical Features of the Big Creek Development," ibid. 13 (1913): 326–36; H. A. Hageman, "The Big Creek Development: Hydraulic and Mechanical Features," ibid., pp. 228–47

[35] "Merger of California Hydroelectric Systems," *Electrical World* 68 (1916): 1134–35.

[36] "A Notable Transmission System," p. 375.

No.	Name	Operating voltage	Freq., cycles	Present	Ultimate	Beginning of operation	Hydraulic or steam	TURBINES Horse-power	Head, feet	Revs. per min.	Shaft hor./vert.	Turbine Manufr.	GENERATORS Kilovolt-amperes	Kilowatts	Power factor	Voltage	Gen. Manufr.	STEP-UP TRANSF. Kilovolt-amperes	Phases	Low ten.	High ten.	Conn. to ground	Total kVA connected to transmission lines	STEP-DOWN Manufr.	Low tension	High tension	Low ten.	High ten.	Termini	Distance, miles
1	Pacific Light & Power Co.	150,000	50	99,500	300,000	1913	Hyd.	2x10,000	1,780 / 1,900	375	Hor.	Al.-Ch.	17,500	14,875	.85	6,000	W. G.E.	5,833	1	Δ	Y	D.	70,000	W. G.E.	72,000 / 15,000	150,000	Δ	'	Big Creek-Los Angeles, Cal.	241
2	Au Sable Electric Co.	140,000	60	19,000	86,500	1912	Hyd.	3,500 / 4,150	28/33/40	120/164/180	Hor.	Al.-Ch.	2,200/3,333/3,333	2,000/3,000/3,000	.90	2,500	G.E.	3,000	1	Δ	Δ	No	19,000	G.E.	5,000 / 22,000 / 44,000	140,000	Δ	Δ	Au Sable-Battle Creek, Mich.	245
3	Southern Sierras Power Co.	140,000	60	8,750	40,000	1915	Hyd.	3,300	514 and 258	300/164	Hor.	Handy Pelton Doble	2,250/2,000	2,250/2,000	1.	2,200	C.W. Al.-Ch.	750/2,000	1	Δ	Y	R.	8,250	W.	33,000 to 4,000	138,500 / 140,000	Δ	Y	Bishop-San Bernardino, Cal.	239
4	Utah Power & Light Co.	130,000 to 110,000	60	33,000	84,000	1914	Hyd.	16,500 / 15,000	492 / 140	514/180	Vert.	I.P. Morris	12,222 / 11,111	11,000 / 10,000	.90	6,600	W. G.E.	4,000	1	Δ	Δ	No	24,000	G.E.	44,000	120,000	Δ	Δ	Grace, Idaho-Salt Lake City, Utah.	135
5	Pacific Gas & Electric Co.	110,000	60	20,000 to 25,000	142,500	1913	Hyd.	2x5,500	1,375	360	Hor.	Pelton Doble	12,500	10,000	.80	6,600	W.	4,250	1	Δ	Y	D.	25,500	Al.-Ch.	60,000	100,000	Y	Y	Drum-Cordelia, Cal.	110
6	West Pennsylvania Traction & Water Power Co.	125,000	60	32,000	100,000	1914	Hyd.	12,000	82	144	Vert.	W.S.M.	10,000	8,000	.80	6,600	Al.-Ch.	10,000	3	Δ	Δ	No	40,000	Al.-Ch.	32,000 / 6,600	120,000	Y	Δ	Cheat Haven-Butler, Pa.	106
7	Tennessee Power Co.	120,000	60	15,000	75,000	1914	Hyd.	10,000	250	360	Hor.	I.P. Morris	9,375	7,500	.80	6,600	W. G.E.	9,375	3	Δ	Δ	No	18,750	G.E.	13,200	120,000 / 95,000	Δ	Δ	Cleveland-Nashville, Tenn.	140
8	Connecticut River Transmission Co.	120,000	60	14,400	14,400	About 1914	Hyd.	About 3,300	57 to 64	257	Hor.	W.S.M.	2,000	1,600	.80	2,300	G.E.	3,000	1	Δ	Y	No	18,000	G.E.	13,200	110,000	Δ	Δ & Y	Shelburne Falls-Millbury, Mass.	60
9	Inawashiro Hydroelectric Power Co. (Japan)	115,000	50	42,000	73,500	1914	Hyd.	About 10,000	350	375	Hor.	Voith	7,775	7,000	.90	6,600	Dick Kerr	4,400	1	Δ	Δ	No	52,800	W.	11,000	100,000	Δ	Δ	Lake Inawashiro-Tokyo, Japan.	111
10	Au Sable Electric Co.	110,000	30	9,000	45,000	1906	Hyd.	14,400	40	225	Hor.	Leffel	3,000	3,000	1.	6,600	W.	3,750	1	Δ	Δ	No	11,250	W.	19,000 / 7,200	100,000	Δ	Δ	Croton-Grand Rapids and Muskegon, Mich.	35
11	Hydro-Electric Power Commission of Ontario (Canada)	110,000	25	106,800	175,000	1910	Hyd.	12,500 / 13,400	175	187½	Hor.	Wolth W.S.M.	7,500/8,950/9,700	7,500/8,950/9,700	1.	12,000	W. G.E.	3,500	1	Δ	Y	R.	42,000	W.	26,400/13,200/6,600	110,000	Δ	Y	Niagara Falls-Toronto and St. Thomas, Canada.	135 / 90
12	Lauchhammer, A.G. (Germany)	110,000	50	15,000	20,000	1911	Steam			1,000	Hor.	A.E.G. M.A.N.	14,300 / 6,250		.70/.90	5,000/5,000	A.E.G. S.-S.	6,250	3	Y	Y	No	20,000	S.-S.	60,000 / 16,000	110,000	Y	Y	Lauchhammer Mine-Riesa, Germany.	35
13	Georgia Ry. & Power Co.	110,000	60	50,000	63,000	1912	Hyd.	17,000	580	514	Hor.	S.M. Smith	12,500	10,000	.80	6,600	G.E.	3,333	1	Δ	Y	R.	60,000	G.E.	22,000 / 11,000	110,000	Y	Δ	Tallulah Falls-Lindale, Atlanta, Ga., etc.	170
14	Alabama Power Co.	110,000	60	48,000	300,000	1913	Hyd.	17,500	68	100	Vert.	I.P. Morris	14,000	12,000	.85	6,600	W.	4,667	1	Δ	Y	R.	56,000	W.	22,000	110,000	Δ	Δ & Y	Coosa River-Birmingham, etc., Ala.	130
15	Mississippi River Power Co.	110,000	25	112,500	225,000	1913	Hyd.	10,000	32	57.7	Vert.	I.P. Morris W.S.M.	9,000			11,000	G.E.	9,000	3	Δ	Y	D.	72,000	G.E.	66,000/33,000/13,200	95,000	Δ	Δ	Keokuk, Iowa-St. Louis, Mo.	144
16	Lehigh Navigation Electric Co.	110,000	25	33,750	100,000	1914	Steam			1,500	Hor.	G.E.	12,500	11,250	.90	11,000	G.E.	3,350	1	Δ	Y	R.	30,150	G.E.	22,000	110,000	Δ	Δ	Hauto-Siegfried, Pa.	26
17	Cedar Rapids Manufacturing & Power Co. (Canada)	110,000	60	90,000	135,000	1914	Hyd.	10,800	30	55.4	Hor.	S.M. Smith	6,000/4,500	4,400/3,600	.75/.78	4,000	G.E.	6,250	3	Δ	Y	No	72,000	G.E.	6,600	110,000	Δ	Δ	Cedar Rapids, Canada-Massena, N.Y.	60
18	Mexican Northern Power Co. (Mexico)	110,000	60	26,000		1914	Hyd.	10,000	230	360	Hor.	I.P. Morris	10,000	7,500	.85	6,600	Esch. W.	3,333	1	Δ	Y	R.	22,500	Esch. W.	6,600	110,000	Δ	Δ	Boquilla-Parral, Chihuahua, Mexico.	47
19	Ebro Irrigation & Power Co. (Ltd.) (Spain)	110,000	50	50,000	200,000	1914	Hyd.	18,500 / 18,000	450 / 225	250	Hor.	I.P. Morris Holyoke Mach. Al.-Ch.	12,500 / 10,000	10,000	.80	11,800 / 11,000	I.P. Morris	2,500	1	Δ	Y	No	66,666	W.	13,200	110,000	Y	Y	Segre River-Barcelona, Spain.	105
20	Chile Exploration Co. (Chile)	110,000	50	75,000	160,000	1915	Steam	14,500	69	225	Hor.	Esch. W.	16,666	10,000	.60	6,600	Esch. W.	4,444	1	Δ	Y	No	40,000	S.-S.	25,000 / 6,600	100,000	Y	Y	Tocopilla-Chuquicamata, Chile.	86
21	Sierra Electric Power Co.	110,000			22,500		Hyd.																		5,000	About 100,000	Δ	Y	Mill Creek-Oakland, Cal.	
22	Sierra & San Francisco Power Co.	104,000	60	34,000		1910	Hyd.	11,750	1,500	400	Hor.	Pelton	8,500	8,500	1.	4,000	G.E.	2,233	1	Δ	Y	D.	26,800	G.E.	11,000	104,000	Δ	Y	Stanislaus-San Francisco, Cal.	138
23	Great Falls Power Co.	102,000	60	21,000	125,000	1910	Hyd.	6,000	105	225	Hor.	S.M. Smith	3,500	3,500	1.	6,600	G.E.	1,200	1	Δ	Δ	No	21,000	G.E.	2,500	91,800	Δ	Δ	Great Falls-Butte and Anaconda, Mont.	150
24	Yadkin River Power Co.	100,000	60	24,000		1912	Hyd.	5,900 / 5,200	45	164	Hor.	S.M. Smith	6,000 / 4,500	4,400 / 3,600	1.	4,000	G.E.	6,250	3	Δ	Y	D.	37,500	G.E.	60,000	100,000	Δ	Δ	Blewitts Falls-Raleigh and Lumberton, N.C.	96
25	Colorado Power Co.	100,000	60	10,000	10,000	1909	Hyd.	9,000	170	400	Hor.	I.P. Morris	5,000	4,000	.85	4,000	G.E.	3,333	1	Δ	Y	R.	15,000	G.E.	6,600	90,000	Δ	Δ	Glenwood-Denver, Colo.	152
26	Great Western Power Co.	100,000	60	50,000	100,000	1909	Hyd.	18,500 / 18,000	450 / 225	400	Vert.	I.P. Morris	12,500 / 10,000	10,000	.80 / 1.	11,800 / 11,000	G.E.	10,000	1	Δ	Y	R.	50,000	G.E.	11,000	90,000	Δ	Δ	Big Bend-Oakland, Cal.	164
27	Southern Power Co.	100,000	60	75,000		1911	Hyd.	5,200	69	225	Hor.	Holyoke Mach. Al.-Ch.	3,000	2,550	.85	2,400	W.	4,000	1	Δ	Y	R.	24,000	W.	44,000 / 13,000 / 2,400	90,000	Δ	Δ	Great Falls, S.C.-Durham, N.C.	210
28	Shawinigan Water & Power Co. (Canada)	100,000	60	45,000	80,000	1914	Hyd.	18,500	145	200	Hor.	I.P. Morris	15,000	12,500	.96	6,600	W.	14,000	3	Δ	Y	R.	28,000	G.E.	12,800	85,000	Δ	Δ	Shawinigan Falls-Montreal, Canada.	87
29	Los Angeles Aqueduct	100,000	50	22,500	138,000	1914	Hyd.	14,000	870	300	Hor.	Doble	9,375	7,500	.80	6,600	W.	3,130	1	Δ	Y	D.	28,350	W.	33,000 / 16,500	85,800	Δ	Δ	San Francisquito-Los Angeles, Cal.	47
30	Tata Hydroelectric Co. (India)	100,000	50	32,000	64,000	1914	Hyd.	11,000	1,661 to 1,727	300	Hor.	Esch. W.	10,000	8,000	.80	5,000	S.-S.	3,333	1	Δ	Δ	No	40,000	G.E.	6,000	85,800	Δ	Δ	Khopoli-Bombay, India.	43

Figure X.11. Transmission systems of the world operating at and above 70,000 volts, 1914, ranked according to operating voltage. △ = mesh grouping (three phases joined as equilateral triangle); Y = star grouping (three phases united at a common junction). From table compiled by Selby Haar as a supplement to Electrical World, 25 April 1914, and revised and brought up to date by Haar for the Hydro-Electro Section, National Electric Light and Power Convention, Philadelphia, Pa., 2–5 June 1914. Reprinted in Bureau of the Census, Central Electric Light and Power Stations, foldout facing p. 132. Data on transmission lines not included here.

No.	System	Location	Voltage
31	Phünwerke, A. G. (Germany)	Hamburg–Ludwigshafen, Germany	100,000
32	Società Italiana di Elettrochimica (Italy)	Pescara River–Naples, Italy	88,000
33	Appalachian Power Co.	New River–Roanoke, Va., Coalwood and Bluefield, W. Va.	88,000
34	Rio Janeiro Tramway, Light & Power Co. (Brazil)	Lages River–Rio Janeiro, Brazil	88,000
35	Sao Paulo Electric Co. (Brazil)	Sorocaba–Sao Paulo, Brazil	88,000
36	Tasmania Hydroelectric & Metal Co. (Tasmania)	River Ouse–Hobart, Tasmania	88,000
37	Mexican Light & Power Co. (Mexico)	Necaxa–Mexico City, Mexico	85,000
38	Toronto Power Co. (Canada)	Niagara Falls–Toronto, Canada	85,000
39	Victoria Falls & Transvaal Power Co. (South Africa)	Vereeniging–Johannesburg, South Africa	84,000
40	Northern Power Co.	Hannawa Falls–Potsdam, N. Y.	80,000
41	Energia Electrica de Cataluna (Spain)	Pyrenees Mountains–Barcelona, Spain	80,000
42	Katsuragawa Denryoku Kabushiki Kaisha (Japan)	Komahashi–Tokio, Japan	77,000
43	Southern California Edison Co.	Kern River–Los Angeles, Cal.	75,000
44	Au Sable Electric Co.	Rogers Dam–Muskegon, Mich.	72,000
45	City of Milan (Italy)	Grosseto–Milan, Italy	72,000
46	Società Generale Elettrica dell' Adamello (Italy)	Cedegolo–Milan, Italy	72,000
47	Montana Co.	Canyon Ferry, Hauser Lake–Butte, Mont.	70,000
48	Hidroelectrica Espanola Molina (Spain)	Molina–Madrid, Spain	70,000
49	Pennsylvania Water & Power Co.	Holtwood, Pa.–Baltimore, Md.	70,000
50	Compania Hidro-Electrica Irrigadora del Chapala, S. A. (Mexico)	Santiago River–Guadalajara, et al., Mexico	70,000
51	Societa Elettrica Riviera di Ponente (Italy)	S. Dalmazzo–Novi, et al., Italy	70,000
52	Swedish State Railways (Sweden)	Porjus–Kiruna, Sweden	70,000
53	City of Winnipeg (Canada)	Point Dubois–Winnipeg, Canada	66,000 to 72,000
54	Cie Grenobloise de force et lumiere (France)	Pomblière–Lyons, France	About 70,000
55	Otten-Gosgen Power Plant (Switzerland)	Aswil–Bottmingen, Switzerland	70,000

mercial and municipal, were second only to New York State's.[37] Comparison of California with the leading electric-power-generating states—New York, Illinois, and Pennsylvania—reveals that California had the largest percentage increase in kilowatt capacity and in kilowatt-hour output during the decade 1902–12.

---

[37] Bureau of the Census, *Central Electric Light and Power Stations,* pp. 30, 33, 34, 46, and 50. The statistics are for private and municipally owned central stations supplying the public; see ibid., p. 15.

CHAPTER XI

War and Acquired Characteristics

ENGINEERS and managers engaged in prediction prefer the extrapolation of trends to the formulation of complex scenarios based on likely interactions of trends and contingencies suggested by historical precedents and analogies. Before World War I they undoubtedly anticipated that the trends of the past decade or so would extend into the future. Because of the momentum of the electric supply industry, this was a reasonable assumption. Moreover, the factors influencing the growth—the context in which it occurred—were, by implication, projected into the future. The combination of growing momentum and reinforcing context was expected to overwhelm contingent perturbations. History, of course, raises grave doubts about extrapolations, but the study of history is not required of engineers and managers.

History is a record, among other things, of contingencies, even of unanticipated catastrophes, natural and man-made. In August 1914 a man-made catastrophe struck Europe and later spread to America. The immediate effects and consequences of World War I have been documented by historians. The effects of the war on the evolving electric power systems, however, have been overlooked. These systems, too, were affected by the war and acquired characteristics that survived into peacetime.

The perturbations and lasting changes brought by catastrophes such as war result from forces that are strong enough to disrupt the momentum of systems. Or, to use a different metaphor, a change in the environment selects out different characteristics. World War I, for instance, altered the relative influence of the various contextual factors affecting electric power systems. Engineers and managers who, because of their experience and special competence, were committed to smallness of scale nevertheless acknowledged the primacy of output when personal and national survival seemed to depend on it. Likewise, companies and governments that were economically and ideologically committed to vested interest and authority yielded in the face of the same imperative. A history of power supply during the war will show this.

Because technology is often manifested in material form—machines, processing equipment, structures, and tools—its lasting effects are easily observed. Ideas and events, historians argue, have effects far beyond the time in which they occur, but their effects are less visible. In the case of technology, even the casual observer knows that he is surrounded by things that were made in the past, things that were often made under substantially different circumstances. In a sense, therefore, surviving technology brings to the present the character of the past, a past that imposed its characteristics on the technology when it was first invented, developed, and introduced into use. The technology of electric power systems that was introduced during World War I not only caused perturbations in trends but also carried into peacetime certain aspects of the wartime environment. The extremely large electric generating stations that were built to fill the pressing and unusual needs for electric power during World War I survived the war and became, in a sense, a solution in search of a problem. Another, less obvious case is the large number of interconnections of electric light and power systems that were made during the emergencies of wartime and carried over into peacetime.

This chapter considers how war brought the governments of Germany and the United States to fund the development of power plants of unprecedented size. The reader's attention is also directed to the series of wartime interconnections in Great Britain and the United States that resulted from the unusual, even radical, demands that were made on formerly reluctant utilities by central governments concerned about survival. After noting the creation of wartime artifacts, this chapter further observes the efforts made by the three governments to maintain in peacetime the momentum of their wartime planning and nurturing of technology. The efforts failed in the three cases considered here. The reason for the failure was in part a conservative reaction comparable to that which is often encountered in the political realm after radical changes have been introduced. Once the disruptive force—in this case, war—is removed, the prewar context again prevails. There were other causes, however, and they differed from country to country.

The fact that the exigencies of war caused the accelerated development of certain technological characteristics—in this instance, large size and interconnection—shows again that the rate and direction of technological change can be shaped by nontechnological factors. In other words, technology is not necessarily a simple extrapolation of its past, or a working out of inherent technological implications. The cases considered suggest that war did not so much stimulate the invention and development of new technologies as clear away the political, economic, and other nontechnological factors that prevented or retarded the utilization of existing technologies. The imperatives of war did not reverse the direction of technological evolution nor did they cause mutations; rather, they broke a conservative crust that had restrained adjustments in course and velocity. The wartime history of electric power supply suggests that a society under the influence of even more pressing demands—a moral equivalent of war—could alter the shape of technology even more drastically.

In the United States an outstanding instance of a mammoth artifact built under the emergency circumstances of war, and standing like a large white elephant at war's end, was the nitrate-fixation plant and its partially completed hydroelectric power station and dam located on the Tennessee River at Muscle Shoals, Alabama.[1] The large demand for nitrates used in fertilizer and explosives, coupled with America's tenuous dependence upon Chile's nitrate monopoly, persuaded the U.S. Congress—even before the country entered World War I—to include in the National Defense Act of 1916 provision for the construction, within the United States, of facilities capable of fixing atmospheric nitrogen. Already in Germany, several nitrogen-fixation processes had been developed, including the high-pressure and high-temperature Haber-Bosch process.[2] Such a plant was built in the United States at Sheffield, Alabama, but the government, doubtful of the ability of American engineers and scientists to handle the Haber-Bosch process, also contracted with the American Cyanamid Company for the construction of a cyanamide-processing plant. In comparison to the Haber-Bosch method, the cyanamide process consumed large amounts of energy; thus the plan was to build the Muscle Shoals hydroelectric station and dam and use its cheap power to supply the cyanamide plant.

The Muscle Shoals nitrate facility was completed just before the war ended in November 1918. Close to $70 million had been invested in it, but no nitrates from it had been used during the war. A transmission line had been erected to bring power from a steam power station because the dam and hydroelectric power station were not yet completed. The government continued construction of these until 1925, by which time an additional $45 million had been invested.[3] Government financing provided the capital investment of more than $100 million in the nitrate plant and hydroelectric facility, for private capital was reluctant to invest the large sums demanded by the high first cost but relatively low operating cost hydroelectric plants.[4] As the values and imperatives of wartime faded, Muscle Shoals stood like an organism that had passed into a foreign—even hostile—environment.

Wartime demands for energy forced the Germans to build giant power plants as well. While the United States took a leap forward in the development of hydroelectric power, Germany accelerated the development of electric power plants fueled by brown coal, or lignite. In both cases the energy source was abundant and cheap, and the plants that were built to utilize it were large-scale and capital-intensive. Because of the emergency

[1] For a general study of the Muscle Shoals nitrate plant and power station and the postwar controversy they stimulated, see Preston J. Hubbard, *Origins of the TVA: The Muscle Shoals Controversy, 1920–1932* (New York: Norton, 1961). On the response to the emergency demand for nitrates in the United States, see U.S., War Department, Ordnance Office, Nitrate Division, *Report on the Fixation and Utilization of Nitrogen*, War Department Report no. 2041 (Washington, D.C.: GPO, 1922).

[2] For a description of nitrogen-fixation processes, especially the Haber-Bosch method, see J. R. Partington and L. H. Parker, *The Nitrogen Industry* (New York: Van Nostrand, 1923).

[3] Hubbard, *Muscle Shoals*, p. 7; the government invested $68 million in the nitrate plant. In *History of the Georgia Power Company, 1855–1956* (Atlanta: Georgia River Co., 1957), p. 274, Wade H. Wright states that the hydroelectric power station at Muscle Shoals was completed in 1925 with 184,000 kw. installed at a cost of $45.6 million.

[4] During the war, the private utilities complained of difficulty in raising funds for expansion. The Philadelphia Electric Co., a large utility that was extremely hard-pressed by wartime demands arising from shipbuilding on the Delaware River and from munitions plants, was not able to raise the $25 million it needed to construct the "Delaware Station," which was sorely needed to meet anticipated demand in 1918–19. See Nicholas B. Wainwright, *History of the Philadelphia Electric Company, 1881–1961* (Philadelphia: Philadelphia Electric Co., 1961), pp. 137–40.

nature of wartime, central governments provided the necessary capital. The particular imperatives of war accelerated economies of scale and the use of capital-intensive technology, thereby creating a wartime style of technology. Two giant German plants spawned by the war were Knapsack (later renamed Goldenberg) and Golpa-Zschornewitz. The former, located in the large brown-coal fields near Cologne, became a part of the Rheinisch-Westfälisches Elektrizitätswerk system that supplied heavy industry in the Ruhr valley. Golpa-Zschornewitz, like Muscle Shoals, was constructed in response to a critical wartime shortage of nitrogen.

Some experts predicted that the shortage would paralyze Germany's armies and starve her people soon after the war began. The world's largest importer of nitrogen compounds before the war, Germany had depended especially on Chile's natural coastal deposits. The country's other major source of nitrogen was a by-product of her own coke and gas production. After the British navy cut Germany's supply lines, the German people faced dire shortages of explosives and agricultural fertilizer, both of which required nitrogen. They had to rely on stored nitrogen compounds until new processes for their production could be developed within the country. These processes required unusually large amounts of electrical energy.[5] A major response to this critical problem was the Golpa-Zschornewitz power station.

The history of the Golpa-Zschornewitz station dates back to 1915, when Germany's stored nitrogen supplies neared exhaustion and the nitrogen derivable as a by-product from the manufacture of coke and gas from coal could not match demand. The situation was alleviated only as the Germans brought into operation the new plants for fixing in compounds the nitrogen of the atmosphere. After the government found that private companies were not willing to build plants of sufficient capacity unless the government guaranteed a long-term market, it decided to finance the building of its own nitrogen plants and to turn over to private industry the operation of the plants. One of the largest of these, Piesteritz, was situated in the brown-coal fields about 80 miles southeast of Berlin.[6]

Piesteritz required a giant power station. Early in 1915 the Reich, or central, government persuaded the Braunkohlenwerke Golpa-Jessnitz AG, owner of the brown-coal fields and a subsidiary of the Allgemeine Elektrizitäts-Gesellschaft, to build the Golpa-Zschornewitz station (Fig. XI.1). The government contracted to buy electric power for the nitrogen plant from the station. Under the direction of Georg Klingenberg, the internationally known head of electric power plant design at AEG, the station was completed in the remarkably short period of nine months. The design was for 180,000 kw., and the station went into operation in January 1916 with 128,000 kw. installed. Braunkohlenwerke formed a new company, Elektrowerke Aktiengesellschaft (EWAG), to own and manage the Golpa-Zschornewitz station. The station soon developed technological and economic problems, however, and these were intensified because the guar-

---

[5] Thomas P. Hughes, "Technological Momentum in History: Hydrogenation in Germany, 1898–1933," *Past and Present*, no. 44 (August 1969), pp. 106–32.

[6] Georg Boll, *Entstehung und Entwicklung des Verbundbetriebs in der deutschen Elektrizitätswirtschaft bis zum europäischen Verbund* (Frankfort on the Main: VWEW, 1969), pp. 27–29.

Figure XI.1. *Golpa-Zschornewitz power station at the mine mouth. From* Electrical World *75 (1920): 602.*

anteed price of current for Piesteritz was too low. In May 1917 the situation became more troublesome when a neighboring plant exploded and damaged the Golpa-Zschornewitz station. For a while the energy demands of the nitrogen plant at Piesteritz could not be met. The series of problems led to negotiations for the government to purchase the Electrowerke company and its power station. The Reichstag agreed and on 1 October 1917 the Reich government took over operation of the station.[7]

Supplying the needs of Piesteritz, but faced with an excess capacity after the explosion of the neighboring plant (which had also been a consumer), Elektrowerke, now Reich-owned, began in 1918 to construct a 110,000-volt transmission line to Berlin, where the city government had recently acquired BEW. The establishment of high-voltage transmission to absorb the excess capacity of the Golpa-Zschornewitz station heralded the development of a peacetime transmission system centered on the giant artifact the war had spawned.

Giant power plants were built to meet the increased wartime demand for electricity in Germany and the United States. Interconnection was another means used by the belligerents to raise output; it was a way to increase supply without building new plants. In Britain, where there was a shortage of building materials, interconnection seemed especially attractive. When Britain entered the war, her electric supply industry was shaped by the political power of local governments. The situation was ripe for interconnection. Wartime soon brought shortages of labor, coal, and electricity (the last-named caused, in part, by the other two). A shortage in the supply of electric machinery also developed as the electrical manufacturers turned to more profitable munitions contracts. The government provided exemp-

[7] Ibid., p. 28.

tions for workers in the electric supply industry, but shortages of coal and machinery created critical problems because the growth and modernization of the armaments, or munitions, industry depended upon increased electrification of the means of production. Engineers and managers from various fields of industry were brought into the departments and committees created by the wartime cabinet, and these men tended to see the answer to the problem in further electrification, especially in economies of scale, and in coordination and systematization through interconnection.[8] The Newcastle upon Tyne Electric Supply Company (NESCO) stood as a model for them in British electricity supply.

Several of the committees and departments that were established during the war had particular responsibilities for electricity supply and machinery manufacture. Before 1917 these committees concentrated on short-term responses to shortages; after 1917 they dealt with technological changes that responded not only to immediate wartime needs but also to anticipated peacetime industrial needs. Charles Merz and his partner, William Mc-Lellan, were asked by the government to serve in these departments and committees because of their remarkable success in creating before the war an exception to the British rule, a large regional power system (see p. 446 below). McLellan became the advisor on power supply to the director of production in the Ministry of Munitions created by Prime Minister Lloyd George in May 1915. The importance of power supply was further recognized by the establishment of a Department of Electric Power Supply in June 1916 with McLellan as head. Merz became a member of the Coal Conservation Sub Committee of the cabinet's Reconstruction Committee. This subcommittee was headed by Lord Haldane, who admired the industrial organization and efficiency of the Germans. Lord Haldane delegated most of the committee's work to Merz, and an interim report on electric power supply was presented in April 1917. Merz, who had been unable to reorganize London's supply in 1905–6, now applied his concepts to reorganizing the supply network of the entire country.

As an advisor on power supply, McLellan advocated interconnection. He knew that interconnecting parochial utilities would increase the diversity of load and the load factor, thereby bringing a fuller use of existing capacity. He and his associates had to contend, however, with the deeply entrenched spirit of competition, even hostility, that prevailed between the municipally owned utilities and the private ones. Ideological tensions were compounded by the bewildering variety of supply systems, voltages, and frequencies in Britain, a variety that made interconnection a much more difficult task than it was in the United States. Nevertheless, spurred on by wartime exigencies, the British government and its advisors effected limited agreements among competitive utilities and financed interconnection. At the urging of Merz, the Board of Trade, which had jurisdiction over the British electric supply industry, in May 1916 promised priority in the allocation of labor and materials to utilities that would interconnect. In September 1916 two leading trade associations, the Incorporated Municipal Electrical Association and the Incorporated Association of Electric Power Companies,

---

[8] Leslie Hannah, *Electricity Before Nationalisation* (Baltimore: The Johns Hopkins University Press, 1979), pp. 53–57.

transcended their differences sufficiently to issue a joint recommendation for the establishment of liaison groups to discuss interconnection. The Institution of Electrical Engineers also organized discussions of the interconnection and reorganization of England's electric supply system. The imperatives of war stimulated a radical change in attitudes and actions. Tangible results were not as impressive as giant power plants of the United States and Germany, but the change that was brought about in the direction of development was a more Herculean achievement.

William McLellan and Arnold Gridley, also a consulting engineer and McLellan's successor in the Department of Electric Power Supply, complemented interconnection by arranging for industrial consumers to take electricity from utility systems rather than use inefficient, isolated plants or expand isolated-plant capacity. This policy resulted in an increase in the capacity of the most efficient utility power plants and in economies of scale. McLellan and his staff were disappointed by the unwillingness of the government to insist more often on expansion with postwar reorganization of the nation's power supply in view.[9] However, expansion of the more efficient plants in industrial areas and the interconnection of plants did save coal and allowed the supply industry to increase output at about the same rate as during the prewar years. The charts of these statistics concealed notable changes beneath their smooth curves, however. Aggregate sales in the United Kingdom rose from 2,100 gwh. (gigawatt hours) in 1914 to about 4,000 gwh. by 1918, but the London increase was only 14 percent between 1914 and 1918. The load factors of generating stations increased from a national average of around 23 percent before the war to 30 percent by 1918.[10]

In the United States, power consumption also continued upward in 1915 as the country looked to its defenses. After the United States entered the war in the spring of 1917 and before the war's end in the fall of 1918, more than two million horsepower was added to the electric-generating capacity of the country, an increase of 10 percent in less than two years. Experts nevertheless predicted there would be a dire shortage of electrical power in the winter of 1918–19 if the war continued. Faced by the likelihood of a shortage, Bernard Baruch, head of the War Industries Board formed by President Wilson in July 1917 to mobilize the economy, ordered a survey of the electrical generating facilities in the United States. In order to determine capacity, efficiency, and the extent to which the output was being used by war-related industries and services, army engineers acting for the board examined power stations in the critical regions where war industries were concentrated. The survey not only allowed the board to identify areas of shortage and excess, and to locate new war contracts accordingly, but it also revealed—and this proved important for the long-range development of the electrical power industry—the possibility of using existing power facilities more effectively by interconnecting power stations and utilities that had complementary load and diversity factors.[11]

[10] Ibid., pp. 59 and 61.
[11] Bernard Baruch, *American Industry in War: A Report of the War Industries Board (March 1921)* (Englewood Cliffs, N.J.: Prentice-Hall, 1941), pp. 298–99; U.S., Congress, House, *Emergency Power Bill*, 65th Cong., 2d sess., 1918, H. Rept. 795.

Striving to meet the anticipated power shortage of the winter of 1918–19, the War Industries Board and other government agencies, local and federal, stimulated—even ordered—interconnections of private utility systems in order to raise the load factor.[12] Privately owned utilities had been infrequently interconnected in the United States before World War I, despite the practicality of high voltages for transmission of power between large-area utilities.[13] As *Electrical World* observed during the war, "The incentive for interconnecting electric service systems has been absent where each individual company has been able to meet its local requirements and at the same time maintain a sound financial condition."[14] War changed the local requirements; the utilities were no longer able to meet them; and the "incentive"—at least among those in government and industry who were aware of the impending critical shortages—emerged. Because utilities were regulated by commissions in each of the states and because regulatory conditions varied from state to state, the actions that were needed to stimulate or order the interconnection of utilities in 1917–18 varied. One major interconnection in one state suggests the general character of the wartime endeavor in the United States.

Interconnection in Massachusetts involved the Salem Electric Lighting Company, the Malden Electric Company, and the Revere Suburban Gas & Electric Company. The decision to interconnect these three companies resulted from their complementary load and diversity factors and illustrates how these factors were taken into account. The three companies served districts to the north of Boston. The Salem and Malden companies carried an industrial, commercial, and domestic load, while Revere's load was greatly influenced by the seashore resort district it served. Each company had salient characteristics: Salem had an excess of generator capacity; Malden's generator capacity was slightly below the maximum load it could anticipate; and Revere, the resort district, experienced its maximum load at 8:00 p.m. on summer evenings (the other two stations carried maximum loads during the winter months at about 5:00 p.m., when factories and businesses were still open and homes were lighted for the evening). Interconnection meant that Salem could share its excess capacity with Malden and that Revere, because of its diverse character, could supply power to the other two during winter peaks and draw power from them during the summer evenings. Not only did the interconnection result in increased power where and when needed, but it raised the load factor for each company.[15]

Following the war, the technology of interconnection was used appropriately; interconnections were used or not, as circumstances required. Compared to the financing of the wartime giant-power plants, the capital investment in interconnection was small. By contrast, the central governments in the United States and Germany faced a serious problem in de-

[12] House, *Emergency Power Bill,* pp. 2–3, 6–9.
[13] Editorializing about interconnection, the *Electrical World* wrote: "War accelerates negotiations between interests which in peacetime might work for years with little or no physical co-ordination." *Electrical World* 71 (1918): 1246.
[14] "Interconnection a Boon to New England Plants," ibid. 70 (1917): 422. I am indebted to Thomas Guider of the University of Delaware for calling my attention to this and other references cited on wartime interconnections in Massachusetts.
[15] Ibid., pp. 422–24.

Figure XI.2. *Interconnections raising load factor in Massachusetts. From Electrical World 71 (1918): 1192.*

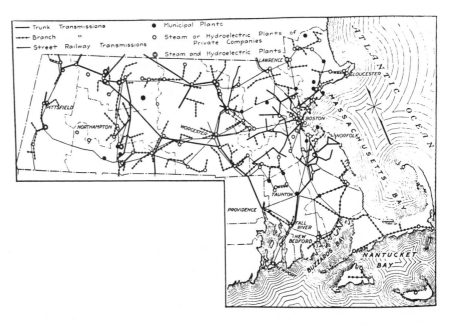

termining the future of the giant power plants. They were, in peacetime, exotic artifacts because of their large size, government financing, and dependence upon nearby munitions plants. For the U.S. government the problem of Muscle Shoals was nearly unsolvable. The National Defense Act of 1916, contrary to the traditional American opposition to government ownership, unrealistically provided that after the war the government should maintain ownership of the plant, using Muscle Shoals to manufacture nitrogen for explosives and fertilizer. Congress, however, defeated the Wadsworth-Kahn Bill, which would have maintained government ownership by means of a corporation whose stock was entirely owned by the government. The new administration of President Warren Harding in 1921 asked for offers from private sources that would guarantee the government a return on its investment. A straightforward and substantial commitment from private enterprise seemed unlikely, however, because the total investment was unprecedentedly large for the electric power field.[16] With continued government operation doubtful and private investment improbable, dramatically resourceful and innovative responses were needed to solve the problem posed by Muscle Shoals.

On 8 July 1921 Henry Ford submitted a bid for Muscle Shoals. In retrospect this commitment is understandable: Ford had proven himself an industrialist of dramatically broad vision in the development of his system for the mass production of automobiles; the innovative character of the man matched the exceptional nature of the problem. Furthermore, Ford was fascinated by opportunities to organize production on a mammoth

[16] As of 1912, the largest-capacity hydroelectric power station in the United States was the 112,500-kw. plant on the Mississippi River at Keokuk, Iowa; it was built at a cost of about $25 million by the Mississippi River Power Co. U.S., Department of Commerce, Bureau of the Census, *Central Electric Light and Power Stations and Street and Electric Railways, with Summary of the Electrical Industries, 1912* (Washington, D.C.: GPO, 1915), p. 123 and foldout facing p. 132.

scale, and Muscle Shoals provided him an opportunity to produce power massively. His financial offer was not remarkably generous; among other stipulations in it was the right to lease the power facilities for 100 years and amortize the government's investment by a modest annual payment of $46,547. This payment, he calculated, would produce the total sum desired by the government if it was invested at 4 percent and compounded semiannually for 100 years, after which time ownership would pass to him.[17]

If Ford's plan for financing was closefisted, his vision of Muscle Shoals's future was expansive. Ford indicated not only that he expected to use Muscle Shoals hydroelectric power to support an industrial complex in the immediate vicinity but that he was also thinking in terms of the regional development of the Tennessee valley. He and his engineers were well aware that the recently established practicality of transmission voltages as high as 110,000—even 220,000—volts meant that electric power from an abundant source like Muscle Shoals could be transmitted economically throughout a region.[18] In seeking support for his offer, Ford, who dissociated himself from the financiers of Wall Street and identified himself as a productive industrialist, relied heavily upon the American faith in technological genius and industrial entrepreneurship; he even brought in his friend and hero, the venerable Thomas Edison, to advise him.

Ford's vision matched the scale of Muscle Shoals, but his offer was frustrated by a coalition that formed against him in the Harding administration and Congress. Nevertheless, by 1924 his offer had stimulated widespread interest in the future of Muscle Shoals, for he had appealed to the public, especially in the South, to whom he promised a new prosperity through regional development. With the widespread interest, new points of view were formulated, and the issue became more complex. As Preston J. Hubbard concluded from his study of the Muscle Shoals project, the problem of the ultimate disposal of this wartime edifice became the core of a major controversy in the power fight of the 1920s involving advocates of private enterprise, contenders for public power, proponents of regional development, and factions within these broad groups.[19] A wartime expedient had demonstrated a technological potential and revealed the inherent political, economic, and social problems associated with it.

More specifically, the parties to the controversy over the Ford offer, whose positions were backed by congressmen and administration officials, included privately owned electric utilities (led by the Alabama Power Company); Wall Street industrial and banking concerns; Andrew Mellon, secretary of the treasury under Presidents Harding and Coolidge; John W. Weeks, secretary of war under Harding; and Senator George W. Norris

[17] In *Muscle Shoals*, pp. 28–146, Hubbard explores the Ford offer and the reaction to it.
[18] For data showing the increase in transmission voltages between 1889 and 1934, see *Electrical World* 83 (1924): 998. According to this source, transmission voltages of roughly 110,000 were practical in 1908, and by 1923 the number had risen to 220,000. However, these estimates were based on pioneering endeavors rather than on common practice. The world's first 110,000-volt transmission line was built from Croton–Grand Rapids to Muskegon, Michigan, in 1906; the second was built from Niagara Falls to Toronto and St. Thomas, Canada, in 1910. Bureau of the Census, *Central Electric Light and Power Stations*, foldout facing p. 132.
[19] Hubbard, *Muscle Shoals*, pp. vii–viii.

of Nebraska, chairman of the Agriculture Committee, a leader of progressive congressmen and an advocate of public power and regional planning. The way in which these various advocates intended to exploit Muscle Shoals reflected not only different economic and political views but the different ways in which the technology of electric power could be developed. The utilities were agreeable to government ownership of the hydroelectric plant at Muscle Shoals, but they wanted the government to sell them the power wholesale for transmission and distribution to their customers; the Wall Street bankers and the well-established energy-intensive manufacturing industries like aluminum and carborundum had the twofold objective of opposing Ford's monopoly of Muscle Shoals power and favoring development of isolated industrial power sites along the Tennessee River; and the advocates of public utilities envisaged the coordinated development of power, agriculture, and industry.

Only the coming of another national emergency could resolve the controversy posed by these technological implications. The imperative for employment and economic development imposed by the Great Depression brought about dramatic and decisive action comparable to the decisions that had created Muscle Shoals. In May 1933 Congress enacted President Franklin Roosevelt's recommendation for the development of the Tennessee valley by means of the Tennessee Valley Authority Act, and the Muscle Shoals dam and power station became the keystone of the new government-constructed, -owned, and -operated system.[20]

In the United States, the existence of the giant, government-owned plant at Muscle Shoals caused an ideological clash that for more than a decade thwarted effective utilization of the plant in a regional system. In Germany, by contrast, the Golpa-Zschornewitz plant, whose origins were similar to those of Muscle Shoals, made the transition to peacetime with relatively little ideological conflict. After the German central government in 1918 constructed a 132-km. high-voltage line from Golpa-Zschornewitz to Berlin, the city-owned utility, the former BEW, had the capacity to supply an aluminum works at Rummelsburg, near Berlin. Another line, 18 km. in length, was built from Golpa-Zschornewitz to Bitterfeld to supply an aluminum works. After the war, EWAG, the Reich-owned utility that operated Golpa-Zschornewitz, continued to supply the nearby wartime nitrogen plant, for it turned from munitions to the production of nitrogen-compound fertilizers. Following the success of the Berlin and Bitterfeld transmission lines, EWAG reached out in 1920 with a 110,000-volt line to Leipzig. EWAG then acquired the brown-coal plant at Trattendorf, which had been built in 1915 by the Swiss electrical manufacturer Brown Boveri to supply electricity to nearby Lonzawerk, an energy-intensive, carbide-producing plant. A 110,000-volt transmission line was then erected to connect Trattendorf with Golpa-Zschornewitz. By acquiring other power plants and utilities and making contracts to supply large industrial consumers and small utilities in bulk, EWAG in the 1920s became one of Germany's largest regional utilities.[21] Because of the absence of paralyzing conflict between govern-

---

[20] The literature on the Tennessee Valley Authority is extensive. One recent study is Thomas K. McCraw's *TVA and the Power Fight, 1933–1939* (Philadelphia: Lippincott, 1971).

[21] Boll, *Verbundbetriebs*, pp. 28–30.

ment and private interests, the transition of wartime technology to peacetime use was smooth.

The giant wartime power plants, with their enormous capacity, sprouted
peacetime transmission systems as they reached out for new markets. In
England, the United States, and Germany, the wartime experience with
centrally planned interconnections influenced peacetime thinking about
systems. The success of emergency wartime interconnections stimulated
the planning of further interconnections, not only in anticipation of postwar
demand, but also in light of the imperatives of future wars.[22] The engineers
and managers who had been called into wartime government service were
especially interested. Beyond a simple increase in interconnections lay the
integration of utilities. A simple interconnection tied together at one junc-
ture or a limited number of points only two utility systems; often the
capacity of the tie line was small, permitting transfer of only a limited
amount of excess power. Control of each system remained independent.
By contrast, integration would bring two formerly independent systems
under common control. No longer would it be possible to distinguish the
boundaries of these systems by their tie lines, for these would become part
of the transmission-line network of the integrated system.[23] Institutional
merger did not follow immediately, however.

In the United States in 1919 Secretary of the Interior Franklin Knight
Lane expressed a carry-over of wartime attitudes when he wrote, "The
enormous development of war industries had created an almost insatiable
demand for power, a demand that was over-reaching the available supply
with such rapidity that, had hostilities continued, it is certain that we should
now be facing an extreme power shortage."[24] For this reason he recommended that Congress study the feasibility of developing a comprehensive,
"superpower" system for the generation and distribution of electricity in
the Boston-Washington industrial region. The secretary's interest was focused on supplying large industry and transportation networks with cheaper
power. Others, like Gifford Pinchot, governor of Pennsylvania, and Morris
Cooke, his adviser on electrical supply, called for unified power systems
that would supply (economically) more general social needs, such as the
need for cheap power in rural areas. The governor maintained that a
unified giant power system, with its improved load factor, would result in
"raising the standard of living . . . eliminating the physical drudgery of life
. . . winning the age-long struggle against poverty."[25]

Only one month after World War I ended, William S. Murray, a consulting electrical engineer, urged the secretary of the interior to prepare

[22] Maj. Gen. William Crozier, "Giant Power and National Defense," *Annals of the American
Academy of Political and Social Science* 118 (1925): 105.

[23] After the war some utility executives strongly advocated interconnection to counter the
loss of company independence that followed integration; they were reluctant to see holding
companies or the government (by regulation) take over the integrated systems. Alex Dow,
head of the Detroit Edison Co., was a case in point. Raymond C. Miller, *Kilowatts at Work: A
History of the Detroit Edison Company* (Detroit, Mich.: Wayne State University Press, 1957), p.
225.

[24] W. S. Murray et al., *A Superpower System for the Region between Boston and Washington*, U.S.
Department of the Interior, United States Geological Survey Professional Paper no. 123
(Washington, D.C.: GPO, 1921), p. 9.

[25] Gifford Pinchot, "Giant Power," *Survey Graphic* 51 (1924): 561.

the ground for this superpower system by supporting a detailed survey of the energy resources of the industrial Northeast from Boston, Massachusetts, to Washington, D.C. The survey would be similar to the one made earlier for the War Industries Board. In transmitting the request to Congress, the secretary of the interior wrote, "The country is now passing through a period of transition, which, I firmly believe, will soon be followed by one of industrial activity and expansion."[26] Congress agreed, and the survey was carried out. Using the results of the survey, Murray and his associates submitted in 1921 a report or plan for a superpower system which he projected could be fulfilled by 1930. In essence, Murray recommended that the large-scale and efficient power plants then owned by northeastern private utilities be supplemented by new superpower plants (60,000–300,000 kw.), both thermal and hydroelectric; transmission lines of 110,000 to 220,000 volts would integrate these. Having been joined by this network, the power plants would feed their output into a power pool, which would be tapped at load centers for distribution by the utilities to consumers. The utilities' existing high-voltage transmission lines (33,000 volts) would be used for distribution to customers in 1930—to industrial, electrified railroad, and smaller consumers. Holding companies funded by investors would own and manage the superpower corporation's facilities— the new large-scale power plants and the high-voltage transmission lines. Utilities would own the lower-voltage distribution network. Murray estimated that, compared with the cost of an unintegrated system supplying the same region, an investment of $1 billion by 1930 would result in an annual saving of more than $200 million because of the higher load factor and other economies of the integrated system.[27] (See Fig. XI.3.)

Giant Power was another idea spawned by wartime conditions. It, too, involved a plan for an integrated system, but it was limited to the state of Pennsylvania. The moving spirit behind the discussions of Giant Power in the early 1920s was Governor Gifford Pinchot (1865–1946) of Pennsylvania, a leading conservationist and progressive follower of Theodore Roosevelt. Pinchot's able technical assistant was Morris Llewellyn Cooke (1872–1960), a consulting engineer with a reputation for liberal social views. Giant Power differed from superpower in technological aspects, but—perhaps more interestingly—also in its social potential. Pinchot saw Giant Power as a means of bringing a social revolution through cheaper electricity, rural electrification, and giant mine-mouth plants.

The history of Giant Power is one of failure. Conceived of under the influence of wartime circumstances, the plan experienced the chill of the postwar reaction, or return to normalcy. Similar bold plans in Britain and Germany also failed to reach fruition. For a time, however, Giant Power attracted considerable professional and public attention; the electrical journals and the popular press gave it frequent and broad coverage. Proponents saw it as a step toward a New Jerusalem; opponents considered it an alien scheme that threatened continued progress in the utilities field. Its most prominent proponent, the governor of Pennsylvania, described it as follows:

[26] Murray et al., *Superpower System*, p. 9.
[27] For a summary of the superpower plan, see especially ibid., pp. 9–26.

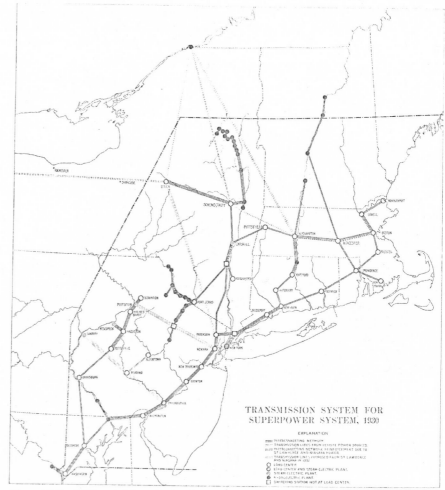

Figure XI.3. *William S. Murray's plan for a superpower system. Murray et al.,* Superpower System, *plate III.*

Giant Power is a plan to bring cheaper and better electric service to all those who have it now, and to bring good and cheap electric service to those who are still without it. It is a plan by which most of the drudgery of human life can be taken from [the] shoulders of men and women who toil, and replaced by the power of electricity.[28]

An opponent, the head of the Pennsylvania Electric Association, which represented 90 percent of the state's utilities, said:

The very clear purpose of this plan is to take from any electric service system the benefits it has thus far accrued by reason of able management, successful financing, painstaking research work, [and] courage in the replacement of

---

[28] *Report of the Giant Power Survey Board to the General Assembly of the Commonwealth of Pennsylvania* (Harrisburg, Pa.: Telegraph Printing Co., 1925), pp. iv–v (hereafter cited as *Giant Power Survey*). The account of Giant Power presented here is taken from Thomas P. Hughes, "Technology and Public Policy: The Failure of Giant Power," *IEEE Proceedings* 64 (1976): 1361–71. See also Jean Christie, "Giant Power: A Progressive Proposal of the Nineteen-Twenties," *Pennsylvania Magazine of History and Biography* 96 (1972): 480–507.

apparatus. . . . Private initiative is to be driven out of the electric service companies and [to] be supplanted by a political plan based upon a socialistic theory and offering all the possibilities of the construction of a state-wide, all powerful, political machine.[29]

The issues and origins of Giant Power were, however, far more complex than the rhetoric of the governor and this industry spokesman indicates. The controversy consisted of far more than a simplistic confrontation between the politician and the capitalist. It involved differing views about the exploitation of newly developed technology in the field of electrical power; it was fired by the enthusiasm of conservationists who had flourished in Theodore Roosevelt's Progressive party; it was heightened in intensity by the waxing power of privately owned holding companies in the United States, and the achievements of publicly owned utilities in Canada and Europe; it aroused public interest in the Northeast because there was evidence that the South and Far West were moving ahead in the development of electric power; and it was influenced by the prior confrontations of the personalities involved. With so many complications, the problems raised by Giant Power could not be solved by rhetoric. Nor could they be settled by technological expertise. For one thing, engineers disagreed among themselves; for another, the variables and the options extended far beyond the technological realm to the political, economic, and social arenas.

These were some of the attitudes and general trends in the context of which Morris Cooke and Gifford Pinchot introduced their Giant Power plan. Cooke's concept can be traced back directly to his wartime experiences. As an adviser on energy and transportation problems, he had recommended that the government lighten the freight load on railroads by financing the construction of large power plants at the mine mouth and by transmitting power from these to load centers by high-voltage lines. Specifically, Cooke wanted mine-mouth plants built in eastern Pennsylvania to supply Philadelphia.[30] In making these recommendations, Cooke raised the possibility of a confrontation with the Philadelphia Electric Company. The war ended before Cooke's recommendations could be acted upon, but his plans reveal his early commitment to one of the central objectives of Giant Power. They also marked the beginning of a struggle between Cooke, Philadelphia Electric, and other utilities. The passions invoked by Giant Power and even the character of the plan itself cannot be comprehended, however, unless we first review the history of Cooke's activities in relationship to Pennsylvania's utilities. Long-term trends were manifest in Giant Power, but the role of individuals like Cooke cannot be ignored.

Cooke studied engineering at Lehigh University, but he gained prominence more as a reformer than as a technical expert.[31] Early in his career

[29] Pennsylvania, General Assembly, *Giant Power: Proceedings before the Committee on Corporations of the Senate and the Manufacturers Committee of the House of Representatives, being a Joint Hearing on Senate Giant Power Bills Nos. 32, 33, 34, 35, 36, and 37. Extraordinary Session of 1926,* pp. 34–35 (hereafter cited as *Joint Hearing on Giant Power*).

[30] Cooke to Robert Woolley, 15 April 1918, Morris L. Cooke Papers (Group 40, Box 200, file labeled "Cheap Electric Power"), Franklin D. Roosevelt Library, Hyde Park, N.Y. (hereafter cited as Cooke Papers).

[31] The biographical sketch of Cooke is drawn from Edwin T. Layton Jr.'s *Revolt of the Engineers* (Cleveland: Press of Case Western Reserve University, 1971), pp. 154–71.

he became an advocate of scientific management and an associate of Frederick Taylor, the originator of the movement. Cooke, however, did not make a name for himself by applying principles of scientific management to industrial problems; he used scientific management as the ideological foundation for his advocacy of social reform as well as the reform of the engineering profession which social reform entailed. Cooke wanted to see experts, especially engineers, solving social problems through the application of engineering and rational management principles: he was convinced that this would never come to pass, however, unless the professional engineer could distance himself from the influence of private business and finance. Since most engineers were hired by large manufacturers and utilities, Cooke believed that it was difficult for them to serve the public interest, for public and business interests conflicted. Because Cooke was an effective publicist with influential contacts in the publishing world, his ideology alone was enough to threaten business interests.

Cooke suited his actions to his ideas by using his political skills to construct a power base for himself and other like-minded engineers within the American Society of Mechanical Engineers (ASME). He believed that the ASME and the other professional engineering societies could promote the public interest much more effectively than individual engineers could. He saw the electrical utilities as the major obstacle to this goal. He considered domination of the American Institute of Electrical Engineers (AIEE) by the electrical utilities a notorious fact; the utilities' domination of his own ASME was less obvious but no less certain to him. Because of his close association with Frederick Taylor (1856–1915), who became president of the ASME in 1906, and because of a reform movement among younger members of the ASME, the utilities had to take Cooke seriously in professional-society politics.

As if that were not enough for the utilities, Cooke engaged in a hotly contested rate case involving one of the country's largest utilities, the Philadelphia Electric Company. Rudolph Blankenberg, the successful reform candidate for mayor of Philadelphia, named Cooke director of public works. His appointment may have resulted more from his family background, cultural style, and political sentiments than from his engineering expertise, for Cooke, like many other reform politicians of his day, came from a genteel and moneyed family dedicated to reform carried out in the public interest by professional elites. Whatever the reasons for the appointment, Cooke, as a result, was in a position to obtain information about utility services and to publicize his interpretation of that data. In the case of Philadelphia Electric, which he brought to court, he was convinced that the utility was overcharging the city for street lights and residential customers for general service. The overcharging resulted, Cooke believed, from the utility's inflating the value of its assets.

Because the state's regulatory body, the Public Service Commission, allowed utilities to earn a percentage of their assets as profit, inflated valuation resulted in increased profits. Cooke managed to have Philadephia Electric bring in outside engineering experts to make a new evaluation, but the results, he insisted, were biased by the pro-utility prejudice of the company's principal consultant. Because that consultant was Professor Dugald C. Jackson, head of the electrical engineering department at M.I.T.

(pp. 148–49 above), and because Cooke alleged that a number of utility consultants from the academic world were prejudiced like Jackson, Cooke's reputation as a radical grew. In the case of Philadelphia Electric's rates, however, Cooke succeeded in getting the company to reduce them, both for the city and for residential customers.

Cooke's authorship of the Giant Power recommendations was certainly enough to cause the utilities to view the plan apprehensively, and Governor Pinchot's sponsorship of it did nothing to relieve the tension. Pinchot had a long history of reform activities motivated by his conviction that public and business interests were not synonymous. A formidable advocate of the causes he chose to champion, he was characterized as enthusiastic, ambitious, and self-righteous.[32] Like Cooke, Pinchot's family was affluent and had close contacts with the high-minded progressive reformers of his era. Furthermore, and also like Cooke, he was an effective publicist. After studying forestry in Europe, he became chief forester of the U.S. Forestry Service. His program called for the scientific management of timberlands and for the conservation of these lands as a national resource for his and future generations. He had seen Europeans husband their limited resources (forest and other), and he was determined to stem the thoughtless exploitation of the resources in America that he knew to be finite, popular opinion to the contrary.

Pinchot's activities drew him into a close political alliance with his long-time progressive friend Theodore Roosevelt. They also sparked a controversy with politicians and financial-business interests that he believed were intent upon exploiting the coal-rich lands in Alaska and the hydroelectric power sites in the American West that he wanted reserved for government control. His stand brought him into conflict with, among other powerful persons and corporations, the Morgan-Guggenheim syndicate. Because of his reputation for making enemies, he carried a legacy into the Giant Power controversy that only increased its intensity. He was motivated in his advocacy of Giant Power by his conviction that he was waging a struggle against selfish and avaricious monopolies organized by powerful financial interests. Among these was the old foe, the House of Morgan, which, according to Pinchot, controlled the Philadelphia Electric Company, among other utilities. (Pinchot and Cooke, it appears, had much in common.)

In 1922, having gained a reputation for his reforms and his objectives, Pinchot won the Pennsylvania gubernatorial election. Shortly afterward, in 1923, he established the Giant Power Survey Board, among whose prominent members was Philip P. Wells, deputy attorney general of Pennsylvania. An advisory committee to the board included Samuel Gompers, Arthur E. Morgan, Henry L. Stimson, and William Allen White. Morris Cooke was director of the board. In 1925 the board submitted its report to the General Assembly of Pennsylvania (see Fig. XI.4).[33] The report was widely circulated by the governor so as to draw nationwide attention to the principles of Giant Power; it also included proposals for legislation.

Numerous newspapers and journals, popular and technical, carried sum-

[32] Gifford Pinchot, *The Power Monopoly: Its Make-up and Its Menace* (Milford, Pa.: Privately printed, 1928), p. 5.

[33] See *Giant Power Survey.*

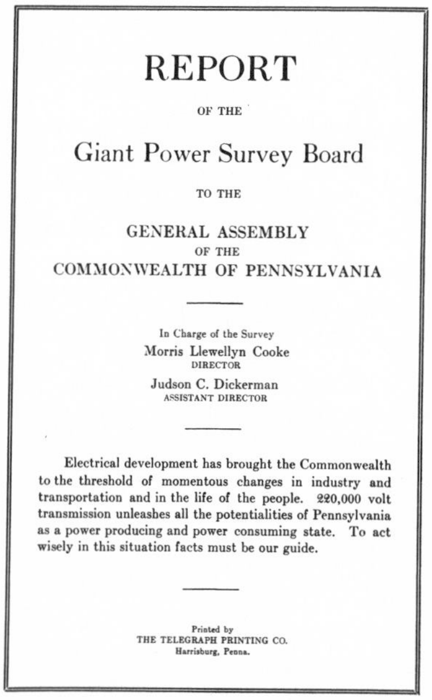

REPORT

OF THE

Giant Power Survey Board

TO THE

GENERAL ASSEMBLY
OF THE
COMMONWEALTH OF PENNSYLVANIA

In Charge of the Survey
Morris Llewellyn Cooke
DIRECTOR
Judson C. Dickerman
ASSISTANT DIRECTOR

Electrical development has brought the Commonwealth to the threshold of momentous changes in industry and transportation and in the life of the people. 220,000 volt transmission unleashes all the potentialities of Pennsylvania as a power producing and power consuming state. To act wisely in this situation facts must be our guide.

Printed by
THE TELEGRAPH PRINTING CO.
Harrisburg, Penna.

Figure XI.4. Title page from Giant Power Survey.

maries of the plan. *Electrical World* printed a matter-of-fact summary, but did so under the heading "Pinchot Takes Radical Stand."[34] This signaled the reaction of the utilities, but before the opposition could mount an effective campaign, the governor and Cooke were receiving considerable attention for their program. In essence, the Giant Power plan, as developed in the report and as later incorporated in legislation, called for the establishment of giant mine-mouth power plants (each with a generating capacity of at least 300,000 kw.) in western Pennsylvania; the construction of a network of high-voltage transmission lines capable of carrying 100,000 or more volts; and the adaptation of the distribution systems of private, municipal, and cooperative utilities to deliver electricity from the high-voltage network to industrial, residential, traction, and other consumers. The mine-mouth power stations were to be integrated with plants that processed raw bituminous coal to derive gas, benzol, ammonium compounds, and other chemicals before delivering low-grade fuel material to the power plants.

The Giant Power concept appealed to utility engineers and managers because its bold use of advanced technology, large-scale power plants, and wide-area grids of transmission lines promised efficiency and economy. However, once Pinchot and Cooke's proposals concerning regulation, licensing, ownership, rates, corporate structure, and securities were comprehended, the bloom was off the rose. Taken together, these aspects of the plan, as legislative hearings and public debate revealed, could, in the opinion of the utilities, be labeled government interference, even a prelude to government ownership. As issues were defined and the controversy developed, it became clear that in the case of Giant Power, the question was not so much whether or not to adopt the new technology as who would own and control the technology and in whose interest.

Regulation by the state was to be achieved through both the established Public Service Commission and a new, permanent Giant Power Board. The latter would authorize incorporation of Giant Power generating companies and Giant Power transmission companies.[35] Since all major power plants and transmission facilities in the future were to be of the new and large kind, the state would determine not only the kind of technology used in general but also the regulatory details of the charters issued to the new, privately funded Giant Power corporations. Existing utilities could apply for Giant Power charters, but other investors would have the opportunity to apply for the right to incorporate for the purpose of constructing and operating the generation and transmission facilities as well. The generating companies were to be given authority by the Giant Power Board to appropriate by condemnation enough mining lands to supply fuel for not more than fifty years.

The transmission companies were to be common carriers of electricity purchased at wholesale prices from the Giant Power generating companies and from surviving utility power plants when they had a surplus above the needs of their own distribution systems. The routes of the transmission

[34] *Electrical World* 85 (1925): 421–22.
[35] The following description of the Giant Power plan is based primarily on *Giant Power Survey* and *Report of the Giant Power Board to the Governor of Pennsylvania, December 7, 1926* (Harrisburg, Pa., 1927).

lines would be determined by the Giant Power Board. The transmission companies would sell power to distribution companies, the majority of which, at least for the foreseeable future, would be the existing utilities. Because authorization for new power plants would be limited to Giant Power companies, the existing utilities over the years would be gradually transformed from multipurpose enterprises to simple distribution companies.

As if the prospect of this transformation were not enough for the owners of the powerful utilities to ponder, there was another bitter pill they would have to swallow if Giant Power legislation was enacted. After fifty years, the state would have the right to take over the Giant Power facilities and operate them upon payment of the money the Giant Power companies had prudently invested during that period. During the fifty years, the companies would have paid an annual fee to amortize the cost of the land mined or used as right of way for transmission. (The right of way would have been acquired by the companies through the state's power of condemnation or by simple purchase.)

The Giant Power companies (both generating and transmission) were expected to make a profit. As in the case of existing utilities, the profits of the new companies would be regulated by the state. What was new was the rate base. Morris Cooke had seen in the valuation procedures of the utilities an opportunity for them to make excessive profits despite their rates being regulated on the basis of percentage of investment. Under Giant Power laws the rate base would not be replacement value of the plant at the time of rate change; it would be the amount prudently invested in facilities. The rate of return on investment was to be adjusted to allow sufficient profit to attract needed capital. Through the Public Service Commission the state would also control the issuance of securities, the par value of which had to correspond to the amount of capital invested.[36]

To an American utilities industry already alarmed by nationalization legislation in Great Britain, Germany, and elsewhere in Europe, Giant Power signified ultimate government ownership. Anxieties were further aroused by Pinchot and Cooke's commitment to rural electrification. Only 6 percent of the power generated by utilities served farms in 1925. According to the Giant Power plan, the Public Service Commission would be empowered to require extension of distribution systems to unserved territory; furthermore, the formation of rural mutual and district distribution companies was authorized and encouraged. Mutual companies were to be formed by voluntary associations of consumers; district companies would be created upon the favorable vote of a majority of the district's inhabitants and of the owners of "a sufficient majority of the acreage." The district companies would have the power to tax, assess benefits and damages, and so on. These companies would buy power from the Giant Power transmission companies in the same manner as utilities.

Another part of the Giant Power plan (subtle, but not lost on the holding companies) pertained to the interstate transmission of power. Pinchot and Cooke deplored the opportunity that large utilities and holding companies

[36] *Giant Power Survey,* pp. 4–8.

had to operate outside the jurisdiction of the state and federal governments. The federal government had not yet enacted legislation regulating the interstate flow of electricity, and the states found it frustrating to regulate a utility or holding company whose activities extended beyond their political boundaries. Among other provisions, the proposed Giant Power legislation called for Pennsylvania to make pacts with other states to regulate interstate electric transmission and thereby avoid the question of federal jurisdiction.[37]

The Giant Power plan had to be presented to the Pennsylvania legislature in the form of bills, a process that would publicize the utilities' opposition. Pinchot probably anticipated the opposition; he may even have expected defeat in the legislature; but he looked beyond it to the national impact the publicity attendant upon the introduction of the legislation and the publication of the lengthy *Report of the Giant Power Survey Board* would have.[38] (Pinchot considered Giant Power to be a national issue, and he himself was thought of as a likely candidate for national political office). Since both Pinchot and Cooke were experienced publicists, the presentation of Giant Power to the legislature—and to the people—was impressive in its persuasiveness. The arguments went beyond technical detail and beyond the dry statements of legislative advocacy.

In his message of transmittal of the Giant Power report, Pinchot compared the electrical supply revolution to the steam revolution. "Steam," Pinchot wrote, "might well say of electricity, 'One mightier than I cometh, the latchet of whose shoes I am not worthy to unloose.' "[39] Steam power had changed the face of the earth, the governor believed, but its potential for the good of all the people had been lessened by the conflict between capital and labor and by the unplanned character of the industrial revolution. The United States especially, he continued, had benefited from steam, gaining the highest ratio of mechanical power to citizen in the world, but if electricity were used for all the people, it would bring an even higher standard of living. Pinchot wanted every home to have electric light, heat, and labor-saving appliances; he wanted every farm to have electricity for milking, feed cutting, wood sawing, and a multitude of other onerous tasks. He called for electricity to be generated in such a way and in such places that the cities would be free of the nuisance of smoke and ash. He forecast that electricity would effect the substitution of garden cities for slums, small industrial communities for gargantuan, chaotic cities, and would bring about a return to the countryside.

Pinchot believed that these developments would be possible only if the spreading net of electric power could be brought under public control. Pointing out that the interconnection of utilities had already made regional systems a reality, he predicted the development of a nationwide grid. Grids integrated power under central control as steam never had been able to. Pinchot foresaw the growth of technological systems; he also predicted that centralized financial and business systems would control and monopolize

[37] Ibid, pp. 9–10.
[38] Pinchot to Cooke, 18 February 1925, and Cooke to Pinchot, 20 April 1925, Cooke Papers (Group 40, Box 35, File 391).
[39] *Giant Power Survey*, p. iv.

the technology. On the good side, electric power was capable of showering upon the people "gifts of unimaginable beauty and worth." "On the bad side," Pinchot observed, "it is as though an enchanted evil spider were hastening to spread his web over the whole of the United States and to control and live upon the life of our people."[40]

The evil spider was the monopolistic utility financier, and the net was the spreading holding company. Governor Pinchot had no doubt that a unified electrical monopoly extending throughout the nation was inevitable and close at hand. The only question was would a selfish few control it or would the people regulate it. If the utilities would not cooperate in public regulation, then control through public ownership was the only alternative. Giant Power, the governor argued, provided for public regulation; it accepted private ownership. "As Pennsylvania and the Nation deal with electric power," he concluded,

... so shall we and our descendants be free men, masters of our own destinies and our own souls, or we shall be the helpless servants of the most widespread, far-reaching, and penetrating monopoly ever known. Either we must control electric power, or its masters and owners will control us.[41]

Morris Cooke, also no mean master of rhetoric, let the governor paint with a broad brush; Cooke masterfully described the technology and economics of Giant Power so that the intelligent layman concerned about public policy could understand and support it. Cooke probably saw the task of winning over the utility engineers and managers as an exercise in futility. He settled instead for the Henry Stimsons, the William Allen Whites, and the Charles W. Eliots. Cooke appealed to the intellectuals and the elitist reformers and took little heed of the impact of his recommendations upon vested interests in the utility field.

In his section of the 1925 report, Cooke recommended that the existing utilities eliminate their generating plants as these became obsolete and that they ultimately limit themselves to distributing electricity bought wholesale from the Giant Power companies. Pinchot said this in general terms; Cooke was succinct and blunt. As for the generating stations then operating, if they were efficient, they would be used for peak loads; if they were inefficient, they would simply be eliminated. (Cooke did not indicate the fate of those who had expected a long-term return on their investment in these power plants.)[42]

Cooke also emphasized by-product recovery from coal at mine-mouth plants (see Fig. XI.5). Boldly, he recommended the use of 20,000 tons of bituminous coal each day by combined by-product recovery plants and 500,000-kw. electric power plants. As a precedent, he could cite only the largest coal distillation plant in the United States, the Clairton high-temperature distillation plant of the U.S. Steel Corporation, which provided, among other products, coke for blast furnaces and gas for steel-working furnaces. He assumed a demand for the large amounts of ammonium sulfate, benzol, tar, gas, coke, and other substances that the Giant Power plants would pour upon the market. Because consumption of the world's

[40] Ibid, p. xii.
[41] Ibid, p. xiii.
[42] Ibid, pp. 36, 42.

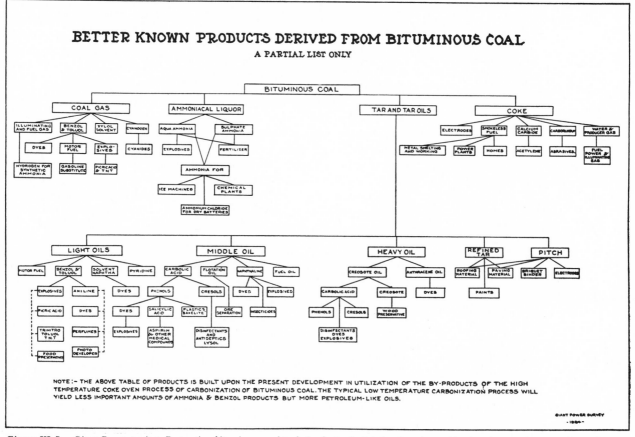

Figure XI.5. *Giant Power project: Pretreating bituminous coal to derive by-products before burning residue in mine-mouth power plants. From Giant Power Survey, p. 104.*

petroleum resources was expected to be rapid, he also suggested that oil and gasoline be derived from coal distillation.[43] Cooke may have heard of the decision of the I. G. Farben Company in Germany to invest heavily in a process for producing gasoline from coal; both he and I. G. Farben anticipated petroleum shortages, unaware as they were of the immense oil fields that would be found in the southwestern United States within several years.

Cooke anticipated the objection from utility engineers that the coal-mining regions lacked an adequate supply of cooling water. Generally it was agreed that an electric power station needed at least 400 tons of cool water to condense the steam exhausted from turbines for each ton of coal used in producing the steam for the turbines. The customary way of obtaining cooling water in Pennsylvania in the twenties was to situate power plants on large rivers or on tidewater. The Allegheny was one of the few rivers in the coal-mining districts of eastern Pennsylvania that had sufficient summer flow to cool Giant Power plants. Cooke recommended that dams be built to provide this water in slack times. The dams, he argued, would also serve as a means of flood control and would provide a more regular supply

[43] Ibid, p. 33.

of water for waste and sewage disposal. In addition to dams—and here he went far beyond American practice—he urged the construction of mammoth cooling towers, which allowed a given amount of water to be circulated through the condensers, cooled in the towers by the atmosphere, and circulated once again through the condensers. He called attention to the practice in Europe, especially at Golpa-Zschornewitz and Trattendorf in Germany, where power plants with cooling towers needed water only to make up for loss due to evaporation.[44]

As a young student of forestry, Pinchot had observed how Europeans husbanded their limited forests, and he urged similar ways upon resource-rich and profligate Americans. It is not clear whether Cooke realized that by referring to European precedents in electric power practices such as the cooling towers he, too, was asking that the techniques of a resource-short region be adapted in one accustomed to an abundance of resources. He did not explicitly acknowledge that in Europe, especially in Germany, mine-mouth plants and cooling towers were located where low-heat-value coal, brown-coal (lignite), was being exploited, a fuel that was too low in calories to be transported because of the high weight-to-caloric-content ratio.[45] The power plant moved to the fuel; the fuel did not move to the site of power-plant cooling water and load. Cooke presented European precedents—just as he referred to the American precedent in high-voltage transmission in California—simply as advanced technology showing the way of the future. The subtlety of suiting technology to factor endowment may have been lost on him.

Cooke, like Pinchot, was greatly concerned about the farmers' lack of access to electricity (see Fig. XI.6). Cooke showed that of 202,250 farms in Pennsylvania, only 12,452 were estimated to have public utility service. (His estimate, he wrote, was higher than that made by the utilities.) Nearly 20,000 farms had either their own electric plant or gas plant, but that left 171,581 farms without modern means of illumination in the second decade of the twentieth century.[46] (Some of these farms were Amish, however, and in those instances electricity was refused for cultural reasons.) The major reason so many farms went unserved was that the utilities would not invest in the distribution lines needed and would not levy rates similar to those they charged their industrial customers. Because farms were widely scattered, investment in the necessary distribution facilities was relatively high compared to the cost of serving urban customers, and partially because of this the farmers were charged rates like those of residential or lighting consumers in the cities. Like Pinchot's plan, Cooke's proposal for solving this problem included mutual companies and district utilities, but it also called for utilities to allocate as much as 5 percent of their current annual capital expenditures to rural electrification. In addition, Cooke wanted lower rates which would encourage increased consumption. Cooke's recommendations became more significant with hindsight for he was ap-

[44] On the problem of cooling, see also August Ulmann, Jr., "The Influence of Water on the Location of Giant Power Plants," *Annals of the American Academy of Political and Social Science* 118 (1925): 141–48.

[45] In 1924 Cooke sent O. M. Rau, an engineer who was assisting him on the Giant Power project, to Europe to observe the electric supply practices followed there.

[46] *Giant Power Survey*, pp. 37 ff.

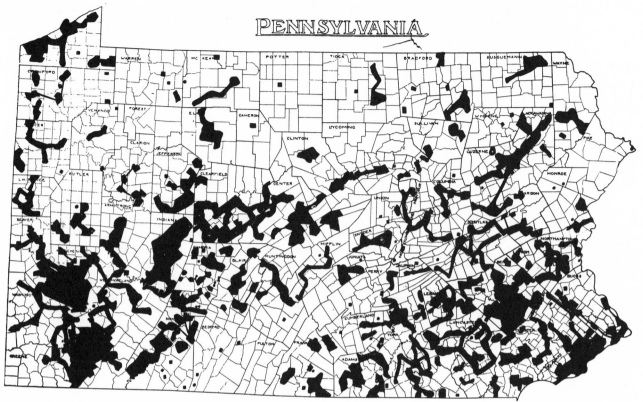

Figure XI.6.   *Areas of Pennsylvania within economical range of existing electric supply, c. 1920; only 50 percent were served. From* Giant Power Survey, *p. 39.*

pointed head of the Rural Electrification Administration under the New Deal.

Some of the utilities' responses to the idea of Giant Power were shrill. One engineer labeled Cooke's ideas communistic.[47] Representative of a more rational reaction from the utilities was a 1925 address by Charles Penrose.[48] In contrast to Cooke, whose technical achievements were limited, Penrose, an electrical engineer from Princeton, had worked as an electrical engineer for the Philadelphia Electric Company and was its engineer in charge of erection of Schuylkill No. 2 Station in 1914–1915. From 1920 he had been assistant general manager of the prominent Philadelphia consulting engineering firm of Day & Zimmermann. He was, in 1925, chairman of the Public Relations Section of the Pennsylvania Electric Association (the trade association of the state's utilities) and a member of the Public Relations Executive Committee of the National Electric Light Association (the trade association of utilities nationwide).

[47] W. S. Murray, letter to editor, *Electrical World* 85 (1925): 513. Murray, a consulting engineer, had another scheme for large power stations and high-voltage transmission. He called his plan Superpower. It was to cover the northeastern United States.

[48] Charles Penrose, "Power in Pennsylvania" (Address before the Hydro-Electric Conference, Philadelphia, Pa., 10 March 1925); offprint in Cooke Papers (Group 40, Box 199, file labeled "Propaganda").

Addressing engineers in conference in Philadelphia, Penrose struck two responsive chords: (1) an experienced engineer opted for reasonable progress from a sound base of experience; and (2) engineers had the responsibility of informing the public about sound policy in order to save it from the reckless ventures of politicians and other laymen treading on the engineers' ground. He also insisted that technological change should not play fast and loose with vested interests, which to him were economic ones, and that industries should not subsidize farmers and residential consumers by allowing them to be charged rates comparable to what the industries paid.

In a revealing remark, Penrose labeled Giant Power technology "radical." To an experienced and prudent engineer it was radical because the length and voltage levels of the transmission lines proposed had not been thoroughly proven in practice and because the 500,000-kw. power and by-product plants were unrealistically large for the mine-mouth. Penrose also questioned the wisdom of making industrial Pennsylvania power plants dependent upon the mining district's limited supply of cooling water (dams probably could not compensate for droughts) or upon cooling towers. Artificial cooling in America had been limited, he reported, to installations less than one-tenth the size of Giant Power plants. He also doubted the common sense of a proposal that would have all of eastern Pennsylvania relying on 300-mile-long transmission lines. (Reliability of service meant more than lower rates to many industries.) The electrical utilities in Pennsylvania, Penrose insisted, were moving ahead at a reasonable rate in their effort to interconnect systems, raise transmission voltages, and improve diversity and load factors. Pennsylvania, he observed, was a prosperous state, and Philadelphia was "the Workshop of the World."

Pennsylvania's utilities also attracted capital because they were recognized as being prudently managed and profitable. Penrose clearly spoke for those who had invested in the utilities when he said:

When we reflect upon the immense investment represented in generating capacity operating today in the electric public utility systems in Pennsylvania, we will appreciate why such a proposal must merit the closest economic scrutiny. This is self-evident because of the intimate relationship, in Pennsylvania as elsewhere that exists between the financial stability of the electric utilities and the financial stability of Industry, Manufacture, and indeed Business itself with the State. You may be the judges as to whether the proposal (I quote), "our service companies will be relieved of the necessity for providing generating stations," is to say the least startling![49]

He also asked shrewdly if the authors of the Giant Power plan had considered what might be the reaction of industrial corporations that had already invested heavily in supplying the coal by-product market that Giant Power plants would enter.

After attaching the label "another utopia" to Giant Power, Penrose ended his address by asking the engineers to take responsibility for the future of the state of Pennsylvania by seeing to it that the public "shall know what are the facts." There seems to have been no doubt in his mind that his informed analysis, based as it was upon prudent engineering practice and sound economic principles, provided those "facts."

[49] Ibid., p. 4.

The history of the Giant Power proposals in the Pennsylvania legislature involved committee hearings and reformulations. One major change was a retreat from the initial position that generating, transmission, and distribution companies should be separate entities. The fate of the bill was finally decided in January and February 1926 at a joint hearing held by the Committee on Corporations of the Senate and the Manufacturers Committee of the House of Representatives. Advocates and opponents of Giant Power were on hand to testify.[50]

The strategy of the bill's opponents had been indicated by Penrose; others offered variations upon his themes. To substantiate the argument that Giant Power was unrealistic, radical technology, three well-known professional engineers, Professor Arthur E. Kennelly, Professor Paul M. Lincoln, and Harold Buck, were called to testify. They had been sent to the hearing by the American Engineering Council at the request of the joint committee. The three said that they did not represent the council but were on hand as experts to offer opinions and present facts. Their facts and opinions turned out to be just about the same.

Harold Buck spoke first and stated the consensus.[51] A consulting engineer from New York City, he declared that he represented no one but himself. In the past, however, he had been employed by the General Electric Company and had served as an electrical engineer for the Niagara Falls Power Company before entering private practice and engaging in the design and construction of large hydroelectric and power transmission systems throughout the country. Echoing Penrose, Buck told the legislature that to support Giant Power was to tamper recklessly with a successful industry, the public utilities in Pennsylvania. That which was sound about the Giant Power plan was already being put into practice by the utilities, he noted: interconnection for the exchange of surplus power; economically prudent rural electrification; load-factor increases; and improvements in the economics of generation. Other aspects of Giant Power technology had not been and would not be adopted by utility engineers because, Buck argued, they were preposterous. He called the legislators' attention to the electrical systems of Canada and Great Britain as examples of the baleful influence of government control and regulation. (Others at that time held up the publicly owned Ontario regional system as a model for the United States to follow.) After hearing Buck's remarks, the spokesman for Giant Power asked Buck if he had ever appeared as an expert witness in a public controversy other than for a private company. Buck recalled two such occasions.

Professor Arthur Kennelly was well known even to laymen. He had been a principal assistant to Edison at the West Orange Laboratory for about seven years and had served General Electric and other companies as a consulting engineer, residing for a time in Philadelphia. Subsequently he had become professor of electrical engineering at the Massachusetts Institute of Technology and at the time of the hearing he was at Harvard. After seconding Buck's remarks and amplifying some of them, Kennelly sounded

---

[50] See *Joint Hearing on Giant Power.*

[51] Ibid., pp. 64–74.

a recurring theme: "In my opinion the present system is working well."[52] He, too, expressed the view that government interference retarded the utilities industry abroad; he feared that interference by the state government—in contrast to state regulation as generally practiced—would dry up the sources of needed capital.

Professor Paul Lincoln, director of the School of Electrical Engineering at Cornell, had earlier worked for Westinghouse and the Niagara Falls Power Company. He repeated many of Buck's pronouncements and went on to stress his concern about the stability of a 220,000-volt transmission line carrying more than 125,000 kva. Like Buck, he urged the continued transport of coal (energy) by railroad to power plants rather than the transmission of electricity (energy) from such plants to the load via high-voltage lines across hundreds of miles. Not only was rail more reliable but it was more economical, he testified. He referred to high-caloric—not brown—coal. In cross-examination, Lincoln was asked if he had ever prepared a study recommending the use of large mine-mouth plants to transmit power from the Pennsylvania coal fields to the industrial East. As was expected in view of his testimony, Lincoln said never. Counsel for the bill then laid before the committee a study done by Lincoln in 1919 in which he advised that there was no engineering obstacle to building in the Pennsylvania coal fields a 500,000-kw. plant capable of 220,000-volt transmission. The witness admitted that his memory was poor in this instance; he also said that in 1919 he had not known what he knew in 1925 about the stability problem.[53]

Proponents of the Giant Power bill presented no engineers from private utilities or electrical manufacturers, no engineering professors, and only one consulting engineer—a specialist in cooling towers, which was still an insignificant field in America. However, they did have an impressive witness in T. A. Panter, assistant chief engineer of the Los Angeles Water and Power Bureau, a municipally owned utility. Supplying the city of Los Angeles with about 70 percent of its total energy, the bureau purchased a substantial portion of this energy from the privately owned Southern California Edison Company's transmission system. Southern California Edison drew much of the power it sold to the municipal utility from hydroelectric plants in the Sierra Nevadas, 250 miles away. Panter testified that on the basis of his experience with service from this transmission line, he anticipated there would be no serious difficulties with Giant Power transmission. He also stated, contrary to Lincoln, that he thought 125,000 kva. could be transmitted without stability problems on a 220,000-volt line.[54]

In general, the testimony and cross-examination at the hearing proceeded unemotionally and matter-of-factly. The exception was an oration given by a member of the U.S. Congress for many years, James Francis Burke of Pittsburgh. Burke began by saying that a more dangerous and destructive set of measures had never been presented to the legislature. He then warned that if the legislature should ever become sufficiently socialistic to enact Giant Power into law, it would also have to provide a "burial place for the 750,000,000 dollars already invested in good faith in

[52] Ibid., p. 81; for Kennelly's testimony in its entirety, see ibid., pp. 79–83.
[53] Ibid., pp. 75–79.
[54] Ibid., pp. 88–92.

the industry." Giant Power, he said, not only smelled of socialism but was "pregnant with the vicious elements of confiscation." It was "false bait for the farmer" and would lead to one-man control of the vast industry by the governor. He closed by asking, "Is Pennsylvania to lead America in following Russia into the dismal swamp of commercial chaos and financial disaster? Are we to forget our Franklins, our Westinghouses, our Edisons, our Steinmetzes, and all the geniuses whose names light up the horizon of industrial progress?"[55] (The fact that Steinmetz had run as the Socialist party's candidate for New York State Engineer seems not to have disturbed Burke.)

It is difficult to understand why Burke was called to appear before the committee—his appeal differed so dramatically from the general temper of the hearing. Perhaps the answer lies in knowing more about the members of the committee. In any case, whether moved by Burke, more reasoned discourse, or other considerations, the joint committee rejected the Giant Power legislation, reporting the bills negatively on February 8. The Senate subsequently concurred; the House never bothered to report.

The history of Giant Power is instructive. At issue was the attempt by an "external force," the Pinchot administration in Pennsylvania, to change the direction of the development of the electric light and power industry in the United States. The effort failed, but the proponents of Giant Power do not seem to have understood the deeper reasons why. Pinchot blamed the power of a great and growing monopoly masterminded by giant financial interests; Cooke shared this view and also attributed the failure to the conservatism of the engineering establishment, an attitude he had struggled against as a member of the American Society of Mechanical Engineers. Both men gave more credit to the influence of human will—perhaps they would have preferred to say human machinations—than to long-range trends or factors.

The historian can focus on trends, factors, and the momentum of a large and growing industry. Philip P. Wells, Pinchot's deputy attorney general, was nearer to the mark when he told the joint committee that in fact the utilities opposed Giant Power not on an engineering basis but because of the "regulation and the pooling and the farmers' bills."[56] Charles Penrose and Arthur Kennelly were much closer to the roots of the opposition when they called attention to the heavy investment of capital that had been made in the existing technology by the utilities and to the possibility that a scheme like Giant Power might drive away the investors who were financing the orderly development of the industry. They would have been even nearer the mark if they had gone on to say that Giant Power would shift the authority and responsibility of presiding over the growth of electric light and power systems from the engineers, managers, and owners of the existing utilities to the new men of the Giant Power system. Pinchot and Cooke were not simply proposing a radical technology; they were proposing radical change in the deepest sense of the word. They were calling for a shift in power, an economic revolution.

[55] Ibid., pp. 54, 56, 59.
[56] Ibid., p. 30.

By the mid-1920s the superpower and Giant Power movements had lost momentum. The wartime spirit of imperative innovation had given way to a more conservative attitude. America's utilities proceeded with interconnection, but not in accordance with a master government scheme for an entire state or region. The vision of planned social revolution through technology gave way to the long-standing confidence that private enterprise and American technological genius would bring profit and progress.

The plan for Giant Power in the United States had provided for government planning and control at the state level, and the resulting ideological conflict had been between state government and private enterprise. In Germany after the war another bold plan for government reorganization of electric supply was advanced. There the ideological conflict developed between the central and state governments, not between private and public power. The central, socialist-dominated government passed an electrification law in December 1919, when the disruptive, unconventional, extreme environment of wartime was still motivating radical actions. On 8 August 1919 the Reich Treasury Minister Wilhelm Mayer had presented to the German National Assembly at Weimar the government's proposal for establishing a unified electric supply system for all of Germany. Mayer made his proposal under the general provision of the Socialization Law passed by the National Assembly on 23 March 1919. Paragraphs 2 and 4 of this law empowered the Reich government to socialize, with compensation, undertakings that exploited natural resources, especially coal and water power, to develop energy from them. These undertakings were to be integrated into a national economy.[57]

The Reich government thought that a unified German system of supply would solve several pressing problems. By increased utilization and expansion of existing central German brown-coal and south German hydroelectric power plants, and through additional construction of these, the government hoped to relieve a postwar coal shortage resulting from the loss of hard-coal (*Steinkohle*) mines under the terms of the Treaty of Versailles and from shorter working hours and lower productivity in the mines. The government, as noted, also faced the problem of finding a market for the large-capacity power plants that had been built during the war to meet special wartime manufacturing needs, such as aluminum refining and nitrogen fixation. Construction of high-voltage transmission systems would make possible in peacetime the distribution of electricity from these very large plants over a wider market area. A large area of distribution would also bring a diversified load and the essential high load factor for these large, capital-intensive plants.

Intent upon utilizing the brown-coal deposits of Germany, the govern-

[57] The discussion of the 1919 proposal for the socialization of electric supply in Germany is based on Treasury Minister Wilhelm Mayer's address of 8 August 1919; see E. Heilfron, ed., *Die Deutsche Nationalversammlung im Jahre 1919 [und 1920]*, vol. 7 (Berlin: Norddeutsche Buchdruckerei und Verlagsanstalt, n.d.), pp. 509–21. The account of the effort to establish an all-German system is based on Thomas P. Hughes, "Technology as a Force for Change in History: The Effort to Form a Unified Electric Power System in Weimar Germany," in *Industrielles System und politische Entwicklung in der Weimarer Republic*, ed. H. Mommsen, D. Petzina, and B. Weisbrod (Düsseldorf: Droste, 1974), pp. 153–66.

ment, for reasons of economy, favored high-voltage transmission supply. Because it is high in calorific value, hard, or stone, coal—the principal source of thermal energy used by German power plants before World War I—is relatively economical to transport by railroad. For this reason, hard-coal electric power plants in the Ruhr region and elsewhere were built near urban or industrial centers where the demand for electricity was concentrated and heavy. By contrast, brown coal, having a low caloric content, is uneconomical to transport. For this reason, power plants using brown coal were built at the mine site and the energy had to be transmitted to urban centers via high-voltage transmission lines.[58]

A shortage of hard coal, the problem of large power plants deprived of their wartime markets, the availability of brown coal, and the existence of water-power resources in southern Germany were some of the special circumstances that shaped the Reich government's plan. The general objectives of higher load factor and harmonious optimization of varied plant characteristics also were conditioning factors. For instance, the government assured south German hydroelectric utilities that interconnection with thermal plants elsewhere in the Reich would assure a regular supply of electricity in the south during times of relative drought. Correspondingly, electricity generated by thermal plants in other regions could be supplemented by excess cheap hydroelectric supply at the point of origin. The government's proposal was presented as a highly rational means of using German resources to supply electricity efficiently and cheaply to all of Germany.

The Reich's plan for nationalization was described as moderate and flexible. Emphasis was placed on the construction of a high-voltage network of transmission lines by the Reich government; all lines above 50,000 volts would be Reich-owned. Electricity would be fed into this all-German pool, or network, by power plants throughout Germany. Electricity would be distributed at lower voltages to local centers in all of Germany and from these to consumers—urban, industrial, rural, and railway. The Reich would gain centralized control over all major power plants by acquiring majority stock in them or by purchasing them outright, but state and local governments would be urged to share in stock ownership as well. Through centralized control, the Reich would expand and interconnect individual plants as well as existing or envisaged regional systems. In 1919 the latter included the Rheinisch-Westfälisches Elektrizitätswerk (RWE) system, a mixed private and local-government enterprise; Elektrowerke AG (EWAG), a Reich system; and the projected Bayernwerk, a Bavarian state enterprise. In his argument before the National Assembly, Treasury Minister Mayer emphasized the Reich government's desire to share ownership with the provincial governments and to be advised by them in fulfilling the plan. He compared the electrification plan to the proposal by Friedrich List, a century earlier, for a unified railway system in Germany; Mayer confidently predicted that the acceptance of the plan by the National Assembly and its successful fulfillment would demonstrate to the world the vitality of the

[58] Great Britain, Foreign Office and Ministry of Economic Warfare, Economic Advisory Branch, *Economic Survey of Germany, Section D: Fuel, Power, and Public Utility Services* (London, HMSO, 1944), p. 45.

new Germany. The law providing for the socialization of electricity was enacted on 31 December 1919.

The new law was never implemented, however. An advisory committee was charged with the responsibility of carrying out its provisions by preparing detailed plans and procedures. The committee, which included members from the Reich and state governments, as well as from mixed-ownership and private companies, met without practical result. The inactivity of the committee frustrated the plans of the various electric supply undertakings, for these, not knowing what course of action later might be decided upon by the committee, were reluctant to act at all. Under these confused conditions, Dr. A. Julius Curtius of the Economic Ministry responded in 1926 to Oskar von Miller's suggestion that the Reich government authorize and support the preparation of a technical report on the present and future condition of electric supply in Germany. Von Miller believed that a report filled with information about the technology and economics of systematic electric supply might prove to be a basis on which a unified electric supply structure could be erected. An eminent Bavarian engineer, von Miller, who had successfully planned regional electric supply systems (including the Bayernwerk in Bavaria), had proven himself adept in formulating plans that responded to the particular interests of private and government-owned utilities and to the principal objectives of a unified system.

Von Miller's report, *Gutachten über die Reichselektrizitätsversorgung,* was published in 1930. It was based on an analysis and projection of detailed information he had obtained by questionnaire from German electric supply undertakings. Von Miller envisaged a high-transmission ring, or *Ringleitung,* of 200,000–380,000 volts gathering and disseminating electricity throughout the country (see Figs. XI.7 and XI.8). The *Ringleitung* would bring the benefits of higher load factor, optimization of varied power-plant characteristics, and coordination through central control. Von Miller circumspectly advised that the objective might be achieved in several ways, such as legislation or voluntary cooperation. His report, however, brought no practical result during the Weimar years; his biographer writes that the regime did not have the "power and will" to overcome the opposition to it.[59]

Other explanations were given for the failure of plans for an all-German grid, and they, too, emphasized German parochialism. In 1935 a publication of the Wirtschaftsgruppe Elektrizitätsversorgung der Reichsgruppe Energiewirtschaft der deutschen Wirtschaft (Electric Supply Section of the German Energy Group of the German Economy) (WEV) asserted that the fruitless endeavors to form a unified electric supply system in Germany came to an end only in 1933 with the triumph of the National Socialists, a regime, the WEV believed, which had the strength to overcome parochialism, or particularism, and other obstructionist factors.[60] More recently,

[59] Wilhelm L. Kristl, *Der weiss-blaue Despot* (Munich: Pflaum, n.d.), p. 274.
[60] Wirtschaftsgruppe Elektrizitätsversorgung der Reichsgruppe Energiewirtschaft der deutschen Wirtschaft (WEV), *Die Elektrizitätswirtschaft im deutschen Reich, 1935* (Berlin, n.d.), p. 11.

Figure XI.7. *Oskar von Miller's proposed 220,000-volt grid system for Germany. From Oskar von Miller, Reichselektrizitäts-Versorgung (1930), plan 21.*

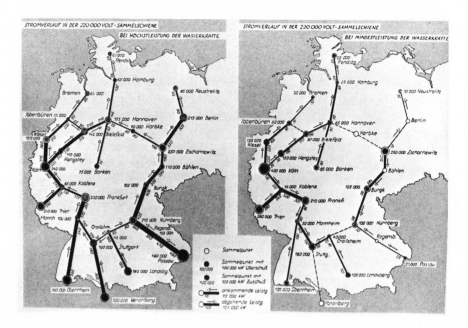

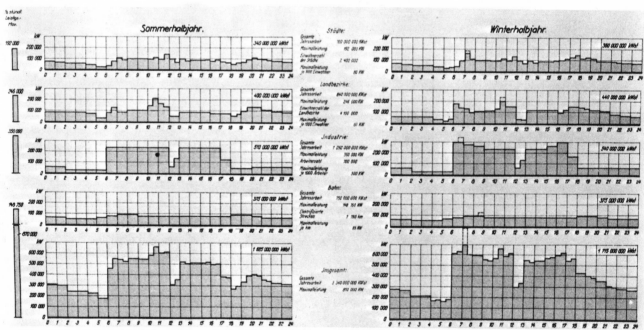

Figure XI.8. *Estimated loads for summer and winter days in Bavaria (right bank of the Rhine), 1935: Cities* (Städte); *rural districts* (Landbezirke); *industry* (Industrie); *railway* (Bahn). *From Oskar von Miller,* Reichselektrizitäts-Versorgung *(1930), plan 2.*

a German source attributed the Weimar Republic's failure to the diverse contending interests of a middle-class state, a particularism that was sustained by a complex economic structure involving private, county, state, and Reich ownership of power plants.[61]

During the Weimar period, the relative importance of government-owned and mixed government- and privately owned power plants grew rapidly. In 1928, 57.3 percent of Germany's power-plant capacity was publicly owned; 28.9 percent was controlled by mixed government- and privately owned companies; and only 13.8% was privately owned.[62] The government owners included the states, cities, counties, and the Reich. Proponents of a unified system generally agreed that the non-Reich government owners and the private capitalists were at the root of the particularism that undermined their efforts. An American authority also concluded that this ownership structure, supportive as it was of regionalism, was a principal factor blocking the fulfillment of such plans for unification as von Miller's.[63] As a consulting engineer and as a member of the Bavarian parliament, Oskar von Miller had struggled on behalf of large, unified, efficient systems of supply for many years. During that time he had become intimately acquainted with the diverse interests that might be, or feared they would be, adversely affected by the creation of a unified supply system. Writing in 1926 on the problems of the German electric supply industry, he argued that a rational plan for efficient supply to all areas of Germany, rural and urban, industrial and agricultural, needed to be carried out irrespective of the interests of various political parties, local governments, and competing private enterprises. Undoubtedly, when thinking as an engineer, he tended to be impatient with those whose goal was other than efficiency. Von Miller, it seems, considered the objective of supplying all of Germany with low-cost, efficiently generated electricity to be a rational and nonpolitical goal.[64]

Dr. Ing. Oskar Oliven, member of the *Vorstand* (Executive Committee) of the Gesellschaft für elektrische Unternehmungen Ludwig Löwe, AG, Berlin, also attributed to narrow political interests the opposition to large, unified systems of electric supply. Oliven was well known for his advocacy of a unified system that would extend beyond Germany to include neighboring European countries. In 1930 Oliven believed that the opposition of locally owned German utilities was so great and their political power in the Weimar Republic was so strong that the best tactic for those, like himself, who advocated large, rational systems was to seek, through compromises, small, progressive steps. Oliven, too, believed that his goal of efficiency was

---

[61] See Helga Nussbaum, "Versuche zur reichsgesetzlichen Regelung der deutschen Elektrizitätswirtschaft und zu ihrer Überführung in Reichseigentum, 1909 bis 1914," in *Jahrbuch für Wirtschaftsgeschichte* (Berlin, 1968), pt. 3, pp. 120–121, who cites Hans-Joachim Hildebrand, *Wirtschaftliche Energieversorgung* (Leipzig: Deutscher Verlag für Grundstaffindustrie, 1965), 1: 45–49.

[62] Robert A. Brady, *The Rationalization Movement in German Industry* (Berkeley: University of California Press, 1933), p. 214.

[63] Ibid., p. 229.

[64] "Ausführungen des Sachverständigen Dr. Oskar von Miller über die derzeit wichtigsten Fragen der Elektrizitätswirtschaft," in *Ausschuss zur Untersuchung der Erzeugungs- und Absatzbedingungen der deutschen Wirtschaft: Die deutsche Elektrizitätswirtschaft . . .* (Berlin: Mittler, 1930) p. 145.

nonpolitical. Furthermore, he assumed that a technology was progressive if it contributed to large-scale efficiency.[65]

The electric supply system in the state of Württemberg in 1928 manifested the characteristics that underlay Germany's parochialism. The region was characterized by numerous distribution systems in close proximity. In the lowlands, large and small cities predominated, and therefore communal (local-government) stations of varied characteristics prevailed. Overland power stations, however, supplied the rural regions between the small cities. Because of the agricultural economy and the absence of heavy industry such as chemical plants, a widely dispersed population used electricity in such a way that the load factor was low in comparison to the more heavily industrialized regions of the country. Efforts were made to consolidate the distribution systems of the large stations and the communal power stations, but in the late twenties the electric supply system in Württemberg was "sehr zersplittert."[66]

The owners of Württemberg (and other) autonomous utilities, whether local-government or privately owned, resisted rational reorganization because they foresaw various consequences the progressive engineers had overlooked.[67] As a result of centralized control, owners and managers of existing power plants could anticipate a loss of authority. Owners of small and inefficient power plants could reasonably expect their plants to be retired from service, a step that would cause them a financial loss, loss of authority, or both. Persons with vested interests in power plants that had a very high load factor because of fortuitous local circumstances could look forward to having to share this advantage with plants that were not so fortunate. Similarly, power plants that utilized energy which was unusually convenient or cheap would probably have to supply this very cheap power to the general system in order to lower the overall costs of drawing power from the system. The end result would be higher costs for local customers, whose new rates would be determined by system costs, not by the costs of the local station. Similarly, the system would probably supply power to rural regions, where power generation would be costly because of the widely distributed load; in effect, the large urban and industrial-area stations could end up subsidizing rural consumers. These are but a few of the changes that could be anticipated; these few, however, illustrate how the existing electrical supply technology was the foundation for a host of political and economic dependencies that would be altered by the substitution of a new technological foundation for the existing one. Such was the nature of parochialism.

Consideration of these vested interests suggests a generalization about the nature of technology: existing technologies give rise to binding nuclei for a host of dependent political and economic interests. Interwoven with

[65] "Ausführungen des Sachverständigen Dr. Oliven über die Wirtschaftlichkeit der deutschen Elektrizitätsversorgung," ibid., p. 195.

[66] "Ausführungen des Sachverständigen Dr. Siegloch über die Elektrizitätsversorgung des Landes Württemberg," ibid., pp. 373–75.

[67] The explanations for particularism are derived by analogy from Oskar von Miller's addresses in support of the Walchenseewerk and Bayernwerk as printed in *Stenographisches Bericht über die Verhandlungen der Kammer der Reichsräte des bayerischen Landtages*, 1908–18; a copy of this stenographic record is in the Bavarian State Library, Munich.

political and economic interests of particular kinds, technology is far from neutral. Like other political and economic forces, it can be labeled as conservative or liberal—or even radical. Furthermore, acute politicians, businessmen, and other decision-makers can sense—even if they do not articulate their perceptions—that a conservative technology will maintain the existing structures and trends and that liberal ones will bring changes in the direction of societal development. Manifestly, numerous persons with political and economic power in the Weimar Republic foresaw the changes that a unified electrical system would bring and they opposed these changes on the grounds of their personal or corporate interests. The technology of unified electric supply systems was by no means a neutral technology. The dice were loaded in favor of Reich authority and a national economy.

World War I also stimulated radical thinking in Britain about the peacetime reorganization of electrical supply under government direction. The British had an added incentive for bold schemes because many believed that their supply industry was backward compared to the supply systems of Germany and the United States. Even before the war ended, parliamentary committees began to propose guidelines for peacetime electricity supply. A Ministry of Reconstruction committee headed by Lord Haldane (with considerable input from Charles Merz) and two Board of Trade committees, one headed by Charles Parsons and the other by Sir Archibald Williamson (with Merz as an influential member), made similar proposals for the reorganization of Britain's electric supply industry.[68]

The Haldane-Merz Report took the offensive against parochial British interests:

Parliament was apparently convinced that the generation and supply of electricity must be dealt with in a big way, though how important this would become they perhaps hardly foresaw. They were, however, apparently *afraid* to insist on the amalgamation of the existing lighting enterprises which, as has been shown, were and are still each limited to a few miles of area instead of covering, as they should, a few counties.[69]

Merz and others battling for regional supply had to find a means of overcoming Parliament's fear.

The report demonstrated Merz's skill in appealing to British pride.[70] Experience had taught him that reorganization of the electric supply industry was a political problem, so he appealed to the mood of a people concerned about national survival. He equated the wartime struggle for survival with the struggle that would inevitably follow in peacetime—the ceaseless struggle to maintain industrial supremacy. The essence of his logic

[68] Great Britain, Ministry of Reconstruction, Reconstruction Committee, Coal Conservation Sub-Committee, *Interim Report on Electric Power Supply in Great Britain* (London: HMSO, 1918) (hereafter cited as Haldane-Merz Report); Great Britain, Board of Trade, Electric Power Supply Committee, *Electric Power Supply* (London: HMSO, 1918).

[69] Haldane-Merz Report, p. 15; italics added.

[70] For another example, see Charles H. Merz, "Electrical Power Distribution," *Engineering* 102 (1916): 262–63, where he wrote: "Proceeding on these lines [high-voltage distribution for the compact industrial districts of England] we shall not be merely copying America or Germany—we shall be doing something that is right for England because it is England, because England is radically different from other countries as regards the technical development and layout necessary to secure cheap power."

was syllogistic: the world's leading industrial power exploits the most advanced energy technology; the most advanced energy technology is regional electricity supply; and therefore, in order to lead, England must adopt regional supply. The report illustrated and provided evidence for this argument by referring to the higher level of use of electric power per worker in the United States; the vital new industrial processes involving electricity in Norway, Sweden, and Germany; and the larger supplies of hydroelectricity available to other industrial nations. The conclusion to be drawn was obvious: parochial interests had to give way to national priorities.[71]

For its part, the Board of Trade appointed Charles Parsons, the steam-turbine inventor and manufacturer, to head its "Committee on Electrical Trades after the War." The responsibility of this committee was to report on the state of Britain's electrical manufacturers and supply utilities. The committee concluded that the backward state of the latter had caused the depressed condition of the former. The Parsons committee, like Haldane and Merz, viewed the electric supply industry—both the utilities and the manufacturers—as a key to national strength.[72] A major reason for the failure of the British supply industry to develop as successfully as its American and German counterparts was thought to be vested interests and political constraints resulting from the "conventional conceptions of public privilege" represented by Parliament and the local authorities.[73] The public privileges to which the committee referred included protection against monopoly and against environmental nuisance and hazard, such as overhead wires. Considering both the manufacturing and supply ends of the industry, the committee declared that the war had demonstrated that nothing less than the well-being of the empire depended on the large-scale employment of electricity in industrial production. Referring often to Germany as a model, the committee concluded that the absence of a close alliance between Britain's industrial banks and its large-scale electrical manufacturers (a major explanation for the advanced state of electrification in Germany) had handicapped the electric supply industry in Britain.[74]

In 1917 the Williamson committee took heed and concluded that Britain needed to do more than interconnect existing utilities; it needed to develop a comprehensive system for the generation of electricity and therefore to reorganize the entire supply system.[75] The committee called for the establishment of district electricity boards to consolidate electricity generation and provide regional transmission by purchasing the most efficient of the existing power plants and by constructing new large-scale plants and high-voltage transmission systems. Distribution was to be left to the existing utilities, whether privately or publicly owned. The boards were to be non-profit, mixed-ownership corporations set up outside the civil service and

[71] Haldane-Merz Report, pp. 4, 7, and 8.

[72] Great Britain, Board of Trade, Departmental Committee, *Report of the Departmental Committee Appointed by the Board of Trade to Consider the Position of the Electrical Trades after the War* (London, HMSO, 1918), p. 5.

[73] Ibid., p. 4.

[74] Ibid., p. 7.

[75] Board of Trade, *Electric Power Supply*, p. 1.

owned in most cases by public bodies, such as local governments, but also by private enterprises, such as the private power companies.[76] Britain's six hundred or so electric supply utilities would in time be coordinated, or in many instances eliminated, by the new district systems. The committee also envisaged the creation of a small board of electricity commissioners to take over regulation of the supply industry for the Board of Trade.

The recommendations of the Parsons and Williamson committees, which represented a cross section of established political and industrial views, were not acted upon until after the signing of the peace in 1918. Peace, however, brought a muting of the values the exigencies of war had stimulated. Vested private and public interests that had been willing to relinquish authority to the national government in the interest of central coordination looked once again to the nurture of private and local autonomy. Even though the schemes of the central government had originated with technical and managerial professionals like Charles Merz and William McLellan—in short, with private industry—in the minds of many they now represented government intervention. One dissenter from the chorus of committee members calling for district electricity boards anticipated the conservative reaction to a radical reorganization of Britain's electric supply industry when he insisted that state control would lead to state management, and that state management was inherently unsound.[77]

The legislation that emerged in response to the Haldane-Merz, Parsons, and Williamson reports represented a complex compromise between innovative wartime concepts and peacetime reaction. The Electricity Supply Act of 1919 left fulfillment of the general reorganization scheme to voluntary action on the part of existing utilities and to the power of persuasion the electricity commissioners could bring to bear. The idea of district electricity boards took the form of joint electricity authorities (JEAs, an organizational concept borrowed in part from German practice). These could be established under the 1919 law if and when the authorized undertakers (private and public utilities) in a district chose to do so. The utilities within a district could turn over to the JEA all or any one of the powers the utilities held for supplying electricity. Furthermore, the utilities could transfer to the JEA their plant, transmission, and distribution facilities. In short, the JEA would be given the responsibility for coordinating and developing the district's supply of electricity if the existing utilities so decided.[78]

On the national level, five electricity commissioners were to be appointed by the Board of Trade, the government department that had regulated and authorized electricity supply since the 1880s. The Electricity Supply Act of 1919 stated that the electricity commissioners should exercise the powers conferred upon them by the law, but that they "shall act under the general directions of the Board of Trade." Some reformers who advocated more radical reorganization had wanted the commissioners placed under the jurisdiction of a department other than the Board of Trade, with which the history of Britain's electricity supply was closely associated. The general charge given the commissioners was promotion, regulation, and supervision

[76] Hannah, *Electricity Before Nationalisation*, pp. 65–66.

[77] Ibid., p. 66.

[78] Electricity Supply Act of 1919, 9 & 10 Geo. 5, ch. 100, § 6.

of the supply of electricity as provided by the several electric lighting acts. The law specified that three of the commissioners should be full-time officers and three should be selected for their practical, commercial, and scientific knowledge.[79]

The third major provision of the 1919 act stipulated that the electricity commissioners could designate electricity districts after holding a local inquiry and providing ample opportunity for advice from and consultation with all parties concerned—public, private, and large consumer. The intent was to establish supply regions that had good load diversity, higher load factors, an economic mix of generating stations, and a coordinated and economical system of transmission and distribution. These regions would not be coextensive with existing electricity districts. The electricity commissioners would have the power to require that utilities facilitate coordination by standardizing the type of current, frequency, and voltage they used. The Board of Trade, however, could exempt utilities when it decided that compliance with such an order entailed unreasonable expense.[80]

Even though the 1919 legislation left ultimate power over the utilities in the hands of the Board of Trade and its general mode was voluntary action and persuasion, the reaction against it was immediate and substantial. The local-government utilities had habitually opposed reorganization plans that would create electricity supply on a scale larger than the geographical limits of their political authority. In the case of the 1919 act, however, there was some ambivalence among them. They realized that a JEA could usurp the power enjoyed by an autonomous utility; on the other hand, a JEA, if controlled by the local authorities in a district, could frustrate the rise of private regional power companies that had hitherto languished because the municipalities refused to buy power wholesale from them and thereby left them the more expensive task of supplying the less industrialized and less populated districts.

By the mid-1920s the optimism expressed in the early reports of the electricity commissioners had given way to the admission that their efforts were meeting with limited success. Interminable public inquiries and negotiations among undertakings, in accordance with the 1919 act's insistence on persuasion and voluntary cooperation, were part of the problem. Utility engineers and managers resisted the closing of inefficient stations and the transfer of technological control of joint electricity authorities when this meant loss of authority and status. Small companies, both public and private, especially those associated with a local government jealously protecting its autonomy, fought absorption by larger ones. Municipal utilities continued to refused to buy bulk or wholesale power from private power companies. Ingenious arguments were presented to defend the generation of power within small, local-government-owned utilities. It was said that governments could borrow money at lower interest rates than private power companies could, and that they thereby offset the technological efficiencies the large power companies might claim. In general, "[the] divisive tendencies inherent in the ownership structure of the industry had effectively reasserted themselves."[81]

[79] Ibid., § 1.
[80] Ibid., § 5 and § 24.
[81] Hannah, *Electricity Before Nationalisation*, p. 85.

The history of electric supply in the Greater Lancashire area after World War I illustrates all the inertia, divisive tendencies, and parochialism that frustrated the newly appointed electricity commissioners. As in London, electric supply in Greater Lancashire was notoriously fragmented. Large and small municipalities guarded their autonomy. The Power Securities Corporation, a holding company, controlled the bulk supplier in the area, the Lancashire Power Company. Wartime brought some interconnection, and the engineers in the area agreed in 1917 that central control and planning for the area were needed. Thirty-five supply authorities, however, rejected the engineers' proposal because mixed ownership would have resulted. The advantages of mixed ownership as practiced in Germany, especially by the Rheinisch-Westfälisches Elektrizitätswerk, did not override British objections. After 1919 the electricity commissioners advocated a joint electricity authority for the area, but the utilities rejected the proposal, preferring to remain independent. In the end, only a weak advisory board and piecemeal interconnections were achieved.[82]

To observers considering the long history of tense and frustrating efforts to reorganize Britain's electric supply industry, the remarkable turn of events that led to the establishment of the Grid in 1926 came as a distinct surprise. The first question that arises is, What confluence of trends and events countered the existing inertia and changed the course of development of electricity supply in Britain? The question takes on added significance when one recalls that many persons considered reorganization of this industry to be the key to Britain's recovery of industrial strength and leadership. The story of this reorganization will be told in the following chapter.

[82] Ibid., p. 82.

CHAPTER XII

Planned Systems

T HE radical change in needs and attitudes brought by World War I stimulated bold concepts for peacetime projects, but as has been seen, the postwar return to normalcy dampened enthusiasm for government-sponsored technology in the United States and technology controlled by the central government in Britain and Germany. Private ownership in the United States, parochial control in Britain, and local or regional development in Germany regained their prewar momentum. There were new technological opportunities to utilize, however, and the more traditional owners and managers of electric supply facilities sought to do so. As a result, innovations were made, but in the United States and Germany the changes were less radical than Pinchot and Cooke's Giant Power proposal and Germany's Socialization Law of 1919. By contrast, in 1926 a major change in direction was brought about in the development of power supply in Britain.

The developments that took place in electric supply systems in the 1920s are comparable to those that occurred in railway systems in the second half of the nineteenth century. Major railroad systems were then interconnected and standardized with respect to gauge and equipment. As a result, the major traffic centers and routes of regional and national systems were identified. The principal routes were then upgraded, traffic nexus and switching yards were laid out, and the trunk lines of the regional and national systems were further developed.

The post–World War I rationalization of electric supply took many forms, but it is generally categorized as having been either evolutionary or planned. In a subsequent section, consideration will be given to the way in which the well-established, large-area utilities with relatively long histories evolved into regional systems integrated by high-voltage transmission networks. In this section, emphasis is placed on the new systems that were established according to master plans. The purpose of these new systems was to knit together, on a regional scale, utilities that had formerly evolved independently. The planned networks, or grids, usually took the form of high-voltage lines ringing a supply region, or polygon, the sections of which met at major load centers. They differed from the evolving networks, which, because of the generally less orderly character of historical change, usually

324

were more complex in form. The planned grids represented the pooling of energy from utilities that preserved their legal identities, primarily as distributors of the pooled energy. In some instances, a separate corporate entity owned and managed the grid; in others, the utilities presided over the grid, or pool, using a committee structure. Because the participating utilities preserved their corporate identities and because they often negotiated decisions about the operation of the power pool, such grids were analogous to confederations of nation-states. Under some plans, the utilities or power companies fed the pool from their own power plants. Under others, the grid took power both from its own plants and from the plants of participating utilities. Here we will focus on three variations upon the essential grid, or pool, concept: the Pennsylvania–New Jersey Interconnection (PNJ), the center of which was Philadelphia, Pennsylvania; the Bayernwerk in Bavaria; and the National Grid in Great Britain. In the case of both the Pennsylvania–New Jersey Interconnection and the Bayernwerk, large hydroelectric plants fed the grid's pool of power. Utilization of such large blocks of hydroelectric power was part of the original plan for each of these systems; in contrast, peacetime utilization of the giant networks at Muscle Shoals, Alabama, and Golpa-Zschornewitz was an afterthought.

The PNJ was clearly a more conservative response to technological opportunities than Giant Power had been. The history of the PNJ pool and related hydroelectric power plants also demonstrates that the builders of these giant electric power systems faced organizational and financial problems that demanded solutions as original as the technology itself, or even more so. Managers and financial entrepreneurs would make commitments that were as bold as those made by engineers.

One such commitment was made on 20 February 1926, at about the time Pinchot and Cooke's Giant Power project was voted down in the Pennsylvania legislature. The timing was not entirely fortuitous. On 20 February the Philadelphia Electric Company, long an adversary of Cooke's, obtained a license from the Federal Power Commission to build a hydroelectric plant on the Susquehanna River at Conowingo, Maryland. The plant was to become one of a number of plants that fed power to the interconnected Pennsylvania–New Jersey transmission grid. The state of Pennsylvania had lost the opportunity to lead the nation in the development of government-sponsored electric supply systems, but a privately owned Pennsylvania utility did take the lead in devising an innovative mode of exploiting interconnection and high-voltage transmission. The FPC license authorized the Philadelphia power company to build on the Susquehanna River at Conowingo the nation's second-largest hydroelectric project; the construction of Conowingo set the stage for the formation, by interconnection, of the nation's largest power pool.

The Conowingo project provided a means of controlling and utilizing the forces of nature, which is the traditional function of technology. For more than half a century, various schemes had been formulated for harnessing the Susquehanna as it gradually dropped about 150 feet in the last twenty-five miles before it emptied into the head of the Chesapeake Bay.[1]

---

[1] C. W. Kellogg, "The Conowingo Hydro-Electric Project," *Stone & Webster Journal*, April 1926.

In 1912 the Pennsylvania Water & Power Company had erected a hydroelectric plant at Holtwood to harness 60 feet of the fall, but an annual power output of about 1,280,000,000 kwh. remained to be developed.

If the flow of the river had been regular, the water could have been regularly drawn off to energy-intensive industries attracted by cheap electricity. The flow had been recorded for years, however, and seasonally and unpredictably it varied greatly. At the time of the famous Johnstown flood of 1889, 750,000 sec.-ft. were measured; on another occasion, a low of 2,200 sec.-ft. was recorded. These natural variations suggested the need for a power system that would connect a hydroelectric plant at Conowingo with existing steam plants. In times of low water or drought, the thermal plants would carry the base load, and Conowingo would carry the peak; in times of high river-flow, the base load would be transferred to Conowingo, and the thermal plants would absorb the peak.

The Philadelphia Electric Company, which in 1921 was the nation's seventh-largest utility, planned to feed the power of Conowingo into a grid formed by interconnecting Philadelphia Electric with two neighboring utilities, the Public Service Electric & Gas Company of New Jersey and the Pennsylvania Power & Light Company (PP&L), which operated in central and eastern Pennsylvania. Not only would the market for Conowingo be expanded but the diversity of load within the load area with which it was connected would increase. (See Figs. XII.1 and XII.2.) The technology for the Conowingo plant and the interconnection scheme was available; the project's planners introduced management and financial innovations to nurture their particular adaptation of the technology.

The interstate character of the project complicated the planning, however. The dam and power plant would be located in Maryland; the upper half of the reservoir, in Pennsylvania; and the transmission system, in Pennsylvania, New Jersey, and Maryland. The dam and power plant required action by the public service commissions (the regulatory agencies) of both Maryland and Pennsylvania. Because the Susquehanna River was navigable, the hydroelectric installation also needed a license from the Federal Power Commission. The Philadelphia Electric Company responded to the regulatory and licensing challenge by forming the Philadelphia Electric Power Company, a Pennsylvania corporation, to hold title to the Pennsylvania property outside Philadelphia Electric's domain. The Susquehanna Power Company, a subsidiary and a Maryland corporation, would hold title to the dam, powerhouse, and Maryland portions of the reservoir and transmission system. Because they would not supply consumers directly, these power and transmission facilities would not be subject to state regulation, as were the utilities.

The joint hearing held by the Pennsylvania and Maryland public service commissions and the Federal Power Commission was eventful. Opponents of Conowingo focused on the financial aspects of the project. Most vociferous was Thomas E. Mitten of the Philadelphia Rapid Transit Company, a company that bought power from Philadelphia Electric. Mitten's lawyer contended that Philadelphia Electric had paid an exorbitant price for the franchise on the Susquehanna and that interest rates and other costs would burden Philadelphia Electric's customers. Mitten insisted that his primary concern was to avoid the resulting increase in rates for his company and

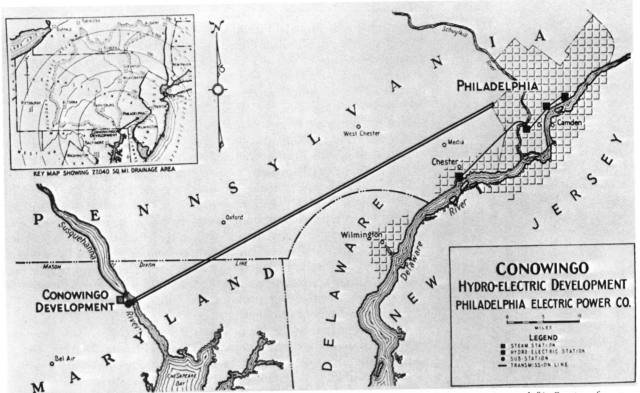

Figure XII.1. *Transmission system, Conowingo Dam to Philadelphia. Note the Susquehanna River drainage area (upper left). Courtesy of Philadelphia Electric Co.*

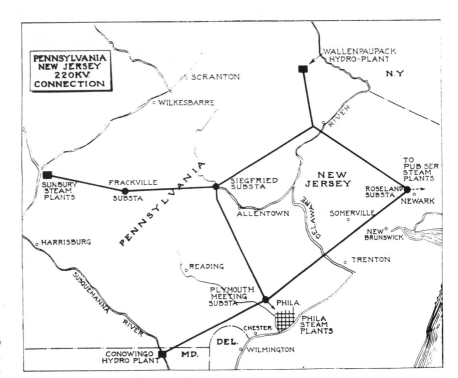

Figure XII.2. *The Pennsylvania–New Jersey Interconnection. Courtesy of C. W. Watson, Philadelphia Electric Co.*

other major customers; others, however, argued that his objective was to gain control of Philadelphia Electric by discrediting the company's management. Mitten did solicit Philadelphia Electric voting proxies for himself and his associates while Philadelphia Electric was under attack.[2] The alleged takeover attempt failed. The Conowingo project was supported by Philadelphia Electric's stockholders and was licensed to the extent necessary by the regulatory agencies, but Mitten's charges relating to "interest rates and allied matters were largely sustained by the Commissions."[3]

Drexel and Company, stockbrokers of Philadelphia, then formed a syndicate with Brown Brothers and Harris, Forbes & Company to market the Philadelphia Electric Power Company's bonds, which were valued at $36 million. The allocation of the subsidiary's preferred stock to Philadelphia Electric shareholders raised an additional $12 million. Drexel and Company, however, had acquired the option on the Conowingo property several years before transferring it to Philadelphia Electric for financial considerations. Critics of Drexel and the utility pointed out that by acting prudently and decisively, Philadelphia Electric could have acquired the option directly. Mitten's supporters made the "totally unwarranted" allegation that Drexel had bribed a vice-president of Philadelphia Electric to facilitate Drexel's involvement in the acquisition.[4]

Gifford Pinchot, whose Giant Power project had been voted down "by vigorous opposition of the companies" days before this hearing, also challenged the Conowingo project.[5] He feared that the Federal Power Commission would allow Philadelphia Electric to issue bonds in excess of the net amount invested in the project. Then, he believed, the company would base its profit and regulated rates on an inflated valuation of Conowingo. Pinchot also opposed allowing control of the project by Philadelphia Electric Company owned common stock, favoring instead control by the nonvoting preferred stock held by investors. In a letter to the Federal Power Commission he stated categorically that voting power and the major investors' risk should not be separated. The executive secretary of the Federal Power Commission informed Pinchot on 27 February that his agency had recommended limitation of the total par of securities to the amount of the net investment in the construction of the project and had denied the issuance of no-par common stock. Pinchot called the decision "a great step in public utility regulation for Pennsylvania."[6]

With the hearing over and the project licensed, the engineer-managers at Philadelphia Electric must have been relieved to proceed beyond what they pejoratively called "politics." As was customary for private utilities, Philadelphia Electric turned to consulting engineers for assistance in carrying out a project that was beyond the technical expertise of the company. Stone & Webster, the Boston consulting engineering firm, designed and

[2] Nicholas B. Wainwright, *History of Philadelphia Electric Company, 1881–1961* (Philadelphia: Philadelphia Electric Co., 1961), p. 175.

[3] Ibid., p. 177.

[4] Ibid., p. 176.

[5] Pinchot and Federal Power Commission correspondence, 13 February to 8 March 1926, printed in *Report of the Giant Power Board to the Governor of Pennsylvania, 7 December 1926* (Harrisburg, Pa., 1927), p. 13.

[6] Ibid., pp. 100–103; quotation on p. 103.

supervised construction of the hydroelectric facility, and Day & Zimmermann, another consulting firm, installed the transmission lines. The Arundel Corporation of Baltimore constructed the main body of the solid concrete dam. Construction of the hydroelectric installation began in March 1926; on 1 March 1928 power was transmitted to the Plymouth Meeting substation northwest of Philadelphia, a major facility in the new power pool.

The principal engineer for Conowingo—the man who had enthusiastically spearheaded the undertaking—was W. C. L. Eglin. After studying engineering in Glasgow, in 1889 he joined the Philadelphia Edison Company, a predecessor of Philadelphia Electric. Described as "testy, impulsive, and very sure of himself," Eglin is remembered as "an aggressive, brilliant worker" who modernized and unified Philadelphia Electric's system.[7] As early as the spring of 1922 he undertook preliminary studies and commissioned special reports on the Conowingo site. He encouraged Drexel and Company to take the option on the franchise when his own company declined the offer. During the construction phase, as chief engineer he coordinated the contractors and pressed to keep the work on schedule. Only three weeks before the project was completed, however, he died at the age of fifty-seven. Having begun as a dynamo-tender, he was vicepresident and director of Philadelphia Electric at the time of his death.

Construction of the Conowingo plant was presented to the public as a heroic manifestation of America's powerful building instincts and technological strength (see Fig. XII.3). The removal of a railroad, the rerouting of the main Baltimore-to-Philadelphia highway across the crest of the dam, and the displacement of the village at Conowingo—changes made necessary by the damming of the Susquehanna—were major challenges. Moreover, a temporary town had to be built to house the thousands of construction workers. Publicity releases boasted that if set up on Pennsylvania Avenue in Washington, the dam would fill the avenue from curb to curb, reach the height of a ten-story office building, and extend from the Capitol grounds to the Raleigh Hotel.[8] The size of the power plant building matched that of the Equitable Building in New York City, in which 15,000 persons worked; the turbines in the power plant were the largest ever built. Even Philadelphia, then called the "Workshop of the World," would have had a hard time absorbing Conowingo's annual power output.

Conowingo was comparable to the widely publicized Niagara Falls and Muscle Shoals projects (see Table XII.1), but it needed a regional load and a power pool. Eglin used load curves to make this point. In the first, which was drawn for winter conditions, the crosshatched area represents the base load Conowingo would carry. The lower load curve, the demand of Philadelphia Electric, dips below Conowingo's capacity in the early morning hours. Thus, the capacity of Conowingo indicated by double crosshatching would be lost. If, on the other hand, Conowingo were integrated into a power pool (whose load curve also is shown), then all of Conowingo's capacity could be used. Another set of load curves shows the advantages of interconnection during the summer months, when, because of slower

---

[7] Wainwright, *Philadelphia Electric*, p. 69.

[8] Kellogg, "The Conowingo Hydro-Electric Project," p. 2.

Figure XII.3.   *Conowingo Dam under construction, 1927. Aerial photo by Victor Dallin. Courtesy of the Eleutherian Mills Historical Library, Wilmington (Greenville), Del.*

river flow, Conowingo would carry only the grid's peak load. Even in the months when water flow was low, Conowingo could be utilized more fully as part of a utility pool than if it were operated with only a load from Philadelphia Electric.[9]

As noted, by means of a high-voltage grid, the systems of Philadelphia Electric, Public Service Electric & Gas of New Jersey, and Pennsylvania Power and Light were interconnected. In 1923, when there was much talk about superpower and Giant Power and an "atmosphere of State and Federal involvement,"[10] the three utilities initiated planning for the hookup.

[9] W. C. L. Eglin, "Symposium on Interconnection: Conowingo Hydroelectric Project, with Particular Reference to Interconnection," *AIEE Transactions* 47 (1928): 380.

[10] "Comment on 'The Osgood Report,'" an undated memorandum in the Philadelphia Electric Co. Archives, Philadelphia, Pa., p. 1.

TABLE XII.1. CONOWINGO IN COMPARISON

Plant	Owner	Installed Horse Power
Niagara Falls	Niagara Falls Power Co.	452,500
Conowingo	Philadelphia Electric Power Co.	378,000
Muscle Shoals	United States Government	260,000
Holtwood	Pennsylvania Water & Power Co.	158,000
Keokuk	Mississippi River Power Co.	150,000

	Conowingo	Muscle Shoals
Area of watershed	27,000 sq. mi.	30,500 sq. mi.
Average river flow	40,000 c.f.s.	51,900 c.f.s.
Length of dam	4,700 ft.	4,500 ft.
Area of reservoir	8,600 acres	8,640 acres
Height of dam (river bed to top of bridge)	105 ft.	140 ft.
Concrete	660,000 cu. yd.	1,350,000 cu. yd.
Spillway length	2,385 ft.	3,050 ft.
Flood capacity	880,000 c.f.s.	950,000 c.f.s.
Initial installed capacity	378,000 h.p.	260,000 h.p.
Number of units initially	7	8
Capacity of each unit	54,000 h.p.	32,500 h.p.
Foundation	granite	blue limestone

Source: Kellogg, "The Conowingo Hydro-Electric Project," pp. 7 and 9.

They employed Professor Malcolm MacLaren of Princeton University to make a study of load diversity in the area each utility served. MacLaren's favorable report led the three companies to form a committee to study interconnection. The committee issued its report on 1 August 1925, and on 16 September 1927, while Conowingo was under construction, the three companies signed an agreement establishing a new entity, the PNJ Interconnection. Thus, the world's largest integrated, centrally controlled pool of electric power (1.5 million kw.) came into being.[11]

The agreement called for a ring of 220,000-volt trunk lines 210 miles in length with two transmission lines from each participant's system to the other two. The trunk lines were of high capacity, which distinguished the interconnection from those in which small tie lines only loosely connected systems. Construction and costs were apportioned among the utilities. The Public Service Electric & Gas Company of New Jersey constructed a switching station at Roseland; a 220-kv. line from Roseland to Bushkill, Pennsylvania; and another from Roseland to the Pennsylvania–New Jersey line near Lambertville. Philadelphia Electric built a substation at Plymouth Meeting, to which transmission lines from Conowingo were extended, and a trunk line from Plymouth Meeting to Towamencin Township, Pennsylvania; it also constructed a 220-kv. trunk line to Lambertville to connect with Public Service. Pennsylvania Power & Light erected a trunk line from Towamencin Township to its Siegfried substation. An existing Pennsylvania Power & Light trunk line extended to Bushkill, the site of PP&L's Wallenpaupack hydroelectric plant (see pp. 440–41 below). The ring began service in 1930. The last leg, between Roseland and Bushkill, was delayed by right-of-way difficulties; the Plymouth Meeting–Siegfried section, the first to be completed, began operating on 10 February 1928.

[11] Eglin, "Symposium on Interconnection," p. 381.

Power from the hydroelectric and thermal generation stations of the three utilities fed into the interconnection ring at the various substations (Conowingo, for example, at Plymouth Meeting). The utilities also took bulk power from the interconnection at the substations; they agreed not to tap the ring directly for customers. The agreement also called for the extension of the Plymouth Meeting–Towamencin-Siegfried trunk line to PP&L's Frackville substation, which would make it possible for the ring to draw power from the thermal station that was planned at the coal mines near Sunbury. The ring and its connections therefore combined diverse but complementary power sources.[12]

Load dispatching was of central importance to the power pool, as it had been to the separate utilities earlier. The Depression frustrated plans for a PNJ control center at PP&L's Siegfried substation, so Philadelphia Electric dispatched the power generated by the three companies. The central, or PNJ, dispatcher was authorized to direct the three utility dispatchers when switching was required. Philadelphia Electric kept load records and from this and other information predicted load variations. Prediction facilitated quick responses to changing conditions as well as optimum use of the steam and hydroelectric plants feeding into the transmission system. "Only the most alert and skilled load dispatching," Eglin observed, "will realize all of the possible benefits [from interconnection]."[13]

The integrated and centrally controlled PNJ power pool differed from the mergers and utility holding-company structures that were then proliferating. In concept, it can be compared to the confederations of nineteenth-century railroad systems that were formed to facilitate cross-country traffic cooperation. The utility managers and engineers who operated the power pool began to see the PNJ as electrically one company, but financially and organizationally a committee of peers negotiating planning and operations.[14] The PNJ brought the economic benefits of a large system and at the same time preserved the utilities' corporate identities.

Operations by committee introduced problems, but according to utility executives familiar with the history of the PNJ, the record has been one of constructive, realistic negotiations within a broad context of general understanding and trust founded on performance. This interpretation has been sustained by the continued growth of the system, which in 1956 was renamed the Pennsylvania–New Jersey–Maryland Interconnection and which in 1981 coordinated the bulk power supply of eleven "investor-owned electric utilities." With a capacity of 45,693,200 kw., the pool comprised roughly 8 percent of the nation's total capacity in 1975.[15]

The general economic guidelines for operating the interconnection were

---

[12] Copy of agreement entered into on 16 September 1927 by the Public Service Electric & Gas Co., the Philadelphia Electric Co., and the Pennsylvania Power & Light Co. A copy showing the changes that went into effect on 1 January 1936 is on file in the Philadelphia Electric Co. Archives.

[13] Eglin, "Symposium on Interconnection," p. 377.

[14] Interview with Ted Fetter and Charles Watson of the Philadelphia Electric Co., 18 May 1976.

[15] *Pennsylvania–New Jersey–Maryland Interconnection*, a booklet prepared by the management of the pool (1976?); and memorandum of Wilmer Kleinbach, PJM manager, 30 December 1981.

relatively easy to articulate. A major benefit arose from load diversity. Throughout the year the Pennsylvania Power & Light Company had peak loads in the morning, for it supplied coal mines and other industries; the other two utilities, the "seaboard" companies, experienced their largest loads in the evening. Not only did the daily curve differ, but PP&L experienced its yearly peak on an October morning, while for the other two utilities it came on a December evening at about 5:00 P.M., when factories were still operating, lights were coming on, and commuters were on their way. Each day, therefore, diversity in loading, or in peaks, encouraged the exchange of power in order to maximize the use of capacity. The seaboard companies could dispatch less capacity during the December peaks, and the PP&L could use less to cover October mornings. Despite the early emphasis on the benefits of load diversity, however, the operators of the pool soon found the benefits arising from the application of principles of economic mix, or the exchange of economic power, to be as—or more— notable.

The equitable allocation of interconnection benefits among pool members demanded patience and mutual interest. Immediately following the signing of the PNJ agreement in September 1927, each utility named one representative to an operating committee. Initially, the committee included a general manager from Public Service Electric & Gas, an assistant chief engineer from Philadelphia Electric, and the general superintendent from Pennsylvania Power & Light. In its meetings over the next decade the committee focused on forecasting interconnection loads and planning generating capacity and construction.[16] The negotiators, or planning engineers, established the "scheduled," or predictable, diversity and discussed the reaction to and use of "call," or unpredictable, diversity. They determined what charges would be made for bulk power exchanges. They also agreed to the allocation, over a period of years, of scheduled diversity. Only with this knowledge could the individual utilities plan the construction of new plants to provide the capacity each needed to meet and maintain a reserve for reliability. Compromises based on mutual interest were necessary; unilateral tactics could have resulted, for instance, in each demanding power from the others to meet its own peaks and thereby avoid investing in new plants. Instead, the utilities relied on predictable diversity among pool members. The price charged for bulk power exchanges was set both to compensate the utility which had the excess capacity and to provide savings for the utility which drew from the interconnection to meet its deficiency. Because of their different operating principles, the utilities also had to agree on the amount of reserve capacity that was necessary to ensure the reliability of the interconnection and on the cost-accounting methods that would be used to determine the price of power exchanges. The success of the operating committee in solving these and other extremely complex problems and in demonstrating the economies to be derived from interconnection served as a model for other utility confedera-

[16] Minutes of the Interconnection Operating Committee, 1927–56, and related correspondence, the Philadelphia Electric Co. Archives. I am indebted to the management of Philadelphia Electric, especially Mr. Wayne Astley, vice-president, for the opportunity to consult these minutes.

tions and for cooperation among peers in general.[17] Contemporaries recall that during the early years of its operation, managers and engineers from the world over came to inspect the PNJ grid.[18]

These managers and engineers also took note of the Walchenseewerk, Europe's largest hydroelectric plant, when in 1924 it began to power the Bayernwerk, the German transmission grid, which, like the PNJ, was presided over by a corporate entity and involved the cooperation of several utilities. This development was noteworthy not only because the technology involved was complex and advanced but also because a destructive war had only recently ended and the effects of disruptive inflation had just begun to subside. It was of considerable interest as well because Bavaria, the province being electrified, was not a highly industrialized region compared to the Ruhr, the northeast coast of England, or the northeastern seaboard of the United States. The 1924 interconnection marked the culmination of activities associated with two phases of the history of electric power systems—the time of intensive development of hydroelectric plants and the subsequent period of the power pool, or grid. The Walchenseewerk project was first conceived as part of a plan to bring the water power of the Bavarian Alps to heavily populated industrial transportation centers like Munich. It can therefore be compared to the earlier projects that carried the power of the Sierra Nevadas to the California coast. The plan to integrate the Walchenseewerk into the Bayernwerk, a regional transmission grid, in order to obtain an economic mix of power sources and a good load factor ranks alongside the similar project, the Conowingo Dam and the Pennsylvania–New Jersey Interconnection. Furthermore, both the PNJ and the Walchenseewerk-Bayernwerk interconnection were planned regional electrification projects, not regional electrification systems that evolved over a long period of time, as was the case in Rhineland-Westphalia, eastern Pennsylvania, and on the northeast coast of England (see Chapter XIV).

The role that consulting engineers and their firms played during the era of planned regional electrification was critical. By focusing on the activities of Oskar von Miller, one can uncover the complex origins and history of the Walchenseewerk-Bayernwerk interconnection.[19] Like some other leading consulting electrical engineers, von Miller was an engineer-entrepreneur. His experience and competence extended beyond designing, developing, and constructing devices, machines, and works. He was an expert

[17] See pp. 370–71 above for a general discussion of the economic benefits of interconnection—the advantages that accrued to power pools and enterprises as a result of mergers or acquisition.

[18] Letter to the author from M. D. Hooven, consulting engineer, Public Service Electric & Gas Co., 11 February 1971. I am also indebted to Professor John Brainerd of the Moore School, University of Pennsylvania, for information about the early years of the Pennsylvania–New Jersey Interconnection. Bayla Singer of the University of Pennsylvania is completing a dissertation on the history of the Pennsylvania–New Jersey–Maryland Interconnection (formerly the PNJ).

[19] Biographies of Oskar von Miller include Wilhelm L. Kristl, *Der weiss-blaue Despot* (Munich: Pflaum, n.d.); Walther von Miller, *Oskar von Miller: Nach Eigenen Aufzeichnungen, Reden und Briefen* (Munich: Bruckmann, 1932); Ludwig Nockher, *Oskar von Miller* (Stuttgart: Wissenschaftliche Verlagsgesellschaft, 1953); and Theodore Heuss, *Oskar von Miller und der Weg der Technik* (Munich: Deutsches Museum, 1950). I am indebted to Rudolf von Miller for the many enlightening conversations I had with him about his father.

in such endeavors, but he also presided over the activities that brought them into use; he was an innovator as well as an engineer. In order to innovate, or bring technology into use, von Miller solved multifaceted problems involving geography, economics, politics, institutions, social values, and many other factors. The character of the problems varied endlessly. Thus, the expression "engineer-entrepreneur" is used to characterize him.

The history of von Miller's connection with the Walchenseewerk and Bayernwerk projects extends from 1903 to 1924. During part of these two decades, von Miller was an appointed member of the Bavarian Upper House and a state commissioner for electrical development in the post–World War I Bavarian government. He functioned, therefore, at the interface between technology and politics, a highly advantageous position for an entrepreneur of large-scale projects. In these years, he also managed an expanding consulting engineering firm that received a number of large electrification commissions.[20] Inventor-entrepreneurs like Thomas Edison were needed when the essentials of an electric lighting system were being invented and developed; in the United States, manager-entrepreneurs like Samuel Insull played leading roles around the turn of the century, when the management of complex, integrated, centrally controlled urban electric light and power systems was the crucial problem. Von Miller, with his encompassing hedgehog style, was their successor.

Because of his bold and effective advocacy of long-distance power transmission and regional electrification, von Miller had earned an international reputation. Born in 1855, the son of Ferdinand von Miller, founder and sculptor for the Wittelsbach dynasty of Bavaria, Oskar von Miller studied civil engineering at the Munich Polytechnikum. His first professional position was as an apprentice civil engineer in the Bavarian state service, but a visit to the Paris International Electrical Exhibition in 1881 fired his imagination, as it did that of so many others, and precipitated a lifelong commitment to electrical engineering. When he was only twenty-seven, von Miller played a leading role in the organization of the Munich International Electrical Exhibition, which was modeled after the one in Paris. He displayed a special interest by arranging for Marcel Deprez (1843–1918), the French electrical inventor and engineer, to demonstrate the feasibility of long-distance power transmission. Deprez, a pioneer in transmission, used 1,500–2,000 volts of direct current to transmit power from Miesbach over 57 kilometers to the exhibition's Glass Palace in Munich. At Miesbach a 1.5-h.p. steam engine drove a generator; at Munich a .4-h.p. Gramme electric motor drove a pump to create a dramatic waterfall.[21] If von Miller and Deprez had been able to drive the generator at Miesbach by water power, the exhibit would have been even more striking—water power to water power by electrical transmission. Showing remarkable prescience, von Miller joined Emil Rathenau in 1883 in Berlin as a manager of the Deutsche Edison-Gesellschaft (see p. 67 above). In 1890 he returned to

---

[20] On the history of Oskar von Miller's consulting engineering firm, see *Oskar von Miller: Im Dienste der Energiewirtschaft* (Munich: Ingenieurburo Oskar von Miller GMBH, 1955).

[21] *Offizieller Bericht über die im Königlichen Glaspalaste zu München 1882 . . . Internationale Elektrizitäts-Ausstellung*, ed. W. von Beetz, O. von Miller, and E. Pfeiffer (Munich: Autotypie Verlag, 1882), esp. pp. 99–107.

Figure XII.4.   *Oskar von Miller and Thomas Edison. Courtesy of Rudolf von Miller.*

Figure XII.5.   *Oskar von Miller. Charcoal drawing by F. A. Kaulback, 1911. Courtesy of the Deutsches Museum, Munich.*

Munich to establish his consulting engineering firm, Technisches Bureau Oskar von Miller. In 1891, on the occasion of the International Electrical Exhibition in Frankfort on the Main, he again demonstrated his enthusiastic commitment to electrical power transmission (see p. 131 above).

The Munich and Frankfort demonstrations were point-to-point transmissions of power. Von Miller also took the lead in introducing regional electrification by means of network, or ring, systems. Between 1910 and 1912 he planned the Pfalzwerk, a power company that supplied electricity throughout the German province of the Pfalz by means of a 110,000-volt system. At the heart of the system was a new and large coal-fired generating station in Homburg. Existing small generating stations also fed into the transmission system. The Pfalzwerk AG was an early example of a mixed-ownership, regional transmission system organized about a large, efficient, coal-fired power station.[22] The German word for this type of station was *Überlandzentrale*.

With his experience in power transmission and regional electrification, von Miller turned his attention in 1910 to a project he had envisaged for years—developing the water power of the Bavarian Alps. The project demanded consideration of numerous factors and contingencies other than purely technical ones. The most obvious and pressing was the geographical environment. In 1903 von Miller had described the geographical constraints of the region.[23] The northern slope of the Alps, including the part located within the then Kingdom of Bavaria, had for centuries provided at-the-site water power. Fast-flowing streams and rivers rushed down the precipitous incline into the Isar, Danube, and other German rivers. Wooden water wheels housed in mill buildings supplied the power to grind grain, push bellows, raise hammers, saw wood, and perform other tasks. These power plants, however, had to be located on the streams themselves, and their energy had to be used there. Writing in 1903, after electric power transmission had been proven, von Miller proposed transmitting this power to sites where industry and population were concentrated.

Drawing upon information gathered by the Bavarian Hydrotechnical Bureau, which had been established in 1899, von Miller presented a tabular and cartographic analysis of available water power on the northern slope of the Alps. Specified stretches of streams and rivers were measured for fall in elevation; the average water flow was established as precisely as possible; and from this and other data the usable water power was estimated. For example, a 35-kilometer stretch of the small river Loisach, from Partenkirchen to the Kochelsee, fell 120 meters; the river's average flow was 6 cubic meters each second; thus, according to von Miller's calculations, its usable water power amounted to 7,700 h.p. The estimate was qualified according to seasonal climate. As a planning document, the report anticipated a twentieth-century genre.

The most attractive water power site was the Walchensee, a natural lake whose surface was 200 meters above the great plateau extending 60 kilometers northward from the Alps to Munich. Only a rock wall separated the Walchensee from the flat land; the bottom of the lake behind this natural barrier reached to the level of the plateau. The situation invited development, for a tunnel could be dug through the rock at a point a small distance below the surface, and large pipes, or penstocks, could then carry the water rushing down the slope to turbines in a power house nestled against the slope on the plateau. To replenish the Walchensee, streams high in the Alps could be partially diverted. To complete the cycle, once the water passed through the turbines, it could be fed into the Kochelsee, which was located on the plateau near Benediktbeuern, and from there into the Isar River, which rushed across the land to Munich and eventually to the Danube.

[22] *Oskar von Miller: Im Dienste der Energiewirtschaft,* p. 13.
[23] Oskar von Miller, "Die Wasserkräfte am Nordabhange der Alpen," *Zeitschrift des Vereines deutscher Ingenieure* 47 (1903): 1002–8.

Others besides von Miller envisaged harnessing the Walchensee. In the summer of 1904 a civil engineer from Darmstadt, Rudolf Schmick, and a Swiss engineer, Jeanjaquel, made proposals. The same year a Prussian army officer, Major Fedor Maria von Donat, approached the Bavarian government with another Walchensee project. Possibly stimulated by Donat's proposal, the Bavarian government in 1908–9 sponsored an international competition to find the best way to use the Walchensee and still preserve the natural beauty of the region.[24] Throughout the history of the project, the Bavarians showed highly sensitive concern for the impact of technology on the landscape.

Other geographical constraints included population settlements and the location of industries. The water power sites on the northern slope of the Alps were in southern, or Upper, Bavaria; the heaviest concentrations of population and industry were northward, not only in relatively nearby Munich but also in the more distant Nürnberg, Augsburg, Fürth, Erlangen, Aschaffenburg, Regensburg, Würzburg, Bayreuth, Ansbach, and Bamberg. The northern region depended on coal. Because of the immense water power of the Bavarian Alps, von Miller and others realized that transmission could extend to and beyond Munich, not only to cities but to rural Bavaria, where 75 percent of the region's population lived. Oskar von Miller insisted that Alpine water power was a resource for all Bavarians.

The burden of history also had to be weighed when von Miller began to plan his combined hydroelectric plant and regional grid. The makers of blueprints could not begin with clean slates. By 1914, for example, Munich already had electric light, power, and traction requiring up to 19,000 kw. at peak load and 60 million kwh. during the year.[25] Moreover, the electric light and power utilities serving Munich and Upper Bavaria had plans of their own for expanding capacity. Different frequencies, voltages, and other characteristics added to the complexity and difficulty of planning a coherent system for all of Bavaria. With the need for standardization and coordination in mind, but before finally adopting a plan for all of Bavaria, the state, in 1913, in return for the utilities' acceptance of regulation and the possibility of eventual acquisition by the state, began making agreements with large utilities to provide them with the use of state land for transmission lines and a sizable area for exclusive supply. This was intended to encourage the growth of the large companies and to discourage parochialism.

Von Miller took into account not only geography and the legacy of the history of the electrical supply industry in Bavaria but also the state of existing technology in the industrial world. Consulting engineers like von Miller usually do not themselves invent and develop technology, but instead

---

[24] *Water Power Exploitation in Germany: Special Publication for the Session of the Second World Power Conference, Berlin 1930*, with the assistance of the Government Departments of the Reich and the German states (Berlin: Deutscher Wasserwirtschafts- und Wasserkraftverband E.V., 1930), pp. 215–20; and Theodor Koehn, "Über einige grosse Europäische Wasserkraftanlagen und ihre Wirtschaftliche Bedeutung," *Elektrotechnische Zeitschrift* 30 (1909): 966 ff.

[25] These figures were calculated from data for the years 1913–16. See Oskar von Miller, *Bayernwerk zur einheitlichen Versorgung des rechtsrheinischen Bayern mit Elektrizität: Projekt Oskar von Miller, Februar 1918: Erläuterungs-Bericht*, appendix, item 12; copy in the library of the Deutsches Museum, Munich.

depend on manufacturers to fulfill design specifications. Furthermore, large projects like the Walchenseewerk and the Bayernwerk are often designed to take advantage of the state of the art rather than to advance far beyond it. The challenge for von Miller and the other planners and designers of the Bavarian project was artfully to adapt available international technologies to particular local conditions. They were, in short, involved in transfer and adaptation.

The planning and execution of the Walchenseewerk and the Bayernwerk also required political sensitivity and power. Many of the major hurdles to be surmounted were political and legislative rather than technical in nature. After the Prince Regent named von Miller to the *Reichsrat*, or Upper House, of the German Parliament in 1909, he was able to respond to these challenges. Before many other engineers and scientists (Charles Merz had in London), he discovered that large-scale engineering was a political matter. He believed that projects the size of the Walchenseewerk and the Bayernwerk needed the financial resources of the Bavarian government and required its legislative authority. In the *Reichsrat* his professional credentials gave authenticity to his arguments. His forensic gifts, personal magnetism, and decidedly Bavarian characteristics served him well in the *Reichsrat*. Even today, for many Bavarians he still symbolizes traditional values such as geniality (*Gemütlichkeit*), family loyalty, and love of the land. His speeches and debates in the *Reichsrat* are models of the clear exposition of technical subjects. The record suggests that he himself remains a model of the politically sensitive and effective engineer, comparable in many ways to his English contemporary Charles Merz.

In 1909 several important matters pertaining to electrical supply occupied the Bavarian government. The Ministry of the Interior established a Bureau for the Utilization of Water Power and the Supply of Electricity, and the lower house of the Parliament approved in principle the construction of an electric power company to exploit the Walchensee. On the floor of the upper house, von Miller played a prominent role in the debate that developed over legislation arising from government involvement in this relatively new field of technology. In 1910 he made several addresses that established his general approach not only to water power development but also to the general question of the state's role in electrical supply. Quickly he tackled the Walchenseewerk problem, for even though the project had been approved in principle, the state's plans were nebulous and the go-ahead had been qualified by the requirement that a market be identified for the more than 50,000 h.p. waiting to be tapped at the Walchensee. Given Bavaria's stage of industrial development, some thought that this qualification would frustrate the project. Von Miller's strategy for the Walchensee followed the precedent established with the Pfalzwerk: feeding the immense power from the Walchensee power station into a regional network, the Bayernwerk. The transmission network would be supplied not only by the Walchenseewerk but also by other large and efficient generating stations in Bavaria, fossil-fuel as well as hydroelectric. Von Miller realized, however, that the uncoordinated development of electric supply would increasingly threaten coherent regional planning and projects even if this approach were finally approved by the legislature. He knew that local governments and private corporations were building generating plants and distribution systems without considering how these might relate to an over-all plan for Bavaria. He also saw the various government ministries advocating conflicting energy policies. The Transportation Ministry wanted state ownership of generating stations like the Walchensee, but only if they would supply electrified, state-owned railroads; the Finance Ministry was planning steam electric stations, which would be redundant, von Miller pointed out, when Walchensee was built; and the Ministry of Interior intended, without considering how these might relate to a general system of supply, to grant isolated concessions for water power development to private enterprises. Von Miller wanted a rational, systematic government policy. His strategy was to transcend the various ministries by recommending the establishment of a single centralized administrative organ for Bavarian electrical supply.[26]

Von Miller eventually presented a detailed plan for unified electrical supply in Bavaria, but in the meantime members of the legislature raised many other questions about it and the Walchenseewerk. Von Miller's responses from the floor of the legislature are still relevant, for the issues persist. The wide-ranging character of the questions from the legislators resulted from the many vested interests and social, political, and economic principles involved. Some members believed that private enterprise, not the government, should finance, construct, and operate the Walchensee power station. Others argued that the government should own and operate it, but only if the power generated were supplied exclusively to state-owned, electrified railways. Von Miller's attitude on state involvement was pragmatic and flexible: Government funds should be invested in the project, at least to the extent that the Walchenseewerk generated power for government-owned railways, but construction and operation of the station should be contracted to private companies, for von Miller knew from experience that private management was efficient.

Von Miller further raised the level of complexity and sophistication of the debate by suggesting that even though the generating facilities should be operated by a privately owned company, the transmission and sale of electricity should nevertheless be left in the hands of the Bavarian government. He wanted to keep profits from becoming the sole determinant of distribution and price, for other criteria, especially social ones, he believed, should override profit. He had in mind subsidized distribution for economic development and other social needs. When pressed, von Miller took a categorical stand on the government's basic responsibility for the use of the Walchensee: It was a large, scarce, and nearly unique state resource, and nature's bounty should benefit all of the people. The undeveloped region, the isolated farmer, the small artisan, as well as heavy industry, should benefit from the Walchensee, even though heavy industry could more easily justify costly transmission and distribution facilities because of its high and concentrated level of consumption of electricity.[27]

The prospect of Walchenseewerk also raised the issue of environmental

[26] *Stenographisches Bericht über die Verhandlungen der Kammer der Reichsräte des bayerischen Landtages*, 18 March 1910, p. 123; 30 June 1910, p. 314; and 9 August 1910, pp. 542–43. These references are to the dates of von Miller's addresses; the page numbers are for the respective volumes in which the speeches appear.

[27] Ibid., 18 March 1910, p. 123; 31 May 1912, pp. 88–89; 13 June 1914, pp. 322, 324.

aesthetics or amenities. The rise and fall of the Walchensee according to season and demand for electricity might flood the shoreline vegetation at times and lay bare unsightly lake shores at others. Deeply attached to the Alpine land, the Bavarians did not take this consideration lightly. Von Miller and others involved in planning the Walchensee project countered with an agreement to restrict the variation of water level to specified limits. This restriction would impose limitations on water storage and power utilization. In Bavaria, however, "trade-offs" between environment and economy were matters of policy; even the 1909 international competition for designs for the Walchensee had specified as a prime desideratum the preservation of the region's natural beauty. Von Miller also assured the members of Parliament that the levels of the streams and rivers feeding the Walchensee would be maintained.

Artists, especially, feared that the Walchenseewerk would mar a beautiful landscape. Von Miller argued, however, that they should be more concerned about the proliferating automobile than about the Walchenseewerk. Already, he asserted, the auto had made the Alps smoky and noisy; furthermore, he insisted, automobiles, unlike power plants, did not contribute to economic development. (In 1912 the luxury passenger automobile—not the truck or the commuter's small car—dominated the scene.) Electric motors and trains powered by the smokeless Walchenseewerk were preferable, he insisted, to coal-fired steam engines and internal-combustion motors.

Electrification of the railways was another issue raised during discussion of the Walchensee project. Supply of electricity to the railways in Upper Bavaria was, for some, justification for the investment of state funds in the Walchenseewerk. The railways of Bavaria, steam-driven at the time, were state owned and operated. Since coal was a scarce resource, the Ministry of Transportation considered Walchensee power desirable on economic grounds (white coal would replace black). Furthermore, Switzerland, Italy, and other parts of Germany had tested and proven railway electrification. An interesting complication, however, was that the military opposed mainline railway electrification in Bavaria.[28] It feared that concentrating power generation in a single plant, like the Walchenseewerk, rather than widely dispersing it in many locomotives, would greatly heighten the vulnerability of vital transportation links.

Ritter von Maffei, a locomotive builder with a seat in the upper house of Parliament, argued against the construction of the Walchenseewerk as a source of power for railway electrification. He acknowledged that steam railways were being electrified in the United States, but he argued that America could better afford miscalculations than Bavaria could. Von Miller, who had made a tour of technology in America and had followed closely the rapid development of electricity there, pointed out that America was industrially strong because she did boldly exploit her natural resources, like the water power at Niagara (300,000 h.p. compared to the 60,000 h.p. anticipated at the Walchensee). "America was not created by God—nor simply by steam engines," von Miller quipped to his more conservative peers.[29]

[28] Ibid., 9 August 1910, pp. 543–45; 31 May 1912, p. 90.
[29] Ibid., 31 May 1912, pp. 89–90.

Despite the Ministry of Interior's ultimate advocacy of a coherent Bavarian system based on von Miller's proposals, and despite the legislature's approval of the Walchensee project in principle, construction on neither the station nor the Bayernwerk grid began before the outbreak of World War I in 1914. Von Miller's foreboding proved correct; private enterprise and local government continued to build small electrical supply systems and in so doing made eventual coordination increasingly difficult. Dissension within the government and in the legislature frustrated the adoption of interim legislation strong enough to prevent troublesome variations in the characteristics of supply and a duplication of facilities. Despite passage of a 1913 law designed to regulate natural monopolies, disorderly growth continued.[30]

The exigencies and tribulations of war did not cause von Miller to waver in his determination to organize Bavarian electric supply, however. He volunteered his professional services for developing detailed plans and specifications for the system, submitting the plan for the Walchenseewerk in 1915 and that for the Bayernwerk in the spring of 1918. He envisaged the Walchenseewerk and Bayernwerk as projects to be pursued at war's end, when returning soldiers and released industry could stimulate economic recovery. Von Miller's vision and commitment transcended immediate circumstance and demand. The proposal he addressed to the Bavarian Ministry of the Interior in October 1915 revealed a continuing preoccupation with the ultimate goal of a unified Bavarian electric supply system.[31]

Only two of the nine sections of the Walchenseewerk proposal dealt with the Walchensee project per se; the others pertained to the hydroelectric project as a component of the Bayernwerk. The technical character of the installation was described succinctly: six 10,000-h.p. turbines; generators supplying polyphase current; a nearby switch-and-transformer house to carry the power to the transmission lines; and the various structures needed to lead water to and from the generating plant. The report provided detailed instruction for maintaining the water level of the Walchensee as well as of the rivers and streams supplying it. This requirement necessitated using the Walchenseewerk during the day, when loads peaked, and conserving its stored water during the night hours, when demand fell. Von Miller envisaged the Walchenseewerk as a peak-load plant in the unified system; hydroelectric and coal-fired plants elsewhere in Bavaria would carry the base load.[32] The other hydroelectric stations, including low-head, or running-water, plants, would be built on the Isar River near Munich.

Despite the devastating war and all the preoccupations and shifts in values consequent to it, Oskar von Miller still managed to bring the Bavarian Parliament to consider and act favorably on his plans. Completed in March 1918, the Bayernwerk proposal was approved by both houses in June, only

---

[30] Ibid., 13 June 1914, pp. 322–24.

[31] *Vorschläge des Herrn Oskar von Miller in München über den Ausbau und die Verwertung der Walchenseekraft an das Hohe Königl. Bayer. Staatsministerium des Innern, München, den 9 Oktober 1915* (copy in the library of the Deutsches Museum, Munich). The fifteen-page report is, among other things, a treatise on the economics of power systems. See also, Oskar von Miller, "Die Verwertung der Walchensee-Wasserkraft für ein Bayernwerk," *Elektrotechnische Zeitschrift* 37 (1916): 85–89, 102–5.

[32] *Vorschläge Walchenseekraft*, pp. 3–5.

Figure XII.6.   *Painting of Walchensee and the power plant, 1924. Courtesy of the Deutsches Museum, Munich.*

a few months before the armistice.[33] The Bayernwerk was described as a regional grid of transmission lines that would operate at 100,000 volts. Its jurisdiction would extend to (but would not include) the generating plants and distribution systems of connected utilities, including the Walchenseewerk, which was defined as a separate entity owned by the Bavarian state. The existing utilities generating and distributing electricity were variously owned; the Walchenseewerk would be the largest-capacity generating facility. Step-up transformers would raise the voltages from the level at the power plants to transmission voltage, and step-down transformers at distribution points would lower the voltages to 40,000–50,000 volts. In the first stage of Bayernwerk development, the 100,000-volt transmission lines would extend 1,020 kilometers, and for 600 kilometers would be double.

[33] Von Miller, *Bayernwerk: Erläuterungs-Bericht;* C. Zell, "Das Bayernwerk," *Elektrotechnische Zeitschrift* 39 (1918): 361–63.

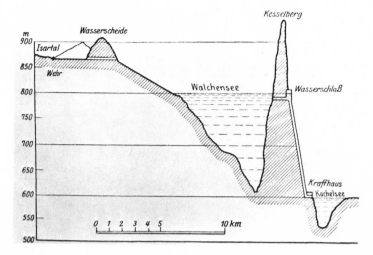

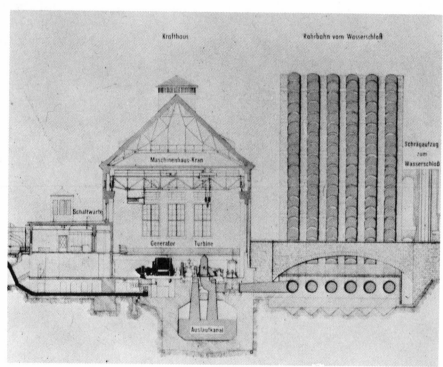

In the second stage, at least 250 kilometers of transmission line would be added. When complete, the transmission system would take the form of a ring; this, von Miller explained, would ensure that distribution centers along it would receive supply from two sides.[34] Such an arrangement would greatly increase the reliability of the system at a time when large-area supply systems were still prone to instability and lightning damage.

In order to determine the overall capacity of the system, von Miller had to anticipate demand. Because critics of the Bayernwerk (and the Walchenseewerk) had insisted that the market for the system would be insufficient, he resorted to a detailed analysis, or model, of economic development in Bavaria. The Bavarian Department for Waterpower Utilization and Electricity Supply helped by circulating a questionnaire designed to allow forecasting of load demand in various sectors of the Bavarian economy. Von Miller's analysis of this data enabled him to anticipate not only the system's total load but also the most appropriate locations for the distribution points, or load centers, where transformers and switches would carry the electricity from the transmission ring into the lower-voltage distribution systems.[35]

Von Miller also had to integrate the existing equipment of the cooperating utilities into the overall system of generation, transmission, and distribution. As he had anticipated, the problem had been complicated by the construction of dissimilar local systems of supply. Like other engineers who were integrating small systems into larger ones after the war, von Miller utilized coupling apparatus, such as rotary converters (a.c.-d.c.). He did not eliminate diversity by making a homogenized system; he systematically maintained and linked diverse elements. When deciding which of the existing generating plants and distribution lines to include in the grid, he generally chose the most efficient ones. Adaptability to the Bayernwerk would be the governing consideration behind any decision to construct new plants and distribution systems.[36]

In the section of his proposal regarding the corporate structure of the Bayernwerk, von Miller suggested an ownership compromise. He recommended a mixed-enterprise corporation: 51 percent ownership by the state (Bavaria) and 49 percent ownership by the various utilities (local-government and private) that were to be integrated into the system. He also recommended that the state of Bavaria should have a majority on the Bayernwerk *Aufsichtsrat* (advisory committee). Such an ownership structure would, he believed, provide the advantages of business, or corporate, organization and management, and would also, through government involvement, protect the general interests of the people, especially the interests of Bavarians living in rural areas where there was little electricity or water power. He predicted that the Bayernwerk would be completed within two or three years after the war and that it would promote postwar economic development.[37]

In his report, von Miller reiterated the economic principles of large,

---

[34] Von Miller, *Bayernwerk: Erläuterungs-Bericht*, pp. 1–3, and appendix, items 6–9.

[35] Ibid., p. 2, and appendix, items 1–5.

[36] Ibid., p. 4, and appendix, items 10–11.

[37] Ibid., pp. 17–18.

interconnected systems. He predicted that interconnection would permit hydroelectric plants to carry 80 percent, and steam plants 20 percent, of the annual load on the Bayernwerk; if the same plants were not interconnected, the steam plants would have to carry 40–45 percent of the total. In coal-poor Bavaria this was a major consideration. Von Miller also stressed that interconnection would permit judicious use, or scheduling, of hydroelectric plants on a daily and seasonal basis, so that low-head hydroelectric plants connected to the grid would carry a heavy load when rivers and streams ran full and a light load in winter when scant rainfall and ice prevented utilization. The Walchensee-stored water would be used for peak loads. Coal-fired plants connected to the system would be drawn upon heavily when the low-head, or running-water, hydroelectric plants and the Walchenseewerk were short of water. In summary, as varying conditions of supply and demand dictated, the low-head plants would carry the base load of the system, the Walchensee would carry most of the peak loads, and the efficient coal-fired plants would carry the remainder. The savings resulting from this type of load management would be substantial.[38]

In his nineteen-page report, which was accompanied by thirty-one pages of highly detailed and informative tables and graphs, von Miller also pointed out that the sum of the peak loads in the system would be less than the sum of the peak loads of the various participating plants if those plants remained outside the system. The result would be less installed capacity and less need for reserve as well. Since the electrical supply industry is a capital-intensive one, this would mean a substantial reduction in unit costs. The report also showed that the rural and more northerly regions of Bavaria would generally demand more from and supply less to the system than Munich and the other urban areas where electrical supply was already well developed. Von Miller's commitment to rural electrification, even if it meant subsidization, was obvious.[39]

Within two weeks of the armistice, von Miller announced that he was prepared to supervise construction of the Walchenseewerk and the Bayernwerk. He insisted that these projects not only would provide work for returning soldiers, but would also, at a time of social unrest, draw together various classes in support of a common constructive endeavor. He advertised for workers for the Walchenseewerk; began to assemble construction equipment at the site; and supervised the start of construction by contractors at the Walchensee (see Fig. XII.9). For the Bayernwerk, he concluded contracts with electrical manufacturers for high-voltage transmission towers; finished surveying and made detailed drawings; and proposed plans for the high-voltage transformer stations.[40]

On 22 January 1919 Interior Minister Erhard Auer of Bavaria's new government named von Miller the *Staatskommissar* (state commissioner) for the Walchenseewerk. Because the government was a republican one, replacing the Wittelsbach dynasty and the conservative wartime ministerial government, von Miller's conservative friends and associates lamented his

---

[38] Ibid., pp. 5–7, and appendix, items 12–29.

[39] Ibid., pp. 2, 8–9, and appendix, items 11, 12, 18, and 25.

[40] Kristl, *Der weiss-blaue Despot*, pp. 185, 198; *Elektrotechnische Zeitschrift* 40 (1919): 303; ibid. 41 (1920): 277.

Figure XII.9.   Transporting a section of penstock to the Walchensee power plant, 5 May 1924. Courtesy of the Deutsches Museum, Munich.

Figure XII.9.   Transporting a section of penstock to the Walchensee power plant, 5 May 1924. Courtesy of the Deutsches Museum, Munich.

association with it.[41] The new government's ability to act decisively and expeditiously in matters concerning the Bayernwerk and Walchenseewerk, however, was the criterion by which von Miller, the engineer, judged it. He had survived years of delay caused in part by divided ministries and conflicting parliamentary interests. A government that was able to exercise power in support of engineering projects in a time of social turbulence and economic confusion was an obvious ally for von Miller.

Von Miller needed the authority of his position as commissioner in order to organize and manage the projects. He was unwilling to entrust the success of his highly technical enterprises to the advice and directives of persons he considered to be uninformed—politicians, bureaucrats, and hastily formed committees acting on the basis of majority rule. His apolitical attitude and policies elicited criticism and opposition during his first year as commissioner. Some argued that the state commissioner and the manager of the projects should not be the same person; for economic and aesthetic reasons, others proposed different routes for the transmission lines; still others believed von Miller was too cautious and extravagant in the measures he took to ensure the reliability of the system and the large generating capacity of the Bayernwerk. At the same time, there were dire predictions of failure for the complex project. There were also those who believed von Miller was acting without adequate professional consultation and advice from others.[42]

[41] Kristl, *Der weiss-blaue Despot*, p. 184; W. von Miller, *Oskar von Miller*, p. 94.

[42] To explain his policies, von Miller wrote a lengthy letter on 18 November 1919 to Minister of Interior Fritz Endres. The letter includes a summary history of the two projects. It is quoted at length in Kristl, *Der weiss-blaue Despot*, pp. 195–99; see especially, pp. 195–96.

In von Miller's opinion, much of the criticism was uninformed or represented the views of persons with special interests. To those who thought him unresponsive to suggestions, he pointed out that he often consulted two distinguished electrical engineers, Professors H. Prinz and Waldemar Peterson. He assured those who supported him that forty years of presiding over innovations had taught him that doubts and criticism were inevitable, but that they usually proved to be exaggerated. He had found that, although unforeseen difficulties usually arose, the experienced innovator generally responded resourcefully.[43] He should have added that an entrepreneur with his commitment and dogged perseverance often outlasted an opposition that was not won over by persuasion or overwhelmed by momentum.

By the end of 1919, however, harassed by his opponents and witnessing the erosion of his authority, von Miller was ready to resign. He did not do so, however, before calling to the government's attention his achievements in the face of manifold difficulties. He had engaged in month-long negotiations over contracts and tariffs; brought about agreements among such diverse interests in electric light and power as the state, cities, rural governmental authorities, and private enterprise; and he had maintained the support of the legislature. He informed the minister of interior that without power commensurate with his responsibility, he would have to consider resignation.[44]

Von Miller did not resign precipitately. He held on to his position as construction on both projects progressed, but in the spring of 1921 he finally resigned as state commissioner. An immediate cause for his withdrawal was the failure of the government to support his position in a disagreement with a consortium of Bavarian cities and several large power utilities. Representing urban, as contrasted to state, government, the consortium wanted the Walchenseewerk to be a part of the mixed-ownership Bayernwerk instead of being purely state owned and controlled, as von Miller had planned. Von Miller countered by proposing that the Bayernwerk, like the Walchenseewerk, be solely state owned. The Bavarian government compromised by forming three corporations for electrical supply in Bavaria: one for the low-head hydroelectric stations of the middle Isar River near Munich, another for the Walchenseewerk, and a third for the Bayernwerk. Furthermore, the government provided that these corporations could raise funds on the public market.[45]

When the three state corporations were formed and the new organizational structure, with its executive and advisory committees, began operating, the position of *Staatskommissar* became superfluous. The government hoped to placate von Miller by naming him to various advisory committees and appointing him general adviser on energy policy. In March 1923 he gave up even this post, however, when the government allowed another utility, the Grosskraftwerk Franken AG, to develop hydroelectricity independently in the industrially advanced region around Nürnberg. This agreement removed from the Bayernwerk a market for electricity that

[43] Kristl, *Der weiss-blaue Despot,* pp. 196, 198.
[44] Von Miller to Minister of Interior Fritz Endres, 18 November 1919 (see note 42 above).
[45] Kristl, *Der weiss-blaue Despot,* pp. 206–10.

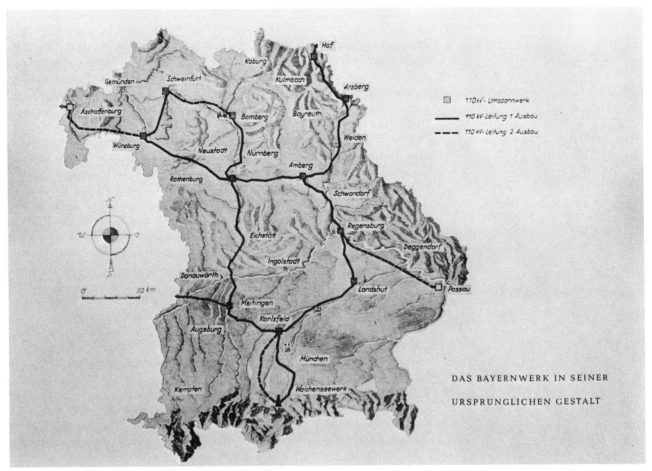

Figure XII.10. *Original plan of the Bayernwerk: Transformer substations,* □; *110-kilovolt transmission line (first stage),* ——; *110-kilovolt transmission line (second stage),* – – –. *Courtesy of Rudolf von Miller.*

would have related well to the rest of the system and contributed to its economic strength.[46]

Yet the Bayernwerk, though limited in scope, and the associated Walchenseewerk were completed, and on 26 January 1924 the Bayernwerk began transmitting power from the Walchenseewerk (see Fig. XII.10). Fifty years later, approximately half the electricity supplied by utilities in Bavaria originated in the Bayernwerk.

In establishing the Walchenseewerk-Bayernwerk system, von Miller displayed remarkable conceptual powers, technical competence, and organizational ability. He also showed by his decisions and policies the manner in which large-scale advanced technology could be introduced into a primarily agricultural region. Characteristic of his approach was an awareness of the unique geographical, political, economic, and social factors that were shaping the particular situations he encountered in Bavaria. His willingness to respond convincingly to Bavarians who feared the defacing of the land-

[46] Elektrizitätswerke E.V., *Elektrizität in Bayern, 1919–1969* (Munich: VBE, 1969), pp. 16–17.

scape is one example of his awareness and flexibility. His unshakable conviction that large-scale development of water power and systematic, planned, regional electrification were appropriate and imperative for Bavaria manifested his vision and his awareness of technological trends. Note should also be taken of the sense of timing and situation he exhibited as he, a political representative of the old regime, seized upon the willingness of a radical postwar regime to act dramatically in a time of great change, when the forces of tradition and routine were weak.

Von Miller is representative of the relatively rare engineer who is able and willing to become deeply involved in politics in order to preside over complex change. He was an engineer-entrepreneur who took part in the political process that determined the fate of his projects and who took into account and provided for the financing, organization, and management of his projects. Furthermore, he attempted to steer a course through the rocks and shoals of the bureaucratic sea and to sail by the changing winds of public opinion. In order to explain the coming of an advanced system of technology, one must take into account the role of individuals like Oskar von Miller.

In 1926, only a few years after the Bavarians established a transmission grid, the British finally enacted legislation establishing a grid and a more unified system of supply. This achievement, in a country whose electric supply had long been parochial, was more political than technological and economic. No individual played as salient a role as von Miller had in Bavaria, but the influence of Charles Merz was substantial. The confluence of events that brought the change in organization of Britain's electric supply industry can be interpreted variously, but the now endemic British concern about the loss of national prestige and power appears to have been the major underlying reason for forming a consensus to break the crust of conservatism. This concern figures prominently in Great Britain's national consciousness today and was expressed, if less urgently, as early as the second half of the nineteenth century. The anxiety has often resulted in a clash of opinion and a paralysis of action, but in 1926 the consensus was that at the core of the problem lay the failure of the country to develop the newest of energy technologies. As a result, Britain's politicians generally agreed not only about the nature of the problem but about the remedy as well. Failure to electrify industry and transportation in the age of electrification was seen as a root cause of the country's failing industrial strength. Low-cost energy and the more efficient industry that would follow upon the establishment of large-area electric supply were considered the solutions. The fact that the reorganization of Britain's electric supply did not bring a dramatic recovery of national strength should not obscure the commitment that was made in the 1920s to a promising and untried solution.

Looking ahead to peacetime policies, Britain's wartime parliamentary committees had stressed the need for reform, but their recommendations, when embodied in the Electricity Supply Act of 1919, had been greatly weakened by the removal of compulsory powers for the electricity commissioners and the substitution of voluntary action and persuasion (see pp. 321–22 above). If the economic situation in Britain had not deteriorated in the political and labor unrest following the war, modest reform might have been tolerated, as modest change had been for decades. But the country emerged from the war with a domestic and foreign debt that was estimated to be ten times higher than that of 1914; coal miners struck in 1919, disrupting the nation's basic energy supply; and in 1921 an unemployment figure of almost a million led to passage of the Emergency Unemployment Act. After the Liberal party split, the Labour party became for the first time the major opposition party, and in 1924 Ramsay MacDonald formed the first Labour cabinet. The Conservatives returned to power the same year under Stanley Baldwin, but they had to contend with the great general strike of May 1926. It was Baldwin's government that presided over passage of the Electricity Supply Act of 1926. This act reorganized Britain's electric supply industry.

The government, and the engineers and industrialists who advised it, knew that the problem of bringing about change in the electric supply system was primarily political. One of Britain's leading consulting engineers and advisers to the government on the reorganization of electric supply, J. M. Kennedy, said that the planned change "does not contemplate the adoption of any policy as regards generating station construction or interconnection which has not been in successful operation either in this or in other countries."[47] By the mid-1920s, the international community of electrical manufacturers, engineering societies, and leading individual engineers and managers was so closely knit by publication and personal communication networks that the planners could assume the availability of a pool of technologies from which to draw.

To organize a community of common interest in Parliament was more difficult. The major circumstances favoring adoption and implementation of government-sponsored reorganization was the presence of a Conservative government that favored it and a Labour opposition that supported it as well. The Conservative commitment was conditioned by the widely held belief that reorganization of a leading sector of the industrial economy would enhance Britain's power—electric, industrial, and international. The Federation of British Industries, which represented a broad sampling of private enterprise, deserted the rigid stance of opposition to government involvement in the industrial economy and settled for token amendments to the eventual electricity bill. After initial resistance, the Institution of Electrical Engineers also supported the bill. The publicity given the various reports of parliamentary committees in popular journals emphasized the promise of a labor-saving electric age and marshaled public support for the measures. The prediction of engineers that the reorganization would result in immense energy savings for the nation and restore its industrial strength also influenced public opinion.[48] The committee report on which the 1926 act was based included, for example, a table showing in simplified form the "consumption per head of population" of electricity in various regions of the world. Great Britain fell at the end of a list of regions, cities, and countries headed by California, Chicago, and Canada and including

[47] J. M. Kennedy, *Report on the Operations of the Proposed Central Electricity Board*, prepared for the Ministry of Transport (June 1926), p. 8; copy in the Electricity Council Archives, London.

[48] Leslie Hannah, *Electricity Before Nationalisation* (Baltimore: The Johns Hopkins University Press, 1979), pp. 96–97.

Tasmania, Shanghai, and Sydney. Former colonies and small countries seemed far better prepared for the electrical age than the nation that had once been the world's undisputed industrial power. The advantage enjoyed in several of these regions because of abundant hydroelectric power was not stressed.[49]

Prime Minister Baldwin dedicated his energies to organizing support for the reorganization plan in Parliament. He could count on Labour support because many of the Labourites saw the establishment of regional supply as a step toward eventual nationalization of electric supply. Baldwin named Lord Weir of Eastwood chairman of a committee to consider the national problem of electric supply and to recommend legislation. Weir, a prominent Scot industrialist and head of a Scottish engineering company, had made a name for himself during World War I as director general of aircraft production and then as secretary of state for air. He made it clear that he believed the electric supply industry was in need of reorganization, and he promised to form a committee that would be small and speedy and that would report in favor of "a new electrical energy policy involving in its application courage and possibly a considerable financial investment."[50] Knowledgeable about committee temporizing, Weir also told Baldwin that he should not expect a report that simply furthered development along existing lines.

When the Weir report was released in 1926, it served as the document of reference for reform advocates. The condition of electric supply in Britain was described as "very disquieting." But there was a remedy—interconnection—not only for the large areas identified by the legislation of 1919, but for the areas between them; interconnection would create a national system. The Weir committee discounted the country's lack of hydroelectricity sites at a time when water power was being rapidly exploited and widely celebrated throughout much of the world, because, it ventured, Britain could exploit coal, had the advantage of a high level of urbanization, and had in close proximity industrial load centers. Great Britain, Weir's committee reasoned, "is in many respects an ideal electrical area, and is far more compact than other countries."[51] Commenting upon the report, Charles Merz said, nonpejoratively, that Britain's grid would be more like a high-voltage distribution system than a transmission system. He was using the distinction commonly made between long-distance transmission and short-distance distribution. As he had said earlier,

Proceeding on these lines [high-voltage distribution for the compact industrial districts of England] we shall not be merely copying America or Germany—we shall be doing something that is right for England because it is England, because England is radically different from other countries as regards the technical development and layout necessary to secure cheap power.[52]

---

[49] Great Britain, Ministry of Transport, *Report of the Committee Appointed to Review the National Problem of the Supply of Electrical Energy* (London: HMSO, 1926), p. 6. The committee was headed by Lord Weir of Eastwood. Hereafter cited as the Weir Report.

[50] Weir to Wilfred Ashley, 1 December 1924, quoted in Hannah, *Electricity Before Nationalisation*, p. 90.

[51] Ministry of Transport, *Review of the Supply of Electrical Energy*, p. 8.

[52] Charles H. Merz, "Electrical Power Distribution," *Engineering* 102 (1916): 262–63.

In suggesting that a proud nation could move from a backward status to an advanced one without simply following the leaders, Merz exhibited psychological acuity.

The Grid was the essence of the technology proposed by the Weir committee and its consulting engineers. Charles Merz gave the new system this name after observing that the network of high-voltage transmission lines that would crisscross the countryside took the shape of a great gridiron imposed upon a map of the country. The gridiron of transmission lines was to interconnect selected stations and to connect with existing and future distribution systems. The Grid was presented to the public as a national system, but in fact it was first organized as a number of regional transmission systems with, between the systems, relatively small coupling transmission lines whose central controls were located in regional headquarters.

The benefits that were expected to accrue were known in industrial regions of the world where large-area supply already existed. The Weir committee report, which was written for members of Parliament and for interested persons who lacked technical training, summarized the "technical considerations" as follows: "generation in large stations, favourably situated as regards fuel, water and load with units of comparatively large capacity; the minimum legitimate amount of stand-by plant; and the highest obtainable load factor, to secure which stations should be inter-connected with one another."[53] According to the report, the existing load factor in Britain was "unreasonably low." The juxtaposition of load factor and national strength showed that Britain was no longer in the era of James Watt; it had entered the era of Charles Merz.

Merz was a principal adviser to the Weir committee; the prominence given in the report to the testimony of Samuel Insull, head of the Chicago enterprise Commonwealth Edison, was not anticipated. Insull was cited several times as a (if not the) leading authority on electric supply. Lord Weir characterized him as "the greatest authority on electricity."[54] Insull's name, prominently featured in the report, provides a reminder that Britain was looking to America for technological leadership and that Insull was an unusually articulate, if sometimes colloquial, spokesman for the concepts of the electrical era. It should also be recalled that Merz and Insull were personal friends and professional associates whose ties constituted a technology-transfer bridge. The Weir report quotes Insull's "striking statement" that the "only limit to the amount of business [electric supply] . . . if the proper amount of brains is devoted to the engineering of selling the energy," is the amount of capital invested. Insull had also told the committee that "any man who refuses it (interconnection)—and I will not exclude England—does not understand the fundamental economies of the business."[55] The British in 1926 were not averse to learning the business of electrical technology from their American cousins, and Insull was easier to accept as an adviser because he was British-born. The fact that Charles Merz had often served as a consultant to Insull also boosted national pride.

[53] Weir Report, p. 7.

[54] Weir to Wilfred Ashby, M.P., Weir Correspondence (9/1), Churchill College Archives, Cambridge, England.

[55] Weir Report, pp. 5 and 8.

The Electricity Supply Act of 1926, which incorporated the recommendations of the Weir committee and its technical advisers, received the royal assent on 15 December 1926. The legislation provided for the creation of a Central Electricity Board (CEB) with executive powers to construct and operate the Grid. The electricity commissioners had already begun planning and were to serve as juridical advisers to the Central Electricity Board in technical matters. The board concept was more original than the technical aspects of the Grid. A public corporation that was able to raise its own funds, the CEB was organized and administered by government-salaried managers and engineers who were not part of the civil service. The salaries of its leading personnel were higher than those of persons with comparable experience and responsibility in the civil service. The CEB was eligible for financial guarantees from the Treasury to aid it in raising funds, but these came from the sale of interest-bearing, nonvoting stock to the public. The CEB never used the financial guarantee, but raised money entirely on its own credit. The chairman and eight board members were named by the minister of transport and were subject to dismissal (also by the minister of transport) only if absent from their duties for six months or more. Their terms ranged from five to ten years. As a corporate body, the board had the status of an authorized electrical undertaking (utility) which owned and operated the Grid on behalf of the government.[56] Conservatives were satisfied because the CEB was not a nationalized authority or government department, but an organization managed by "practical" men who were in close touch with industry. Although the concept was original in many respects, it was not as original as some of its proponents and some historians believed. The public-corporation concept had been embodied in the organization of the Bayernwerk in Bavaria and in the theory of Giant Power defined by Governor Pinchot and Morris Cooke in Pennsylvania (see pp. 334–50 and 297–312 above).

In effect, the chairman of the CEB became the managing director of the Grid. His was the only full-time position on the board. The minister of transport chose as chairman Sir Andrew Duncan, a Scot and a member of a firm of industrial lawyers. Duncan had attracted public attention as a skilled negotiator who was knighted after resolving the coal dispute of 1920. Forty-two years old in 1926, Duncan had a reputation for Celtic charm, tact, and patience.[57] For the chairman of the CEB, these qualities proved more important than intimate knowledge of the technology of large-area supply. Duncan chose Archibald Page as the chief engineer and general manager for the CEB. Page had been an electricity commissioner before leaving that post to become general manager of the County of London Power Company, which was organized to supply the utilities of east London. Two senior engineers were named to assist him: Harold Hobson, who had worked for Merz & McLellan; and Johnstone Wright, who had trained with the Newcastle upon Tyne Electric Supply Company (NESCO). These three engineers remained with the CEB until the mid-1940s, and each

---

[56] Great Britain, Ministry of Transport, *Memorandum on the Electricity (Supply) Bill, 1926* (16 & 17 Geo. 5, ch. 51) (London: HMSO, 1926), pp. 1–4; Hannah, *Electricity Before Nationalisation*, pp. 100–101.

[57] Hannah, *Electricity Before Nationalisation*, p. 102.

served a term as chairman. Charles Merz had been a likely candidate to head the CEB, but he was considered too close to commercial interests in the industry, and John Snell, technical expert and head of the electricity commissioners, wanted to administer the Grid, but he was considered not tactful enough as a negotiator. According to his own testimony, Samuel Insull had been asked by the prime minister to head the CEB, but had refused.[58]

The legislation providing Page and his associates with the authority to reorganize the nation's electric supply industry resulted from innovations introduced by a committee representing manifold interests. The technological and organizational changes brought forth by the 1926 act were complex and made palpable the multitude of private and corporate interests represented in Parliament. Furthermore, not all of the subtleties of concern could be made explicit in the legislation and the procedures derived from it. Therefore, judgment and political skill had to be used to bring about the intended changes.

The Central Electricity Board, the electricity commissioners, and the advisory consulting engineering firms avoided the temptation of assuming that technical expertise could solve the problems. They realized that decisions, in contrast to solutions to problems, were needed; one best answer was not to be found. Their authority, they also wisely acknowledged, was limited. Page, an experienced negotiator, understood that persuasive skill was of the essence, for the decisions the board members had to make required that they venture beyond the limits of their defined powers. In short, Page and his associates had to function as politicians as well as engineers.

Typical of the challenges the board faced and the political tactics it employed were the decisions made about "selected" stations. The law provided that, with the advice of the electricity commissioners, the CEB should select the power stations to include in the Grid. Quantitative guidelines were to be applied. If an operating power station generated electricity at a cost greater than what it would cost to purchase power from the Grid, this station was not "selected." The estimated cost of supply from the power station was calculated without including capital costs, for these were "sunk," or not subject to change. On the other hand, the comparative cost of electricity supplied from the Grid included capital costs such as expenditures to be made in constructing the Grid. This formula for calculation, which involved more complexities than those noted, meant that the operating costs of existing stations were to be compared to the operating-plus-capital costs of the Grid. The approach was comprehensible. The electric supply industry had grown rapidly throughout the world because the operating-plus-capital costs of new technology often were less than the operating costs of old technology. For this reason, old plants could often be abandoned before being amortized.

Page and his associates understood, however, that the cost calculations failed to take into account not only a number of economic variables but noneconomic substantive considerations as well. Power station owners insisted, for instance, that they intended to introduce technology that would

---

[58] Forrest McDonald, *Insull* (Chicago: University of Chicago Press, 1962), p. 276.

dramatically lower their costs. In addition, the operating and capital costs of the Grid were dependent upon unpredictable variations in material and labor prices. Just as important were the noneconomic factors that the calculations ignored, but that Page, the politician, could not. Among these was the pride the managers and engineers and private and municipal owners of power plants took in their organizations and technology and in the authority they held. This, as has been seen, was a constituent of parochialism. Because of vested interests, then, the board selected more stations than the Weir committee recommended. The report called for the number of "selected" stations to be limited to 60; the electricity commissioners identified 118; and the Central Electricity Board designated 140.[59]

As defined in the Electricity Supply Act of 1926, the relationship between the new Central Electricity Board and its predecessors, the electricity commissioners, was complex and left much to interpretation and the guidelines of experience. The act gave the electricity commissioners responsibility for inaugurating planning for the Grid because they had had the experience under the act of 1919 of encouraging regional schemes. Ultimate responsibility for owning and operating the Grid, however, would be lodged with the CEB. During planning and construction both the electricity commissioners and the CEB would have the authority to retain consulting engineers. Britain's leading consulting firms—Merz & McLellan, Kennedy & Donkin, and Highfield & Roger Smith—were retained to advise the board on such matters as the routing of high-voltage lines and the selection of power stations. After the Grid began operating, the commissioners continued to function as the board's technical advisers and as a quasi-juridical body hearing appeals from utilities about decisions of the CEB and operations of the Grid. The electricity commissioners, under the general direction of the Ministry of Transport, remained the regulatory authority for supply companies, or utilities, so the CEB, as an authorized supplier owning and operating the Grid, was subject to the commissioners' regulatory decisions.[60]

As is expected in situations where complexities, present and future, defy the laying down of rules and procedures, conflicts arose between the commissioners and the board. Leslie Hannah, the historian of electric supply in Britain, concludes that "there was rarely any great warmth in the relations between the two bodies [board and commissioners]."[61] The electricity commissioners, headed by the engineer Sir John Snell, preferred technically neat solutions; the CEB opted for choices that were more heavily weighted by relatively imprecise economic considerations and influenced by the vested interests whose cooperation was sought. Snell also had differences with

[59] Hannah, *Electricity Before Nationalisation*, pp. 112–13.

[60] An excellent analysis of the functioning of the CEB and the commissioners is found in a report Merz & McLellan prepared for Consolidated Edison Co. of New York, *Report on Electric Service in New York, London, Paris, and Berlin* (London: Merz & McLellan, 1937); I am grateful to Dorothy Ellison for calling this report to my attention. The plan for the Grid that Merz & McLellan prepared for the Ministry of Transport was entitled *National Electricity Supply: Technical Scheme* (May 1926); J. M. Kennedy's report for the same ministry was entitled *Report on the Operations of the Proposed Central Electricity Board* (June 1926); copies of both are in the Electricity Council Archives, London.

[61] Hannah, *Electricity Before Nationalisation*, p. 108.

Charles Merz. Both men enjoyed substantial prestige in engineering circles, so their differences were not easily resolved. Snell had served with distinction as an engineer after completing his studies at King's College, London. While Merz was making his name as an adviser to and director of the privately owned NESCO, Snell was making his as an engineer for municipally owned utilities, including one on the northeast coast of England that remained independent of the NESCO network. Snell, however, shunned the parochial attachments that fostered small-scale supply and advocated large-scale operations for municipally owned utilities.

The differences between Merz and Snell came to a head after Merz's consulting firm accepted the assignment from the Ministry of Transport to advise the commissioners in initiating planning for the Grid. Merz & McLellan was responsible for planning several of the regional systems that would make up the Grid. This required that the company conduct a survey to ascertain the output and load curves of existing utilities. The consulting engineers also identified "selected" power stations and recommended transmission-line routes. Merz soon decided, however, that the guidelines Snell had laid down would result in a flawed plan for the CEB. Anticipating that the CEB would call upon his firm's services during the construction phase, Merz decided to withdraw from the planning in order to avoid being in the position to carrying out a plan with which he did not agree. Snell complained that the work of Merz & McLellan was moving too slowly and at too great an expense compared with that of other consultants. After several meetings, Snell accepted the firm's withdrawal. A compromise allowed Merz & McLellan engineers to complete the planning while working as individuals. Subsequently the CEB did retain Merz & McLellan as consulting engineers, as has been noted.[62]

The Weir report correctly predicted that construction of the Grid would not involve unprecedented technology. British engineers had solved many of the most difficult technical problems of the electrical era, but the Grid was a large-area transmission system that had been introduced earlier elsewhere. The 132,000 volts decided upon for transmission was well below the 220,000 used in the United States and Germany by 1926, and the transmission distances were shorter. Thus, Britain's engineers concentrated on a host of relatively routine problems that were complicated more by their number, a tight schedule, and economic dimensions than by technical challenge. The economic aspects were of the kind best considered by engineers like Charles Merz and his consulting engineering colleagues, who were accustomed to making decisions based on an optimum mix of costs and technical efficiency.

Standardization of voltage, phase, and frequency was one of the economic and technical decisions Grid engineers made. Standardization was an especially messy problem in Britain because of the diversity embodied in its

[62] Correspondence in bound volume entitled "Electricity Supply Act (1926): Memoranda and Important Correspondence," Merz & McLellan Co. Archives, Amberley, Killingworth, near Newcastle upon Tyne, England. See especially J. Snell to Merz, 21 August 1926; Snell to Merz, 29 March 1927; Merz to Snell, 30 March 1927; notes on a meeting of Merz and Snell, 5 July 1927; notes on an interview, Snell, Merz, McLellan, and Leggat, 7 July 1927; and memorandum from Leggat (of Merz & McLellan) to Merz, 14 July 1927. See also Hannah, *Electricity Before Nationalisation*, pp. 111–12.

small-scale technology. Standardization of electricity supply was like a Chinese box—problems within problems. On the level of high-voltage transmission, the question of voltage and phase was not complicated (in much of the country) by existing commitments and extant equipment. Most high-voltage transmission utilized new equipment. Moreover, the transformation of voltages from generating plants to low-voltage distribution systems was a well-established practice. Transformers could be wired on one side for local conditions and on the other for the standard 132,000 volts decided upon for the main transmission lines. The transformer was a coupling device that permitted the integration of subsystems with different voltage characteristics.

In 1926, choosing a standard type of current did not present serious problems either, because polyphase, or three-phase alternating, current had become a world standard for lighting and power. Direct current survived in large cities where highly efficient direct-current Edison stations prevailed until after the turn of the century, but it was understood that the transition to polyphase current would be made. In the meantime another coupling device, the rotary converter (a.c.-d.c.), allowed the direct-current areas to be fed by polyphase currents and encompassed in systems that were primarily polyphase.

Standardization of frequency, however, was not as readily solved, for frequency converters were not as efficient and reliable as rotary converters. Therefore, equipment had to be rewired if the frequency was other than the Grid's standard 50 cycles. This problem was especially serious in the northeast coastal region supplied by NESCO, the regional supply utility that had expanded impressively under the guidance of Merz & McLellan. NESCO had previously standardized at 40 cycles (see pp. 458–59 below).

Beginning in the spring of 1927 and ending in August 1931, plans for the various regional transmission systems that would together constitute the National Grid were adopted at the rate of about two each year. Construction of the Grid ended on schedule in about six years. In 1933 the Central Electricity Board began operating as a power transmission utility in central Scotland and east-central England, and by 1936 the Grid was in full commercial operation in all regions except the northeast coast of England, the location of NESCO and the major standardization problem; operations began there in 1938 (see Figs. XII.11 and XII.12). The initially planned Grid cost approximately £26.7 million. During the six years of its construction 150,000 tons of steel were used, mostly for 28,000 transmission towers. The Grid's overhead lines required 12,000 tons of aluminum and 200,000 porcelain insulators. British manufacturers were heavily engaged in the construction of transformers and switchgear, metering, and control equipment. The expansion of existing power stations and the construction of new ones also kept turbine and generator manufacturers busy and gave them experience that had formerly been concentrated among the American and German manufacturers.[63] Both the construction and the operation of the Grid stimulated the post–World War I economy of Great Britain.

[63] Great Britain, Central Electricity Board, *The Grid* (booklet published by the CEB in 1946[?]), pp. 4–7. Charles H. Merz described the Grid in "The National Scheme of Electricity Supply in Great Britain" (Paper read before the British Association for the Advancement of Science, South Africa, 1929); copy in the Merz & McLellan Co. Archives.

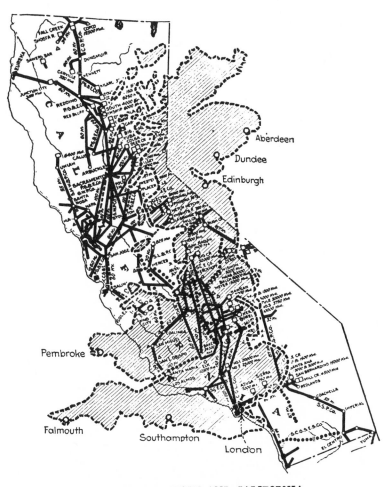

Figure XII.11. Simplified map of proposed English 132-kv. Grid, 1933. Adapted from the CEB's Annual Report 1932.

GREAT BRITAIN AND CALIFORNIA

The former has 40,000,000 more inhabitants than the latter and comprises one of the greatest industrial districts in the world. Its electrical output, however, is no larger than California's, and while Great Britain is planning restricted power districts made up of interconnected systems, California already possesses an interconnected network stretching from one end of the state to the other. The maps are drawn to the same scale.

Figure XII.12. Superimposed map of Great Britain showing the British Grid to be of no greater extent than California's interconnected power network. From Electrical World 82 (1923): 1056.

A better understanding of the complex nature of the Grid can be derived from consideration of several of its constituent regional systems. In the London area, the Grid, like the Pennsylvania–New Jersey Interconnection in the United States, was a power pool (see pp. 325–34 above). The Central Electricity Board left unchanged the ownership of the "selected" private and local-authority-owned power stations there, but took over the direction of operations and the planning of expansion. The CEB also assumed payment of the capital and operating costs of these selected plants. The entire electrical output of the stations was at the disposal of the Grid, and most of it was fed into the transmission network. The supply authorities, or utilities, then purchased from the Grid all the electricity they needed for their distribution systems. The price charged by the CEB represented the cost of production adjusted for load factor plus an additional charge for transmission. The price was guaranteed to be lower than would be incurred if the selected station generated for itself.

By the time the Grid began operating, there were thirteen privately owned companies in the county of London. The JEA concept that was established under the Electricity Supply Act of 1919 had encouraged consolidation, but the idea did not gather momentum until passage of the 1926 act in anticipation of the Grid. All thirteen companies were distributors, but nine had turned over generation to the London Power Company, the stock of which they fully owned. Of the other four, the City Company had a major power plant at Bankside, and the County of London Company had major power plants at Barking "A" and Barking "B." Consolidation had also occurred on the distribution side. Six of the companies had placed their common stock under the control of a holding company, the London Associated Electricity Undertakings. Several other companies had come under the control and management of the Metropolitan Company and the County of London Company. The prices charged customers by the private companies varied and allowed for profit, which was regulated by the electricity commissioners. London's thirty-two local-authority-owned utilities remained completely independent as distributors of electricity, but in terms of generation they were coordinated by the Central Electricity Board. All "selected" power stations, private or local-authority-owned, were directed by the board. Other stations made agreements with the CEB about operations; some, for instance, operated as peak-load plants. In 1935 the total capacity of a 50-cycle generating plant in London was 2,313,000 kw. and its output was 4,418 million kwh., 115 million of which was for traction.[64]

The operation of the Grid can also be examined by considering one of the regional schemes. The seventh, which was announced in 1930, pertained to southwest England and Wales. The region had a population of about six million and an area of slightly over 17,000 miles. Included in it were highly industrialized areas, mines, and seaports. Before the Grid was established, 165 utilities owned 107 generating stations. Six of these stations were chosen to become part of the Grid, and two new stations were constructed. The CEB had the authority to arrange for a utility in the area of supply to provide a new station. If no such arrangement could be made, the electricity commissioners could authorize the CEB or some other company or person to construct the station. The new station would then be operated according to Grid procedures.[65] Two of the selected stations, one at Bristol and another in South Wales, were designated as base-load plants. The transmission system for the region consisted of 623 miles of high-voltage cable and 25 transformer stations. Erection of these stations was opposed on aesthetic grounds by landed squires.[66]

In operation, the Grid did not by itself lastingly resuscitate the British economy, all the bright predictions of the parliamentary committees notwithstanding. More-muted expectations were stated as early as 1927 (after the 1926 act had been safely passed) by Archibald Page. Sensitive to the

---

[64] Merz & McLellan, *Report on Electric Service.* The description of the London Grid presented here is based on this report (pp. 55–91), which describes the situation around 1936.

[65] Electricity Act of 1926, § 6.

[66] L. W. W. Morrow, "Great Britain Electrifies," *Electrical World* 96 (1930): 252–53.

self-esteem of the managers and engineers of existing companies, whose cooperation was needed, Page discounted the more extreme statements made earlier about the backwardness of Britain's electric supply industry. He pointed out that Britain had a well-developed gas industry, whose service reduced the demand for electric lighting, as well as a long-established practice of using steam power in industry without intervening electric power transmission. He observed that although the British steam plant might be less efficient than its electric counterpart, the economics of replacement did not always favor conversion. Page also raised doubts about the significance of per capita consumption of electricity, a statistic the committees promoting the Grid had stressed. He said that per capita consumption of coal and oil might be better indicators of industrial development, for in Britain the consumption of coal subsumed the generation of most electricity and took into account other energy supplies, such as gas. He rightly pointed out that the share of the cost of electricity that was attributable to distribution, in contrast to transmission, amounted to between 20 and 75 percent of the total according to location, and that distribution was in the hands of the utilities. Page also cautioned that savings from the Grid would not be impressive in areas where modern power plants already operated. He may have misled his audience, however, when he said that legislation had not been as responsible for the delay of large-area supply in Britain as the rivalry between large undertakings and an "overzealous display of local patriotism by the smaller municipal undertakings" had been.[67] The earlier legislation in fact protected the parochialism.

Even though the Grid did not revolutionize British industry, but instead brought changes that were more in line with Page's realistic anticipations, the reorganization of Britain's electricity supply did result in notable achievements. Because interconnection permitted the use of smaller, reserve plants, capital expenditures were lower than if supply had been from independent plants. By 1938 the CEB used a safety margin of approximately 10 percent reserve capacity compared to the 43 percent estimated reserve for 1925. As a result, the CEB estimated that £33 million of capital expenditure on spare plants was saved. It also estimated that in 1935, operating costs for specified power stations were 11 percent less than the cost of running the same stations under independent ownership in 1932, before the Grid began operations. Despite a pricing policy that by 1939 resulted in a surplus of at least £2.75 million (after paying interest on securities and providing a sinking fund for these), the CEB had reduced the price of electricity in bulk (at the busbars, where it was fed into the transmisson system) to .34d per unit in 1939 compared to a cost of 1.098d in 1923. The cost of electricity was also reduced in the United States and Germany during this period, but as historian Leslie Hannah concludes,

It [reorganization and the Grid] was, on the whole, a happy experience and one with which the industry could be well pleased. In the space of fifteen years between 1925 and 1940 the national grid system enabled the British supply undertakings to overcome their previous lag in development and inaugurate a vigorous expansion of sales based on low prices for the consumer. By the late

[67] Archibald Page, "Inaugural Address," *IEE Journal* 66 (1927): 1–11.

1930's Britain's electricity consumption per head of population equaled that in countries with similar income levels.[68]

An even more impressive tribute is paid to the originators and developers of the National Grid by those who are persuaded that it was a major reason for the climb in the 1930s of Britain's industrial economy from the lower ranks among industrial nations to a position behind only the Soviet Union and Germany in rate of growth.[69]

[68] Leslie Hannah, "A Pioneer of Public Enterprise: The Central Electricity Board and the National Grid, 1927–1940," in *Essays in British Business History,* ed. Barry Supple (Oxford: Clarendon Press, 1977), pp. 224–25. In this article, Hannah cites D. A. Wilson, "The Strategy of Sales Expansion in the British Electricity Supply Industry Between the Wars," in *Management Strategy and Business Development,* ed. Leslie Hannah (London: Macmillan & Co., 1976); Harry W. Richardson, "The New Industries Between the Wars," *Oxford Economic Papers* 12 (1961); and idem, *Economic Recovery in Britain, 1932–39* (London: Weidenfeld & Nicolson, 1967).

[69] I am indebted to Leslie Hannah for calling my attention to this argument.

CHAPTER XIII

The Culture of Regional Systems

THE planned power pools and grids constructed in the United States, Germany, and England heralded the era of regional electric supply systems, which linked cities, towns, countryside, and remote industrial sites. Alongside the planned systems, such as the Bayernwerk and Britain's Grid, the evolving systems of mostly privately owned utilities flourished. As will be seen in Chapter XIV, the latter had a long history and were characterized by incremental growth. In the 1920s, however, engineers, managers, and financiers realized that a technological form—the regional system—had fulfilled inherent implications and could be considered conceptually mature. (Actual growth, of course, would continue.) This realization was manifested in the engineers' and managers' conceptualization of underlying technical and economic relationships; their deft identification of critical problems and efficacious solutions; their introduction of organizational forms that were well suited for expansion; and their development of organized knowledge (science of technology) about regional power systems. Just as the the universal system had gathered momentum in the 1890s, so regional systems gathered momentum and flourished in the 1920s. Moreover, just as the earlier discussion of the momentum of the universal system suggested the concept of the culture of technological systems—in this case the values, ideas, and institutions that arose in response to the needs and opportunities for growth that were seen and defined by professional engineers, managers, and system builders—so, too, a culture of regional power systems is discernible. Supportive of the development of the technical core, these contextual elements were themselves shaped by the technical characteristics of the expanding power systems.

The technical core of the regional power system involved the widespread use of steam turbines in power plants, which after 1900 resulted in the unprecedented and unexpected spread of power systems. Station engineers and managers originally introduced steam turbines to replace the monstrously large reciprocating engines used in heavily populated areas, where the price of real estate was high. Turbines, however, proved to be more efficient than reciprocating steam engines, and they also lowered installation, operating, and labor costs. The largely unforeseen consequence of the introduction of the turbine was the sharp acceleration of the quest for load sufficient to fulfill the economy-of-scale potential of a large, efficiently loaded turbine. The turbines were, in effect, supply in search of demand. A superficial analysis might present the load-seeking turbine as an instance of technological determinism, but closer study will show that the force of the technology was circumscribed by a host of nontechnological factors which, together with the technology, shaped the events and trends that followed the introduction of the turbine.[1]

The large, load-seeking steam turbines found their complement in far-flung high-voltage transmission lines which followed the population shift to the suburbs. If the turbine was located in an urban center like Chicago, high-voltage transmission lines carried electricity out into the surrounding region. If the turbine was located at a mine mouth or brown-coal pit, transmission lines spread the power over the region and distributed it to industries, transportation systems, commercial enterprises, and private residences. In the 1920s and early 1930s, when regional systems proliferated, voltages of 100,000 and higher became state-of-the-art, so transmission over hundreds of miles became economical, particularly for hydroelectric systems (see Fig. XIII.1 and Table XIII.1). Furthermore, point-to-point transmissions grew into networks, grids, or rings as the long lines were punctuated by major switching stations where other generating stations fed power into the system or power for nearby loads was taken out.

Economic considerations pushed the introduction of larger turbines and larger, diversified supply areas. Electric supply was capital-intensive technology, and the overhead stemming from capital costs was a substantial part of the cost of a unit of electric power. Advocates of large, regional systems argued that the unit cost of generation was an inverse function of the size of the turbines and generators. They appealed to the long-accepted proposition that capital cost per unit of power generated decreased with the increase of the physical size and capacity of machines. The reason for this decrease was, in part, a matter of geometry, for heat and magnetic-field losses increased with surface area, and the ratio of surface area to capacity decreased with the increase in physical size of the turbine and generator. Others pointed out that the large, regional structure improved a system's load factor. Moreover, since capital charges were paid on a system's full capacity, a higher load factor meant a reduction in unit cost.

The growth of regional systems not only presented interesting technical problems to engineers but it also challenged managers and financiers. Technical, organizational, and financial inventions resulted. The turbine, for instance, proved to be an engine that enhanced organizational growth. Before World War I, the geographical extent of most utilities did not provide a load to match the supply potential of turbine-driven central stations. Utilities had obtained their franchises, or concessions, during an era of small-area, direct-current distribution and reciprocating-steam-engine generation. With technical change, this organizational structure became outmoded, and utilities began to reach out for new territory and new customers and to invent new organizational structures. Until the Grid,

[1] Thomas P. Hughes, "Technological History and Technical Problems," in *Science, Technology, and the Human Prospect: Proceedings of the Edison Centennial Symposium,* ed. Chauncey Starr and Philip Ritterbush (New York: Pergamon Press, 1980), pp. 148–49.

Figure XIII.1. *Increase in transmission voltages.* From Giant Power Survey, *p. 27.*

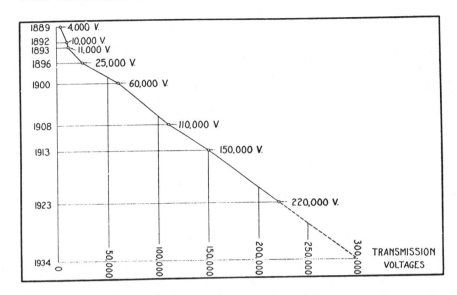

TABLE XIII.1. TRANSMISSION-SYSTEM MILEAGE IN THE UNITED STATES, 31 DECEMBER 1932

Voltage	Total System Mileage	Percentage
6,600–22,000	83,995	33.41
22,000–110,000	124,466	49.51
110,000–220,000	27,665	11.00
Not designated	15,282	6.08
Total	251,408	100.00

Source: U.S., Federal Trade Commission, *Utility Corporations* (S. Doc. 92, 70th Cong., 1st sess., 1928), Part 72A (1935), p. 41.

England was an exception to this trend because local governments confined power systems to their jurisdiction.

The introduction of the turbine, with its high-voltage transmission lines, generated unprecedented capital requirements, too. During the era of Edison, the cost of distribution lines almost matched the cost of generating equipment. In the 1920s the cost of a high-voltage transmission system was layered on low-voltage distribution costs. The transmission system included overland steel transmission towers, copper or aluminum lines with costly insulators, and a complex array of lightning-protection apparatus, circuit breakers, switches, transformers, and synchronous generators. The capital demands of the American utilities during the 1920s, when many regional systems were under construction, exceeded those of the railroads during the decades of their most rapid expansion.[2]

Consulting engineers also played a highly important, though still largely unacknowledged, role in furthering the growth of regional systems. The technical, organizational, and financial problems inherent in regional sys-

[2] U.S., Department of Commerce, Bureau of the Census, *Historical Statistics of the United States: Colonial Times to 1957* (Washington, D.C.: GPO, 1960), pp. 432–33 (ser. 95) and p. 589 (ser. 146). See also Forrest McDonald, *Let There Be Light: The Electric Utility Industry in Wisconsin, 1881–1955* (Madison, Wis.: The American History Research Center, 1957), pp. 186–87.

tems demanded the entrepreneurial spirit, comprehensive experience, and talent possessed by the consulting engineering firms and their principals. When political problems entered the mix as an important component, the consulting engineer's broad view and capacity for a multifaceted response became even more important. The consulting engineer assumed the mantle of the entrepreneur ready to preside over technical change in a complex environment. Demonstrating their entrepreneurial talents impressively, Merz & McLellan and the firm of Oskar von Miller responded to the politics-laden problems of planned government grids; in the United States, consulting engineering firms like Stone & Webster and Electric Bond & Share, would demonstrate a masterly ability to cope with the heavily managerial and financial problems that arose in the American electric supply system in the 1920s.

The regional systems of the twenties constitute the third stage in the evolution of light and power systems. The Edison direct-current systems of the first stage were characterized by homogeneity of supply and load. Generators and components of a similar kind were housed in a single location. Reciprocating steam engines were usually of the same kind or had the same characteristics—high or low speed, direct or belt drive. The distribution system took a standard voltage from the central station and delivered a standard voltage to consumers. The load on the system was almost entirely incandescent lamps, the motor being a rarity.

The systems of the second phase were more heterogeneous. They were characterized by the concept of transmission and distribution introduced as the "universal system" at the Chicago exposition of 1893. Generators with substantially different characteristics in different power stations were interconnected. Reciprocating steam engines and steam turbines were often used in the same power station. Varying outputs from the different kinds of generators (some alternating current, some direct) were combined in a single transmission system through the use of couplers—transformers (including Scott-connected transformers), synchronous generators, motor generators, frequency converters; couplers also connected different kinds of load, lighting and power (stationary and traction).

The regional systems of the 1920s can be categorized as the third stage in the evolution of electric supply systems. Again, the hallmark is increased heterogeneity. Whereas in the universal system of the second stage, different kinds of loads were systemically joined according to the concept of load factor, in the third stage, different kinds of energy sources were combined according to the more recently articulated concept of economic mix. The introduction of turbines and high-voltage transmission made the third stage possible, but they were not the essence of it. Turbines and high-voltage transmission stimulated the construction of far-flung systems, and the spread of these was so extensive as to include natural resources of various kinds. The engineers and managers of utilities took advantage of the presence of such varied energy sources as hard coal, bituminous coal, brown coal, high-head water, and low-head water in their supply areas to obtain an economic mix. The capacity to encompass complexity had increased.

The combination of diverse energy sources in the system allowed for an economic mix, a technical-economic concept on the supply side that was comparable in importance to load factor on the load side. The origins and evolution of the concept and practice are exceedingly difficult to document, however. Probably the concept followed the practice. Having combined, say, a coal-fired plant and a hydroelectric one for obvious economic reasons, engineers and managers responsible for system growth then thought through the general implications of their empirical probes. Because both Germany and the United States had coal and water power within economical transmission distance, systems organized according to the principles of economic mix spread rapidly in the two countries after World War I. In water-power-short England, waste-head power stations were connected with coal-fired plants, a variation on the same theme.

An economic mix is an interconnection of power plants whose energy sources are utilized complementarily. For example, an ideal economic mix would include an urban power plant using hard or bituminous coal, a lignite-fired plant at an open-pit site, a high-head hydroelectric plant drawing from a natural lake at high elevation, and a low-head plant using the running water of a river. These might be separated from one another by fifty to a hundred miles, possibly more. The operation of the plants would be carefully scheduled to take advantage of their complementary characteristics. The advantages derived were measured by an increase in economy and reliability; the means to achieve these were technical.

Scheduling the use of power plants in an ideal mix was a challenging and extremely interesting problem for utility engineers and managers. At their disposal they often had a variety of responses to changing, or transient, natural and man-made events. The high-head hydroelectric plant could be used sparingly, if at all, for peak load during the winter freeze. The run-of-river, or low-head, plant was less useful during periods of drought. The high-head plant was scheduled for maximum use during spring thaws; the running-water plant might be fully utilized for base load then too. The plant fired by hard coal was scheduled for heavy use during the winter freeze and probably during periods of drought, as was the plant fired by brown coal. Utilization of brown-, bituminous-, and hard-coal plants depended on comparative fuel costs. Strikes by hard-coal miners or an increase in the cost of rail transport of hard coal would increase the demand for brown coal, for instance. Layered on the challenge of scheduling the utilization of natural resources economically was the problem of efficient use of the plants, whose operating costs varied, in part, according to their age. In some actual systems, old thermal plants were used only during times of peak demand and when the more economical units were being serviced.

The economic-mix system demanded complex controls for scheduling power plant utilization. Increasingly complex information networks and remote controls were needed in the control, or dispatching, center. To facilitate control, engineers and mathematicians wrote equations describing the essential parameters, variables, and functions of the different power systems, and they attempted to design analog computers to solve the most tedious of these. In addition, load forecasting was helpful to dispatchers in scheduling power plant use. Load forecasting required the keeping of historical records, especially load curves, and information about changes in population, transportation, industrialization, and social patterns, and the weather. All of these directly affected the magnitude, location, and variation of the load. Utilities with hydroelectric plants in the mix also used historical hydrographic data and kept a running record of rainfall, snowfall, runoff, and other relevant details. These extensive data-gathering, record-keeping, and data-processing activities were carried out as long as their cost did not exceed the benefits measured by lower operating expenses, better load curves, and reliability of service.

The regional electric power system stimulated utility managers and engineers to think in terms that are now associated with systems engineering and information-and-control science. Engineers in other fields—petroleum refining, for example—also were thinking systematically, but electrical engineers made the more substantial contribution to systems engineering. Decades earlier, the managers and designers of railway systems had experienced the problems of scheduling traffic where and when it was needed, switching to meet that need, and forecasting loads. Production-line engineers had concerned themselves with routing and flow, but electrical engineers dealt with a system in which storage was uncommon—storage batteries were used sparingly—and the product flowed at the speed of light. So challenged, the electrical engineers began working out a science of information and control. In the 1920s, engineers increasingly used concepts such as "coordination," "integration," "control," "flow," "concentration," "centralization," and "rationalization." This is the language of systems. Among these concepts, "rationalization" is of unusual interest.[3]

Rationalization was much discussed in the 1920s. In Germany its proponents, both expert and popular, were unusually articulate and enthusiastic. They gave *Rationalisierung* varied and complex definitions. As applied to industry in general, the concept was decidedly influenced by developments in electric supply, and associated with it were the terms used by the engineers and managers who were building regional electric systems: *coordination, integration, stability, plan, order, control,* and *system.*[4] The rationalization movement in older industries, such as coal and iron, involved reorganization of both the physical plant and the management structure. This resulted in increased plant specialization and the alignment of technically related specializing plants. Badly located and equipped plants, mines, and businesses were closed. More efficient and larger-scale technology and management methods were introduced. Engineers who had been trained in scientific analysis and managers who were familiar with cost accounting, labor efficiency, and rational schemes of organization were employed. Standardization and research were encouraged. The rationalizers, whether managers or engineers, wanted to eliminate material waste, mechanical friction, and human confusion. The most ambitious of the rationalizers intended to extend their scope beyond a single plant, business, agency, or

---

[3] On the earlier development of information-and-control systems see Stuart Bennett, *A History of Control Engineering, 1800–1930* (New York: Peregrinus, 1979); Otto Mayr, *Zur Fruhgeschichte der Technischen Regelungen* (Munich: Oldenbourg, 1969); and Thomas P. Hughes, *Elmer Sperry, Inventor and Engineer* (Baltimore: The Johns Hopkins Press, 1971).

[4] Robert A. Brady has provided a deeply informed analysis of rationalization in *The Rationalization Movement in German Industry: A Study in the Evolution of Economic Planning* (Berkeley: University of California Press, 1933); see especially his Introduction for a definition of the term. See also Merle C. Turney, "The Rationalization of Electricity Supply in the United States" (Ph.D. diss., University of Illinois, 1937), esp. pp. 70–122.

industry to encompass the entire economy. They directed their attention to the inefficiencies of older and heavier industry, but they were inspired by their colleagues in the new and high technology of electricity supply.

Rationalization was a governing strategy for regional utilities in the 1920s. It carried many shades of meaning, but generally it signified obtaining the optimal combination of economic gains with a minimum input of economic resources, including capital and labor. In practice it brought, among other policies, the optimum utilization of available generation equipment. New and more efficient units carried steady base loads, while the older, less efficient units were brought into operation only during peak loads. Obsolete generating plants were removed from service and replaced by transformer substations that drew power, via extended transmission lines, from large, modern plants. The layout of transmission and distribution networks was carefully planned to obtain the most appropriate existing and future loads. Economists and engineers made detailed studies to enable them to choose between extending transmission lines or building new power stations to supply a remote load. Rationalization also called for labor-saving equipment and economies of scale. Because the level of capital investment was high, inventions and innovations resulting in the more efficient use of capital and labor were cultivated. The general engineering and management strategy emphasized the economical use of large, capital-intensive technology.

The spread of a system also gave its operators the opportunity to fulfill further principles of load factor. Economic mix was the newer concept, but load factor also provided guidance in lowering costs. In the 1920s, emphasis was placed on developing, by making complementary, a diversity of load in rural agricultural and industrial activities such as mining. From an economic viewpoint, however, the isolated small farm remained an unattractive load because the cost of distribution to a single farm required a capital investment that was unlikely to be repaid through use. As some private utilities noted, in order for them to supply isolated farms at an affordable price, they would have to charge other consumers rates which in effect would subsidize the farmers. The electrification of rural areas would be further stimulated in the 1930s, however, by regulatory requirements and by the cooperative ventures of farmers buying in bulk from private power companies or government agencies. On the other hand, through sales campaigns for home appliances such as stoves and irons, the utilities encouraged increased consumption by small consumers in towns and cities.[5]

The spread of regional power systems stimulated engineers and managers to exploit diversity among widely separated communities, or load centers, ingeniously. Differences in living and working habits arising from variations in the time of sunrise and sunset, contrasts in temperature and rainfall in highland and lowland regions—even differences in social customs stemming from varied ethnic composition and history—were used to shape the load curve. In the United States a few utilities could take ad-

[5] On rural electrification in the United States, see D. Clayton Brown, *Electricity for Rural America: The Fight for the REA* (Westport, Conn.: Greenwood Press, 1980). On the promotion of appliances to increase load in the 1920s and 1930s, see Leslie Hannah, *Electricity Before Nationalisation* (Baltimore: The Johns Hopkins University Press, 1979), pp. 186–212.

vantage of the difference in time zone within one area of supply, for the east side experienced the late afternoon peak load an hour earlier than the west. In Germany, a utility extending from one historic province to another combined advantageously the diverse demand that arose from different religious holidays and other patterns of culture involving the use of appreciable amounts of energy. Diversity, however, was fully exploitable only if predictable.

The concept of reliability and related reserve capacity was further honed and applied by the system builders. A system was judged reliable if the reserve generating capacity equaled the largest likely failure in the system, whether of a generating unit or a transmission line. Where several systems interconnected with a tie line of sufficient capacity, the interconnected system then needed to carry a reserve capacity equal to the probable hazard in the combination. The assumption was made that only one of the major hazards would occur at one time within the interconnection. If three independent systems, for example, each had a likely failure of 100,000 kw., their reserve would total 300,000 kw. Interconnected, they would carry a combined reserve of only 100,000 kw. During periods when a given system's units were shut down for maintenance work, the excess capacity of the other systems within the interconnection would be drawn upon. Such operating concepts and procedures are routine today, but in the 1920s they were just being introduced.[6]

Practitioners of regional systems technology articulated their concepts and strategies during the 1920s. The concepts were more inclusive than those defined by Samuel Insull for Chicago in an earlier era (see pp. 216–21 above). The synthesis of design and operating principles expressed by electrical engineers and managers in numerous technical articles and books in the United States, Germany, Great Britain, and elsewhere can be briefly summarized as follows:

1. Obtaining economies of scale with large generating units (steam and water turbines);
2. Massing the generating units near load centers or economical sources of energy and near cooling water (giant power plants);
3. Transmitting electricity to load centers (high-voltage transmission lines);
4. Cultivating mass consumption by charging low and differential rates and allowing supply to create demand;
5. Interconnecting power plants to optimize their different characteristics;
6. Interconnecting loads to take advantage of diversity and thereby raise load and demand factors;
7. Centralizing control of interconnected loads and power plants (establishing dispatching, or system-coordinating, centers);
8. Forecasting load requirements in order to achieve optimum operations within the interconnected system;

[6] For an excellent description and analysis of the operation of modern power systems, see Wallace E. Brand, "Northeast Electric Bulk Power Supply," *Public Utilities Fortnightly*, 9 June 1966, pp. 65–88; and W. C. Astley, "The System," in *Electric System Operation*, ed. B. G. A. Skrotzki (New York: McGraw-Hill, 1954), pp. 1–54.

9. Lowering installed and reserve capacity and coordinating maintenance shutdowns through the exploitation of power plant interconnections;
10. Accepting government regulation in order to establish a natural monopoly;
11. Earning a regular and adequate return on investment in order to obtain capital at reasonable interest.

These various policies and technologies were interdependent in their functioning and effects. In the twenties all were considered to be integral to the achievement of optimum load factor and economic mix. Not only was the physical technology of regional systems complex, but the theories of system management that evolved were the most sophisticated innovation since the formulation of complex railroad management concepts in the nineteenth century.[7]

The articulated principles were guidelines for the achievement of lower costs and greater reliability of supply, the primary objectives of the regional-system builders. Most of the private and government-owned utilities building regional systems held to this goal; the private utilities embroidered it by adding the profit motive. Not all utilities sought to lower costs, however. For example, small urban utilities were content to obtain a return on their investment; lowering costs by increasing size and consumption was not their goal. Small municipal utilities were also satisfied to supply electricity at prices that assured a load appropriate to the capacity of the plant, and to earn returns that were acceptable to the taxpayers (ratepayers in Britain).

Having defined operating concepts and objectives, the regional utility builders evaluated the performance of the components of the system—the power stations, substations, transmission and distribution lines, etc. The physical variables of each component affecting the achievement of the builders' objectives were quantified as costs, and the system engineers then sought means to vary the characteristics of a component in order to lower costs. The effect of a variation on the other components and on the connectors then had to be ascertained. For this reason it was desirable to model or to develop equations for the system. This step was not difficult, for the engineers were accustomed to stating the physical characteristics of the system quantitatively, and cost accountants and managers could then translate these characteristics and related efficiency data into costs.

Such analyses revealed imbalances or bottlenecks within the system. The imbalances were seen as drags on the movement of the system toward its goals, especially those of lower costs and larger size. Desiring to remove an imbalance or bottleneck, the engineers and managers then defined the reverse salient in the system as a problem or set of problems that, when solved, would remove the impediment. The solution to the problem for one component often reverberated through the system causing imbalance in other components or connectors (transmission lines for instance), and these were then attacked as problems.

Changes in outside forces such as demand and supply also affected the

[7] These principles were articulated by Samuel Insull, Charles Merz, and A. Koepchen, for instance. Their articles, reports, and speeches have been cited throughout this study. See also Thomas P. Hughes, "Regional Technological Style," in *Technology and Its Impact on Society: Tekniska Museet Symposia No. 1* (Stockholm: Tekniska Museet, 1979), p. 218.

achievement of system objectives. System costs often increased because of changes in supply prices or because of conditions, such as regulatory requirements, originating outside the system and beyond the control of its operators. These effects were usually localized within particular system components or connections, however, and thus became identifiable as problems. The solutions to these problems in turn often caused imbalances and the need for further correction. As a result of the on-going process of engineering and cost analysis and problem solving, or invention and development, the regional system usually remained in a state of dynamic imbalance. The changing environment and internal growth prevented the achievement of an optimal dynamic equilibrium. Only if the system had been closed—not influenced by external forces—might an equilibrium have been achieved once growth ended.

System imbalances constituted reverse salients, and their identification as critical problems often led to technical solutions (see pp. 14–15 above). Managers and cost accountants turned to engineers, inventors, and scientists for these solutions. Reverse salients, critical problems, and solutions were last considered in this study in the discussion of the period of transition from direct current to alternating current in the late 1880s. Because the critical problems of the d.c. system could not be adequately solved, the new, alternating-current system developed on the basis of a solution that could not be encompassed in the old system. In contrast, during the 1920s, the era of the mature regional system, critical problems were solved within the context of the existing system, reverse salients were corrected, and the system's momentum was maintained. Controversies arose about alternative solutions, and different regions incorporated technologies with different characteristics, but these differences were not considered to be so great as to herald the emergence of a new system.

The nature of the critical problems solved by engineers and industrial scientists in the 1920s can be indicated by consideration of some of the outstanding ones that were common to the United States, Germany, and Great Britain. Control of the heterogeneous regional system, with its diverse loads and mixed energy sources, was an embracing critical problem. Inadequate control, or failure to continuously match supply to demand, resulted in instability in the system, which manifested itself as variations in voltage and frequency. The dispatching center was the locus of the problem and the site of the solution. The typical control, or dispatching, center of the 1920s differed from the control room of an urban system operating a decade earlier (see Figs. XIII.2 and XIII.3).[8] The earlier center might have had a marble-paneled switchboard with an ornamental iron framework. Attendants manually operated switches mounted directly on the board. The earlier control room was probably located in the power house to allow personnel direct access to the machinery operators. Contact with other

[8] This description is based on Georg von Siemens, *History of the House of Siemens*, trans. A. F. Rodger, 2 vols. (Frieburg/Munich: Alber, 1957), 2: 142–45. See also Frank Gillooly, "Reinforcing System Operator's Memory," *Electrical World* 74 (1919): 347–50; and *System Operation*, a pamphlet issued by the New York Edison and Allied Companies in July 1927. I am indebted to Ms. Dorothy Ellison of the Consolidated Edison Company of New York for information on the history of load dispatching at Consolidated Edison and for the use of several illustrations.

Figure XIII.2. *Evolution of the load-dispatching, or systems-operations, center of Consolidated Edison of New York: The system operator's board in 1902. Generators, transmission and distribution feeder lines, and ties between central stations are diagrammed. Tags hung on pegs indicate the state of system components. The first system operator's board used by Consolidated Edison was introduced in 1898, but no photo of it survives. Courtesy of the Consolidated Edison Co. of New York.*

Figure XIII.3. *Evolution of load dispatching: Consolidated Edison's semi-automatic system operator's board, 1912. Courtesy of the Consolidated Edison Co. of New York.*

power plants was by telephone. With the advent of the high-transmission voltages of the 1920s, circuit breakers and switches had to be located a safe distance from the control room operators, either in a neighboring room or, better, in an open-air switchyard. Then, using remote-control electric signals to energize electromagnets or small electric motors, the control room operated the oil-filled circuit breakers and switches. By telephone instruction, if there were operators present, the dispatcher maintained control over circuit breakers and turbogenerators at locations tens or even hundreds of miles away. The central control room for the system received telegraph signals from remote sensors indicating the state of various unattended units and sent signals to control the performance of these units. The central control room received voltage, amperage, and frequency readings and information about the physical state of turbines, boilers, dam water-flow, circuit breakers, and switches in the far-flung transmission network. Large electrical manufacturers and small companies specializing in instrumentation—for instance, Leeds & Northrup in the United States—developed transmitters, receivers, and displays to provide central control rooms with information by telemetering.[9] Control centers in turn constructed large models of the ever-changing systems. Lights on the models indicated which transmission lines, generators, bus bars, and transformers were in use; the lights for closed switches were one color, those for open switches another. Surrounded by his instruments and controls, the dispatcher became a popular symbol of high technology in the twenties.[10]

Serving as a reference for the control of systems, the frequency meter was an extremely important instrument for the dispatcher. For formerly autonomous systems to merge or for systems to exchange power by interconnection, precise standardization and regulation of frequency was required. In the United States, 60 cycles was the standard frequency for most systems; 50 cycles prevailed in Germany; and Britain was in the process of abandoning its plethora of variations and settling on 50 cycles. A system controller knew that if the frequency of the system varied within tolerable limits around the standard, the supply of energy into the system would equal the energy taken by load and losses. If the frequency fell considerably, it was his job to signal the power stations to increase energy supply by fueling boilers more rapidly in thermal plants or by increasing water flow in hydraulic installations. If the frequency rose significantly, he would call for a decrease in the flow of energy into the system. The frequency meter gave a direct reading of the frequency of the waves of alternating current, but more importantly provided an indirect indication of the speed of rotation of the turbogenerators. The rotors of the generators felt the load on the system in the strength of the magnetic field within and against which they turned. The frequency of a power system was analogous to the revolutions of the early Watt steam engines, the governors (controllers) of which increased or decreased steam flow to maintain a balance between energy input and output. Inventors and engineers designed automatic feedback controls, for frequency regulation and quick response were critical problems, but because the automated system was so complicated, manual control of generation was the general expedient in the 1920s.[11]

The focus on problems of system control is revealed by the list of topics dealt with in papers published by the American Institute of Electrical Engineers from 1929 to 1934. The topics differed radically from those of papers published a generation earlier. Of the twenty-four new topics introduced, more than half pertained to power systems and transmission.

---

[9] Leeds & Northrup has deposited its papers in the archives of the Eleutherian Mills Historical Library in Wilmington (Greenville), Del.

[10] Similar control centers were developed from other industries. See, for example, Eduard Schmäing, "Process Control Rooms," *BASF Review* 24 (1974): 66–71.

[11] Bennett, *A History of Control Engineering*, pp. 166–67, 171.

Those that were written about most frequently were "interconnected power systems and stability"; "automatic stations, telemetering, and supervisory control"; "A-C distribution networks and relays"; "automatic control of frequency and voltage regulation"; "transformers—tap changing under load and surge-proof"; "mercury arc rectifiers and inverters"; "oil circuit breakers—theory, design, and testing"; "impulse voltage—generation, measurement, and testing"; and "new types of fuses, lightning arresters, protective gaps, arcing horns, switchgear, etc."[12]

The list of subjects also reveals another major "critical problems" area pertaining to regional systems, insulation. Surge-proof transformers, oil-filled circuit breakers, impulse voltage, and new types of fuses and lightning arresters all relate to high-voltage transmission and especially to the insulation made necessary by it. Surges are high, transient voltages caused by short circuits or lightning; circuit breakers present serious insulation problems; impulse voltages are generated in experiments to test the insulation used in power systems; and protective gaps and arcing horns are designed to carry off, or interrupt, overloads caused by lightning before the extreme overload damages equipment; mercury arc rectifiers provide a means of converting alternating current to direct current in high-voltage, universal systems. These were the critical electrical problems of the rapidly evolving regional systems.

The need to develop insulation that would adequately protect equipment and transmission towers from high voltages, surges brought on by short circuits, and the transient extremes caused by lightning was so general that electrical engineers and scientists in the 1920s analyzed, categorized, and quantified their experience and experiments to establish a science of insulation. How academics scientifically approached the high-voltage problem of corona about the turn of the century has already been explored (see pp. 158–60 above). In the twenties, some of the same researchers and many others turned to the study and design of insulation. Porcelain insulators of complex design and unusual material properties were needed to suspend the transmission lines from the great cross-country towers; bushings were needed to carry lines into transformers; and circuit breakers had to be appropriately insulated with oil. These were some of the specifics that fell under the general rubric of insulation.[13] As had been the case with instruments for system control, some companies, such as Ohio Brass in the United States, specialized in solving the critical problems of insulation (see Fig. XIII.7, p. 380, below).[14]

[12] Charles F. Scott, "The Institute's First Half Century," *Electrical Engineering* 53 (1934): 666. The critical problems of transmission were discussed in Frank Baum, "Some Power Transmission Economics," *AIEE Proceedings* 26 (1907): 743–57; P. W. Sothman, "Problems of High-Tension Transmission Lines," ibid. 33 (1914): 201–14; and P. Junkersfeld, "Distribution of Electrical Energy: Report of the Sub-Committee on Distribution," ibid., pp. 235–41.

[13] Georg Boll, "Nationale und internationale Verbundwirtschaft auf Höch- und Höchstspannungsleitungen," in *Das Zeitalter der Elektrizität* (Frankfort on the Main: VDEW, 1967), p. 78; and idem, *Geschichte des Verbundbetriebes* (Frankfort on the Main: VWEW, 1969), pp. 129–38.

[14] I am indebted to Joseph T. Lusignan, formerly an engineer at the Ohio Brass Company, for calling my attention to the contributions of A. O. Austin and Ohio Brass to the development of the modern cap-and-pin suspension insulator for high-voltage transmission lines. Among

Figure XIII.4. Network analyzer, 1937. Courtesy of the General Electric Co., Schenectady, N.Y.

A reverse salient of the ever-expanding, complex regional systems was the inability of engineers to precisely analyze and define them with equations that showed functional relationships. Efforts to write these equations resulted in complicated mathematical problems, the solution of which was tedious and time-consuming, if not impossible. Without a precise and clear understanding of the systems, engineers had to rely on empirical, or cut-and-try, methods to improve system performance. Engineers who were thoroughly trained in mathematics defined this reverse salient as a problem calling for new modes of equation solving and more complex analog models of the systems. In the United States, engineers at Westinghouse Electric and General Electric assembled miniature analog power systems in order to investigate the stability problem (see Fig. XIII.4). At the Massachusetts Institute of Technology, Professor Vannevar Bush (1890–1974) of the electrical engineering department and two of his graduate students, Harold Hazen and Hugh Spencer, also embarked on the design of a device to simulate a complex power system. They used small transformers as generators and motors, variable resistors as loads, and artificial transmission lines. The simulator had three generating stations, 200 miles of line, and six load centers encompassed in a space of 50 square feet. Drawing on advice from General Electric engineers and others, the M.I.T. group completed the device, the "network analyzer," in 1929. During the next two decades numerous utilities and electric manufacturers used the analyzer in solving system problems. Similar devices called network calculators and a.c. calculating boards were built elsewhere.[15]

Network analyzers made experimentation possible by varying the arrangement and size of generators, transmission lines, and transformer banks. Measuring the effects of variations in one or several components on the performance of other components, these analyzers tested the system's stability. The results obtained only approximated the real events in the full-scale system, however, for the actual events were too complex for exact simulation.

Engineers—and physicists—also wanted to be able to write general system

---

the articles written by Austin are "Insulation of Some of the Higher Voltage Lines" (Paper presented at the International High-Tension Congress, Paris, 18–27 June 1931); and "A Laboratory for Making Lightning and Other Tests upon Full Sized Insulator Strings and Towers" (Paper presented at the International High Tension Congress, Paris, 6–15 June 1929). On the development of transmission towers and suspension insulators, see also H. W. Buck, "Some New Methods in High-Tension Line Construction," *AIEE Proceedings* 26 (1907): 981–87; and E. M. Hewlett, "A New Type of Insulator for High-Tension Transmission Lines," ibid., pp. 975–79.

[15] I am indebted to Professor Karl Wildes of the M.I.T.'s Department of Electrical Engineering for the information on which this account of the network analyzer is based. Professor Wildes is writing a history of the electrical engineering department at M.I.T. Other sources for the background and invention of the analyzers are Charles L. Fortescue, "Method of Symmetrical Coordinates Applied to the Solution of Polyphase Networks," *AIEE Transactions* 37 (1918): 628–716; the paper by H. L. Hazen, O. R. Schurig, and M. L. Gardner, "Abridgment of 'The M.I.T. Network Analyzer,'" *AIEE Journal* 49 (1930): 872–75; H. A. Travers and W. W. Parker, "An Alternating-Current Calculating Board," *The Electric Journal* 27 (1930): 266–70; and Eric Gross, "Network Analyzer Facilities Keep Pace with Industry Growth," *Electric Light and Power* 34 (1956): 82–85. The last-named article lists the location and date of installation of forty network analyzers between 1929 and 1956.

equations and to solve these in a reasonable amount of time. Vannevar Bush summarized the frustration of those who were attempting to solve equations for the system:

Electrical engineering, for example, having dealt with substantially linear networks throughout the greater part of its history, is now rapidly introducing into these methods elements the non-linearity of which is their salient feature, and is baffled by the mathematics thus presented and requiring solution. Mathematical physicists are continually being hampered by the complexity rather than the profundity of the equations.[16]

The inability to analyze adequately, solve equations expeditiously, and design and predict well was obviously threatening to retard the further development of regional systems.

Between 1927 and 1931 Bush and his associates at M.I.T. invented a continuous integraph for solving second-degree equations and the differential analyzer (a device with mechanical components and an electrical drive) for solving differential equations. The differential analyzer, an analog device, not only proved useful in solving power system problems but also stimulated the development of general-purpose analog and digital computers (see Figs. XIII.5 and XIII.6). In the thirties, engineers at the University of Pennsylvania built a differential analyzer to help solve system stability problems. During World War II, physicists and engineers there invented and built the ENIAC, the world's first general-purpose digital computer. The work on the ENIAC was influenced by the earlier experience with the differential analyzer.[17]

Regional systems problem solving occurred mostly within institutional settings. In the 1880s individual inventors had made an effective contribution to the emerging fields of technology such as electric supply, but by the 1920s the mature systems had accumulated institutions that tended to exclude individuals unsupported by collegial activity and expensive equipment designed specifically to respond to the problems of a well-defined system. During the twenties, invention, research, and development continued to be carried out in the academic world (as work on the differential analyzer illustrates) and in the laboratories of the electrical manufacturers.[18]

The high-voltage laboratories that flourished in the 1920s were among the newest of the research organizations to form close ties with the regional power systems. Set up by electrical manufacturers, universities, and in-

---

[16] Vannevar Bush, "The Differential Analyzer: A New Machine for Solving Differential Equations," *Franklin Institute Journal* 212 (1931): 448.

[17] See Thomas P. Hughes, "*Eniac*: Invention of a Computer," *Technikgeschichte* 42 (1975): 147–65; and Herman Goldstine, *The Computer from Pascal to von Neuman* (Princeton: Princeton University Press, 1972).

[18] The discussion of critical problems has been limited in this section to the electrical system per se, but engineers and inventors were also seeking solutions to critical problems in turbines and boilers for regional systems. These are discussed by Donald A. Wilson in "The Economic Development of the Electricity Supply Industry in Great Britain, 1919–1939" (D. Phil., Oxford University, 1951), p. 234. Wilson provides an informed analysis of British technical practices, as compared to American practices, in the field of electricity supply. He concludes that British technical decisions, although less impressive in isolation, were almost always appropriate to Britain's economic circumstances.

Figure XIII.5. *Vannevar Bush (left) and Harold Hazen (second from right) at the product integraph, a step in the development of analog computers which preceded the differential analyzer. Courtesy of the M.I.T. Museum and Historical Collections, Cambridge, Mass.*

Figure XIII.6. *Vannevar Bush with the differential analyzer completed in 1931. Courtesy of the M.I.T. Museum and Historical Collections, Cambridge, Mass.*

dustrial associations, these laboratories were part of the long-standing trend
that led engineers with a scientific approach and scientists with industrial
interests to turn their attention to the solution of the critical problems of
evolving power and light systems. The trend was easily discernible by the
turn of the century, when the problems of polyphase systems were being
addressed by engineers who drew on electrical theory, tested hypotheses
in laboratory and field experiments, derived quantitative equations, and
obtained data expressing the relationships among complex factors influencing the performance of electric systems.

In Germany in 1921, twenty utilities and ten manufacturing firms established the Research Center for High Voltage Apparatus (Studienge-
sellschaft für Höchstspannungsanlagen E.V.) The center concentrated on
problems associated with insulation, lightning, short circuits, grounding,
electrical interference from high voltage transmission with electric communications, and ice and wind loading on transmission towers and lines.
Through fundamental research, scientists and engineers sought to understand the phenomena. The center also became known for its development
of measuring devices.[19] Moreover, AEG and Siemens, the German manufacturers, established facilities for investigating high-voltage transmission
and for developing system apparatus.[20]

In 1926 Stanford University took the lead among U.S. universities in
establishing a high-voltage laboratory. The university's location not far
south of San Francisco, California, had stimulated the growth of its electrical
engineering activities, and California's utilities, with their pioneering high-
voltage systems, helped support the research program of the high-voltage
laboratory. Harris Ryan, an academic who had done major research on
high-voltage corona effects (see p. 158 above), headed the laboratory, which
was later named for him. In 1926 the laboratory concentrated on the
characteristics of air as a high-voltage insulator, oils as insulators in transformers and switches, and transmission-line insulators and insulation materials working at 220,000 volts and upward. Another goal of the Stanford
center was to study the nature of lightning. An antecedent research facility
at the university had brought collaboration between engineers from the
Pacific Gas & Electric Company and the Southern California Edison Company and faculty members of Stanford's electrical engineering department
in research and experimentation that culminated in the development of
insulators for the state's 220,000-volt transmission lines, one from the Sierras to San Francisco and the other from the Sierras to Los Angeles. By
1929, graduate students on fellowships also were engaged in the Stanford
laboratory's program.[21]

One of the world's most highly publicized high-voltage laboratories was
located at the Schenectady plant of General Electric (which is not to be

---

[19] Boll, "Nationale und internationale Verbundwirtschaft," p. 78.

[20] Von Siemens, *House of Siemens*, 2: 105–11.

[21] Two information releases dated Stanford University, 17 September 1926, and a report
on the laboratory's activities dated 15 February 1929, Harris J. Ryan Collection (SC 25, Misc.
Box 4), Stanford University Archives, Stanford, Calif. I am indebted to Professor Hugh H.
Skilling, former executive head of Stanford's Department of Electrical Engineering, for information about Harris J. Ryan and high-voltage research.

Figure XIII.7.   *High-voltage laboratory of the Ohio Insulator Co. (Ohio Brass Co.): Outdoor impulse gaps* (left and right); *test section of 220-kilovolt transmission towers* (left). *From A. O. Austin, "A Laboratory for Making Lightning" (paper presented at the International High Tension Congress, Paris, 6–15 June 1929), p. 6.*

confused with the General Electric Research Laboratory). The publicity promoting popular consumption of electricity identified the famous Charles Steinmetz as the presiding genius of the high-voltage facility, but F. W. Peek, Jr., carried on research that not only attracted the attention of industrial and academic engineers but influenced the design of GE's apparatus as well (see Fig. XIII.8). As a case history, Peek's research sheds light on the high-voltage experimentation that was done for regional power systems as well as the evolution that had taken place in the relationship between science and technology since the early years of power transmission (see pp. 156–58 above).

Peek, like Ryan, had attended a university that in its early years won renown for its course in electrical engineering. He graduated from Stanford with an A.B. in engineering in 1905. The highly important developments in high-voltage transmission that were then taking place in California, and that undoubtedly made a strong impression on the young student, help explain Peek's commitment to the field. His formal education was com-

Figure XIII.8.   F. W. Peek, Jr. (far right, seated), *with associates and visiting scientists at GE's Pittsfield works, 12 April 1933. Courtesy of the General Electric Co., Schenectady, N.Y.*

pleted some years later when he received the M.E.E. from Union College while working for the General Electric Company in Schenectady, New York.

Peek joined General Electric in 1905, and even though the importance of engineering research on high-voltage transmission was only beginning to be recognized, he found the means to pursue his interest, rigging "his first apparatus in an old box car in a remote corner of the Schenectady plant . . . experimenting, recording, studying, and rebuilding." In 1909 he was one of the first engineers to join Charles Steinmetz's consulting engineering department at Schenectady. It appears that he was encouraged by Steinmetz in the high-voltage research, but was not under his close supervision. A GE engineer, noting the difficulty of assigning credit when closely associated investigators work simultaneously on similar problems, recently observed that Peek was an independent investigator who carried on much in his own way, "worked out his circuits (impulse generators) more or less independently, although it would be reasonable to suppose that Steinmetz, as head of the laboratory, would be informed as to the work Peek and the other members of the laboratory were performing and the apparatus that they were using.[22]

Peek gained a substantial reputation for his research on high-voltage phenomena, especially the corona effect and lightning. The scientific char-

[22] W. W. Lewis to J. H. Hagenguth, 16 May 1968 (a letter loaned to the author by Lewis). The following pages on Peek are based in part on Thomas P. Hughes, "The Science Technology Interaction: The Case of High-Voltage Power Transmission Systems," *Technology and Culture* 17 (1976): 654–59.

acter of his research is attested to by his publication of twenty scholarly papers in *AIEE Transactions* between 1911 and 1931. The articles, especially the series of four papers on "The Effect of Transient Voltages on Dielectrics," seriously question the doubtful proposition that engineers—and Peek was employed as an engineer and was so titled—do not publish. Peek also wrote a book, which is still used, entitled *Dielectric Phenomena in High-Voltage Engineering*.[23]

Peek's education and publications suggest that his approach to engineering problems was scientific. However, the primary influence on his choice of problems was not scientific. From his experience Peek believed that economics—not science—called the technological tune. The increases in transmission voltages were decided upon, he wrote, "purely from the economic viewpoint." The real and anticipated demand for large amounts of power at greater distances from major power sites dictated the striving for higher voltages, and the higher voltages brought the technological problems, such as coronas. In his early engineering research Peek focused on the corona problem (see pp. 159–60 above), but in about 1913 his interest shifted to the effects of lightning upon high-voltage transmission lines and he continued to work in this field until his untimely death in an automobile accident on 26 July 1933.

The availability of a scientific instrument, an impulse generator, was a major reason why Peek shifted his attention to lightning rather than to another reverse salient in the expanding high-voltage transmission front.[24] The machine was available because Peek had designed and constructed a 200-kilovolt impulse generator to investigate coronas.[25] In order to use the machine for lightning research, he first determined the voltage and wave form of real lightning (inclination of wave front, tail, etc.). Establishing the characteristics of lightning so that these could be expressed precisely, even quantitatively, was a research project that would certainly have been called science, even natural science, if Peek's motives had not been application, or the solution of a technological problem.

Peek's motives were not those of pure science, however. His primary reason for investigating the nature of lightning was to create precise simulations of the various forms of it with the impulse generator. These simulations were needed to test the various components of transmission lines.

---

[23] The four lengthy papers, all titled as above, appeared in *AIEE Transactions* 34 (1915): 1857–1909; 38 (1919): 1137–64; 42 (1923): 940–47; and 49 (1930): 1456–69. The book was published by McGraw-Hill in 1915.

[24] It is also possible that Steinmetz and the company's management influenced Peek's choice of the lightning problem.

[25] F. W. Peek, Jr., "Lighting. 1," *General Electric Review* 32 (1929): 604; idem, "The Effect of Transient Voltages on Dielectrics," *AIEE Transactions* 34 (1915): 1857; idem, "The Effect of Transient Voltages on Dielectrics. III," ibid. 42 (1923): 940. W. W. Lewis, also a General Electric research engineer, writes that the 200-kv. transformer was used by Peek before he published his 1911 paper "Law of Corona and the Dielectric Strength of Air," ibid. 30 (1911): 1889–1988. According to Lewis, the "so-called impulse generator consisted of the same 200-kv. transformer used in the corona tests" (W. W. Lewis to J. H. Hagenguth, 16 May 1968). Lewis's letter does not specify the modifications that took place as the 200-kv. testing transformer metamorphosed into the "impulse generator," but Peek's dating the introduction of the impulse generator as "about 1913" ("Lightning. 1," p. 604) and Lewis's dating the first use of the transformer as 1911 suggest such modifications.

During two decades of lightning research, 1913–33, Peek increased the power of his impulse generators to the order of a million kilowatts in order to approximate "voltage and other conditions [resulting from lightning] that usually occur on transmission lines.[26] As a result, he tested full-scale components. Peek also relied on the action of natural lightning on operating transmission lines to provide data, but the natural event was not as manipulable a tool as the man-made one.

In 1929 Peek wrote with obvious satisfaction that progress in obtaining knowledge of lightning had been so rapid that "the subject may now be said to be on an engineering basis," the indication of this being the expression of lightning phenomena quantitatively in volts, amperes, and units of time.[27] Furthermore, Peek stated the following general characteristics of lightning: voltage, on the order of 100 million; current, on the order of 100,000 amps; energy, on the order of 4 kwh.; power, on the order of a thousand billion h.p.; time, on the order of a few microseconds; gradient at breakdown, 100 k.v. per foot; and the charge formed by the action of air currents on raindrops.[28] Beyond this and other quantitative data, Peek pictured the wave shape of lightning on a cathode-ray oscilloscope, determined the time required for a cloud to discharge, and found the attenuation of lightning waves traveling on a transmission line.[29]

Armed with increasing knowledge, Peek made subtle and persistent efforts to apply it. In the application, the science-technology relationship is starkly revealed. Initially, in his AIEE paper of 1915, Peek discussed the general effects of transient voltages on dielectrics, lightning being only a special case of transient voltages, and insulators and conducting lines being special cases of capacitors with atmospheric dielectrics. The major finding announced in this paper was that the behavior of dielectrics when subjected to continuously applied alternating current differed from their behavior when exposed to transient voltages, especially lightning. He introduced the term *impulse ratio,* which he defined as the ratio of the impulse sparkover voltage to the continuously applied sparkover voltage. Peek had found that sparkover across an air gap—over an insulator for example—occurred at a lower voltage for the continuously applied, or normal, high voltage than for the transient voltage wave. The importance of this finding—especially the values of the impulse ratio that Peek determined for different transmission-line components—was obvious for the design engineer attempting to protect the line against transients.

Within a few years of his discussion of the laws governing the rupture of dielectrics by transient and nontransient voltages, Peek addressed the question of practical applications.[30] His research permitted him to provide for the design engineer formulas and tables predicting the effects of lightning voltages on the lightning-arrester gaps, insulators, and bushings used in transmission lines and transformers. His simulated lightning had proved

[26] Peek, "The Effect of Transient Voltages on Dielectrics. III," p. 940.
[27] Peek, "Lightning. 1," p. 602.
[28] Ibid.
[29] F. W. Peek, Jr., "Lightning: Progress in Lightning Research in Field and Laboratory," *AIEE Transactions* 48 (1929): 436.
[30] F. W. Peek, Jr., "The Effect of Transient Voltages on Dielectrics," ibid. 38 (1919): 1137–56.

that these components would fail (permit sparkover) at lower voltages with regular currents than with transient voltages such as lightning. This information was especially valuable to designers of lightning arresters, for an arrester that would ground destructive regular currents might fail to do so with transients. Peek provided detailed information on a fascinating variety of components, such as different forms of protective lightning arresters and combinations of forms involving spheres, points, and horns. He found that "spheres discharge the very steep waves, the horns the moderate ones, and the points continuous high-frequency waves, of slanting front and static."[31]

Peek also supplied design data for the grading shield for insulators, transformers, and transformer insulation. A synthesis of his research was the statement he made in 1929 on lightning-proof transmission lines and the coordination of line and transformer insulation.[32] "A line of moderate height," he concluded, "protected with ground wires and properly insulated [using his data], could usually be made lightning proof against induced voltages [lightning]." The engineer's sense of balance can also be discerned in Peek's recommendation that in an entire transmission system the transformer insulation should be slightly stronger than the bushing flashover voltage, and that this voltage in turn should be higher than the flashover voltage of the line insulators. "It is obviously not good engineering to make the transformer (the most expensive) the weakest link in the insulator chain," he said.[33]

Peek's articles on high-voltage transmission and his painstaking and bold experimentation were responses to problems that arose as high-voltage transmission evolved. There is no substantial evidence that scientific discoveries stimulated engineers like Peek and Ryan to embark on their investigations. These men depended upon the corpus of science, but science was not an immediate cause of their activity. Having worked for large industrial corporations that were heavily committed by investments and competence to the expansion of electric light and power systems, especially high-voltage transmission, Peek was closely attuned to technological events. The events that particularly interested him were those manifesting reverse salients on expanding technological fronts.

The work that Peek accomplished involved a method that should be called scientific. The authorities cited and the periodicals for which he wrote represented organized knowledge; he used and attempted to formulate general statements or laws; mathematics was an analytical tool and a language for him; hypotheses were formulated; and experiments were designed for the laboratory or in nature to test these hypotheses. Such an approach is usually labeled scientific.

Technology and institutions such as high-voltage laboratories tend to evolve together. Edison and his associates formed an urban utility at about the same time they launched their research-and-development program for developing an electric lighting system. When the universal supply system was introduced, utilities metamorphosed into light and power companies,

[31] Ibid., p. 1137.
[32] Peek, "Lightning: Progress in Research," pp. 447–48.
[33] Ibid., p. 447.

and when power transmission became feasible, newly formed power companies presided over its development. Regional power systems also needed new organizational structures. Thus, in addition to high-voltage laboratories, multipurpose consulting engineering firms and a new generation of holding companies came into being. The consulting engineering firm and the holding company did not suddenly appear in the 1920s, for each had a longer history, but they reached fruition then, as evidenced by differentiation of form and function.

The growth of organizations that related systemically to regional power technology added to the momentum of this technology. Large numbers of inventors, engineers, workers, scientists, managers, and entrepreneurs with expert knowledge of, and vested interest in, the technology became committed to its growth. In many cases the new institutions merged with the utilities to form a suprasystem incorporating technological, business, and financial components. In other cases, looser ties were formed, but these ensured that a high-voltage research laboratory, for example, would give priority to the critical problems of a particular utility or that a consulting engineering firm would be retained for an extended period.

Historians of science and technology have long recognized the importance of research scientists and industrial laboratories in the evolution of electric supply, but the role of the consulting engineering firm has usually been overlooked. The consulting engineers emerged in the 1920s as entrepreneurs of regional systems, integrating the systems' technical, economic, and political factors and presiding over the growth process. Earlier, their function had been to advise utilities and others who purchased electrical equipment and constructed facilities. In time, the more successful consulting engineers established firms and took over for some utilities responsibility for designing and constructing power plants. This was far more economical for the utilities than maintaining a permanent staff of experts on large-plant construction. The leading consulting engineering firms in Germany and England—for example, those of Oskar von Miller and Merz & McLellan—also became involved in advising government authorities on electric supply matters because in these countries, in contrast to the United States, government-owned utilities were common. In the United States, where government was not a major source of funds for the utilities, consulting engineering firms found an important role in assisting their clients with the problems of financing. And as such advantages of the large-scale utility as economies of scale in purchasing and in supply became clear, some of the leading consulting engineering firms took over the management of groups of smaller utilities, thereby passing along to them some of the benefits of large-scale operation. Involved in the design and construction of systems adapted to local conditions, providing management, clearing away political obstacles and lobbying for legislation, and obtaining financing, the consulting engineering firms were impressive in the role of entrepreneurs presiding over technological change. Often their entrepreneurial resources were brought to bear in projects that in size rivaled the major railway projects of the late nineteenth century.[34]

[34] For instance, in 1854 a total of $10 million was invested in the Western Railway of Massachusetts. In 1860, at least $30 million was invested in the New York Central Railroad's

Figure XIII.9. *Charles A. Stone (left) and Edwin S. Webster, 1888. Courtesy of the M.I.T. Museum and Historical Collections, Cambridge, Mass.*

Figure XIII.10. *Stone & Webster's first office sign. Courtesy of Stone & Webster, New York, N.Y.*

The firm of Oskar von Miller presided over the planning and construction of the Walchenseewerk and the Bayernwerk in Bavaria, and the Merz & McLellan firm took part in the planning of the British Grid. Subsequently (see pp. 443–60 below), Merz & McLellan took responsibility for the growth of the Newcastle upon Tyne Electric Supply Company (NESCO), Britain's largest privately owned regional utility. Here, however, we will focus on the history of a firm—an institution—rather than the history of a project. The history of the American firm Stone & Webster is one of evolution and differentiation of function and is, therefore, an example of organic growth comparable to that of the utilities it helped to design, construct, finance, and manage.

Charles Stone (1867–1941) and Edwin Webster (1867–1950) founded the Massachusetts Electrical Engineering Company of Boston in 1890, only two years after graduating from the Massachusetts Institute of Technology and the institute's newly established course of study in electrical engineering. The Massachusetts Electrical Engineering Company—the name was changed to Stone & Webster in 1893—was one of the earliest U.S. consulting electrical engineering firms. It offered to advise those intending to invest in the new and rapidly developing field. It was ready to design, estimate, and superintend construction of electric light and power plants; it was also prepared to determine the economy and efficiency of electrical equipment in its small testing laboratory. In a letter to prospective clients, signed as always "Stone & Webster," the young engineers promised to bring to bear "the scientific knowledge and practical experience, both of the company and of experienced engineers in all branches of the profession."[35] In 1890, on the recommendation of President Francis A. Walker and Professor Charles Cross of M.I.T., the firm undertook its first major project: to design and construct a 400-h.p. hydroelectric plant and a short transmission and distribution system for a New England paper mill.

The panic of 1893 deeply influenced the character of the firm. Earlier, during the 1880s, the electrical manufacturers, including Edison, Thomson-Houston, and Westinghouse, took the securities of the newly founded utilities as payment for equipment and rights to use the patented com-

---

physical assets, and by 1883 the total investment in the company was close to $150 million. By 1873 the Pennsylvania Railroad system represented an investment of $400 million. These investments greatly impressed contemporaries and today impress historians studying the managers and financiers who were responsible for them. It should be noted, however, that the cost of the Bayernwerk, for which Oskar von Miller's firm was consultant, was estimated to be 32 million marks, and that the cost of the British Grid, for which Merz & McLellan and several other British firms served as design and construction consultants, was estimated to be £ 250 million. For the cost of the Bayernwerk, see Oskar von Miller, *Bayernwerk zur einheitlichen Versorgung des rechtsrheinischen Bayern mit Elektrizität: Projekt Oskar von Miller, Februar 1918: Erlauterungsbericht* (copy in the Deutsches Museum, Munich). For the cost of the Grid over a fifteen-year period, see Great Britain, Ministry of Transport, *Report of Committee Appointed to Review the National Problem of the Supply of Electrical Energy* (London: HMSO, 1926), p. 18. The information on the railways is from Alfred D. Chandler, Jr., and Stephen Salsbury, "The Railroads: Innovators in Modern Business Administration," in Bruce Mazlish, *The Railroad and the Space Program* (Cambridge, Mass.: M.I.T. Press, 1965), p. 129.

[35] General letter headed "Dear Sir," 1 January 1890, in the files of Stone & Webster, New York City (hereafter cited as S&W files). I am indebted to C. F. Overton for guidance in using these files and to the firm for permission to do so.

ponents of the manufacturers' systems. The new utilities, many of them in small towns, could seldom persuade local investors that their securities, especially the common stocks, were reasonable risks. Until the financial panic of 1893, the flaws in the arrangement between manufacturers and utilities were not fully revealed. The panic, however, caused banks to demand payment on loans made to the manufacturers, who found it impossible to raise cash to meet the demands by marketing the utility common stock they held.

Under the circumstances, J. P. Morgan and other financiers interested in the utilities, the loans, and the stock rescued General Electric, newly formed by the merger of Edison and Thomson-Houston, by paying cash for the utility stocks and establishing a trust, or syndicate, to manage or dispose of the utility property acquired. At this stage, the recently formed firm Stone & Webster became involved. The managers of the syndicate, the Street Railway & Illuminating Properties Trustees, engaged the young partners, who had excellent references in Boston's financial and engineering circles, to appraise the properties acquired. As a result, Stone and Webster had a remarkable learning experience that helped define the character of their company.

Through analysis of the utilities, Stone & Webster obtained a rich store of information about the utility industry, gaining insights into its financial, managerial, and technological problems. Furthermore, they developed a keen sense of how some of the outstanding problems might be solved. Many solutions were of a general kind, being abstracted from common circumstances in the various utilities. They also learned of practices in one company that could be applied—or should be avoided—in another. The experience of Stone & Webster in 1893 might be compared to the case history approach used in business schools.

Not only was Stone & Webster's expertise as a consulting engineering firm greatly enhanced by this experience, but the partners invested profitably in the utilities they analyzed. An anecdote long remembered and often told at Stone & Webster concerns one property examined during the panic of 1893.[36] Recalling his visit to the Nashville (Tennessee) Light & Power Company to examine that property, Charles Webster wrote:

After visiting Nashville and examining the property and making a report, I came to New York and Mr. J. P. Morgan, Sr. asked me to tell him what I thought of the possibilities for the future development of that and other enterprises. I was enthusiastic about them and told him that I thought, if the assets were carefully conserved and the property was wisely developed, it would result in a great property. . . . He told me that, if I felt so confident about the future of these things, he thought I ought to buy them.[37]

Stone & Webster borrowed the money, bought control for a few thousand dollars, "struggled with it, managed it, and built it up" over several years, and sold it for a profit of over $500,000.

Reorganizing utilities for bankers and others led to the firm's acquiring financial interests in other utilities and to its independently offering re-

[36] An unpublished 6-page typescript, "Memorandum on 'Entrepreneur Business' of Stone & Webster," S&W files. This item was reputedly authored by Charles Stone.

[37] Ibid.

organization or consulting services to utilities.[38] Over the next decade Stone & Webster developed a set of interrelated services and interests. These were institutionalized into a coherent concept and structure that anticipated holding-company functions usually associated with the 1920s. By 1906 Stone & Webster was providing centralized management, engineering, and financial services to twenty-eight independent power, light, gas, and traction utilities throughout the United States. The firm had financial interest in the stocks of the various companies, but each company had its own officers, board of directors, and bank accounts. Therefore, Stone & Webster did not function as a holding company or trustee for the companies. It did, however, exercise considerable control over the policies of the companies.

Stone & Webster usually named its executives as managers in charge of the utilities it serviced. These managers reported to a Stone & Webster district manager and to a "sponsor" in the Boston office. There were six sponsors, and together they constituted the executive committee of Stone & Webster. Charles Stone and Edwin Webster did not sponsor particular utilities, but provided a broad oversight of the entire business. They called themselves "the firm." Until World War I, when circumstances placed Stone in the New York office and Webster in Boston, they sat at desks side by side, consulted each other at the end of each day about decisions made and pending, and signed letters and memoranda as "Stone & Webster."

Besides their line function in supervising the managers in the local utilities, the members of the executive committee headed the staff departments in the Boston office. The principal departments were: engineering, purchasing, auditing, corporation, statistical, securities, and library. Of these, Stone & Webster considered its statistical department "one of the unique, and at the same time one of the most important, features of the organization."[39] Each month the statistical department received two reports from the managed companies. The first covered finances, the second the operating statistics. With these the department analyzed each company's condition. Graphs and tables were prepared that not only facilitated the projection of trends but also allowed comparisons of each company with its prior record and with other companies. Among the many statistical studies made over the years was one on the performance of steam-generating power plants, a study that resulted in the increased standardization of equipment and operations as well as higher efficiencies. In about 1905 a comparison of power station performance in managed companies revealed wide variations in practice and performance. As a result, Stone & Webster assigned experienced operators to the power plants for a considerable length of time. Combustion methods, maintenance schedules, and other operations were rationalized, and savings of 10–40 percent in operating costs were achieved as a consequence.[40] The statistics department was so central to the Stone & Webster system that it had general responsibility for

[38] Memorandum by Charles Stone on the early history of Stone & Webster, 30 December 1902, S&W files.

[39] "The Stone & Webster Organization, and the Properties It Manages," *Street Railway Journal*, 7 July 1906; reprint in the S&W files.

[40] "Historical Notes," prepared in April 1930 by L. B. Nash, S&W files.

training new men for managerial posts in the field and for departmental positions in the central office.

The engineering department also kept detailed records and drew on these as it encouraged standardization of the equipment used by the managed companies. Since the reorganization and development plans drawn up by the firm, the sponsors, and the executive committee often called for modernization and expansion to increase the profitability of the various companies, the engineering department prepared plans for remodeling power houses or laying out new transmission and distribution systems. Construction might be supervised by the local company from Boston. Stone & Webster, however, was developing its own construction facilities and personnel as well.

Stone & Webster also increased the marketability of its clients' securities. At a time when the electric light and power utilities were rapidly expanding, financing this capital-intensive industry was a critical problem. The small company found the problem especially severe. If, however, it could identify with Stone & Webster, an immediate advantage followed; furthermore, the Stone & Webster securities department acted as fiscal representative of the managed companies by placing their securities. (Some of the securities were purchased by Stone & Webster, probably some in payment for services.) For many years the firm had relied heavily upon Lee, Higginson & Company of Boston in marketing the securities, so much so that there was concern that the firm would simply be thought of as engineering consultants for the Boston brokerage house. Over time, however, Stone & Webster increasingly placed securities independently. The "black book" contributed to this objective. Begun in 1902 and continued until 1929, this annual report covered the operations of companies under Stone & Webster's supervision. The "black book" contained a condensed statement of outstanding securities, rate of dividend, earnings, and other financial information. The establishment of a publicity department in 1903 also enhanced the reputation of the managed companies.

The library department was not without an important role in the organization as well. Besides storing and retrieving agreements, contracts, reports, and similar material, it prepared "current literature sheets" by abstracting articles in the technical and business literature that were of special interest to Stone & Webster personnel. "This service is not intended to take the place of personal reading of the current literature on the part of individuals," however, a contemporary cautioned.[41] It was not presumptuous of the librarian to expect the personnel to keep abreast of the literature, for by 1912 almost half of the firm's 600 officials and employees were college graduates, and of these, close to 100 came from the Massachusetts Institute of Technology and 50 came from Harvard University.

Stone & Webster departments metamorphosed into Stone & Webster subsidiaries.[42] In 1906 the engineering department incorporated under

[41] "The Stone & Webster Organization," p. 13.
[42] The following description of the Stone & Webster organization is based on a reprint in pamphlet form of an article prepared by *Electrical World* for distribution at the National Electric Light Association convention in Seattle, Wash., 11–14 June 1912; a copy of the reprint is in the S&W files.

the name of the Stone & Webster Engineering Corporation. In 1907 the management function performed by the executive committee, sponsors, and various departments coalesced into the Stone & Webster Management Association, a corporate organization. The Engineering Corporation broadly extended its range of activities. Among its works during its first decade were the expansion of the Boston Elevated Railway Company; the design and construction of the dam, power house, and transmission lines of the Mississippi River Power Company at Keokuk, Iowa, one of the world's largest hydroelectric installations; and the Big Creek–to–Los Angeles transmission system. The firm also designed the Conowingo hydroelectric project. The securities department of Stone & Webster merged with the investment banking house of Blodget & Company and took corporate form as Stone & Webster and Blodget, Inc. By the eve of World War I, the overall organization of Stone & Webster, with its subsidiaries and managed companies, had become a system of mutually reinforcing components. (See Fig. XIII.11.) It was entirely appropriate that the organization took as its symbol the triskelion designed by the original partners. The sign, with its three arms enclosed in a triangle, represented the interconnection of financing, engineering, and construction.

Despite the proliferation of utility holding companies, Stone & Webster, while performing most of the functions of a holding company, delayed its own transformation. The first step toward becoming a holding company was taken in 1925 when it participated in the formation and financing of the Engineers Public Service Corporation, which acquired control of the Virginia Railway & Power Company (later Virginia Electric & Power Company) and about twenty other utilities receiving management services from Stone & Webster. The new holding company performed the services Stone & Webster had formerly provided for the acquired companies. A company officer explained the resort to the holding company form as a way of avoiding absorption of the managed companies by other rapidly spreading holding companies. In 1929 Stone & Webster became Stone & Webster, Inc., of Delaware, a holding company for the Stone & Webster group. By an exchange of stock, it acquired 90 percent of Engineers Public Service Corporation, which by 1933 had invested $95,116,675 in six major groups of light, power, and street railway companies in fourteen states. Stone & Webster, Inc., also held the Sierra Pacific Electric Company, another subholding company. In aggregate, the Stone & Webster group by 1932 consisted of forty-three companies. In addition, it continued to manage companies outside the group. The group generated a little more than 2 percent of the electricity supplied by utilities in the United States.[43]

The transformation of Stone & Webster into a holding company was part of a trend in the electric supply industry, and the consulting engineering firm was but one nucleus around which a holding company could form. Eventual transformation into a holding company was no more foreseen by founders of many of the other predecessor organizations than it

[43] U.S., Federal Trade Commission, *Utility Corporations* (S. Doc. 92, 70th Cong., 1st sess., 1928), Part 66, (Washington, D.C.: GPO, 1934), pp. 564, 565, 661–63; idem, *Electric Power Industry: Control of Power Companies*, 69th Cong., 2d sess., 1927, S. Doc. 213, pp. xxxiii, 177–87 (hereafter cited as S. Doc. 213 [1927]).

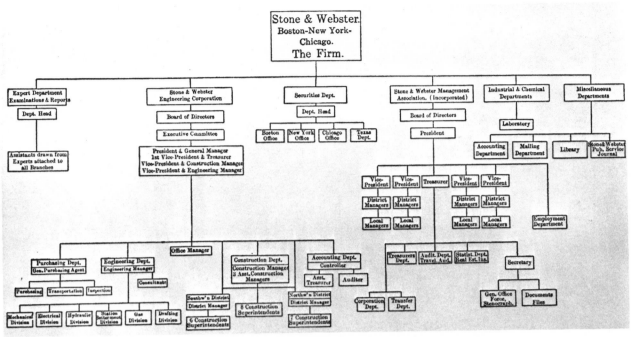

Figure XIII.11. *Stone & Webster organizational chart, 1912. Courtesy of Stone & Webster, New York, N.Y.*

was by Charles Stone and Edwin Webster. In the 1920s, holding companies dominated the privately owned electric utility industry. They grew as the size of all utilities increased.[44] In 1914 eighty-five utility corporations—some of them holding companies—controlled about 70 percent of the total installed capacity of the entire public electric utility industry; of these corporations, thirty-five controlled one-half of this total, sixteen controlled one-third, and ten controlled one-fourth.[45] By 1924, having superimposed themselves on the large—and small—utilities, the large companies that dominated the utility industry had become holding companies (see Fig. XIII.12). In that year, holding companies controlled two-thirds of the gen-

[44] Statistics on the growth of U.S. electric utilities are contained in various government agency reports. One 1911 survey was limited to hydroelectric installations. The information for 1911 was published in U.S., Bureau of Corporations (predecessor to the Federal Trade Commission), *Water Power Development in the United States* (Washington, D.C.: GPO, 1912). On the state of the industry in 1914, see U.S., Department of Agriculture, Forestry Service, *Electric Power Development in the United States,* 64th Cong., 1st sess., 1916, S. Doc. 316. On the status of holding companies in 1924, see S. Doc. 213 (1927), which became, in effect, the first volume of U.S., Federal Trade Commission, *Electric Power Industry: Supply of Electrical Equipment and Competitive Conditions,* 70th Cong., 1st sess., 1928, S. Doc. 46, a history of the electrical manufacturers. For data on holding-company activities in 1932, see U.S., Federal Trade Commission, *Utility Corporations* (S. Doc. 92, 70th Cong., 1st sess., 1928) (Washington, D.C.: GPO, 1928–35). This multipart and multivolume study has two summary volumes, one on utility propaganda and publicity (Part 71A) and the other on holding companies in general (Part 72A). An index for the 95-volume study is given in Parts 81A and 84D. An *aperçu* summary of the state of large corporations and holding companies in the electric utility field can be found in Part 72A, pp. 34–37.

[45] Forestry Service, *Electric Power Development,* quoted in Federal Trade Commission, *Utility Corporations,* Part 72A (1935), p. 35.

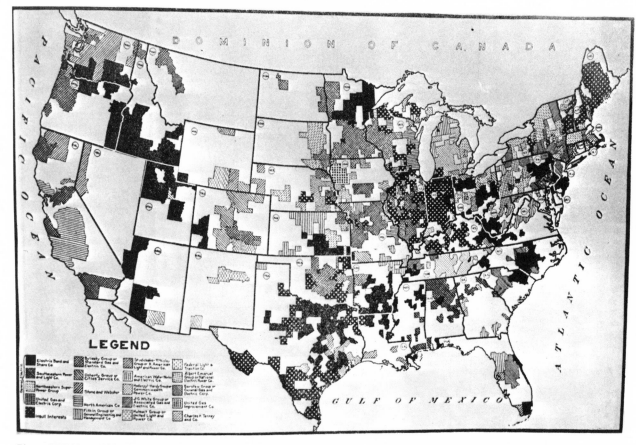

Figure XIII.12. *Fields of operation (counties) of major U.S. electric utility holding companies, 1925. From U.S., Federal Trade Commission, Utility Corporations, Part 72A, p. 56.*

erating capacity of the industry; seven holding-company groups controlled 40 percent, and other holding companies controlled 25 percent. With 13 percent, the holding company formed by General Electric, Electric Bond & Share Company, ranked as the largest; the Insull group, Middle West Utilities Company, held between 8 and 9 percent. For a nation long warned to be wary of trusts and monopolies, these were provocative statistics. Between 1924 and 1932 the General Electric group was reorganized and a new giant, the United Corporation, was formed. Holding-company domination persisted, with sixteen large holding-company groups controlling more than three-fourths of the privately owned industry. For all holding companies, the extent of control was over 78 percent.[46]

Because of a series of government studies in the United States, the adverse publicity given the holding companies by advocates of government ownership, and the notoriety surrounding the Insull holding company after its financial collapse in 1932 and after its founders were tried for mail fraud in 1934, holding companies became a much-discussed and misunderstood

---

[46] Federal Trade Commission, *Utility Corporations*, Part 72A, p. 37.

aspect of the electric utility industry.[47] The mounting tension culminated in 1935 with passage of the Public Utility Holding Company Act.

There was misunderstanding about the origins and function of the electric-utility holding companies. Proponents and opponents debated the advantages and disadvantages of the holding company for the consumer. Because the holding companies came to public attention at a time when investment in stocks in general and in utility stocks in particular was mounting, the public often saw the holding company as the creature of bankers and stock brokers, not of engineers and managers. Furthermore, the public associated the large holding companies and trusts of the past with financial wizardry or skulduggery. The technological origins and functions of holding companies were far less sensational and thus were not discussed or understood.

Contrary to popular opinion, the origins and development of several leading electric-utility holding companies are to be found rooted more deeply in technology and management history than in finance. Capital formation was the major problem of the large utilities to which the holding companies responded in the 1920s, but this need for capital arose from the concentration of large-scale technology. Engineers and technically trained managers dominated the early history of major holding companies. S. Z. Mitchell of Electric Bond & Share, Charles Stone and Edwin Webster, H. M. Byllesby of H. M. Byllesby & Company, and Samuel Insull all were trained as engineers or had extensive experience in engineering and management before heading holding companies. In their management of the holding companies, these men insisted on the introduction of systems technology and management. In order to introduce capital-intensive and large-scale regional technology, however, they had to find ways to finance large-scale investments.

The major holding companies grew large by serving and presiding over the merger of small utilities. As has been seen, the urban utility consolidated and spread out over the large cities until it prevailed as a regulated monopoly. From this activity many economies and rational structures pleasing to the engineer, the manager, the investor—and often to the consumer— emerged. Until the advent of efficient, high-voltage transmission and water and steam turbines, the small cities, towns, and rural areas were supplied— if at all—by utilities lacking the advantages of large-scale, centralized technology, management, and financing. Their relatively inexperienced and untrained managers and engineers used the less advanced technology, for in the United States, manufacturers invested their intellectual and physical resources in large-scale challenges.[48] With the advent of turbines and high-voltage transmission, however, it became possible for regional systems to integrate small communities and supply some of the rural areas. The holding companies then provided these systems with centralized engineering,

---

[47] On the history of Insull's holding companies, see Forrest McDonald, *Insull* (Chicago: University of Chicago Press, 1962), pp. 237 ff.

[48] This impression needs to be examined through research into the policy and product of the major manufacturers. Unfortunately, Harold C. Passer's *Electrical Manufacturers* (Cambridge, Mass.: Harvard University Press, 1953) carries the history of U.S. manufacturers only to 1900.

management, and financing.[49] The holding companies either merged the utilities or organized the small utilities into confederations related by physical or financial interconnections. A government report concluded that the holding company's financing, engineering, and management of utilities serving small communities was probably the chief benefit to be derived from this organizational concept.[50]

With the onslaught of the Depression, the drop in holding-company stock values, and the spread of dissatisfaction among investors in holding companies, however, doubts increased about the advantages of the electric-utility holding company. Ensuing government investigations and reports helped define the functions of the companies as technological, managerial, and financial, but abuses were found in the financial activities.[51] The principal benefit accruing to the operating companies and to consumers, if the service was not abused, was, in the opinion of the Federal Trade Commission, the holding company's financing of the growth and improvement of small, undeveloped operating utilities. Also viewed positively was the holding company's role in facilitating interconnection and large-scale output. The holding company provided the small operating company with the experts' solutions to the problems of generation, transmission, distribution, and utilization of electricity. Moreover, under the aegis of the holding company, "service was improved and extended, consumption increased, and costs of production were reduced with a consequent possibility of, and tendency toward, lower rates, notwithstanding accompanying financial practices which had an opposite tendency as to rates."[52]

In the opinion of the Federal Trade Commission, the list of practices that "had an opposite tendency" was long: the pyramid of utilities formed by the hierarchical holding-company structure often involved the issuing of excessively speculative securities; the holding company often charged excessive fees for consulting services; the writing up of the values of utility securities often placed pressure on the utilities to charge high rates to obtain a profitable return; by means of security-market manipulation, utility securities were sold to the public at highly inflated values; the organizational and financial structure of the holding company frequently permitted a few men with a small investment to gain practical control of immense operations; and interstate contracts, operations, and other relationships made difficult the control of holding companies by state regulatory commissions. Like so many other complex organizations, however, the holding company cannot be adequately defined by abstractions or categories of advantages and disadvantages. The history of several of them will reveal some of the complexities and irregularities of this organizational construct. In the following section we will consider the Electric Bond & Share Company, a holding company established by a manufacturer, and the United Corpo-

[49] Federal Trade Commission, *Utility Corporations,* Part 72A, p. 107.
[50] Ibid., pp. 834, 881.
[51] This summary discussion of the advantages and disadvantages of holding companies is from ibid., pp. 831–82. This document, with its many parts and volumes, resulted from a Senate resolution calling for a consideration of "the value or detriment to the public of such holding companies, owning the stock or otherwise controlling such public-utility corporations, immediately or remotely" (p. 831), and for legislative recommendations.
[52] Ibid., p. 881.

ration, a holding company founded by an investment and banking concern. We have already noted Stone & Webster's transformation from a consulting engineering firm into a holding company. The Pacific Gas & Electric Company (see pp. 276–80) and the Rheinisch-Westfälisches Elektrizitätswerk (see pp. 423–29) are examples of utilities that became holding companies by acquiring the stock of the utilities they absorbed. Reference has been made, as well, to Samuel Insull's formation of the Middle West Utilities Company, which became one of the largest U.S. holding companies.

The history of Electric Bond & Share is linked with the history of General Electric and the career of S. Z. Mitchell (1862–1944). In the 1920s Mitchell and Samuel Insull were the most widely known holding-company entrepreneurs. It is estimated that during this period Mitchell's company controlled more of the utility business than Insull's, for in 1924 Electric Bond & Share controlled at least 10 percent of the energy generated by the industry, while Insull's Middle West Utilities Company controlled only about 8 percent.[53]

The origins of the Electric Bond & Share Company, which was formally established in 1905, can be traced back to 1890. In that year the Thomson-Houston Electric Company, the manufacturer that in 1892 would merge with Edison General Electric to form the General Electric Company, established a subsidiary called the United Electric Securities Company of Boston. The function of the new company was to help small, independent electric light and power companies finance the purchase of generators and other equipment manufactured by Thomson-Houston. Local sources of capital considered neither the bonds nor the stocks of the small, often indifferently run, electric light companies to be a good risk.[54] As a result, the companies could not expand in response to a market that Mitchell calculated was growing at a rate of about 10 percent annually. The problem was more serious for these small utilities than for many small enterprises because their capital investment in relation to income was unusually high; Mitchell estimated that for these utilities it took an investment of from $4 to $6 to produce a gross annual revenue of $1.[55]

The United Electric Securities Company financed Thomson-Houston's small customers by taking their nonmarketable bonds and using them as collateral for the bonds issued by United Electric Securities, which were marketable because of the company's size, management, and relationship to Thomson-Houston.[56] Despite the considerable market for these bonds that developed over the years, however, the problem of financing local utilities was not solved. The panic of 1893 saw many small utilities sink into bankruptcy, and those that survived faced an even less responsive investment market. Even investors who had been willing to buy the bonds of the small-town utilities (if not the stock) insisted after the panic that the utilities raise more capital by issuing common stock. In short, the bond buyers

[53] S. Doc. 213 (1927), p. 41.

[54] Sidney A. Mitchell, *S. Z. Mitchell and the Electrical Industry* (New York: Farrar, Straus & Cudahy, 1960), p. 60 (hereafter cited as *S. Z. Mitchell*).

[55] Mitchell, in an address to the Association of Edison Electric Illuminating Companies in 1920, quoted in *S. Z. Mitchell*, p. 118.

[56] S. Doc. 213 (1927), p. 70.

demanded more owners and fewer creditors. They wanted the capital-hungry utilities to cover with mortgage securities no more than one-half to two-thirds of the cash cost of the power plant and other facilities.[57]

General Electric, the successor to Thomson-Houston in 1893, inherited the United Electric Securities Company of Boston and subsequently formed the Electrical Securities Company of New York primarily to fulfill its bond-purchasing function. Under the leadership of Charles Coffin (1844–1926), its resourceful president, General Electric advised small utilities to sell more junior securities, or stock, but found that "merely pointing the way was not effective."[58] Because General Electric had acquired over the years many shares of unmarketable, small-utility stock in payment for equipment, it, too, had a general interest in solving the problems of the utilities. In 1905 an inventive solution came out of deliberations between Coffin and Mitch-ell.[59] Mitchell was called on for assistance because he had been dealing with small utilities in the Northwest, companies whose electric service was poor and whose financial structures were crude—companies that were, in the vernacular, "cats and dogs."[60]

Mitchell had prepared for a career in engineering at the U.S. Naval Academy. Finding the prospects of the navy bleak in 1885, however, the newly commissioned Mitchell resigned and accepted employment with the Edison Company in New York. Earlier, in 1883, he had been attracted to electrical engineering while installing incandescent lighting on the U.S.S. *Trenton*. Mitchell worked at the Edison Machine Works, the manufacturer of dynamos, and he also helped to wire the distribution system of the Edison Electric Illuminating Company in New York City. Within a year, however, he seized the opportunity to take over the exclusive Edison agency in the Northwest after Henry Villard, Edison's friend and financial adviser, gave it up. From 1885 to 1905 Mitchell sought and found financial backers, obtained franchises, installed equipment, set up operating and manage-ment procedures, and cultivated the growth—or at least ensured the sur-vival—of the region's small electric light utilities. The call Mitchell got from Coffin to deliberate about "cats and dogs" is therefore understandable.[61]

The solution proposed by Mitchell and Coffin was the formation of the Electric Bond & Share Company. The new company turned over its stock to General Electric in exchange for the stocks of power and light companies held by General Electric having a par value of $2,782,150, bonds with a par value of $1,476,000, and $1,300,576.90 in cash. Electric Bond & Share converted the acquired securities into marketable assets by resuscitating and invigorating the utilities that had issued them. This was done—as had been the case with Stone & Webster—by providing management, engi-neering, and financial services. Electric Bond & Share did not purchase a majority of stock in the client companies, but exercised control through a combination of means, including minority stock and contractual, fee-based

[57] Ibid., p. 71.
[58] Ibid., p. 93.
[59] *S. Z. Mitchell*, p. 61.
[60] Ibid., pp. 61, 64.
[61] This biographical sketch of Mitchell is based on ibid., pp. 19–42.

services.[62] In 1925 General Electric divested itself of direct control of Bond & Share by distributing the company's stock to GE stockholders.

Stressing management services, Bond & Share introduced a policy that varied from Stone & Webster's. Bond & Share recommended that the directors of the client company elect a Bond & Share executive as a non-salaried officer or member of its board. This officer—like the executives serving other client utilities—would remain in Bond & Share's New York office; a Bond & Share man would not be named resident manager as was customary with Stone & Webster.[63] From its large staff of managers and advisers, Bond & Share would, like Stone & Webster, designate a "sponsor" for each client utility. The sponsor would visit, observe, inspect, and correspond with that company. He would apply his general knowledge of the utility business to the problems and opportunities of the specific utility company. Furthermore, he would have at his disposal the large Bond & Share staff of specialists in insurance, taxes, rates, public relations, statistics, and other management functions. The contract also called for a "sponsor engineer"; he, too, would stay informed by means of visits and correspondence. Besides the management and engineering specialists, there were to be "sponsor accountants" in Bond & Share's New York office. The problems that these sponsors addressed were expansion of the business, utility interconnection, rates, capital structure, and financial programs.[64]

Electric Bond & Share soon began organizing holding companies to control the client utilities through majority ownership of their stock. Bond & Share itself continued to avoid direct control of the utilities through ownership of a majority of their shares. It established a new contractual relationship with its holding companies rather than with the operating utilities. By 1925 Electric Bond & Share held large blocks of voting stock in, and had various servicing arrangements with, five holding companies. Because it did not have a majority stock interest in any of these, however, it consistently maintained that it was not itself a utility holding company. The Federal Trade Commission and "the investment world" generally, however, believed that Electric Bond & Share held de facto control because the other shares of the five companies were widely dispersed or blocks of them were held "by friendly and cooperative banking interests."[65]

The holding companies often acquired their controlling stock in the operating companies when Electric Bond & Share turned over to the holding companies utility stock obtained by General Electric in exchange for equipment. The holding companies also acquired utility stock by purchase, using income from the sale of holding-company securities, or by exchange of these securities for operating-company securities.[66] Because Stone &

[62] Ibid., pp. 64–65; S. Doc. 213 (1927), p. 73.

[63] In 1906 Stone & Webster named local managers; see *Street Railway Journal*, 7 July 1906, p. 2. In a service contract used in the 1930s, Stone & Webster Service Corp. expressed readiness to "recommend" local executives (Stone & Webster Form no. A706, S&W files).

[64] This analysis of the Electric Bond & Share service is drawn from a 1929 contract reprinted in Federal Trade Commission, *Utility Corporations*, Part 72A, p. 672.

[65] James C. Bonbright and Gardiner C. Means, *The Holding Company: Its Public Significance and Its Regulations* (New York: McGraw-Hill, 1932), p. 104. Bonbright and Means drew upon S. Doc. 213 (1927) in preparing their study.

[66] Bonbright and Means, *The Holding Company*, p. 104.

Webster sometimes obtained stock in exchange for its engineering and management services, it seems likely that Electric Bond & Share received utility securities in this way as well.

During the period of rapid expansion of regional utilities, Electric Bond & Share and the related holding companies emphasized capital formation and found means by which the utilities could dispose of their difficult-to-market junior securities. Seeing every citizen as a customer, Mitchell, who was still head of Electric Bond & Share, told utilities that they needed $30–$80 for each individual added to the U.S. population.[67] The standard financial structure designed by Electric Bond & Share to meet the financial needs of the utility companies controlled by the holding companies called for 60 percent capitalization through the sale of bonds to the public, 20–25 percent capitalization through the sale of preferred stock to the public, and 20–25 percent capitalization in common stock that was to be taken by a holding company.[68] For the holding company, the advantage of keeping the amount of common (voting or controlling) stock low was obvious; control would be less costly. Furthermore, as will be explained, the arrangement could mean for the holding company a high return on investment. The inventiveness of the financiers was comparable to that of the inventors and engineers who had designed the technological structure. Following this plan, Electric Bond & Share participated in the policy making of the five major holding companies: the American Gas & Electric Company (formed in 1906);[69] the American Power & Light Company (1909); the National Power & Light Company (1921); the American & Foreign Power Company, Inc. (1923); and the Electric Power & Light Corporation (1925).[70] The National Power & Light Company also had a major holding company as a subsidiary—the Lehigh Power Securities Corporation (see pp. 439–41 below).

Because American Power & Light received various services from Bond & Share, its history is representative of the concepts and actions of S. Z. Mitchell and his associates. American Power & Light was formed as a holding company to finance two Kansas utility companies that were too small to interest investors in their senior securities and had no market for their common stock. The owner of the two utilities had asked Mitchell to form the holding company to take their common shares. American Power & Light then raised the money for the stock by selling its own securities. Subsequently, the holding company acquired other utilities in the immediate vicinity, and Mitchell merged them into the Kansas Gas & Electric Company, a utility that was large enough to interest investors in its preferred stock and bonds. American Power & Light controlled the Kansas Gas & Electric Company through common stock, and the holding company entered into the customary contractual relationship with Electric Bond &

---

[67] *S. Z. Mitchell,* p. 118.

[68] Federal Trade Commission, *Utility Corporations,* Part 72A, p. 87.

[69] American Gas & Electric Co. originated before Bond & Share had built up its staff of experts, so American Gas & Electric supplied its own managers, engineers, and accountants to the operating companies whose common stock it held. American Gas & Electric relied on Bond & Share for financial services, however. In 1924, Bond & Share owned about 7 percent of the voting stock of American Gas & Electric. See S. Doc. 213 (1927), p. 7.

[70] Ibid., p. xxxvii.

Share. In 1910, directed by Bond & Share, American Power & Light acquired properties in Washington and Oregon. In subsequent years it obtained control of other groups of companies in Texas, Minnesota, Florida, and Nebraska. By 1924 the operating companies held by American Power & Light generated about 2.4 percent of the total commercial output of electricity for the entire country.[71]

When the companies acquired were in the same geographical area, as was the case in Kansas, they were united by transmission lines into a continuous system and were often merged to form larger operating companies. As a result, small, inefficient plants could be shut down, load factor could be improved, varied energy sources mixed, and diversity exploited—as they had been during the era of merger and consolidation. But Mitchell went beyond exploitation of supply and diversification of load to introduce the diversity principle on a higher level of abstraction; he called his idea "financial diversity." Because of this aspect of their policies, Electric Bond & Share's holding companies were categorized as the "diversified investment type."[72]

The principle of financial diversity, as fulfilled by the American Power & Light Company and other Bond & Share holding companies, was explained by Owen D. Young, head of General Electric, as follows:

If one owns a light and power plant in a single community, his investment and his earning power is [*sic*] subject to the risk of that community. Floods may come and wipe it out; cyclones may hurl it down; crops may fail; business depressions there may be acute. The capital invested in that plant, if owned by a single man, is subject to those contingencies. But if men combined their investments in a large number of plants, widely diversified geographically, the floods will never come to all at once; the failure of crops will never come to all at once; a depression in business is unlikely to come to all at once, if the diversity is widely made. Therefore, a given investment in a group of plants is much safer than an investment in a single plant of similar amount. Not only is the principal safer but the continuity of return is better insured.[73]

According to Young and Mitchell, not only were the risks spread in the holding companies so that the investor's principal was safer and the continuity of return better, but the return might well be much larger. Mitchell explained the concept in a simplified example. He said that $100 invested by a holding company in the common stock of a utility should bring a return of $9 as a dividend on the utility's common stock. Assuming that the holding company had raised the $100 by marketing $60 of its own debentures (nonmortgage, or collateral, bonds), $20 of its preferred stock, and $20 of common stock of its own, then on the debentures it would have to pay 6 percent ($3.60) and on the preferred stock 7 percent ($1.40), which would leave $4 from the $9 earned on the utility common stock to declare as a dividend on the holding company common stock or to be invested in the operating companies. As Mitchell observed, "For that possible 20 percent return the holding company common stock should be

[71] Ibid., pp. 93–108 and xxix.
[72] Federal Trade Commission, *Utility Corporations*, Part 72A, p. 84.
[73] Quoted in *S. Z. Mitchell*, pp. 84–85.

saleable."[74] He might have added that the organizers of holding companies were often ready to purchase the common stock themselves.

In a report in 1927, at the height of the holding-company era, the Federal Trade Commission observed that promoters often retained ownership of the common shares of their holding companies and obtained rates of return ranging from 19 percent to 55 percent.[75] High returns could be achieved by increasing the proportion of low-interest debentures and preferred stock among the holding company's securities. For instance, if the holding company in the example above had increased its debentures to 65 percent and its preferred stock to 25 percent, the return on the $10 of holding-company common stock would have been 33.5 percent. S. Z. Mitchell recognized the possibility of high—even excessive—returns on investments, but he characterized his role as that of a resourceful financier of utility expansion rather than a reaper of large profits.

Management, design, and construction services were always central to the organizations erected by Mitchell and by Stone & Webster. As financing increased in importance with the steady growth of the capital-intensive utility industry, however, holding companies were formed whose activities were almost entirely financial. Chartered in 1929, the United Corporation was the creation of three investment banking houses—J. P. Morgan & Company, Drexel & Company, and Bonbright & Company, Inc.[76] Within a few years, the investment bankers had practical control of a network of utilities extending (with one major break) from the Great Lakes and the St. Lawrence River to the Gulf of Mexico. Those who were informed about United Corporation labeled the control "practical" because United Corporation owned only minority stock in the companies in its group. Minority control probably was congenial because immense amounts of capital were involved and because government regulation and public accounting seemed less likely if the corporation remained superficially an investment company rather than legally a utility holding company. Furthermore, the prestige of the three banking houses was so great among the utilities, and the interlocking of the boards of directors was so extensive, that control could be exercised without majority holdings.[77]

In 1931 United Corporation virtually controlled five major holding companies and, through these, numerous utilities. Among the holding companies it controlled was the Niagara Hudson Power Corporation. Unlike many other holding companies, with their widely dispersed properties, Niagara Hudson controlled utilities that were contiguous and interconnected. Organizationally, however, Niagara Hudson was a complex pyramid of subholding and operating companies. Also a part of the United Corporation and organizationally on the same level as Niagara Hudson was the Commonwealth & Southern Corporation, which held utilities in Georgia, Alabama, and Tennessee and which was later involved in a classic

[74] Ibid., p. 80.

[75] S. Doc. 213 (1927), p. xxiv.

[76] This account of the United Corp. is drawn from Bonbright and Means, *The Holding Company,* pp. 127–38; and from Federal Trade Commission, *Utility Corporations,* Part 72A, pp. 75–77 and 111–16.

[77] Bonbright and Means, *The Holding Company,* p. 134.

struggle with the Tennessee Valley Authority. The United Gas Improvement Company, in which United Corporation held a 26.1 percent interest in 1931, was probably the oldest utility holding company in the electric supply field. Originally incorporated in Pennsylvania in 1882 to introduce improvements in the gas industry, the enterprise sought control of gas companies in order to establish a market for its patented method of manufacturing water gas. In time, electric utilities were added to the list of companies controlled by the firm.

With its immense holdings of interconnected—or likely to be interconnected—utilities in the Northeast and Southeast, the United Corporation's status approached that of the superpower system much talked about by Morris Cooke and others immediately after World War I. According to rumors in financial and engineering circles before passage of the Public Utility Holding Company Act of 1935, the investment bankers would preside over the physical interconnection of their utility empire.[78] Despite opposition arising from the natural preference of utility managers to maintain operating autonomy instead of being homogenized into a superimposed system, and despite other manifestations of particularism, the rumors took on credence with the formation of the Pennsylvania–New Jersey Interconnection, which involved the Philadelphia Electric Company and the Public Service Electric & Gas Company of New Jersey, two of the largest utilities in the United Corporation group (see pp. 331–34 above). Furthering the complexity of the situation was the fact that the third member of the Pennsylvania–New Jersey Interconnection, the Pennsylvania Power & Light Company, was in the Electric Bond & Share group, which had harmonious relations with the United Corporation by virtue of the close ties of the two superpower holding companies with the investment banking house of Bonbright & Company and because Electric Bond & Share owned stock in some of the companies of the United Corporation group.[79] Anticipating the emergence of a superpower system, some investors and utility cognoscenti took as a portent the name of another holding company, the American Superpower Corporation, which was formed by Bonbright & Company and had close investment ties with the United Corporation.

The onset of the Depression wrecked the frail pyramidal structure of several U.S. holding companies, including Samuel Insull's, and raised doubts about the advantages of all of them. The mounting doubts culminated in 1935 with passage of the Public Utility Holding Company Act, which sharply curtailed and reversed the spread of the holding company by outlawing holding companies with utilities that were not physically (technologically) connected. The anticipated rise of superpower holding companies that would preside over regional systems covering the entire Northeast or West, or even the whole nation, was not to be. Instead, the private utilities and the federal government found themselves locked in a long-term struggle for regional-system control and ownership, a struggle manifested in the fight over TVA and in frustrated efforts to create other TVAs. America,

[78] Bonbright and Means doubted that the physical integration would occur, because of "the petty jealousies of local politics in New York City and Philadelphia and the rivalries between New York, New Jersey and Pennsylvania" (ibid., pp. 128–30).

[79] Ibid., p. 133.

like London and England as a whole decades earlier, saw technical implications contradicted by what Lloyd George labeled "politics."

In Britain, holding companies dominated the privately owned sector of the supply industry, but compared to the United States, this sector held a small percentage of the nation's generating capacity, and this small sector was influenced by foreign enterprise. American manufacturers had long exported electrical machinery to Britain; in the 1920s, American holding companies exported their practices as well. Samuel Insull's holding company acquired interest in the British Power & Light Corporation, a small British holding company. A much larger involvement by Americans was obtained through the Greater London and Counties Trust (GLCT). The Utilities Power & Light Corporation of Chicago, a small U.S. holding company that controlled less than 1 percent of the generating capacity in the United States, owned the GLCT. GLCT's capital, £300,000 in 1926 and £400,000 in 1927, was used to acquire small British utilities with distribution systems that could draw from the Grid's transmission network. Most of the capital came from the United States, but to relieve some of the concern about the penetration of foreign capital, the company named Britons as executives. GLCT concentrated its acquisitions efforts in southern, rural England and in 1928 acquired 95 percent of the shares of Edmundsons Electricity Corporation; with twenty-nine utilities, Edmundsons was one of Britain's largest holding companies. By the end of 1929 GLCT controlled fifty-four utilities, most of which were rural and small.

Like the American holding companies, GLCT announced plans to rationalize its holdings through interconnections and management and financial services. These efforts were frustrated, however, first by the dispersed character of the company's holdings and then by the financial difficulties of the American owners in the 1930s. The owners initially held the common stock of GLCT, obtained a high return on their investment, as in the United States, and limited the participation of British capital to fixed interest securities. In the thirties, however, the American parent company, Utilities Power & Light, experienced serious financial difficulties and was no longer able to finance its British subsidiary; thus, through the purchase of common stock, British capital began to displace American. In 1935 a British consortium, including Lazards, the British merchant bank, took control of GLCT from the Americans, liquidated it, and established Edmundsons, the former subholding company, as the parent holding company. Edmundsons then continued to rationalize the distribution companies it held.[80]

The Power Securities Corporation, the largest of the British-owned holding companies in the 1920s, was headed by George Balfour, a champion of holding companies and large-area supply. A conservative member of Parliament and head of a utility management firm, Balfour Beatty, Balfour organized the Power Securities Corporation in 1922. He became chairman and owner along with three large heavy-equipment manufacturers, Armstrong Whitworth, Babcock & Wilcox, and British Thomson-Houston. The new organization took over all of Balfour Beatty's utility interests and used Balfour Beatty's consulting engineers and management expertise for them.

[80] Hannah, *Electricity Before Nationalisation*, pp. 227–34.

The combination of electrical manufacturer and utility holding company was already well known in the United States. The Power Securities Corporation became the largest electric-utility holding company in Britain. From this base and with this interest, Balfour fought the implementation of the Electricity Supply Act of 1919 and, except when legally required to do so, opposed cooperation with the electricity commissioners. His strong voice added to the dissonance among private and public interests and proponents of large- and small-area supply.[81]

[81] Ibid., pp. 71–73, 81.

CHAPTER XIV

RWE, PP&L, and NESCO: The Style of Evolving Systems

THE evolution of the regional systems considered here was a long process compared to the short gestation periods of the post–World War I planned systems discussed in Chapter XII. Thus, the evolving systems were shaped by factors and technology from different stages in the history of electric supply systems. Consider, for example, the lasting imprint of World War I on the Rheinisch-Westfälisches Elektrizitätswerk (RWE) in Germany; the traces of Edison direct-current utilities that could be found in the regional system of the Pennsylvania Power & Light Company (PP&L) in the 1920s; and the influence of the Electric Lighting Act of 1882 and the power bills of 1900 on the Newcastle upon Tyne Electric Supply Company (NESCO) in England.

Moreover, regional systems evolved so complexly that they almost defy coherent description. Some utilities expanded by acquiring others and then by technologically integrating the organizationally merged systems so that electrically they functioned as one. Systems also expanded through interconnection with other utilities. Usually these systems were not integrated electrically and therefore were not operated from a single control center. Each utility retained its corporate identity and negotiated contracts for the exchange of power over high-voltage tie lines. The loosely connected systems were often called "interconnections" in the 1920s.[1]

To lend coherence to the description of three major evolving systems—RWE, NESCO, and PP&L—without eliminating their essential diversity, the concept of regional style will be employed. In general, the regional systems of the industrial world shared critical technical problems and embodied similar solutions. Yet there were specific differences that imparted distinctive styles to these systems.

---

[1] Samuel Insull, "Interconnection and Consolidation Effect Vast Benefits to Consumer," *Electrical World* 90 (1927): 1087–90; Alex Dow, "Interconnections," ibid. 89 (1927): 451–56. Articles by engineers and managers who led the way in interconnection can be found in the "Symposium on Interconnection Proceedings," *AIEE Transactions* 47 (February 1928). Among the articles are W. E. Mitchell, "Progress and Problems . . . Interconnection in Southeastern States," pp. 382–91; P. M. Downing, "Some Aspects of Pacific Coast Interconnections," pp. 393–98; and H. B. Gear, "Interconnection and Power Development in Chicago and the Middle West," pp. 399–404. There is also a bibliography, pp. 405–8.

The similarities can be partially explained by the existence of an international pool of technology from which the industrial nations drew. Manufacturers engaged in international trade, patents were generally licensed for international use, scientific and technological literature circulated to all of the world's centers of learning, courses in engineering schools described and rationalized world experience in electrical technology, and engineers and inventors moved and consulted easily across national boundaries. Technology transfer was not so much from point to point or place to place as from place to pool to place. The common technology of the pool was shaped to suit the place.

The differences found in the evolving regional systems—the essence of style—stemmed mostly from the nontechnological factors of the cultural context. These differences and their causes need to be emphasized because they are often overlooked in our era, which tends to advocate a superior, advanced technology—"the one best way"—a way that transcends regional and national differences.[2] In 1909, Charles Merz, the experienced British consulting engineer, saw the differences in regional technologies and indirectly defined style as follows: "The problem of power supply in any district is . . . completely governed by local conditions."[3] Technological style can be defined as the technical characteristics that give a machine, process, device, or system a distinctive quality. Out of local conditions comes a technology influenced by time and place, a technology with a distinctive style. The local conditions external to the technology can be defined as cultural factors; the technology they shape, a cultural artifact.[4] Among the cultural factors are geographical, economic, organizational, legislative, contingent historical, and entrepreneurial conditions. Earlier in this study, the contrasting histories of electric supply in London, Chicago, and Berlin suggested the concept of technological style and the force of cultural factors. The factors do not operate deterministically by shaping the technology through passive human agents in the role of craftsman, inventor, engineer, manager, and financier; the factors only partially shape technology through the mediating agency of individuals and groups. Natural geographical features, for example, are a given, but engineers adapt to, modify, or variously use them by selecting or inventing an appropriate technology. The factors complexly and systematically interact with technology and with one another. Furthermore, they and the relationships among them change as a power system grows.

Of the circumstantial factors that shape the style of a regional system, geography is the most obviously influential. When one defines geography as both natural and human, as such locational characteristics as climate and

---

[2] Thomas P. Hughes, "Regional Technological Style," in *Technology and Its Impact on Society: Tekniska Museet Symposia No. 1* (Stockholm: Tekniska Museet, 1979), pp. 212–14. Several sections in this chapter are drawn in part from a revision of this essay.

[3] Charles H. Merz, "Power Supply and Its Effects on the Industries of the North-East Coast," *Journal of Iron and Steel Institute*, September 1908, p. 4.

[4] On the style of material culture, see George Kubler, *The Shape of Time: Remarks upon the History of Things* (New Haven: Yale University Press, 1962), p. 129. For a study of the varied responses of utilities to environmental factors recently, see Marc J. Roberts and Jeremy S. Bluhm, *The Choices of Power: Utilities Face the Environmental Challenge* (Cambridge, Mass.: Harvard University Press, 1981).

seasonal variations in daylight, the location and character of rivers, lakes, and seas, the availability of mineral deposits, soil and vegetation, elevations, transportation, industry, and demography, it becomes clear that it decidedly shapes the character of electric power systems. The geographical influence is especially marked in the choice of energy sources made by the engineers and managers of systems, as we have already observed. This influence is also notable in the long-term character of the load carried by a power system.

Geography partially determines both supply and demand for a utility. Supply depends on the availability of coal and other fuels in a region. The infrastructure of the region's transportation system also becomes highly relevant. Elevations and precipitation are determinants of supply from hydroelectric facilities. Utility engineers carefully measure snow and rainfall to predict spring thaws. Moreover, the capacity of a utility to exploit minerals and to use the potential energy of water changes with the development of new technology as the system builders extend their network of transmission lines to acquire an economic mix of energy supplies for the system.

On the demand side, geography, broadly defined, is also manifested in trends and cycles. The character of demand is shaped by the mix of mining, electrified transportation, industry, commerce, and agriculture in the region served by the utility. Different industries load the utility in various ways. An electrochemical plant takes a steady load throughout the twenty-four hours of a day; a steel mill causes a peak when using large motors in its rolling mill. A region with electrified irrigation pumps makes a heavy demand on a seasonal basis. Within the region, the location of the various loads, including the population, shapes the character of the distribution and transmission networks of the power system. Utility managers manage or manipulate the load by charging differential rates and by choosing customers insofar as regulations allow.

Economic and geographical factors are difficult to separate in an analysis of technological style. The engineers and managers of electric supply systems operated their utilities in accordance with economic concepts such as load factor and economic mix, but the policy principles based on these concepts were influenced by regional geography. The loads and supplies of energy were site specific for each utility, so the manager often had to invent means of raising the load factor or improving the economic mix of energy sources. Because of geographical circumstances, the highest achievable efficiency in some of the best-designed and -managed utility systems was lower than the efficiency of systems that were less well run. Samuel Insull recognized this when he pointed out that the combination of cheap power and nearby chemical load at Niagara Falls resulted in a performance that could not be matched by a Chicago utility.

The organizational form of a utility and its style are inextricably interwoven. Managers and financiers organized and reorganized utilities to facilitate the steady expansion implicit in regional power technology. Regional utilities in the 1920s shared a capital-intensive and expansionistic drive, but each had its own particular organizational variation. One classification of organizational form is ownership. In the 1920s, manufacturers, banks, and private investors owned many of the evolving regional utilities in the United States—PP&L for example. Ownership was mixed private and government in some instances in Germany; RWE is an outstanding example. Until the establishment of the British Grid, the major evolving regional system in England was the privately owned NESCO. Consulting engineering firms and holding companies also gave organizational form to the regional utilities, which then differentiated and defined their own style.

As earlier chapters of this study have shown, the evolution and style of regional utilities cannot be adequately explained without reference to politics, especially legislation. Legislation manifests the ideological, economic, and social character of a culture. Regional utilities had repeatedly to acquire franchises or monopolies in order to define their supply areas. A hostile legislature, influenced for instance by competing interests, either private or governmental, sometimes withheld the powers needed by utilities to erect transmission lines over the countryside. As a price for rights and monopolies, the legislative bodies usually laid down regulations governing rates and conditions of supply. These regulations were sometimes restrictive and frustrated investment and growth; often they appeared to be framed in a utility's own best interests—and indeed they sometimes were by pliable legislators. Regulations, or lack of regulation, also shaped style. The map of any regional system graphically represents the history of legislation and regulation.

History is often popularly imagined as the story of such obvious events as wars, revolutions, and shifting alignments of political power. From the point of view of utility managers and engineers, these historical circumstances deeply influence the character of power systems, but because these forces are unpredictable and beyond their control, they are contingent. The utilities risk predicting and planning for long-range demographic and economic trends or projections, but catastrophic historical events are beyond rational calculation. Nevertheless, they help determine the technological style of power systems because different utilities in different regions and countries feel the effects of different contingent circumstances. As noted, World War I variously shaped utilities in Germany, Britain, and the United States. Of the evolving regional systems, RWE shows most obviously the immediate and lasting effects of the war.

Although utility managers and engineers could not control contingent events such as wars and depressions, they nevertheless sought through the organization of regional utilities and their related technical characteristics to extend their control over factors outside the electric supply system. In the 1920s, especially in Germany, they attempted and often succeeded in transforming influences and forces once a part of the system's environment—by definition, beyond its control—into a larger system under control of the managers and engineers. In short, the environment became part of the system. Expanding vertically toward their sources of energy, regional utilities acquired hard- and soft-coal mines. They also acquired the transportation and materials-handling facilities that moved the coal to the power plants. In addition, the regional utilities exchanged power with large industries possessing isolated, or independent, power plants. In other cases, interlocking boards of directors virtually integrated regional utilities and their major industrial consumers. In addition to horizontal expansion through merger, a vertical extension into sources of supply and sections of demand was made.[5] In this way, system builders integrated electric power systems into regional industrial systems. This process of vertical integration, of becoming a component in a regional industrial system, also influenced the regional power system's style.

There were also supraindustrial systems, loose confederations in which regional electric systems were linked to other industrial enterprises not by technology but by such influence couplers as interlocking boards of directors and cartel agreements. For instance, power utilities were loosely interconnected with electrical manufacturers through ownership ties and common board members. Such loose connections substantially influenced the style of utilities because an electrical manufacturer and utility could jointly plan and design equipment that would best suit the characteristics of the electric supply system or could modify supply characteristics to suit a given load.

To show the contrasting style of the RWE, NESCO, and PP&L systems, a profile of the characteristics and hallmarks of each will be presented (see Table XIV.1). A survey of the various factors shaping their style will follow. The influence of these factors during the twenties will be examined in detail. The first to be discussed was the most complex of the three evolving systems. In the 1920s the Rheinisch-Westfälisches Elektrizitätswerk AG (RWE), ranked first in installed capacity in Germany (see Fig. XIV.1). It operated the largest-capacity coal-fired power plant in the world. In 1924 the total installed capacity of the utility's generating plants was 475,000 kw. Notably, the managers of RWE had concentrated 365,000 kw., or 76 percent of the total, in two plants, the Goldenbergwerk near Cologne and the Reisholz near Düsseldorf. The Goldenberg plant, placed in operation in 1914 and later named for its designer, fired brown coal taken from a nearby open mine owned by the Roddergrube AG. The high-voltage network, with its 100,000- and 200,000-volt transmission lines, linked the generating plants and the system's load. In 1929 the extension of a 220-kv. line 250 kilometers southward from Cologne to a hydroelectric plant in the Vorarlberg in the Alps substantially affected the character of the system. RWE thus began operating an economic mix of energy—brown coal, hard coal, and the white coal of the Alps. The utility made much of its ability to exchange power between the mine-mouth plants of the Cologne district, the deep mines of the Ruhr valley, and the water-power sites of the Alps. In January 1930 RWE provided further flexibility to its system by placing into operation near Herdecke one of the earliest pumped-storage power plants, which it named for the plant's designer, Arthur Koepchen.[6]

[5] Alfred D. Chandler, Jr., in *The Visible Hand: The Managerial Revolution in American Business* (Cambridge, Mass.: Harvard University Press, 1977), explores the increased capability of managers to coordinate and control manufacturing and transportation enterprises as they expanded by vertical integration to acquire raw material and distribution facilities. He does not deal with electrical utilities.

[6] I am indebted to Ed Todd, a teaching assistant at the University of Pennsylvania, for research assistance on RWE's history. For summary histories of RWE, see Ernst Schmelcher, *RWE, 1898–1954* (Essen: RWE, 1954?); and *Das RWE nach seinen Geschäftsberichten, 1898–1948* (Essen: RWE, 1948). The 1924 statistics for RWE are from p. 38 of the last-named work. On RWE see also Camillo Asriel, *Das RWE: Ein Beitrag zur Erforschung der Modernen Elektrizitätswirtschaft* (Zurich: Girsberger, 1930); *Electrotechnische Zeitschrift* 47 (1926): 65–67 and 104–

TABLE XIV.1. REGIONAL STYLE: TECHNICAL CHARACTERISTICS OF THREE POWER SYSTEMS, c. 1924

Characteristic	Rheinisch-Westfälisches Elektrizitätswerk	Newcastle upon Tyne Electric Supply Co.	Pennsylvania Power & Light Co.
Area	17,500 sq. mi. (1930)	1,400 sq. mi. (1924)	c. 5,000 sq. mi. (1924)
System Installed Capacity (year)	475,000 kw. (1924)	250,000 kw. (1924)	200,000 kw. (1924)
Rating of Two Largest Power Stations (% of system capacity)	Goldenbergwerk (290,000 kw.) (61%) 1924 Reisholz-Werk (75,000 kw.) (15%)	Dunston (71,000 kw.) (28%) 1924 Carville "B"* (55,000 kw.) (22%) 1924	Hauto (70,000 kw.) (35%) 1924 Harwood (41,500 kw.) (21%) 1924
Energy Sources	lignite (open face) and hard coal	hard coal and waste heat	low-grade anthracite
Cooling System	cooling towers at Goldenberg	river water	river water
Transmission Voltages	66,000; 110,000; 220,000	22,000; 66,000	66,000; 220,000
Transmission Frequency (cycles/phase)	50/3	40/3	60/3
Entrepreneur	Hugo Stinnes	Charles Merz	S. Z. Mitchell
Ownership	mixed: government and private shareholders	private shareholders	holding company (private)

Sources: Footnotes of this chapter.

*Figures are not available for the kw. capacity of NESCO's North Tees Central Station. In 1924 it may have been larger than that of Carville "B."

The geography of the Ruhr region shaped the character of the RWE system. The Ruhr had the largest hard-coal deposits in Germany. Deep hard-coal mines supplied fuel for RWE plants and used power from the RWE system for pumps and other mechanical equipment. The first RWE generating plant was completed in 1900 at the site of the Victoria Mathias mine in Essen (see Fig. XIV.2). Less than 100 kilometers from the heart of the Ruhr were the relatively unexploited brown-coal deposits west of Cologne. After World War I these were increasingly mined using strip, or open-face, technology. Moreover, within economical transportation range of the coal mines were the iron-ore deposits of Sweden and Spain; the Rhine, its tributaries, and related canals made possible the importation of iron ore from abroad.

The presence of coal, the proximity of iron ore, and the availability of economical means of transportation all contributed to the industrialization of the Ruhr and lower Rhine. The rise of the steel industry located major power consumers in the region supplied by RWE. The Krupp Company had its headquarters in Essen, as did RWE. The establishment of heavy-chemical plants along the Rhine provided RWE with outstandingly favorable load characteristics. In sum, steel making, coal mining, heavy engineering, and chemical manufacture turned cities and towns like Essen and Gelsenkirchen into densely populated centers and added diversity to the utility's load. Some of the largest industrial establishments generated their own power in isolated plants, but over the years RWE's management persuaded many of these to exchange power with RWE or even become completely dependent on its system. RWE grew rapidly because its engineers and managers integrated the utility into this supraindustrial system (see Fig. XIV.3).

8; and Gerhard Dehne, *Deutschlands Grosskraftversorgung* (Berlin: Springer, 1928). On the economic geography and history of the Ruhr, see Norman J. G. Pounds, *The Ruhr* (London: Faber & Faber, 1952).

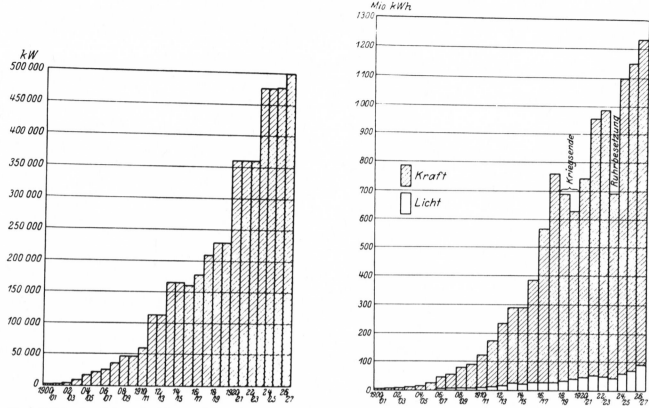

Figure XIV.1. *Installed capacity (kilowatts) (left) and production (kilowatt-hours) (right) of RWE, 1900–1927. From Dehne,* Deutschlands Grosskraftversorgung, *pp. 61–64.*

RWE's managers applied the economic principles of load factor and economic mix to exploit the geographical situation, and thus these principles made their distinguishing mark on the utility's style. As early as 1906, surveys of the developing field of central-station electrical supply in Germany singled out RWE as an outstanding example of supply and load management. At that time, the managers of RWE were integrating the formerly independent, or isolated, generating plants of the Ruhr coal mines into the emerging RWE system. The mines drew supplementary power from the RWE system during their period of heaviest demand, 6 A.M. to 2 P.M.; at other times, the more efficient generating plants at the mines fed their excess power into the RWE system.[7]

The integration of RWE into a widespread industrial system and the application of system economics resulted in part from the influence of Hugo Stinnes and August Thyssen, who in 1902 acquired control of RWE through a consortium. Beginning as an owner of coal mines, Stinnes (1870–1924) became one of the most resourceful system builders in Germany, if not in the world. The first of RWE's power plants was located at his Victoria

---

[7] "Professor Baum," "Beiträge zur Frage der Krafterzeugung und Kraftverwertung auf Bergwerken," *Glueckauf* 42 (1906): 1137–54; *Electrical World* 47 (1906): 811.

Figure XIV.2. *First RWE central station, Essen, 1900. Courtesy of Rheinisch-Westfälisches Elektrizitätswerk, Essen, Germany.*

Mathias mine in Essen. Stinnes conceived of electricity as a way of pooling energy into a generally usable form.[8]

As early as 1902, Stinnes, who became chairman of RWE, associated himself with the economic and technical principles of such bold conceptualizers of electric power systems as Samuel Insull, who envisaged the mass production and marketing of energy being furthered by rationalization of production facilities. As Stinnes wrote in the utility's 1902 report,

> We do not consider it our goal to exploit our monopoly in various districts by limited output at high prices as is the practice with most local suppliers, but believe our responsibility to ourselves and our customers, especially the railroads and industry, is to supply power at the lowest possible price and in largest possible quantities. Through the rational utilization of generating plants and distribution network, we shall regularly establish our rates on the basis of reduced costs.[9]

The organizational form and ownership of RWE evolved complexly. When Stinnes set out to establish a single supply system for the Ruhr region, the company proceeded to purchase other utilities, privately and government owned—for instance, the Elektrizitätswerk Berggeist AG in Brühl and the Bergishes Elektrizitätswerk in Solingen in 1905. RWE negotiated cooperative arrangements with local governments that not only enlarged the utility's area of supply but also provided expansion capital. Furthermore, the arrangements resulted in RWE's systematic involvement with consumers and utility regulators. By 1910, communities such as Solingen,

[8] *Electrical World* 49 (1907): 629.
[9] *RWE Geschäftsberichten*, p. 12; on Stinnes see Gert von Klass, *Hugo Stinnes* (Tübingen: Rainer Wunderlich, 1958).

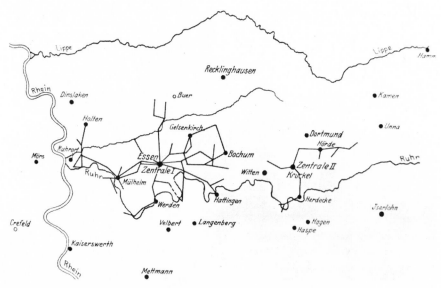

Ruhrort, and the district of Essen were purchasing RWE stock and taking electric supply from the utility. Stock ownership brought the local governments membership on the company's advisory committee (board of directors). At that time, local governments had considerable influence in determining the policy of the mixed-ownership enterprises that supplied electricity to them.[10] Political power and the constitutional arrangements of the local governments in the area of RWE supply facilitated the integration of government-owned and private utilities and encouraged mixed ownership, which was not the case in either the United States or England. This factor greatly affected the characteristics of the RWE system.

During the first two decades of its existence, RWE spread its area of supply into a region centered on the Ruhr valley and evolved into a mixed-ownership enterprise with about half of its shares in the hands of local governments and the other half in the hands of private investors. Negotiations between the utility and local governments in the Prussian provinces of Rhineland and Westphalia (which encompassed the Ruhr region) were facilitated because industrial leaders, allied with Stinnes in matters pertaining to coal, iron, steel, and chemicals, were influential in local government. The ingenuity and complexity of the negotiations that culminated in the mixed-ownership structure is suggested by an episode involving RWE and three utilities owned by local governments. The confrontation was resolved by negotiations during the period 1906–8. In 1906 RWE reached out toward the city of Dortmund and its environs by building a power station to supply the district and obtaining supply concessions. It offered to purchase Dortmund's city-owned utility, but the city refused and, supported by Berlin banks and the electrical manufacturer Allgemeine Elektrizitäts-Gesellschaft, enlarged its own plant. The expansionistic policy of RWE was further frustrated by two other local government utilities in the region, one based at the city of Bochum and the other at Hagen. A solution

[10] *Elektrotechnische Zeitschrift* 30 (1909): 1226; 47 (1926): 107; 29 (1908): 662; 31 (1910): 1262; 47 (1926): 106–7.

to the stalemate was found when Stinnes and his associates and the Dortmund utility agreed to form a new utility to take over the RWE power plant near Dortmund and the concessions RWE had recently acquired there.[11] Both Dortmund and RWE would own stock in the new company.

The growth of RWE was deeply affected by the severe stresses of World War I and the Versailles peace settlement. The year before the conflict erupted, RWE completed a power station at Knapsack in the brown-coal district southwest of Cologne. Equipped with two 15,000-kw. turbogenerators, the plant used brown coal, or lignite, taken by the open-face method from mines owned by the Roddergrube AG, a mining company with which RWE concluded a supply contract. This coal had only one-fourth to one-fifth the caloric value of hard coal. During the war, demand for energy from the war-stimulated manufacturers of aluminum, nitrates, carbide, and other materials caused RWE to expand its Knapsack plant greatly. Six 15,000-kw. turbogenerators were added, and before he died in 1917, the engineer and designer of the station, Bernhard Goldenberg, commissioned the installation of two 50,000-kw. units, turbogenerators that were larger than any others in the world (see Fig. XIV.5). As a memorial, the power station was later named for Goldenberg.

RWE concentrated its capacity at the Goldenbergwerk because of the savings that could be realized in copper, transportation, and labor. The very large units used less copper than a number of small ones supplying the same output; materials-handling equipment moved the brown coal to the power station, thus relieving the railroads of the burden of transporting low-caloric fuel. Endless chains pulled wagons, and gravity bucket-conveyors reduced labor. From 5,000 to 10,000 tons of lignite could be moved daily. Transmission lines carried 100,000 volts of electricity to load centers. Because siting at the coal deposits precluded location of the plant near a natural source of cooling water, a bank of cooling towers was built to service the turbine condensers.[12]

When peace came, RWE continued to expand the Goldenbergwerk. The Treaty of Versailles stripped Germany of 40 percent of its prewar hard-coal supply, and the reparations agreement called for the export of hard coal. Not only RWE, but Germany's public utilities generally, shifted to brown coal. In 1913, 23 percent of the fuel burned was lignite; by 1922, 41 percent was. In 1924 the Goldenbergwerk had a capacity of 290,000 kw., or 60 percent of the total capacity installed by RWE (see Figs. XIV.6 and XIV.7).[13]

When the German government began exporting hard coal to meet the reparations demands laid down at the Spa Conference in 1920, RWE's

[11] Ibid. 27 (1905): 828; *25 Jahre Vereinigte Elektrizitätswerke Westfalen, 1925–1950* (Dortmund: VEW, 1950), pp. 12–13; "Stromversorgung des Rheinisch-Westfälischen Industriebezirks," *Elektrotechnische Zeitschrift* 28 (1907): 243–44.

[12] Dehne, *Deutschlands Grosskraftversorgung*, pp. 57–59, 62; *RWE Geschäftsberichten*, pp. 28–31; "Report on Electric Power Supply Developments in Germany: Zone of British Occupation, 16 July 1919," Merz & McLellan "Reports and Tests" (vol. 1/24, folio 29) Tyne and Wear County Council Archives, Newcastle upon Tyne, England (no author of this typescript is given); and *Elektrotechnische Zeitschrift* 34 (1913): 1388; 38 (1917): 430–31; 41 (1920): 455; 43 (1922): 55.

[13] Dehne, *Deutschlands Grosskraftversorgung*, pp. 50–51, 62; *RWE Geschäftsberichten*, p. 31.

Figure XIV.4. *Bernhard Goldenberg.*
Courtesy of Rheinisch-Westfälisches
Elektrizitätswerk, Essen, Germany.

Figure XIV.5. *The Goldenbergwerk, 1919. Courtesy of Rheinisch-Westfälisches Elektrizitätswerk,*
Essen, Germany.

hard-coal-fired plants failed to receive an adequate supply of the fuel. Ever more dependent on its Goldenberg plant and the nearby supply of brown coal from the Vereinigte Ville and Berrenrath mines owned by the Braun-kohlen-und Briketwerke Roddergrube AG, RWE's managers sought a surer means of guaranteeing supply. They believed that a contractual relationship was not adequate for the future; the needs of the Goldenbergwerk could not be specified over a number of years, and Roddergrube would give priority to the demand for brown-coal briquettes. Under these circumstances, RWE negotiated an *Interessengemeinschaft* ("pooling agreement") with Roddergrube. Signed in 1920, the agreement preserved the legal independence of both parties, but turned over the management of Roddergrube to RWE. The ninety-year agreement also specified how profits to the two parties would be pooled and divided among the respective shareholders. The new managerial arrangement gave priority to supply of the Goldenbergwerk.[14]

Wanting to guarantee a supply of hard coal for other generating stations insofar as reparations and government allocations allowed, RWE also negotiated a similar *Interessengemeinschaft* with the suppliers of its generating facilities in Essen. In 1921–22 RWE further ensured its control over the hard-coal mines by acquiring their stock through an exchange of stock made possible by the substantial increase in RWE capital shares. With its energy base established, RWE looked forward to expanding its supply area.[15]

The RWE Roddergrube *Interessengemeinschaft* facilitated cooperation in matters less obvious than profit and the allocation of brown-coal produc-

[14] *Kölnische Zeitung,* 9 and 14 September 1920.
[15] *RWE Geschäftsberichten,* pp. 30–31.

Figure XIV.6. *Comparative costs of transporting hard coal (Steinkohle), brown coal (Rohbraunkohle), and brown-coal briquettes (Braunkohlenbriketts). From Dehne,* Deutschlands Grosskraftversorgung, *p. 52.*

	ab Werk	bei 100 km Entfernung	bei 200 km Entfernung	bei 300 km Entfernung
Steinkohle	28,6 M	35,6 M / 19,5%	40,6 M / 29,6%	45,6 M / 37,4%
Braunkohlenbriketts	25,1 M	34,8 M / 23,6%	41,9 M / 40,2%	49,1 M / 49%
Rohbraunkohle	12 M	30 M / 60,2%	43,2 M / 72,4%	56 M / 78,5%

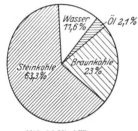

1913: 2,2 Mia. kWh.

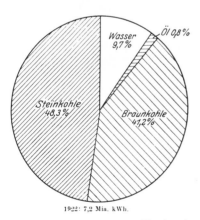

1922: 7,2 Mia. kWh.

Figure XIV.7. *Percentage of hard coal (Steinkohle), brown coal (Braunkohle), and water power (Wasser) in energy use of public electric utilities in Germany, 1913 and 1922. From Dehne,* Deutschlands Grosskraftversorgung, *p. 51.*

tion. It also furthered cooperative technological arrangements. Because the agreement extended over a period of ninety years, capital could prudently be invested in physical facilities that tied the partners together. This meant that more materials-handling equipment could be constructed to move brown coal from the open-pit mine to the brown-coal-consuming power plant. In the 1920s, oversized, dredge-like steam shovels removed the overburden of soil, and endless-chain cutters ripped away the brown coal and dumped it into cars drawn by electrically driven cog-wheel locomotives directly to the power plant.[16] (See Figs. XIV.8–XIV.11.)

Having been in existence for almost a quarter of a century, by 1920 RWE had an identifiable character and a tradition, or momentum. Frugal in its use of energy, taking the craftsman's and engineer's pride in a high level of technical efficiency, effective in raising capital for expansion, flexible and pragmatic in its associations with political authorities, innovative in its organization of mixed-ownership utilities, committed to the mass production of energy and to territorial expansion, and shaped by historical contingencies, RWE had a rich experience from which to formulate general policy principles. RWE's managers articulated a science of economics and technology that was particularly applicable to RWE. These policies, and the technological decisions that stemmed from them, can be observed in the history of RWE in the twenties.

The policy of mass production and large-area supply, associated with Stinnes as early as 1902 and called "rational" by him, was finely honed during two decades of growth and was associated with the "rationalization" movement that took place between the two world wars. The concept, as noted, was generally applied to German industry in the 1920s. After World

[16] *Geschäfts-Bericht . . . Braunkohlen-und Briketwerke Roddergrube Aktiengesellschaft in Brühl,* 1927 and 1929; F. Isermann, "Grossraumförderung in Braunkohlen-Tagebaubetrieben mit Schrägaufzügen," *Zeitschrift des Vereines deutscher Ingenieure* 72 (1928): 1256–62; and Georg Klein, *Handbuch für den Deutschen Braunkohlenbergbau* (Halle: Knapp, 1935), 2: 760–62, 1113–21. I am indebted to the Rheinische Braunkohlenwerke Aktiengesellschaft for these references. See also Robert A. Brady, *The Rationalization Movement in German Industry* (Berkeley: University of California Press, 1933), p. 76.

Figure XIV.8.   *Brown-coal transportation system, open-pit mining, 1912. Courtesy of the Rheinbraun Archives, Pfaffendorf, Germany.*

Figure XIV.8.   *Brown-coal transportation system, open-pit mining, 1912. Courtesy of the Rheinbraun Archives, Pfaffendorf, Germany.*

Figure XIV.9.   *Brown-coal open-pit mining, 1926, at the Vereinigte Ville mine. Courtesy of the Rheinbraun Archives, Pfaffendorf, Germany.*

War I, the policy of rationalization and the principles of its application to electric supply systems were expounded by RWE's management. Rationalization as defined and practiced by RWE—and practiced by American utilities—differed significantly from rationalization as defined by the older, labor-intensive industries in Germany such as mining and iron and steel production. Utilities like RWE were capital intensive and therefore con-

Figure XIV.10. Open-pit mining: Brown-coal transportation system, Vereinigte Ville mine, 1930. Courtesy of the Rheinbraun Archives, Pfaffendorf, Germany.

Figure XIV.11. Brown-coal transportation system, Vereinigte Ville mine, 1938. The group of eight stacks with four behind is the Goldenbergwerk. Courtesy of Rheinische Braunkohlenwerke AG, Cologne, Germany.

cerned about saving capital. Labor was a relatively unimportant factor in the cost accounting of electricity suppliers. (The visitor to a mammoth power plant would observe only a handful of workers.) Only when the production of hard and brown coal was taken into account did the labor factor assume major importance, and the utility engineers who defined rationalization focused on cost accounting as it pertained to the capital-intensive technology under their control—not to the production of fuel. In the high-technology, capital-intensive electric supply industry, rationalization involved an elegance and precision of concept that was unknown in the older, labor-intensive industries, which mixed Taylorism and simple mechanization into an amorphous concept of rationalization.

A. Koepchen, the engineer who in the 1920s as a member of RWE's management committee played a leading role in the formulation of company policy, noted that electrical engineers understood rationalization especially well. He defined the concept as using technology to obtain the economic optimum with a minimum of resources (*Volksvermögen*). He believed the essence of rationalization to be *Verbundbetrieb*, which he defined as the coordination of interconnected power plants supplying a large area. He included in his definition the power plants of a single undertaking as well as those belonging to different utilities if they were interconnected. The goal of rationalization was coordinated operations.[17]

RWE *Verbundbetrieb* stressed not only the interconnection of power plants but also an economic mix of the energy sources used by the interconnected plants. As has been noted, the geography of the Ruhr region provided RWE with the opportunity to mix brown-coal and hard-coal generation; then, during the twenties, the utility took the substantial step of incorporating water power into its economic mix. Prior to this, RWE *Verbundbetrieb* had called for brown-coal plants located at the mines to carry the system's base load because their year-round generating costs were the lowest in the mix, and for the hard-coal plants to pick up any load that exceeded the Goldenberg station's capacity. After hydroelectric plants were connected, RWE used the output of the running-water, or low-head, hydroelectric plants as it became available because the energy could not be stored and because the capital costs of these plants were low. The high-head plants with dams or reservoirs allowed energy to be stored in the reservoirs, so their output was used seasonally according to supply and demand.

In the 1920s, rationalization as practiced by Koepchen and other RWE managers involved load management as well as *Verbundbetrieb*. The company persuaded chemical works in the Ruhr and along the Rhine to abandon their isolated plants and take power from RWE. Chemical loads were especially sought after because they used electricity for heat as well as for electrolysis. These steady loads, taking current night and day, provided a nearly ideal load factor. The iron and steel works of the Ruhr region also loaded the RWE system economically because the rolling mills and electric furnaces were large, steady consumers. From its founding, RWE had improved its load factor by supplying energy for mine pumps that cleared the mines of water during the night hours.[18] During World War I, while building the Goldenbergwerk, RWE bought shares in a company that produced potash lye and soda lye, a company that in turn became a major RWE customer. RWE also supplied the Erftwerke, a newly constructed aluminum works which during and after the war was one of the utility's largest consumers. Dyeworks and metallurgical plants such as the Bergische

---

[17] A. Koepchen, "Das RWE in der deutschen Elektrizitätswirtschaft" (Speech delivered in the Haus der Technik in Essen on 28 March 1930), pp. 3–5; copy in the library of RWE headquarters, Essen. The speech was also published; see A. Koepchen, *RWE Elektrizitätswirtschaft* (Essen: RWE, 1930), p. 10.

[18] Koepchen, *RWE Elektrizitätswirtschaft*, p. 4.

Stahlindustrie Lindenberg in Remscheid and the Krefelder Stahlwerk also became major RWE customers.

In a widely quoted talk given on 28 March 1930 in the Haus der Technic in Essen, Koepchen compared the efficiency and resulting economy of the RWE system with the results achieved by other utilities operating in smaller supply areas and with a less extensive and diverse mix of power plants. He characterized RWE as an outstanding example of rationalization, or *Verbundbetrieb,* and he articulated the principles of the policy. He compared the load curves of RWE, the Hamburg utility, and the Berlin utility (BEWAG). With a better load curve and economic mix of generating plants, RWE reduced the costs to large, industrial consumers below the costs they would incur in their isolated plants, thereby bringing industries with good load characteristics into the system. This further improved RWE's load factor and allowed the utility to reduce the price it charged the small, commercial and residential consumers. Koepchen acknowledged that RWE's expansion eliminated the autonomy of the utilities and isolated power plants that were incoporated into the RWE system, but he argued that the economics of rationalization would prevail over the objections of individuals and small companies desiring to remain *Eigener Herr im Haus* ("master in one's own house").[19]

As described by Koepchen, *Verbundbetrieb* resembles horizontal business combination. There is, however, an important difference between business combinations, such as trusts and cartels, and *Verbundbetrieb.* Business combinations are commonly formed for the purpose of controlling or regulating supply, price, and marketing; *Verbundbetrieb* is a combination for the purpose of increasing the technical efficiency of production. *Verbundbetrieb* is technological and physical combination. Technological combination, however, is often complemented by a business combination. Hence we arrive at the association of *Verbundbetrieb* with business combination.

RWE *Verbundbetrieb* combined similar production units, generating facilities. RWE also entered into technological combinations comparable to vertical business combinations. Units along the line of production from energy resources and raw materials to finished products and available energy were combined. The form of these vertical combinations varied, as with horizontal ones, from firm and formal agreements to loose and informal ties. Much as transmission lines were the technological connectors for *Verbundbetrieb,* materials-handling and energy-conversion technology were the linkages for vertical combination. The business organizations and relationships that were formed to manage the vertical technological combinations included long-term contractual relationships such as that between Roddergrube and RWE.

Taking advantage of the German government's encouragement of *Verbundbetrieb* and the legal status of trusts, cartels, and other forms of combination, RWE not only entered into technological (or tight-linkage) combinations but formed loose confederations as well. While these loose combinations did not involve physical technology, they did usually reflect the relations of energy supply and production—for example, the relationship between the manufacturer of electrical equipment and the utility using

[19] Ibid., pp. 3–5.

the equipment. RWE's management was prone to loose combinations because of the system-building drive of the utility's founder, Hugo Stinnes, and his successor in 1924 as head of RWE, Albert Vögler.

Stinnes favored vertical combination and sought linear arrangements involving raw material (such as coal), manufacturers (such as iron and steel mills), and finished-goods producers (such as machine works). Immediately following World War I, he presided over the largest combination of industrial enterprises yet formed in Germany, the so-called Siemens-Rhein-Elbe-Schuckert Union, which included coal, steel, and electrical-machine manufacture. RWE coordinated its policies with the union through Stinnes, who looked upon electric power transmission as the eventual physical, or technological, connector in this combination. The union did not survive Stinnes's death, but its history stands as testimony to his instincts and policies. Stinnes's public utterances suggest that he viewed his rational industrial creations as substitutes for, and improvements upon, centralized government socialism.

Despite the planning and ordered design that were inherent in rationalization, *Verbundbetrieb,* and combinations, RWE, partly in reaction against planning on the national level and partly out of a commitment to the pragmatic approach, advocated regional-system growth by evolution rather than by long-term planning. After more than twenty-five years of operations, RWE knew its history and characteristics well enough to define its philosophy of growth. In their 1930/31 annual report, RWE's managers insisted that long-term planning forced utilities into a Procrustean bed that prevented response to an unceasing flux of opportunities and challenges. Plans, they argued, take into account only the circumstances that exist or are conceivable when the plans are formalized; therefore, they fail to provide for the technology and circumstances of the future. Plans, they concluded, become especially constraining blueprints in new, rapidly evolving fields of technology like electricity supply. (RWE may have had von Miller's Bayernwerk and plans for an all-German grid in mind.) Driven to expand by their economic motives and by technological potential, RWE's managers favored instead the "natural" course of development through economically prudent, step-by-step responses to concrete opportunities.[20]

The ecotechnical principles of RWE (more generally termed its style) were embodied in the utility's technology in the 1920s. RWE's technology was a cultural artifact, a physical manifestation of the utility's strategy and structure. To ensure that the technology was representative, engineering managers were needed who could integrate politics, economics, and technology. Indicative of this emphasis was the elevation of Arthur Koepchen, formulator of rationalization concepts, to the position of principal manager following the death of Bernhard Goldenberg in 1917. An engineer by training, Koepchen was known for his outstanding managerial abilities. He has been compared to Emil Rathenau, the innovative founder and manager of AEG.[21] Koepchen presided over the design and construction of a 220,000-volt transmission line that RWE extended over 800 kilometers from the

---

[20] Annual report for 1930/31, quoted in *RWE Geschäftsberichten,* p. 50.

[21] Georg von Siemens, *History of the House of Siemens,* trans. A. F. Rodger, 2 vols. (Freiburg/Munich: Alber, 1957), 2: 111.

Figure XIV.12. Prof. Dr. Arthur Koepchen. Courtesy of the Rheinisch-Westfälisches Elektrizitätswerk, Essen, Germany.

brown-coal fields and hard-coal mines of northern Germany to the water power in the south. This technological project exemplifies well RWE's style in the twenties. Koepchen undoubtedly took part in finding the way to finance construction of the line; he also seems to have made the decision about its feasibility. In April 1924, with E. Henke, another member of the managerial committee (*Vorstand*), Koepchen visited the United States to study high-voltage transmission as well as U.S. management and finance procedures.[22] California had led the way in high-voltage transmission. Several years later, however, Koepchen took pains to point out that although California had two 220,000-volt lines before RWE began construction of its great north-south transmission line, the California system grounded its neutral, or *Nullpunkt,* and therefore the voltage between line and earth was only $220,000 \div \sqrt{3}$, or 127,000 volts. The RWE line was not grounded, because the post office and railroads objected to interference with their telegraphic communications.[23]

Both in Germany and the United States, the utilities turned to the manufacturers for detailed designs to fulfill general specifications. The RWE transmission line was one of the major contracts of the decade, so AEG, Siemens-Schuckert, and Felton & Guilleaume competed. RWE specified hollow steel conductors after Koepchen discussed these with the Americans. Siemens-Schuckert drew upon the design of a hollow cable patented in 1919 by Rudenberg, designed new suspension insulators, and planned the construction routine. The hollow steel cable was 60 percent cheaper than solid copper cable. RWE awarded three contracts: 40 percent of the length of the line to Siemens-Schuckert and 30 percent each to AEG and Felton & Guilleaume provided that each used the Siemens-developed cable. Another problem solved by the manufacturers was design of outsized switchgear and transformers; the latter had to be adaptable to rail transport despite their weight of 168 tons.[24]

Another major RWE project of the twenties, a pumped-storage power plant, exemplifies the management's drive to raise the load factor and to utilize fully the capacity of its economic mix of power plants. The plant at Herdecke on the Ruhr River began operations on 28 January 1930 (see Fig. XIV.13). It, along with the 220,000-volt transmission line, was the most impressive of RWE's technological achievements in the 1920s. The concept of Herdecke, later named the Koepchenwerk, was summarized by Koepchen as follows:

The RWE, not wishing to continuously install new sets in its thermal power stations only for the purpose of covering the short winter lighting peaks, thought it advisable to use the generating capacity available at no further capital outlay during the night due to the then smaller load to pump water into an upper reservoir. The power for the night pumping would originate in otherwise underloaded thermal stations, low-head hydraulic plants, or blast-furnace gas power stations. [During the day the stored water would be released from the reservoir to drive water turbines and generators to provide power when needed.] The Herdecke pumped-storage power station, the capital cost of which per

[22] *RWE Geschäftsberichten*, p. 37.
[23] Koepchen, *RWE Elektrizitätswirtschaft*, p. 9.
[24] Von Siemens, *House of Siemens*, 2: 112–13.

Figure XIV.13. Koepchenwerk (earlier known as Herdecke). Courtesy of the Rheinisch-Westfälisches Elektrizitätswerk, Essen, Germany.

kilowatt was not higher than that for a thermal power station of similar size, shows that such a station is thoroughly able to pay for itself; moreover, this storage power station, apart from achieving its function of transforming low-grade power into high-grade power, may also be called on as an instantaneously effective stand-by.[25]

The public generally and some technical persons found absurd the notion of pumping water up to a reservoir from a river and then letting it run down to drive water turbines. RWE's management, well versed in the economics of *Verbundbetrieb,* went ahead and constructed a facility that in 1930 drew as much as 75,000 kw. from the RWE system during periods of low load and supplied as much as 132,000 kw. at times of peak load (see Fig. XIV.14). The reservoir, excavated on a hill about 165 meters above an artificial lake on the Ruhr, stored 1.6 million cubic meters of water. The lake, the Hengsteysee, had been built by the Ruhrverband, Essen, to reduce pollution in the Ruhr River. The RWE power plant had three complete units initially, each consisting of a Francis turbine, a combination motor-generator, and a pump. The motor-generators were rated at 33,000 kw. at 330 rpm; two were of AEG manufacture and two, Siemens-Schuckert. At times of low load, the motor-generators functioned as motors to drive the 33,000-h.p. pumps that lifted the water from the Hengsteysee to the storage reservoir; during times of peak load on the RWE system, they functioned as generators driven by the water turbines at 300 rpm. The water flowed down in four pressure pipes, or penstocks, whose inner cross section tapered from 3.20 meters at the valve control house near the reservoir to 2.55 meters at the power house; the length of the pipes was 292

[25] Koepchen, *RWE Elektrizitätswirtschaft,* p. 7.

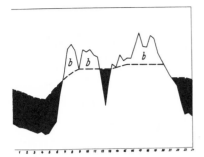

Figure XIV.14. *Inflow of electrical energy of storage water pumps at Herdecke power plant (shaded areas) and outflow of electrical energy from Herdecke water turbines (b) during a twenty-four-hour period. From Koepchen, "Das RWE."*

meters. A fourth pump was to be added to complete the plant's fourth unit. RWE expected the installation in full use to operate as a generating station for 4.4 hours in each twenty-four, which would result in an output of about 580,000 kwh. Construction of the installation involved up to 2,000 workers and took two years. RWE planned the entire layout with the co-operation and construction supervision of the Lahmeyer Company, Frankfort on the Main. When pumping, the Herdecke plant was expected "to flatten essentially the load curve of the entire RWE system" (see Fig. XIV.15).[26]

By the end of 1930, the RWE system had a generating capacity of 800,000 kw., 500,000 kw. of which were concentrated in the Goldenbergwerk, the largest power plant in Germany. In order to achieve the most economic mix of generating facilities in response to the system's ceaselessly varying load, the interconnected power plants and load needed to be centrally controlled and monitored. For this purpose, RWE tied together its spider web of transmission and distribution lines at Brauweiler, in the brown-coal fields west of Cologne. There, in October 1929, the main switching station of the system began operating. Its control panels indicated the state of the system at various points, and the load dispatchers could remote-control the switches and circuit breakers, connecting and disconnecting the various generating facilities (see Figs. XIV.16 and XIV.17). RWE's managers considered the Brauweiler control center to be the greatest concentration of electric power in the world.

The high-voltage transmission line extending through Germany was a dramatic achievement that recalled the great trunk-line railways of the nineteenth century. Pumping water uphill, the Herdecke plant was a wondrous paradox to the technologically naïve. And the Brauweiler switching station symbolized the power concentrations that were possible in the twentieth century. Nevertheless, the most complex problems of these technological projects were political. The history of the continuous expansion of the supply territory of RWE is one of political struggle, as can be seen in the following narrative account of the thrusts and counterthrusts made, the ground gained and the compromises negotiated.

On one level the competition for energy sources and market involved competition among utility holding-companies and among utilities. Beneath the surface, however, the struggle was one for power—political, not technological—among government authorities. Representing local governments and private shareholders, RWE competed for territory with two other large utilities, the Preussische Elektrizitäts AG (Preussenelektra), owned by the state of Prussia, and Elektrowerke AG (EWAG), controlled by the Reich, or central, government. At stake was control of Germany's electric supply—not only dominance of the electric supply industry but the associated political authority and economic power of various levels of government. The interests represented by RWE were the most complicated. Local governments, mostly in the Rhineland and including the Ruhr, voted RWE's controlling stock; but private interests, banks and industrial enterprises,

---

[26] RWE, *Pumpspeicherwerk Herdecke a.d. Ruhr* (Essen: RWE, n.d. [1930?]), a booklet in the RWE Archives, Essen; quote from Koepchen, *RWE Elektrizitätswirtschaft*, p. 7. See also A. Koepchen, "Warum Pumpspeicherwerke?" (n.d.), an essay in the RWE Archives (archive no. XXII/III/2C), Essen, Germany; and "Beseitigung der Spitzenlast," *Deutsche Allgemeine Zeitung*, 31 October 1929.

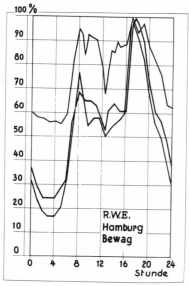

Figure XIV.15.   Representative load curves (c. 1930) for RWE, the public utility in Hamburg, and that in Berlin (Bewag). From Koepchen, "Das RWE."

also had substantial ownership and influence in the company. To complicate matters further, the state of Prussia also held minority shares in the company.[27] Therefore, RWE represented local government, private enterprise, and central government. Some local government authorities were hostile to RWE, however, because it deprived them of the ultimate authority to manage their public utilities.[28]

Early in the 1920s RWE began its concentrated move southeastward to water-power sites in the Alps. It acquired transmission-line rights and power-exchange interconnections by buying controlling stocks in the Elektrizitäts-Actiengesellschaft vorm. W. Lahmeyer & Company. This company had been the principal stockholder in RWE from its founding until 1902, when Hugo Stinnes and August Thyssen bought the controlling shares. Subsequently, Lahmeyer, which was headquartered in Frankfort on the Main, sold its electrical manufacturing facilities to AEG and concentrated on constructing and operating power plants and transmission systems. Lahmeyer was especially active in hydroelectric systems. Among the subsidiary utilities it controlled as a holding company were the Mainkraftwerke AG, which supplied territory on the right bank of the Rhine; the Kraftwerk Altwürttemberg AG, which from a hydroelectric installation on the Neckar River near Heilbronn supplied the regions of Stuttgart; and the Lech-Elektrizitätswerke AG in Augsburg, which operated the largest distribution system in Bavaria. To acquire controlling interest in Lahmeyer and its subsidiaries, RWE had to deal with the Elektrobank (later Elektrowatt) in Zurich, Switzerland, a company that financed electrical undertakings and owned shares in Lahmeyer.[29]

The move southward toward the Saarland also involved the piecemeal acquisition of controlling interest in utilities. RWE acquired shares in undertakings supplying cities, towns, and rural districts preliminary to arranging system interconnections and extensions of high-voltage lines. The utilities with which RWE made such arrangements included those with headquarters at Bad Kreuznach, Meisenheim, Koblenz, Worms, Idar, and Trier. These were mostly in Rhineland-Palatinate. The Saarland, with its hard-coal mines, did not revert to Germany until the plebiscite of 1935, but anticipating the union, RWE extended its high-voltage power lines into the region and interconnected along the way with water-power stations on the Moselle and its tributaries.[30]

During its expansion, RWE encountered the widespread electric supply system of the state of Prussia. In the vicinity of Frankfort on the Main, Prussia had acquired a power station and wanted to supply the city. Because RWE considered the area to be in its sphere of interest, it opposed the move. Anticipating that RWE might attempt to frustrate negotiations with the city, the Prussian government withheld the right of way RWE needed to cross the Main River and move southward with its high-voltage trans-

[27] *Elektrotechnische Zeitschrift* 45 (1925): 1168.

[28] Wilhelm Treue, "Die Elektrizitätswirtschaft als Grundlage der Autarkiewirtschaft und die Frage der Sicherheit der Elektrizitätsversorgung in Westdeutschland," in *Wirtschaft und Rustung am Vorabend des 2. Weltkrieges*, ed. F. Forstmeier and H. E. Volkmann (Düsseldorf: Droste, 1975), p. 138.

[29] *RWE Geschäftsberichten*, p. 33.

[30] Treue, "Elektrizitätswirtschaft," p. 136.

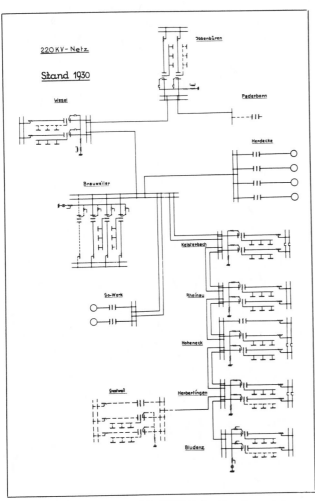

Figure XIV.16. *Schematic of the RWE high-voltage transmission network. From Koepchen, "Das RWE."*

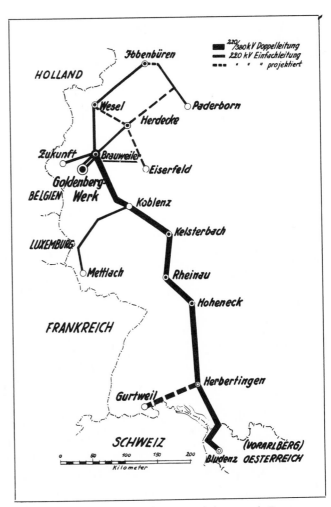

Figure XIV.17. *RWE high-voltage transmission network. From Koepchen, "Das RWE."*

mission system. RWE complained about a state government's using its legal authority to favor its own utility interests over those of a competitor, but RWE had made use of the legal authority of the local governments among its shareholders to acquire franchises and rights of way. In 1927 Prussia and RWE negotiated a demarcation treaty that came to be known as the *Elektrofrieden* ("electrical peace"). Western Germany was defined as the sphere of influence of RWE, and the area from the North Sea coast along the Weser River to Frankfort on the Main was designated the sphere of interest of Prussia. The peace was seen as an interval in the long-term struggle for control of electrical supply in Germany.[31]

Prussia got its supply contract with the Frankfort utility, and RWE crossed the Main. RWE then proceeded to reach an agreement in the province of Baden to extend its major high-voltage transmission line (220,000 volts) to Rheinau on the Neckar River. There an interconnection was made with

[31] Ibid.

the supply network of the Badenwerk, which drew upon water-power stations on the upper Rhine and in Switzerland. In this way RWE's coal-fired plants and the Alpine water-power plants could exchange energy economically. RWE's eventual goal was to extend its 220,000-volt transmission line (through agreements with Württemberg utilities) directly to the Alpine water-power sites of the Vorarlberg district of Austria. In the meantime, RWE and the Bayernwerk agreed to interconnect systems, thereby mixing energy from the brown-coal fields of the Rhineland and the water-power stations of the Bavarian Alps.[32]

The negotiated peace, or demarcation treaty, of 1927 also resolved some differences concerning RWE's western territory. The energy involved in this instance was brown coal. The Prussian government had purchased the Braunkohlen-Industrie AG ("Zukunft"), which mined brown coal and supplied electrical energy in the vicinity of Aachen, less than 125 kilometers from the Ruhr valley, the heart of the RWE system. RWE persuaded Prussia to give up its interest in the "Zukunft" by turning over to Prussia its shares in brown-coal fields in the Braunschweig area in central Germany. RWE also acquired by agreement an interconnection of its system with the electrical supply systems of the brown-coal area of central Germany.[33]

As observed in this study, the technological linkages often needed institutional contexts. The political struggles surrounding the 220,000-volt transmission line were mirrored in organizational forms. Notable among these was the Westdeutsche Elektrizitätswirtschaft AG, which provided for cooperation among the west German utilities tied to one another by the transmission line (see Fig. XIV.18). RWE and the Vereinigte Elektrizitätswerke Westfalen, GmbH, Dortmund; the Kommunales Elektrizitätswerk Mark AG, Hegen; the Braunkohlen-Industrie AG ("Zukunft"), Eschweiler; the Mainkraftwerke AG, Frankfurt-Höchst; the Badische Landeselektrizitätsversorgung AG (Badenwerk), Karlsruhe; the Hessische Eisenbahn AG, Darmstadt; and the Elektrizitätswerk Rheinhessen AG, Worms, participated in the organization, each holding 10 percent of the Westdeutsche shares except the last two, which shared a 10 percent holding. The remaining 30 percent of the shares were reserved for Württemberg and Palatinate utilities, logical choices for expansion because of their proximity to the transmission line.

The Westdeutsche Elektrizitätswirtschaft was in part a response to those who insisted that all German utilities should be tied together technologically and administratively. Oskar von Miller's plan of 1930 was an expression of this conviction (see p. 315). RWE believed that more flexible and pragmatic responses were possible within a loosely structured association such as the Westdeutsche Elektrizitätswirtschaft. Through it the associated utilities could further the technology of interconnection and thereby rationalize their common interest in an energy mix and economical load diversity over a large region of Germany.[34] The Westdeutsche Elektrizitätswirtschaft was an institutional expression of *Verbundbetrieb*. It was also representative of RWE's empirical and evolutionary style.

[32] *RWE Geschäftsberichten*, p. 42; selections from the annual report for 1925/26.
[33] Ibid., p. 45 (a selection from the annual report for 1926/27).
[34] Ibid., pp. 49–50.

Figure XIV.18. *Coupled systems of the Westdeutsche Elektrizitätswirtschaft AG. From Koepchen, "Das RWE."*

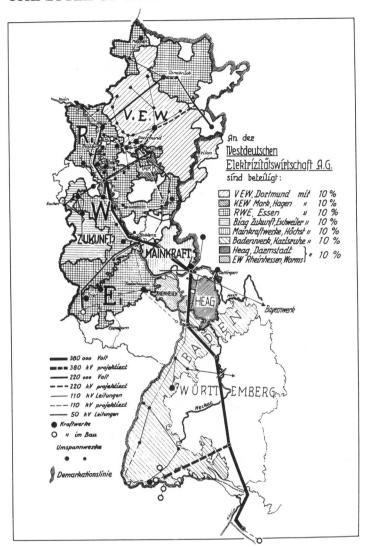

East of the 220,000-volt RWE transmission line, RWE's most powerful competitors also projected a transmission line to connect the brown coal of the north with the water power of the south. Like RWE, the competition sought an economic mix in which the southern utilities would supply inexpensive power when water was abundant, especially in late spring and early summer, and the coal-fired plants of the north would "wheel" power south in the winter, when snow and ice lowered the level of the Alpine lakes. On 16 May 1928 the Elektrowerke AG, the Preussische Elektrizitäts AG, and the Bayernwerk formed the Actiengesellschaft für deutsche Elektrizitätswirtschaft (German Electricity Company) to connect Hamburg with the Bavarian Alps, where Oskar von Miller's Walchenseewerk would feed into the system. Along the way, a 220,000-volt connection was made with the EWAG system of central Germany, which included the Golpa-Zschornewitz brown-coal plant built in World War I.[35]

[35] Gerhard Dehne, "Interconnection in Germany," *Electrical World* 92 (1928): 378–79.

Such competition was not only technological and political; it was financial as well. To compete for such projects as long-distance transmission lines, the German utilities resourcefully drew from both private and government sources of capital. RWE increased its shares and sold them to private investors and local governments. Paradoxically, the Prussian government bought RWE shares as well. In addition, RWE obtained long-term loans from U.S. banking houses, which in the twenties were especially interested in German investments.

RWE and the other large German utilities were unusually heavily capitalized enterprises. By focusing on electrical manufacturers, historians often overlook the leading role electric utilities played in the German economy. In 1927 the most heavily capitalized electric utility in Germany was RWE, whose stock amounted to 155 million reichsmarks ($37 million); the stock of AEG, the most heavily capitalized electrical manufacturer, totaled 186 million reichsmarks ($44 million). The assets of all the German electric utilities combined amounted to more than the combined assets of the electrical manufacturers.[36] By 1930, RWE's stock totaled 243 million reichsmarks, while the share capital of the largest German bank—the merged Deutsche Bank and Diskonto-Gesellschaft—was 285 million reichsmarks.

A substantial share of the financing for RWE expansion came from the National City Bank of New York. In 1925 National City loaned the utility $10 million; in 1927, $15 million; and in both 1928 and 1930, $20 million. In 1931, $7.5 million more was loaned. Converted into reichsmarks, the loans through 1930 amounted to 273 million RM, which was 30 million RM more than the value of RWE shares as of 30 June 1931.[37] The stock of RWE had to be increased not only to provide for the acquisition of other utilities but also because the loans from the National City Bank carried the option to buy the utility's stock. Throughout the decade of heavy financing and large increases in the number of stock shares, the local governments of Germany nevertheless retained the majority of RWE's voting stock.[38]

The financial problems of RWE differed markedly from those of the Pennsylvania Power & Light Company, a comparable regional utility in the United States, but in terms of technical characteristics, the two systems were notably alike in the 1920s. The similarities of these two utilities are attributable in large part to the international pool of technology from which both drew and to the common character of the geography—both human and natural—of the regions they supplied. Their differences stemmed largely from the contrasting politics, economics, history, and organizational forms of the regions in which they evolved.

A contrast in style is suggested by a comparison of the basic characteristics of the two systems (see Table XIV.1, p. 409).[39] Both RWE and PP&L used

[36] Gerhard Dehne, "German Electrical Industry Capitalized at $1,000,000," ibid., pp. 966–67.

[37] *RWE Geschäftsberichten*, p. 52.

[38] Ibid., pp. 34, 35, 52.

[39] Statistics for the Pennsylvania Power & Light Co. are taken from its publication *Pennsylvania Power & Light Company*, 8 vols. in 9 (Allentown, Pa., 1940?). Volume 1 is entitled *Origin and Development of the Company;* volume 2, *Corporate History of the Company;* and volume 3, *Origin and Development of Predecessor Companies.* These three volumes are especially helpful to the historian. All volumes will hereafter be cited as *PP&L History.* I am indebted to Mr. George

mine-mouth plants; PP&L burned low-grade anthracite that could not be economically transported to the market, and RWE burned low-caloric lignite. The two utilities operated at different frequencies, however; RWE at 50 cycles, PP&L at 60 cycles. Both utilities transmitted at 220,000 volts, and both would in a few years reach out to water-power sites to improve the economic mix of their energy sources. On the other hand, RWE was a mixed private and government-owned utility, while PP&L was a private holding company.

Consideration of fifty years of PP&L's history will reveal more of the characteristics—the style—of a regional utility. That history will also provide another example of the way in which context—both regional and national—shapes an evolving system. Especially notable is the fact that PP&L evolved, as did Insull's Commonwealth Edison in Chicago, in the relative absence of government regulation. Also important was the strong drive for geographical expansion that prevailed in the American context (see Fig. I.4, pp. 10–13).

Pennsylvania Power & Light, like RWE, supplied a predominantly industrial load in a coal-mining and steel-producing region. By 1930, 70 percent of its output (kilowatt-hours) went to industrial customers. Of the power supplied to industry, a disproportionate amount (45 percent) went to coal mining. (See Figs. XIV.19 and XIV.20.) PP&L's largest single industrial customer was the Bethlehem plant of the Bethlehem Steel Corporation, the second-largest steel company in the United States.[40]

PP&L's coal-mining and heavy industrial load was concentrated in the Lehigh valley of Pennsylvania, which was, on a smaller industrial scale, like the Ruhr valley in Germany. In addition to the Bethlehem Steel Corporation, smaller iron- and steel-related enterprises were situated in the valley, which extended into the large, anthracite coal-mining region of northeastern Pennsylvania. Allentown, with a population of about 90,000, was the largest city in PP&L's territory, and Bethlehem, a few miles distant, had a population of 60,000. Also concentrated in the valley were the cement industry, slate quarries, and silk manufacturers. To the northwest, about 50 miles from Allentown, were the anthracite fields, the nation's center for the mining of this hard, high-caloric coal. The Pennsylvania Power & Light Company also supplied the agricultural and small-industry district in the valley of the Susquehanna River north of Harrisburg, the state capital, the population of which was about 80,000. In Britain, NESCO served a region that nurtured a disproportionate share of the nation's mining and industry. While less outstanding than NESCO in the national context, the region served by PP&L was notable for including between one-third and one-half of the industry, as measured by wage earners and payroll, in a state whose industrial output ranked second in the nation.[41] To the south and east of

---

Vanderslice, vice-president and comptroller of PP&L, for the loan of volumes 1–3. The *PP&L History* was prepared as required by the Pennsylvania Public Utility Commission and the Federal Power Commission.

[40] *PP&L History*, 1: 132, 165.

[41] This estimate of PP&L's supply of the state's industry is based on statistics for the twenty-eight counties in which PP&L operated. PP&L served only twelve of the twenty-eight counties in their entirety, a large part of eight counties, and small portions of eight others. Ibid., p. 114.

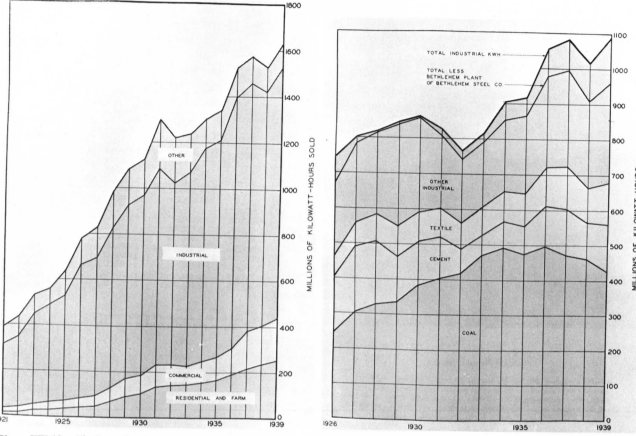

Figure XIV.19. *The Pennsylvania Power & Light Co.'s total load, 1921–39. From* PP&L History, *1: 133.*

Figure XIV.20. *The industrial load of Pennsylvania Power & Light, 1926–39. From* PP&L History, *1: 165.*

Pennsylvania's capital was Lancaster County, one of the foremost agricultural regions in the United States and part of the area PP&L served. The county was known not only for the richness of its soil but also for the assiduity of the Amish people, who were superior farmers. (The Pennsylvania Power & Light Company encountered in the Amish people an unusual problem, however, for they refused, because of their determination to avoid dependence on large-scale technological systems, to use electricity.)

The history of PP&L serves well as an example of the widespread and long-term process of organizational merger and consolidation that has characterized the history of electricity supply in the United States (see Fig. XIV.21). As in the case of many other U.S. regional utilities, PP&L's business history culminated in the 1920s with the consolidation of the merger movement, the formation of a holding-company pyramid embracing the company, its predecessor companies, and other large regional systems. The history of PP&L is especially interesting because its beginnings are associated with several of the earliest Edison utilities. The ultimate consolidation was carried out under the auspices of the Electric Bond & Share Company, the engineering, management, and financial enterprise that originated in the activities of the Edison-based General Electric Company.

Figure XIV.21. *Integration of utilities*
(power supply groups) through new
organizational forms and holding
companies. Note that in 1905 the merger
trend accelerated. From PP&L
History, *1: 61.*

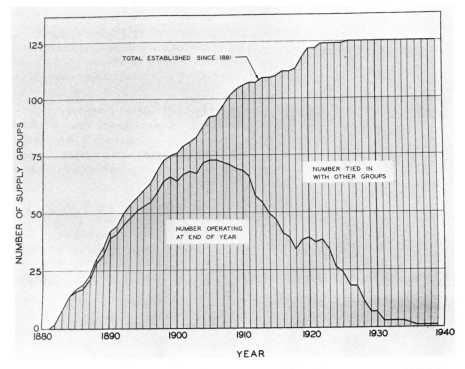

The Edison Electric Illuminating Company of Sunbury was one of PP&L's notable predecessor companies. At Sunbury in 1883 Thomas Edison and his associates introduced the three-wire system of distribution in response to one of the most pressing critical problems of the Edison system. The Edison chemical meter for measuring the consumption of electricity on the customer's premises also was tried out there. The Sunbury project, according to the Edison representatives who organized it, revitalized Thomas Edison's drastically waning fortunes.

According to standard histories of the Edison enterprises, Sunbury provided Edison with the opportunity to try out central-station incandescent lighting in small cities and towns. Edison called these "Village stations." Consumers were more widely dispersed there than in cities like New York, but the three-wire system reduced distribution costs by allowing the voltage to be raised to 220. By connecting lamps across an outside and inside (neutral) wire, 110 volts could be supplied to the customers. (See pp. 83–84 for a discussion of the three-wire technology.) Edison established and financed the Thomas A. Edison Construction Department to build "Village stations" when the financial backers of the parent Edison Electric Light Company failed to support his small-city ventures.[42] Before it was absorbed in the fall of 1884 by the Edison Company for Isolated Lighting, the Edison Construction Department installed central stations in nearly twenty towns. Five of these were predecessor companies of Pennsylvania Power & Light.

Philip B. Shaw of Williamsport, Pennsylvania, the state of Pennsylvania's representative, or manager, for the Edison Electric Light Company, ob-

[42] Harold C. Passer, *The Electrical Manufacturers, 1875–1900* (Cambridge, Mass.: Harvard University Press, 1953), p. 99.

tained funding locally and organized electric lighting companies at Williamsport, Shamokin, Sunbury, Mount Carmel, and Hazleton, Pennsylvania. The five small towns were situated in the anthracite coal fields between the Susquehanna and Lehigh rivers (except for Williamsport, which was located on the Susquehanna farther north). Shaw obtained letters patent, or legal authorization for the companies, between the fall of 1882 and the fall of 1883.[43] His recollections of the system's subsequent history differ from standard Edison histories. Shaw recalled that in the fall of 1882 former Thomas Edison boosters had become critics and that J. P. Morgan, in Shaw's presence, "[had] classed Mr. Edison as an impostor, a fakir and a charlatan" because of problems and delays in construction of the Pearl Street station in New York City. Shaw believed that "Edison's vision of giant castles had crumbled into microscopic units" and that Edison had become a persistent pessimist.[44] Shaw's account then noted his urging the despondent Edison to try his system in small towns, where gas light was expensive and coal inexpensive, rather than in the large cities. Edison then named Shaw the Edison representative for Pennsylvania after Shaw described his prior experience as promoter of patented devices. Shaw raised cash from persons in Williamsport and elsewhere in the region; arranged for the Edison Construction Department to build the plants; and negotiated the exchange of cash and the stock of the local companies for an Edison license and equipment from the Edison Electric Light Company, which held the Edison patents.[45] After construction began on the plants in Sunbury and the other towns, "a wild rush for parent company stock ensued, and the stock with a par value of $100 skyrocketed to $4400 per share." Therefore, it was no idle guess, Shaw ventured, "to assume that to Pennsylvania, and particularly to the towns above mentioned, belongs the credit and distinction of having rescued from total submergence the greatest industry ever conceived."[46]

By drawing on earlier experience with an above-ground transmission system that had been funded on a small scale by the Edison Electric Light Company at Roselle, New Jersey, in 1882, Edison designed the central-station and distribution system for Sunbury.[47] (The three-wire system was not used in Roselle, however.) Edison sent W. S. Andrews and Frank Sprague to Sunbury to superintend construction and to train local men during construction to operate the Edison Electric Illuminating Company. With thousands of spectators on hand, Edison inaugurated the system on 4 July 1883 in a town numbering no more than 6,000 people. A number of

[43] The five utilities ultimately became part of Pennsylvania Power & Light. Shaw also established a company at Bellefonte, Pa. See "Memorandum Data for use by the Publicity Division of the Franklin Institute of Pennsylvania." Unsigned, but authored by Shaw, this memo (on file at the Franklin Institute Museum Archives, Philadelphia, Pa.) was probably written during the early 1930s when the original dynamos and steam engines from the Edison station at Sunbury were on loan to the Franklin Institute. I am indebted to Anne Millbrooke for calling my attention to this document and for permission to refer to her unpublished research paper, "Electricity in Central-Eastern Pennsylvania: Pennsylvania Power & Light Company and Its Predecessors," done at the University of Pennsylvania in the fall of 1976.

[44] Shaw, "Memorandum."

[45] Edison Electric Light Co., *Bulletin*, September 1883, pp. 46–48.

[46] Shaw, "Memorandum."

[47] Edison Electric Light Co., *Bulletin*, September 1883, p. 47.

Figure XIV.22. *Edison direct-current, three-wire-distribution central station in Sunbury, Pa. Courtesy of the Edison Archives, Edison National Historic Site, West Orange, N.J.*

commercial buildings and houses were lit, including the City Hotel. The central station was a small wooden structure housing a Babcock and Wilcox boiler and two "L" type Edison generators belted to high-speed Armington & Sims steam engines (see Fig. XIV.22). The system had a capacity of about six hundred fifty 10-c.p. incandescent lamps.[48] With technical advice and equipment from the parent Edison company continuing to be given in exchange for bonds and stock, Sunbury expanded and by 1888 served more than a hundred customers and about 1,200 lamps.[49] The company was reorganized and involved in mergers several times between 1888 and 1911, when it was absorbed by the Northumberland County Gas & Electric Company and lost its corporate identity. Northumberland was later acquired by Pennsylvania Power & Light.[50]

Small utilities with Edison licenses, like the utilities established by Shaw, began operating in other parts of eastern and northeastern Pennsylvania and eventually were integrated into a regional system by the Pennsylvania Power & Light Company. Thomas Edison and the Edison Electric Light Company took stock from these companies in partial exchange for equipment and advice. Thus began a long history of manufacturer influence in utility operation. The load carried in the early stations was mostly incandescent lighting supplemented in some cases by a complementary arc-lighting plant made up of equipment from one of the early arc-light manufacturers—Thomson-Houston, Excelsior, Brush, United States, American, Schuyler, Sperry, and Weston, for example. A poor load factor resulting from the absence of load other than lighting and a high peak during the evening hours weakened the small utilities, however. Their perilous situation deteriorated further when "the battle of the systems" manifested itself on the local level in the later 1880s with the appearance of competing, alternating-current lighting stations using Westinghouse, Thomson-Houston, and other non-Edison equipment. The competition from alternating-current stations in small towns was especially sharp for direct-current utilities because the competitor's a.c. distribution at 1,000 volts or more increased the area of economical service. The a.c. utilities also held an advantage because arc-lighting and alternating-current, incandescent systems were compatible (both used high voltages). By 1900 the era of duplication of facilities and competition had brought the establishment of no less than sixty-four different companies serving eighty-eight communities in the region that was later to be served by the Pennsylvania Power & Light Company. The technical characteristics of these systems

[48] Information on the early history of the Edison Electric Illuminating Co. of Sunbury can be found in the Pennsylvania Power & Light Co. Papers at the Eleutherian Mills Historical Library in Wilmington (Greenville), Del. (hereafter cited as PP&L Papers). These papers include financial records, such as ledgers and journals, and formal records, such as minute books, of the predecessor companies of PP&L, which was formed in 1920. For instance, among the papers are a photostatic copy of the first Minute Book and the original of the second Minute Book of the Sunbury company. I am indebted to Dr. Richmond Williams and Mr. Hugh Gibb for guidance in the use of these papers. See also W. S. Andrews, "The Story of Sunbury Station," *Popular Electricity* 3 (1910): 391–93; Harry L. and Samuel Keefer, "Installation in Sunbury of First Three Wire System," *Sunbury Daily Item*, 25 May 1916.

[49] Millbrooke, "Electricity in Central-Eastern Pennsylvania," p. 15.

[50] The competition among electric utilities in Sunbury from 1900 to 1911 is succinctly chronicled in *PP&L History*, 3: 30–32.

varied widely in level of voltage and frequency as well as in type of current.[51] The situation was parochial.

From about 1895 to 1910 the effects of "the battle of the systems" attenuated with the resolution made possible by couplers, but the electrification of streetcars and the exploitation of hydroelectric power brought additional challenges for the utilities. Some transit companies used power from existing utilities, but others built their own power plants. The existing utilities could have supplied the streetcar companies either by adding d.c. equipment or by using converters if their generators were the a.c. type, but the transit companies, believing that the transit load was sufficient or that it could be supplemented by obtaining their own franchises to supply lighting, pursued their goals independently. Power companies with hydroelectric stations also competed with the existing lighting companies by obtaining distribution rights instead of selling power wholesale to the utilities. These developments in Pennsylvania reflected what was happening throughout the utility industry.[52] A summary of the history of several utilities in northeastern Pennsylvania—later PP&L's territory—is in many ways representative of these trends (see Fig. XIV.23).

Organized in 1886, the Edison Electric Illuminating Company of Lancaster, Pennsylvania, adapted to change when in 1892 it installed a separate alternating-current distribution system. Nevertheless, "the battle of the systems" made its way to Lancaster in 1893 when Westinghouse interests established a second utility, the Citizens Electric Light, Heat & Power Company, in this small city. The two companies competed for about two years until the Edison Company bought the Citizens Company, abandoned the less efficient, Edison generating plant, and moved some equipment to the newer power plant acquired in the purchase. The rise of hydroelectricity then had local repercussions as a new company, the Lancaster Electric Light, Heat & Power Company, incorporated in 1897, supplied Lancaster from three a.c. hydroelectric plants along Conestoga Creek. The competition so reduced prices that the Edison Company was unable to declare dividends for several years. In 1902, however, the Edison Company agreed to take over the entire power production of the hydroelectric company and its distribution facilities in Lancaster, thereby initiating a process of integration. The integration of thermal and hydroelectric facilities in turn led to the institution of a load dispatcher to manage the diverse power supplies to achieve optimal economies.

The early history of electric supply in Lancaster differed in one respect from utility-industry trends: a transit company did not complicate and stimulate developments. Elsewhere in eastern Pennsylvania, however, transit companies did shape history. In Allentown, a small city in the Lehigh valley,

[51] Ibid., 1: 5–6.

[52] Utility industry trends among hundreds of small utilities in the United States are defined and described by Forrest McDonald in *Let There Be Light: The Utility Industry in Wisconsin, 1881–1955* (Madison, Wis.: The American History Research Center, 1957). McDonald provides an overview for the different phases of utility history in Wisconsin on pp. 3–32, 103–25, 181–201, and 299–322. Thomas C. Martin also provides overviews of trends in U.S., Department of Commerce, Bureau of the Census, *Special Reports of Central Electric Light and Power Stations* (Washington, D.C.: GPO, selected years). See, for example, the *Report* for 1907 (published in 1910), pp. 96–123; and the *Report* for 1912 (published in 1915), pp. 111–76.

Figure XIV.23. *Corporate history of the Pennsylvania Power & Light Co. From* PP&L History, 2: xviii.

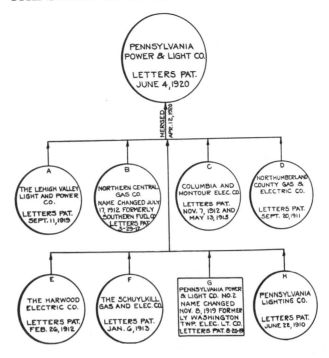

a lighting company was established in the 1880s. About 1890, when the potential load of the Allentown Electric Light & Power Company exceeded the company's ability to finance the additional plant, the Industrial Improvement Company of Boston arranged for the Allentown & Bethlehem Rapid Transit Company, which was already operating a transit network and a power plant, to take over the common stock of the Allentown Electric Light & Power Company and to enlarge its generating plant to serve both the transit network and the distribution system of the Allentown Power Company. In 1903 the situation was further rationalized when a new company, the Lehigh Valley Transit Company, acquired the Allentown & Bethlehem Rapid Transit Company and the Allentown Electric Light & Power Company. In 1913, the Lehigh Valley Transit Company merged and consolidated the Allentown Electric Light & Power Company and a number of small properties it had acquired into the Lehigh Valley Light & Power Company, a principal predecessor of PP&L.[53]

Another stage in the prehistory of the PP&L began with the introduction of the turbine and high-voltage transmission. The new technology stimulated integration in northeastern Pennsylvania as it had in the regions served by RWE and NESCO. The relatively short history of the Harwood Electric Power Company, a predecessor of the Pennsylvania Power & Light Company, provides an example of the process of integration and the formation of district systems. In 1907, the operator of an anthracite mine, Calvin Pardee, financed and organized the Harwood Electric Power Company. The company was designed to preside over a mine-mouth power station in the anthracite-mining district near the small town of Hazleton

[53] *PP&L History,* 3: 2–6.

in northeastern Pennsylvania. Mine-mouth plants were not new; RWE's first plant was built at the Victoria Mathias mine in Essen. Pardee transferred several small turbogenerators and the stock of the Harwood Coal Company to capitalize the new company. The temporary plant had two 300-kw. Westinghouse-Parsons machines and one 500-kw. General Electric turbogenerator, but by 1913 its capacity had risen to 12,000 kw. and the plant had achieved the status of a district station. The plant stood alongside a large coal breaker, where chunks of coal from the mine were broken down, cleaned, graded, and sorted according to size. The power plant's capacity exceeded the mining and industrial load in the franchise area of the new company, so, by stock purchase and merger, Pardee acquired the territory of the Consumers Electric Light & Power Company of Hazleton, which had been formed in 1906 to acquire charter rights to distribute electricity in Hazleton. Using power from the Harwood plant, the Consumers Company rendered service more economically than an older competitor, the Hazleton Electric Light & Power Company, the successor to the Edison Company established by Philip B. Shaw in 1883. Because of its antiquated equipment, some of it the Edison direct-current type, the Hazleton Company was unable to compete and sold out to Pardee in 1912.[54]

In 1913 the Lehigh Coal & Navigation Company—which is not to be confused with the Lehigh Valley Light & Power Company—constructed another district station equipped with turbogenerators only 10 miles from Harwood, at Hauto. One of the nation's oldest industrial corporations, the Lehigh Coal & Navigation Company had been founded in 1822 to exploit the anthracite fields and to build a canal system to transport the anthracite out of northeastern Pennsylvania to the Philadelphia port and market. Over the years the company had acquired extensive coal fields and related manufacturing and transportation facilities, especially in the Lehigh valley.[55] Its ability to move capital from mining to transportation and then to electric power made it a highly adaptable energy and investment enterprise. The company had incorporated the Lehigh Navigation Electric Company in 1912 for the primary purpose of supplying electricity to its coal mines in Carbon County, to cement mills in the Lehigh valley, and to the Bethlehem Steel Company in Bethlehem, about 30 miles from Hauto, where the new power plant was to be situated.[56] The initial equipment at the Hauto installation consisted of three 10,000-kw. General Electric turbogenerators. The Hauto power plant was also a mine-mouth installation.

In 1913 the Lehigh Navigation Electric Company continued the trend toward organizational integration by purchasing from Alfred D. Pardee controlling stock in the Harwood Electric Company (formerly the Harwood Electric Power Company), and then in 1915 maintained a technical trend by interconnecting its Hauto plant with the Harwood plant. The agreement between the Lehigh Navigation Electric Company and the Harwood Com-

---

[54] Ibid., 1: 12.

[55] H. Benjamin Powell, *Philadelphia's First Fuel Crisis* (University Park: Pennsylvania State University Press, 1978), pp. 82–84.

[56] The merger and consolidation agreement by which the Lehigh Navigation Electric Co. was formed was dated 23 December 1912. See Minute Book, Lehigh Navigation Electric Co., 27 July 1913 to 30 July 1919, item 7019, PP&L Papers.

pany, which retained its corporate identity despite Lehigh ownership, included a complex and resourceful arrangement for power exchange.[57] The technical merger was made possible by the use of a coupler in the form of a 6,000-kw. 25/60-cycle frequency changer. The "short form merger act of 1876" of Pennsylvania, which permitted one company to sell its franchises and its property to another, facilitated this and similar mergers.[58]

By 1917 the Lehigh Company had built up demand to a connected load of 24,100 kw., but it was predominantly a large, industrial-customer load. It had only 112 customers and 63 percent of the power it delivered was used by five cement companies. To meet demand, the company periodically borrowed money (until its debt was over a million dollars) from the Lehigh Coal & Navigation Company to finance construction of power plants and transmission and distribution lines. The financial transactions were facilitated by S. D. Warriner, who was president of both companies.

The decision-making process and the modes of development adopted for the Lehigh Navigation Electric Company changed, however, in the fall of 1917. S. Z. Mitchell, president of Electric Bond & Share, was present at the Lehigh Company's board of directors' meeting on 12 September 1917. During the meeting, several members of the board resigned and new ones, including Mitchell, were named. The minutes laconically state that subsequent to his election, Mitchell took part in the proceedings. He not only took part, he became the engineer of change. Thenceforth, Electric Bond & Share, through its holding companies, determined the policy of the Lehigh Navigation Electric Company and much of the utility business in eastern Pennsylvania.[59]

Mitchell and S. D. Warriner attended the board meetings of the Lehigh Company throughout 1918, but not until July 1919 was the impending transformation in company organization revealed. The minutes of the July 29 meeting include a statement of business and technological policy that was used by holding companies to justify the transformation of the electric supply industry throughout the country. The policy was inventive; at least for a time, it solved a critical problem of electric supply that was comparable in seriousness to the technological problems Edison solved when he introduced the high-resistance lamp. Four decades earlier, the critical-problem solver (Edison) was at Menlo Park; in 1919 he (Mitchell) worked out of the New York offices of Electric Bond & Share. Appropriately, Edison, the inventor-entrepreneur, and Mitchell, the financier, presided over different phases of the history of electric supply systems (see pp. 14–17 above).

The usually matter-of-fact minutes explained that the Lehigh Navigation Electric Company could acquire on reasonable terms the funds needed to provide the service demanded of it only if a merger was organized and the combined properties pledged as security, and only if the combined companies increased their earnings through the rationalization of technology and economies of scale. The particular merger the board had in mind was

[57] Meeting of the Board of Directors, 23 June 1915, Minute Book, Lehigh Navigation Electric Co., ibid.

[58] This short history of the Lehigh Navigation Electric Co. is based on *PP&L History,* 1: 15–16 and 2: v–vi.

[59] Meeting of the Board of Directors, 12 September 1917, Minute Book, Lehigh Navigation Electric Co., item 7019, PP&L Papers.

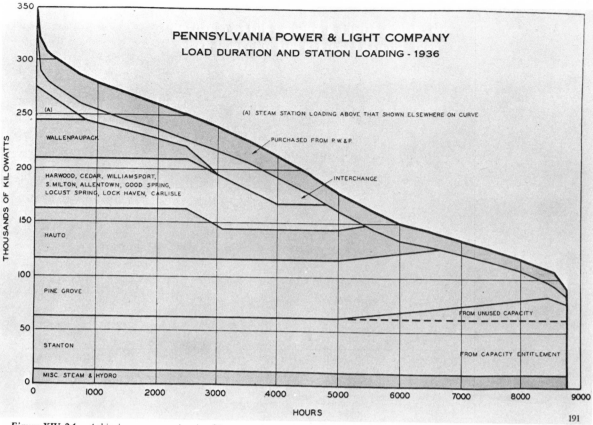

Figure XIV.24. *Achieving an economic mix of power stations to meet varying demand. From* PP&L History, *1: 191.*

that of the Lehigh Navigation Electric Company and the Lehigh Valley Light & Power Company, the latter having been formed in 1913 by the Lehigh Valley Transit Company from small utilities in the Allentown and Bethlehem area.[60]

The two companies merged and consolidated in July 1919 to form "The" [addition of the definite article constituted the only change in name] Lehigh Valley Light & Power Company. The stock transactions involved in the merger suggest the nature of the financial and business reorganization that took place. The complexity of the transactions suggests why many utility heads at the time were known more for their financial ingenuity than for their ability to solve technological and managerial problems. The old Lehigh Valley Light & Power Company had common stock outstanding amounting to $641,700, and for this the new company exchanged $1,500,000 of preferred stock. The Lehigh Navigation Electric Company's $3,037,000 of common stock was exchanged for $2,178,700 of the new company's common stock. The $3,678,700 of outstanding common stock of the two old companies, then, was exchanged for $3,678,700 of stock of the new (some preferred, some common). Share for share, the new preferred stock was valued less highly than the common stock of Lehigh Valley Light & Power, while the new common stock was valued more highly than the old

[60] Meeting of the Board of Directors, 29 July 1919, ibid.

common stock of the Lehigh Navigation Electric Company. "The" Lehigh Valley Light & Power Company thus had authorized common stock valued at $4,150,000 and preferred stock valued at $1,700,000, which in combination equaled the authorized common stock of the two old companies. The old companies had not listed preferred stock. Their bonded indebtedness amounted to $619,500 (LVL&P) and $3,909,000 (LNEC).[61]

Within a month after the state of Pennsylvania issued letters patent establishing the new company, the company's board of directors discussed the raising of $3 million to carry out the technological steps it thought necessary to meet demand. Additional capacity was planned for the Hauto and Allentown plants, and new transmission lines were to be constructed and old ones raised from 22 kv. to 66 kv. In conjunction with these changes, other components of the enlarged system, such as switches and transformers, were to be added or replaced. The interrelationship of financial and technological solutions was clear.[62]

The election of S. Z. Mitchell to the board of directors of the Lehigh Valley Navigation Electric Company had signaled the involvement of Electric Bond & Share. In 1920 the influence of this corporation became more explicit when the Lehigh Power Securities Corporation, a holding company associated with Electric Bond & Share, took over regional supply in northeastern Pennsylvania.[63] This followed the establishment—by further merger and consolidation—of the Pennsylvania Power & Light Company of Allentown, Pennsylvania. The new formation was formally discussed at a meeting of the board of "The" Lehigh Valley Light & Power Company on 17 April 1920. The holding-company rationale was stated explicitly in the minutes, undoubtedly for the historical record, for those present were well acquainted with it.

It is the consensus of opinion that through merger and consolidation of this company with such companies [several other neighboring utilities] on an equitable basis, a much greater flexibility in financing can be obtained, which would be distinctly to the advantage of each company, and especially to the advantage of the public in the territory served by each company, as it would facilitate the possibility of providing a basis for the necessary property extensions and developments.[64]

Again, stock transactions characterized the consolidation and merger agreement.[65] The six operating utility companies that were drawn in (besides "The" LVL&P) were the Northern Central Gas Company, the Columbia & Montour Electric Company, the Schuylkill Gas & Electric Company, the Pennsylvania Lighting Company, the Harwood Electric Company,

[61] Joint Agreement of 23 July 1919, Minute Book of "The" Lehigh Valley Light & Power Co., item 2352, PP&L Papers.

[62] Meeting of the Board of Directors, 1 October 1919, ibid.

[63] Lehigh Power Securities Corp. was identified as one of five holding companies under the substantial or complete control of Electric Bond & Share in 1933. After 1928 it was a subsidiary of the National Power & Light Co., another holding company formed in 1921 and controlled by Electric Bond & Share. See U.S., Federal Trade Commission, *Utility Corporations*, Part 72A (Washington, D.C.: GPO, 1934), p. 616.

[64] Meeting of the Board of Directors, 17 April 1920, Minute Book of "The" Lehigh Valley Light & Power Co., item 2352, PP&L Papers.

[65] Joint Agreement of Consolidation and Merger, 12 April 1920, ibid.

and the Northumberland County Gas & Electric Company. Also named was another, nonoperating, Pennsylvania Power & Light Company, which had been formed earlier simply to hold the name. All the companies were territorially contiguous. The Northumberland Company counted among its predecessor companies the Edison Electric Illuminating Company of Sunbury, the three-wire-station company established in 1883; and among the antecedents of the Pennsylvania Lighting Company was the Edison Electric Illuminating Company of Shamokin, which had been formed under the same auspices as the Sunbury company. The line of development from the Edison Electric Light Company, which licensed these two, to Electric Bond & Share was long, complex, but continuous.

Each of the companies involved received in exchange for its stock the preferred cumulative stock of the Pennsylvania Power & Light Company. The Lehigh Valley Transit Company also received PP&L preferred cumulative stock in exchange for the same number of preferred shares of "The" Lehigh Valley Light & Power Company it had held. The exchanges took the majority of the 35,000 cumulative preferred shares authorized by the new company, but left 65,000 shares of noncumulative preferred stock and 310,000 shares of common stock, all of which went to the holding company, the Lehigh Power Securities Corporation. The Pennsylvania Power & Light Company (Lehigh Power Securities Corporation) could then sell stocks and bonds to the public, but keep control of the voting common stock in its own or other friendly hands.

The Pennsylvania Power & Light Company continued to expand its territory by merger and consolidation, principally in 1923, 1928, and 1930. In 1923 the Lehigh Power Securities Corporation acquired for PP&L the stock of five utilities. Thirty-four more companies operating in, or adjacent to, the region serviced by PP&L were merged with it in 1928. In 1930 Lehigh Power Securities brought twenty-one more utilities into the system.[66] With the financial flexibility and resources acquired, the technology of the system was rationalized and expanded. Further additions were made to the Harwood, Hauto, and Allentown steam stations, and in 1924 construction was begun on the 40,000-kw. Wallenpaupack hydroelectric station (Fig. XIV.25). The hydroelectric station carried the system's peak loads and brought more efficient and economical utilization of the steam plants. In 1925 the company further expanded the capacity of its steam station at Pine Grove, which had been acquired by merger, and in conjunction with the Scranton Electric Company built for base-load operation a 100,000-kw. station of unusually high economy.

Hundreds of miles of high-voltage transmission lines were added to the system, but the most technologically innovative project was the 220,000-volt interconnection of the Pennsylvania Power & Light Company's plants with those of the Philadelphia Electric Company and the Public Service Electric & Gas Company of New Jersey (see pp. 331–32 above). The load diversity offered by Pennsylvania Power & Light complemented especially well the loads of the other two, more urban utilities, for their peak loads came in the late afternoon, while the peak of PP&L, primarily an industrial load, came in the mornings. The three companies thus exchanged power.

[66] *PP&L History,* 1: 27–38.

Figure XIV.25.   Wallenpaupack hydroelectric station. From PP&L History, *1: 47.*

The Lehigh Power Securities Corporation had become a subsidiary of the National Power & Light Company in 1921. It was one of the major holding companies controlled by Electric Bond & Share. Therefore, the Pennsylvania Power & Light Company took engineering, construction, management, and financial services from some part of the National Power & Light Company group or directly from Electric Bond & Share. In 1933, for example, PP&L was listed as paying the largest amount of all the related utilities to Electric Bond & Share for supervisory, or management, services ($353,152). It also paid for engineering and auditing services.[67]

The system builders of the holding-company era redefined load diversity to include investment diversity. Mitchell repeatedly spoke of the advantages of spreading investments, or "risks," so that the loss of load because of a depressed economy resulting from an orange-crop failure in the territory of a Florida operating company would be compensated for by a bumper crop in the territory of another operating company owned by the holding company.[68] He added that "the principle of large volume and diversification as applied to various risks in order to stabilize results through the law of averages has been followed for over 200 years by the great insurance companies of the world."[69] PP&L's system builders expressed a variation on this theme by contending that their energy system allowed risks to be diversified and spread in their region in such a way as to maintain the economic health of the region. Their case in point was the anthracite industry.

---

[67] Federal Trade Commission, *Utility Corporations*, Part 25, pp. 460–63.

[68] Sidney A. Mitchell, *S. Z. Mitchell and the Electrical Industry* (New York: Farrar, Straus & Cudahy, 1960), p. 88.

[69] U.S., Federal Trade Commission, *Electric Power Industry: Supply of Electrical Equipment and Competitive Conditions*, 70th Cong., 1st sess., 1928, S. Doc. 46, p. 172.

The annual production of anthracite coal reached its peak in Pennsylvania in 1917. The cause of the subsequent decline was complex and involved the use of other fuels as well as increased operating costs in the anthracite fields because of the depletion of the most accessible seams in deposits worked since the 1850s. By 1930 the annual production had fallen steadily to 70 million short tons from the 100 million tons produced in 1917. Despite this, through abandonment of isolated plants and electrification of machinery and production processes, the anthracite industry increased its use of PP&L electricity until 1936, when a gradual decline set in. However, the increase in electrification was not the point PP&L's managers wanted to make about the economic advantages of regional electrification. In their advocacy of regional supply they stressed the ability of the integrated utility to shift energy supply from districts of relatively economic decline to regions of economic potential—more specifically, from districts where coal seams were near exhaustion to distant ones as yet unexploited. Even more to the point was the ability of the regional utility to shift energy from declining industries to thriving ones. For instance, the annual production of Portland cement in the company's territory doubled between 1918 and 1927, years of sharp decline in the production of anthracite. Furthermore, the cement industry was an especially desirable customer because of its favorable load curve.

PP&L built up an industrial sales department to stimulate technical changes and economic development, and this in turn brought increased electricity consumption and improved load-curve characteristics.[70] The success of this department depended to no small extent on the rates the company charged, for they were lower for the larger industrial consumers than for residential and farm consumers. In 1930, for instance, while the industrial load amounted to 70 percent of the system's total, the revenue from that load was only 40 percent of the total revenue earned from the sale of electricity. In the same year, while the residential and farm customers used only 9 percent of the kilowatt-hours sold, their share of the total revenue from sales was 28 percent.[71] Disappointed in the performance of the private utilities, proponents of rural electrification pointed out these statistics.

In the years immediately following its organization in 1920, PP&L feared that the pressures of contingent circumstances would get beyond its control. These pressures resulted from a conjuncture of factors and events that can be traced to the wartime experiences and political trends discussed in this study in connection with the Grid in Britain, the Bayernwerk in Germany, and the Giant Power plant in the United States. Giant Power could have impinged directly on PP&L, for it was designed to rationalize production and distribution in Pennsylvania and to emphasize rural electrification.

In the political arena, the privately owned utilities effectively resisted Giant Power. They argued that their policies would achieve, by more practical means, fulfillment of the goals embodied in the Giant Power plan. The construction of the Pennsylvania–New Jersey Interconnection by Pennsylvania Power & Light and two neighboring utilities was subsequently offered as evidence of these practical means of rationalization. PP&L also

[70] *PP&L History*, 1: 163–74.
[71] Ibid., pp. 132–36.

pointed to the progress that had been made in rural electrification since 1926. The number of farm customers had grown from 3,116 in 1926 to 25,519 by 1939. The company estimated that in 1939, 57 percent of the total number of farms in the company's territory were connected and that service was available from existing lines to 66 percent of these. In the United States as a whole, only 28 percent of the farms were receiving electric service.

The company took a leading part in formulating a plan for rural electrification in Pennsylvania. Embodied in General Order no. 28 of the state's Public Service Commission, the plan was worked out by the commission in cooperation with committees appointed by the Pennsylvania Electric Association (an association of electric utilities), and the Pennsylvania State Council of Farm Organizations. The Pennsylvania Electric Association had opposed Giant Power. General Order no. 28 required that the farmer pay a portion of the cost of the extension of service to him, a step that was favored by the private utilities and opposed by leading advocates of subsidized rural electrification. Order no. 28 also created the Joint Committee on Rural Electrification. Consisting of nine farm representatives and seven spokesmen for the utilities, the committee helped work out cooperative arrangements between farmers and the utilities.[72] This tension between the private utilities and advocates of government support for rural electrification may have contributed to the sharp upswing in the number of PP&L's farm customers in 1936, the same year that the U.S. government established the Rural Electrification Administration, an agency that was authorized to make loans to cooperatives and to farmers for rural electrification.

If a slide showing an extremely large territory covered by widely dispersed power plants, each radiating a relatively small-area network of transmission lines and displaying interconnections among these localized subsystems, were projected on a screen, an experienced utility engineer of the 1920s would have little trouble identifying the system as being characteristic of PP&L's style. If another slide showed an extremely large power plant located at an open-face lignite mine, a nearby bank of cooling towers, and a great transmission line extending southward to streams fed by melting snow, an engineer of the 1920s would have little difficulty identifying the style as RWE's. If still another slide showed a large system with a surprisingly low transmission voltage and an irregularly routed transmission network connecting hard-coal and waste-heat-generating plants, the electrical engineer or manager would know it was NESCO. If he were shown more artifacts—NESCO's turbines for example—identification would be even less difficult. NESCO, like RWE and PP&L, had a characteristic style.

The northeast coast of England was industrial like the Ruhr and Lehigh valleys, but the regional system operated there by the Newcastle upon Tyne Electric Supply Company (NESCO) was smaller. Nevertheless, it stood out as an exceptional development in Britain before the construction of the national Grid. By means of a network of transmission lines, NESCO in the 1920s supplied a standardized three-phase current at 40 cycles to a 1,400-

[72] D. Clayton Brown, *Electricity for Rural America: The Fight for the REA* (Westport, Conn.: Greenwood Press, 1980), pp. 29–31.

square-mile area in the region of Newcastle upon Tyne. The system's highest transmission voltage was only 66,000 volts, compared to the 220,000 volts used by RWE and PP&L. In 1924, NESCO's installed capacity amounted to about 250,000 kw. and its annual supply was about one-eighth of the total electricity sold in Great Britain (see Table XIV.1, p. 409). NESCO's name was later changed to North Eastern Electric Supply Company, but the original name will be used here.

The northeast coastal system of NESCO depended on two kinds of energy, hard coal and waste heat. In 1924, 50 percent of its power was concentrated in two stations (Carville "B", 22 percent and Dunston, 28 percent). Though the dozen or so waste-heat plants provided much less capacity than the hard-coal plants, they were notable because they used waste heat from blast furnaces and coke ovens. The large, coal-fired NESCO stations—Carville "B", Dunston, and North Tees—were situated on, and cooled by, rivers. Although these coal-fired plants were smaller than the monstrous Goldenbergwerk and did not have or need the highly developed coal-handling system of the German mine-mouth station, they were more advanced in boiler and turbine design.

NESCO's regional style contrasted remarkably with the character of utilities elsewhere in England. Not only did NESCO's urban competitors elsewhere sell far fewer units but they operated at a significantly lower load factor. In 1918 NESCO sold 578,416,840 units; the Manchester Corporation sold only 184,675,190. Manchester ran second to NESCO in terms of load factor as well; the former achieved only 32.4 percent, while the latter reached 56 percent. After Manchester, the Sheffield and Birmingham corporations, or local governments, were the largest suppliers in England.[73] All were larger urban districts than Newcastle but had smaller supply systems. Further explanations for the contrasting history of electric supply in Newcastle and in urban industrial centers like Manchester and Liverpool will be found in the cities' political, economic, and social histories.

Physical and industrial geography explains many of NESCO's characteristics. Newcastle, an urban port in Northumberland, the northeastern-most county of England, stands on the northern bank of the Tyne, about 8 miles from its mouth at the North Sea. In the early twentieth century, the tidal river carried many ships to Newcastle (in 1911 a city of more than 200,000 inhabitants) and other densely populated manufacturing towns along its banks (see Fig. XIV.26). In the vicinity there were immense stores of bituminous coal, which was mined and exported as well as consumed by industry in the region. Along the northern bank of the Tyne stood such industrial enterprises as Armstrong, Whitworth & Company, as well as the shipyards of Swan, Hunter, and Wigham Richardson & Company. The turbine-manufacturing enterprise of Charles Parsons was situated adjacent to NESCO's Carville "A" power station. Blast furnaces, engineering works, chemical plants, and brick and tile added to the industrial activity of the

[73] The information for Lancashire is from Great Britain, Ministry of Reconstruction, Reconstruction Committee, Coal Conservation Sub-Committee, *Interim Report on Electric Power Supply in Great Britain* (London: HMSO, 1918), pp. 12–13. The information for the other cities is from a memorandum filed in "Electricity (Supply) Bill," item 36 in the C. H. Merz lockbox, Merz & McLellan Co. Archives, Amberley, Killingworth, near Newcastle upon Tyne, England. I am indebted to John M. Burnett of NESCO for 1924 NESCO statistics.

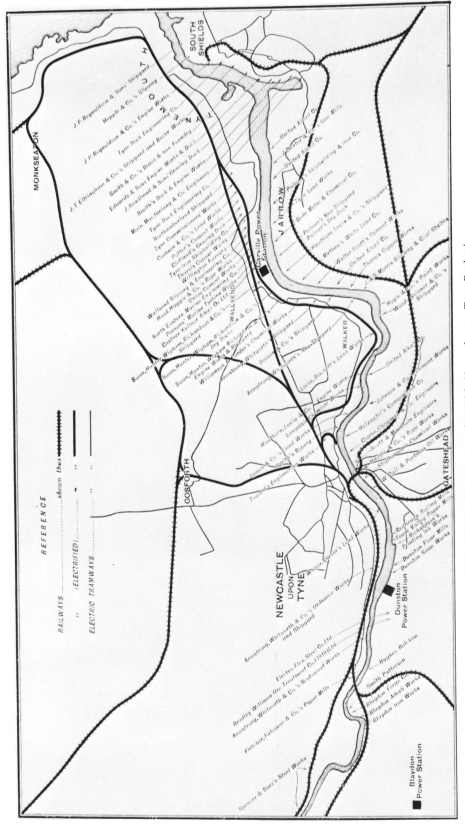

Figure XIV.26. The Tyneside industrial district supplied by NESCO, 1911. Courtesy of NESCO, Newcastle upon Tyne, England.

Figure XIV.27. *Characteristics of NESCO, c. 1923. Courtesy of NESCO, Newcastle upon Tyne, England.*

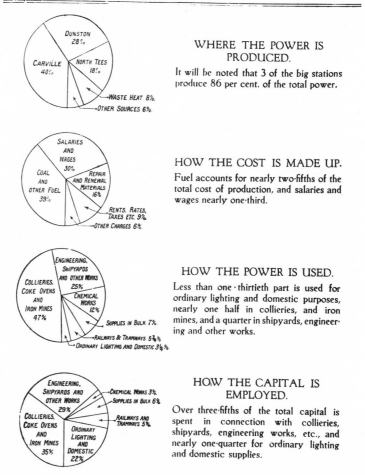

N. E. S. COMPANY LIMITED

WHERE THE POWER IS PRODUCED.

It will be noted that 3 of the big stations produce 86 per cent. of the total power.

Dunston 28%
Carville 40%
North Tees 18%
Waste Heat 8%
Other Sources 6%

HOW THE COST IS MADE UP.

Fuel accounts for nearly two-fifths of the total cost of production, and salaries and wages nearly one-third.

Salaries and Wages 30%
Coal and Other Fuel 39%
Repair and Renewal Materials 16%
Rents, Rates, Taxes Etc. 9%
Other Charges 6%

HOW THE POWER IS USED.

Less than one-thirtieth part is used for ordinary lighting and domestic purposes, nearly one half in collieries, and iron mines, and a quarter in shipyards, engineering and other works.

Engineering, Shipyards and Other Works 25%
Collieries, Coke Ovens and Iron Mines 47%
Chemical Works 12%
Supplies in Bulk 7%
Railways & Tramways 5½%
Ordinary Lighting and Domestic 3½%

HOW THE CAPITAL IS EMPLOYED.

Over three-fifths of the total capital is spent in connection with collieries, shipyards, engineering works, etc., and nearly one-quarter for ordinary lighting and domestic supplies.

Engineering, Shipyards and Other Works 29%
Collieries, Coke Ovens and Iron Mines 35%
Chemical Works 3%
Supplies in Bulk 6%
Railways and Tramways 5%
Ordinary Lighting and Domestic 22%

region. On the eve of World War I, NESCO supplied a region that had 5.1 percent of the population, mined 19.6 percent of the coal, made 36.5 percent of the coke, mined 37.5 percent of the iron ore, produced 37.7 percent of the pig iron, and built 51 percent of the merchant ships in the United Kingdom.[74] In the mid-twenties, almost half of NESCO's output was used in coal and iron mines; one-quarter was consumed by shipbuilding, heavy manufacturing and other works; chemical plants used about one-tenth; and the remainder supplied railways and tramways, bulk consumers, and ordinary lighting and domestic appliances (see Fig. XIV.27). By itself, however, geography is not a sufficient explanation for the differences between NESCO and other English utilities supplying industrial districts.

[74] The population figure is for 1911; the other statistics are for 1913. See A. R. Sloan, "Fuel Economy on the North-East Coast as a Result of Electric Power Supply" (Paper read before the 1916 meeting of the British Association for the Advancement of Science, Newcastle upon Tyne, 5–9 September 1916), p. 2; reprinted in *Engineering* 102 (1916): 293–94.

The success of the Newcastle upon Tyne (later North Eastern) Electric Supply Company in forming a regional supply system can be explained in part by political factors. A series of victories in Parliament brought the supply company the legislative authority it needed to expand its supply area. The victories were achieved on various occasions in competition with the Newcastle & District Company, a rival utility supplying the city, and later with several power companies that intended to supply electricity in bulk to the area. As a result, the privately owned NESCO grew in much the way that utilities expanded in Germany and the United States, where political power and legislation were, respectively, supportive and subservient. Paradoxically, the legislative context within which NESCO expanded was the same as that for other English utilities: a series of parliamentary electric light and power laws and bills. NESCO, however, used these laws to enhance rather than frustrate growth. The utility's history, then, is testimony to the fact that legislation is not a sufficient explanation for the retardation of growth of the British supply industry. A survey of the growth of NESCO in the political context shows how the laws were used.

NESCO obtained provisional orders authorizing it to supply Newcastle in 1889 after the 1888 amendments to the Electric Lighting Act of 1882 were passed. At about the same time, Charles A. Parsons, the turbine manufacturer who had an engineering works situated in the Newcastle area, started a rival electric supply utility, the Newcastle & District Company. The two supply companies agreed to divide the city of Newcastle between them; NESCO would supply the area north of the river, the Newcastle & District Company the area south of it. The two companies also cooperated but failed in an effort to prevent the Newcastle Corporation, the local government, from obtaining parliamentary authorization to provide electric tramways in Newcastle.

The expansion of NESCO was facilitated by Parliament's decision, following the recommendations of the Cross committee in 1898 and the Kitson committee in 1900, to enact power bills enabling companies to supply bulk power over large areas. Most English power companies did not flourish, however, because the enabling bills prohibited supply by a power company to districts that already had an authorized distributor unless the distributor agreed. Moreover, powerful urban political interests caused Parliament to exclude supply within cities from the power bills. NESCO, however, managed to avoid these restrictions. Charles H. Merz notes in his recollections that Sir James Kitson, head of the parliamentary committee that reviewed the power bills, was "a director of the North-Eastern Railway."[75] (This interest in the economy of the region may have led Kitson to support its electrification.) Both NESCO and the Tyneside Electric Company, a power company formed by Parsons, applied for power bills to supply the Tyneside

[75] The early history of NESCO is given in James R. Beard, "NESCO in the Very Early Days," *Nesco Magazine*, no. 1 (n.d.); the quote is from p. 7 (copy in the Merz & McLellan Co. Archives). Beard, a partner in Merz & McLellan, had access to—and quotes from—J. Theodore Merz's "Reminiscences" and Charles Merz's "Notes," both of which are on file in the Merz & McLellan Co. Archives. Beard also presented "Their Achievement—Our Heritage," an address to the North-Eastern Centre, Institution of Electrical Engineers, 14 December 1959 (copy in Merz & McLellan Co. Archives). In addition, there is a succinct history of NESCO in Ministry of Reconstruction, *Interim Report on Electric Power Supply in Great Britain*, app. D.

Figure XIV.28.   *J. Theodore Merz. From Rowland,* Progress in Power, *facing p. 1.*

region. On this occasion (1900), as on many others, Charles Merz proved to be an effective expert witness before a parliamentary committee. NESCO emphasized that as a result of an agreement with the Walker & Wallsend Gas Company, which had commissioned a power station with Charles Merz as a consulting engineer, it already had a power station supplying the area north of the Tyne River. NESCO defeated the Parsons bill, put through its own, and took over the gas company's power station.

Charles Merz then continued to carry out his plans for large-area (eventually regional) power supply. In 1902 NESCO secured passage of another parliamentary power bill authorizing extension of supply north of Newcastle into the coal-mining area of Northumberland. A rival scheme of Parsons's Newcastle & District Company was defeated. In 1904 NESCO reached an agreement with the Durham Power Company to the south to supply it with electricity in bulk and then took over the company by stock purchase. In 1906 NESCO acquired control of the Northern Counties [Power] Company. NESCO was then able to raise the capital needed to develop the supply areas of the power companies it acquired, money the companies individually had been unsuccessful in raising earlier.[76]

NESCO's successes in obtaining parliamentary authority to expand its system also depended on the engineering, business, and financial network of J. Theodore Merz and his son, Charles. The network was made up of relatives, Quakers, and business and professional associates. J. Theodore Merz's wife was a Quaker; her brother, John Wigham Richardson, was a Tyneside shipbuilder (Swan, Hunter, and Wigham Richardson & Company), and her sister married R. Spence Watson, a solicitor with a Quaker background and a leader of the Liberal party in the Tyneside region. The Merz family also associated with leading Tyneside industrial, engineering, and scientific figures because of J. Theodore Merz's education and training. Of German background, J. Theodore's father was born in Manchester but returned to Germany to live in the university town of Giessen. J. Theodore attended the universities of Giessen and Göttingen, concentrating his studies in chemistry and chemical engineering. He returned to Britain in 1868, however, and settled near Newcastle, where he worked for a chemical manufacturer. He founded the Newcastle Electric Supply Company in 1889 and served as its chairman from 1901 to 1917. In addition, he was a director of British Thomson-Houston, the electrical manufacturer, and, most remarkably, the author of one of the major scholarly studies of his time, the four-volume *History of European Thought in the Nineteenth Century.*[77]

These associations with science, engineering, business, and politics were further enriched by J. Theodore Merz's scholarly activities and by the

[76] For this history of NESCO and the firm of Merz & McLellan, I am indebted to Dr. Richard A. Hore and Ms. Kathleen Bramley of Merz & McLellan. Dr. Hore, of the technical and scientific staff, and Ms. Bramley, librarian, arranged for me to use the historical records held by Merz & McLellan at the firm's principal offices in Amberley, Killingworth, near Newcastle upon Tyne, England. They also directed me to the extensive collection of Merz & McLellan records now deposited in the Tyne and Wear County Council Archives, Newcastle upon Tyne.

[77] John Rowland, *Progress in Power: The Contribution of Charles Merz and His Associates to Sixty Years of Electrical Development, 1899–1959* (London: Newman Neame, 1960), pp. 11–14. This is a privately printed history authorized by Merz & McLellan.

Figure XIV.29.   *Charles Merz.*
Frontispiece from Rowland, Progress in
Power.

literary and historical interests of R. Spence Watson, a recognized authority on the history of the Newcastle region. Through his father and R. Spence Watson, Charles Merz met leading British and European intellectuals and politicians. The circle of influence was further widened by the access the Merz family had to leading London banking figures such as Leonard Cunliffe of Cunliffe Brothers. These associations helped prepare Charles Merz to function as a consulting engineer and entrepreneur, roles in which an ability to relate to the worlds of engineering, politics, and financial power was necessary. Oskar von Miller, the leading German consulting engineer, had similar associations because of his family's prominent position in Bavarian society.

The Merz family, their relatives, and their associates were close because Newcastle and Tyneside constituted "somewhat of an isolated unit both culturally and industrially from the rest of England."[78] The leaders of this industrial community used their political and financial power to support the privately owned NESCO while most other urban centers, excluding London, promoted government-owned electric supply. Even NESCO's immediate rival, the Newcastle & District Company, was a private enterprise. As Charles Merz pointed out on numerous occasions, small-scale, mostly government-owned supply companies elsewhere in England incurred higher costs and realized lower per capita consumption.

The organizational structure of NESCO was unusual because of the utility's involved relationship with a consulting engineering firm. This, too, should be taken into account in explaining NESCO's style. Other utilities—Chicago's Commonwealth Edison for instance—regularly used one consulting firm (Sargeant & Lundy in this case), but the relationship between NESCO and Merz & McLellan was so close that the first was rarely spoken or written about in isolation from the other. NESCO provided electricity to the northeast coast of England, but contemporaries attributed the growth of electric supply in the region to Charles Merz and to Merz & McLellan. The power stations of NESCO attracted world-wide attention, but they were designed and constructed by Merz & McLellan. NESCO had its own board of directors and managers—men of reputation such as R. P. Sloan—but Charles Merz, not Sloan or the other company heads, was known as the regional entrepreneur in the world of electric supply. Even histories of NESCO give Merz, rather than company managers and engineers, the prominent place.[79]

The organic relationship that existed between NESCO and Merz & McLellan can be seen in a survey of their history and particularly of the career of Charles Merz (1874–1940). In 1889 Merz was too young to take part in the founding of NESCO by his father, J. Theodore Merz; his uncle, R. Spence Watson; and six others. Charles attended Bootham School in York and Armstrong College in Newcastle, but he did not complete the course of study at Armstrong. He apprenticed at Pandon Dene, the first of NESCO's generating stations. He also apprenticed with Robey of Lincoln on high-speed engines and with British Thomson-Houston. After appren-

[78] Beard, "Their Achievement—Our Heritage."

[79] See, for example, Beard's "Their Achievement—Our Heritage" and "NESCO in the Early Days."

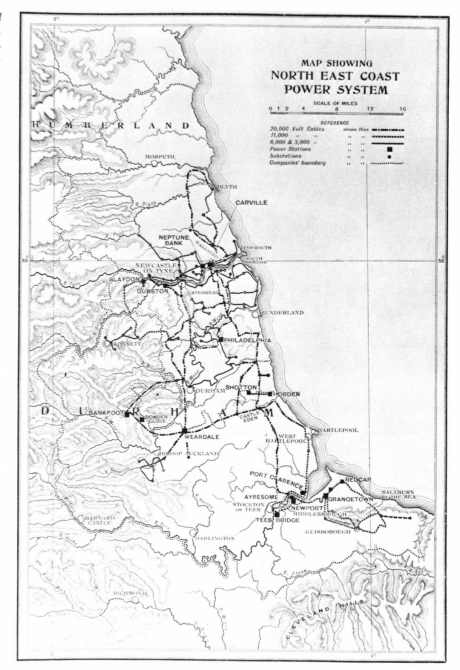

ticeship, he joined BTH and became manager and engineer of its electric supply utility at Croydon. Also for BTH, he supervised construction of an electric plant for tramways and lighting at Cork, where he met William McLellan, who later became his partner in the consulting engineering firm Merz & McLellan.[80]

[80]J. M. Burnett, "Charles H. Merz, 1874–1940," *Papers Presented at the Fourth Weekend Meeting on the History of Electrical Engineering, Durham, 2–4 July 1976* (London: Institution of Electrical Engineers, n.d.), pp. 24–35.

Figure XIV.31.   George Westinghouse,
Lord Kelvin, and Charles Merz in 1903.
From Rowland, Progress in Power,
facing p. 40.

J. Theodore and Charles Merz drew together in their work when in 1899 Charles was asked by the Walker & Wallsend Gas Company to superintend construction of an electric power plant to supply power to industry on the Tyneside. Charles's uncle, John Wigham Richardson, was a director for Walker & Wallsend. Charles was eventually offered a job as engineer for the company, but he refused, preferring to establish his own consulting engineering firm so that he might plan, design, and supervise construction for a number of clients in addition to Walker & Wallsend. The Neptune Bank power station that was built for Walker & Wallsend was the first three-phase supply utility in England and attracted considerable professional attention.[81] No less a scientific figure than Lord Kelvin performed the opening ceremony on 18 June 1901. Kelvin had been an advocate of direct current.

Established in Newcastle, adviser to a power company, and close once again to NESCO and his father, Charles Merz conceptualized a technological and economic strategy that became his hallmark. As noted, NESCO soon purchased the electric supply business of the Walker Gas Company and embarked, with Charles's and his father's guidance, on an innovative policy of developing a regional power company instead of simply an urban utility. Charles then asked the Scotsman William McLellan to join him. McLellan, who was two months younger than Charles Merz, had completed the engineering course at Liverpool University and worked for the electrical manufacturer Siemens Brothers of England.

The details of Merz and McLellan's early plans have survived and give a rare glimpse of the origins of a consulting engineering firm. In 1900 the young men corresponded after Merz asked McLellan to join him in the venture. McLellan was careful to stress that he in no way wanted to trade on his friendship; he preferred that a businesslike arrangement be nego-

[81] Beard, "Their Achievement—Our Heritage," p. 11.

Figure XIV.32. *William McLellan. From* Rowland, Progress in Power, *facing p. 1.*

tiated. Merz said that since he did not intend to give all of his time to Newcastle projects, McLellan should become the resident manager there. The two men informally agreed that McLellan should receive a monthly salary amounting to £300 per annum, an additional £200 at the end of the year, and one-quarter of the firm's net profits each year. Merz's £1,000 salary would be deducted in determining the net. (Merz was at liberty, so long as he did not use the Newcastle staff, to take consulting work elsewhere.) Both men also wanted to turn back into the consulting business three-quarters of the share each took from the net profits.

They estimated that during the first two years the firm's income would be as follows: Neptune Bank power station for Walker & Wallsend Gas Company, £7,500 on a £150,000 project; underground transmission/distribution cable for NESCO, £5,000 on a £100,000 project; and a tramways project, £2,000 on £40,000, for a total of £14,500 in consulting fees. They also estimated that annual expenses for their firm would amount to the £1,500 salary for the two of them, £350 for travel expenses, and £1,850 for office expenses. Since the expenses for two years amounted to £7,400 and since they had already spent £1,400 for their office, the two men expected a net gain of £5,700 in the first two years, of which Charles Merz's share would amount to £4,275 and William McLellan's £1,425. They planned to reinvest as agreed.[82]

In their offices they planned to have two salaried employees (probably engineers) and a chief draftsman, whose annual salary would be £200, the same as the engineers'. Also on hand would be three assistant draftsmen, three junior assistants, four typists, and two office boys. The Merz & McLellan staff would in time cooperate closely with the staff of NESCO. By 1913 the staff of Merz & McLellan had grown to seventy-three and the company had opened a London office on Victoria Street, traditionally the district where engineers had their offices. An office in London facilitated parliamentary associations, testimony before parliamentary committees, and work in the Commonwealth.

Charles Merz traveled throughout the world for Merz & McLellan. In 1901 and again in 1906 he visited the United States, where he met Samuel Insull. A close professional and business relationship developed between the two, surviving even Insull's financial collapse and public disgrace following his indictment on charges of fraudulent practices as a holding-company entrepreneur, charges of which he was eventually declared innocent. Merz organized a private luncheon for the "fallen giant" at the Reform Club, "an action by which Insull was much moved."[83] On several occasions Merz & McLellan also served as a consulting firm for Commonwealth Edison, especially on steam turbine plants and on general development matters.[84] Merz, Parsons, and Insull worked together in 1912 when

[82] McLellan to Merz, 4 August 1900, and draft of an agreement (n.d.), C. H. Merz lockbox, Merz & McLellan Co. Archives.

[83] Rowland, *Progress in Power,* p. 76.

[84] See, for instance, Merz & McLellan, "Commonwealth Edison Company and Associated Companies: Report on the Present and Future Development of the Electrical System," Merz & McLellan Papers (vol. 1/1.8, Folio 29), Tyne and Wear County Council Archives. The firm also prepared a major study of the urban utilities of London, New York, and Paris, among others, for the Consolidated Edison Co. of New York; see Chapter XII, note 60, above.

Insull's Commonwealth Edison Company ordered a 25,000-kw. unit, then the largest in the world. Merz also observed engineering practices, prepared reports, and organized construction projects in Australia, Argentina, South Africa, India, and in American and British cities other than Chicago and Newcastle upon Tyne. He became an agent of technology transfer on a grand scale, taking with him the experience he gained in building the NESCO system on the northeast coast of England and bringing back state-of-the-art ideas from the rest of the world. He also associated with other leaders in electrical engineering and management, such as Georg Klingenberg of AEG. No history of technology transfer would be complete without consideration of consulting engineers like Merz.

We have already noted in this study Merz's effectiveness as a parliamentary witness on behalf of power bills, his contributions to the creation of the Grid, and his leading role during World War I as a member of parliamentary commitees planning the reorganization of electric supply in England after the war (see pp. 319–21 above). He was believed by many to be the most effective expert witness in the engineering world. In presenting his cases he used highly imaginative statistical tables, graphs, and charts prepared by his firm. In World War I he also served as a member of the Admiralty's committee on antisubmarine warfare and then as director of experiments and research for the Admiralty, a position that was used to draw scientists and engineers into wartime research and development. Thus Merz was a pioneer in the trend that would culminate in the forging of the scientific, engineering, and military establishment that performed so effectively in World War II.[85] In 1940 Merz's London home was hit by a bomb and he and his two children were killed; his wife survived.

Charles Merz articulated and disseminated the economic and technological principles of regional systems. He took into account the technological and economic givens and forged strategy and tactics for NESCO. In so doing, he joined the ranks of such men as Samuel Insull, Hugo Stinnes, Oskar von Miller, A. Koepchen, Charles Stone, Edwin Webster, and S. Z. Mitchell. Each offered a particular emphasis in his publications and works. Merz stressed the development of highly efficient steam-turbine power plants to supply large industrial consumers and traction. In 1904, with William McLellan, he published a paper on the design of power stations that became an authoritative statement among engineers. Merz and McLellan foresaw in 1904 that

large power schemes will probably eventually consist of a network of transmission cables, substations and distribution cables, supplying all the power requirements of a neighborhood and drawing a supply from more than one power station, the latter being located where electricity is to be obtained most cheaply from either coal or waste products.[86]

They then went on to discuss the advantages and disadvantages of steam

[85] For correspondence and other material about Merz's relationship with Sir Eric Geddes, first lord of the Admiralty during World War I and later minister of transport, see item 36 in the C. H. Merz lockbox, Merz & McLellan Co. Archives. Leslie Hannah, in *Electricity Before Nationalisation* (Baltimore: The Johns Hopkins University Press, 1979), and John Rowland, in *Progress and Power*, also discuss Merz's government service.

[86] Charles H. Merz and William McLellan, "Power Station Design," *Journal of the Proceedings of the Institution of Electrical Engineers* 33 (1904): 4; reprint in the Merz & McLellan Co. Archives.

turbines, a question that was quite new to most of the engineering world. The first steam turbines had been introduced in a public utility in 1890 at the Forth Banks station of Charles Parsons's Newcastle & District Company, but these were small, 75-kw. units. Other installations had followed in England, but a sizable, 1,000-kw. unit did not result until 1900, in Elberfeld, Germany. Merz & McLellan's early interest in turbines is better understood when one notes that the turbines installed in England were made by Charles Parsons, whose engineering works were located at Tyneside. Merz and McLellan found opinion still divided on the advantages and disadvantages of the turbine as compared to the reciprocating engine, but they pointed out that innovative installations were under way, including Commonwealth Edison's installation at Fisk Street in Chicago. They credited Parsons with the development of the turbine, suggesting that De Laval's turbine was so high-speed that it was practical only in small sizes.

In the 1904 article and later essays Merz stressed the regional character of technology.[87] Because of his familiarity with technology throughout the world, he realized that a universal best way did not exist; instead, a variety of styles prevailed. He believed that local conditions completely governed power supply within a region. Therefore, he chose to write about the relative advantages of high- and low-speed reciprocating engines for driving generators, noting that the high-speed engine had taken root only in Britain. Merz also discussed the differences in system frequency encountered: 25 cycles in the Midlands of England and western Scotland; 40 cycles on the northeast coast of England; 50 cycles in Lancashire and Yorkshire; 50 cycles in France and Germany; 40 cycles in Italy; 25 and 60 cycles in the United States; and 50 and 60 cycles in Japan. (Japan could draw on both European and American technology.) He went on to observe that different energy sources could be exploited advantageously in different regions. He was particularly interested in his own region's use of waste heat from coke ovens and blast furnaces. In 1924 he explained why, compared to the rest of the world, relatively low voltages were used for transmission in England. England was an aged and settled nation where landowners enjoying traditional rights obstructed the granting of wayleaves that would make feasible extended high-voltage point-to-point transmission. Furthermore, England was densely settled, industrially developed, and had numerous coal fields; high-voltage transmission like that used in California was not needed. The British erected inexpensive wooden poles to follow the irregular boundaries of the land, "which do not follow straight lines but are the result of gradual growth from an almost immemorial past."[88] The contrast between this pattern and the wide open spaces in much of the United States was dramatic.

[87] Merz published a number of articles and presented a number of talks, reprints of which are in the Merz & McLellan Co. Archives. Among them are Charles Merz and William McLellan, "Power Station Design"; Merz, "Power Supply and Its Effects on the Industries of the North-East Coast"; Charles H. Merz, "Electric Power Distribution" (Paper read before the Annual Meeting of the British Association for the Advancement of Science, Newcastle upon Tyne, 5–9 September 1919, and published in *Engineering* 102 [1916]: 262–63); and Charles H. Merz, "The Transmission and Distribution of Electrical Energy" (Talk given at the First World Power Conference, London, 1924).

[88] Merz, "The Transmission and Distribution of Electrical Energy," p. 9.

It would be interesting to know the extent to which the author of *The History of European Thought in the Nineteenth Century* shaped the concepts of his son. J. M. Burnett of Merz & McLellan believes that "it was probably J. Theodore rather than his son who first conceived of a regional supply system, the first in Britain."[89] Little evidence of J. Theodore Merz's influence in electrical matters is available, however. His memo of October 1898 entitled "On the Present Position and the Further Development of the Newcastle Electric Supply Station" is an exception. He prefaced his remarks by saying that he did not consider himself an expert on electrical supply but that he had talked to many experts and was an intelligent compiler. He argued that NESCO's system was out of date because the station was small and supplied light to only a small area of the city. Its single-phase alternating-current distribution had been outmoded by polyphase distribution. More important, the company needed to build a large station on the Tyne River, where condensing water was plentiful, and outside the city, where land was inexpensive and smoke would not cause a nuisance. The station would supply industrial and densely settled residential areas primarily. J. Theodore recommended an expenditure of £40,000 for the station and suggested ways of raising the money.[90]

The Merzes coordinated technology, economics, and their social network. While local-government-owned electric supply undertakings elsewhere were extremely sensitive to the demand for electricity made by taxpayers or ratepayers, NESCO concentrated on the industrial power load, much of which was controlled by the Merzes' Quaker kin or friends. After 1900 NESCO's load became increasingly industrial, and by the 1920s, no more than 10 percent of the load could be attributed to ordinary lighting and domestic appliances. Charles Merz used the staff of his consulting engineering firm to sell NESCO power to local industries and then the firm acted as a consultant in the design of the equipment and installation for these customers. Merz & McLellan's designs nicely integrated supply and demand. It is not surprising to find among the major customers of NESCO in 1910 Armstrong, Whitworth & Company and Swan, Hunter, and Wigham Richardson & Company, the directors and managers of which were either related to or closely associated with the Merzes socially and in various industrial activities.[91] Sir Andrew Noble, chairman of Armstrong, Whitworth & Company, for instance, was an interested participant in Charles Merz's effort to organize electric supply in London (see p. 249 above). The Merzes' ties with the shipbuilding firm of Swan, Hunter, and Wigham Richardson have been noted.

As a consulting engineer for NESCO, Charles Merz provided mines and industry with designs that reorganized production on the northeast coast of England. He advised the potential customer in detail about the way in which electric motors could be used to replace steam engines and how the workplace could be reorganized because individual electric motor drive permitted more freedom of location than did machines driven by steam

[89] Burnett, "Charles H. Merz," p. 24.
[90] A copy of the memo is on file in the Merz & McLellan Co. Archives.
[91] *Engineering*, c. 1910–11; reprint in the Merz & McLellan Co. Archives.

engines, with their countershaft and belt-transmission.[92] Merz recalls that once their bias in favor of steam was overcome, the engineers and ship-builders of Tyneside found the new power highly adaptable, and that within a few years they were using far more power in the form of electricity than they had employed earlier in the form of steam.[93]

Power stations designed and built by Merz & McLellan for NESCO embodied the partners' views on the primacy of industrial load, economy of scale, and turbine efficiency. The Neptune Bank station, which was commissioned in 1901, was the first station designed by Merz & McLellan for the Newcastle region. Originally equipped with marine reciprocating engines built by John Wigham Richardson's company, the Neptune Bank station was expanded several years later following the introduction of a 1,500-kw. turbine unit, the largest built by Parsons up to that time. Merz had visited the nearby Parsons works, which NESCO supplied with electricity, to inspect the precedent-setting Elberfeld turbines while they were being built.[94] The Neptune Bank turboalternator was the first large unit to generate three-phase power.[95] The station supplied a universal system providing three-phase and direct current at different voltages. The *Electrician* described it as the first utility in the United Kingdom to supply electric power in bulk.[96]

Two years after the commissioning of the Neptune Bank station, the demand on NESCO had become so great that in 1904 the company began construction of the Carville station, which was "the first large generating station of the modern type" in the world.[97] The station embodied principles enunciated by Merz and McLellan in their paper of 1904. Among these was the "complete unit" system, which meant, in effect, dividing the station into a number of autonomous generating systems. The generating units were entirely turbine-driven; two had an output of 3,500 kw. and two produced 1,500 kw. The large units were more than double the capacity of any turbines that had been built up to that time. Merz and McLellan were now fully committed to turbines, despite some early operating problems at the Neptune Bank station.

In 1910 NESCO opened another station—at Dunston—with three turbines totaling about 30,000 kw. These turbines had been built by Brown Boveri of Switzerland and AEG of Germany. The next major technological advance, however, came in 1916 with the commissioning of a second station at Carville. For many years Carville "B" (Fig. XIV.33) held the record as

[92] Memorandum of C. H. Merz to R. P. Sloan of NESCO, October 1905, Folder 29, C. H. Merz lockbox, Merz & McLellan Co. Archives.

[93] Beard, "NESCO in the Early Days," p. 6. Beard is quoting from C. H. Merz's "Notes."

[94] Charles H. Merz, "Autobiography," chapter entitled "Cork," p. 15, Merz & McLellan Co. Archives.

[95] Beard, "Their Achievement—Our Heritage," p. 11.

[96] "Electric Power Supply at Newcastle-on-Tyne," *The Electrician* 47 (1901): 319–27; copy in Merz & McLellan Co. Archives.

[97] L. Coram, "British Power Stations: Plant Closed Down" (a Merz & McLellan manuscript booklet providing technical data on NESCO stations designed by the firm from 1904 to 1927), Merz & McLellan Co. Archives; Beard, "Their Achievement—Our Heritage," p. 14.

Figure XIV.33. *Carville "B" power station. Courtesy of NESCO, Newcastle upon Tyne, England.*

the most economical power station in the world.[98] An unprecedentedly high steam pressure of 250 psi at a temperature of 650°F was used in the station's five Parsons 11,000-kw. turboalternators. The increase in thermal efficiency brought about by raising pressure and temperature became one of the "critical problems" of the 1920s. Merz & McLellan added to the momentum of the trend by using 450 psi and 700°F at North Tees, the next station it designed for NESCO. North Tees had Metropolitan Vickers 20,000-kw. turboalternators. Begun during the war and completed in 1921, the North Tees station won such a reputation that anecdotes are still told of the ruses tried by American engineers wanting to inspect it despite wartime security regulations. In 1922, W. S. Monroe of the Chicago consulting firm of Sargent & Lundy described North Tees as the most advanced power station in the world. Continuing his tour of European power stations, Monroe was particularly eager to discuss high-pressure and high-temperature design with "our associates" Merz & McLellan. (He found the most important power station in Europe after North Tees to be the new Genn-

[98] Beard, "Their Achievement—Our Heritage," p. 15.

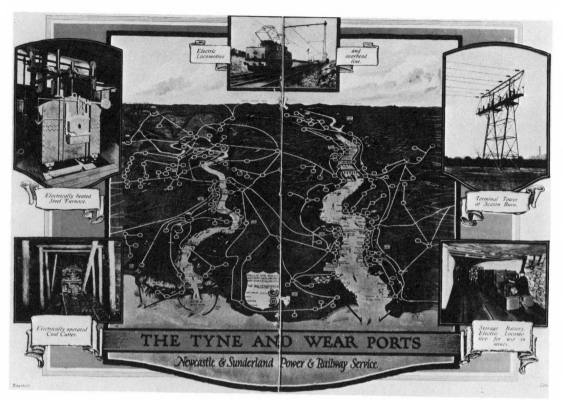

Figure XIV.34. *NESCO's transmission and distribution system, c. 1923. Courtesy of NESCO, Newcastle upon Tyne, England.*

villiers station of L'Union d'Electricité in Paris, a station that had been designed by Ernest Mercier.)[99]

The year 1926 and establishment of the Grid brought a dramatic change in the orientation of NESCO and Merz & McLellan, for they were forced to rethink their plans for regional electrification in the context of a national network. The most immediate and difficult problem for NESCO was the change in frequency. It was an especially vexing problem because Merz and his associates had designed an integrated, standardized-frequency regional supply system that was unique among England's utilities. After 1926, ironically, the system's 40-cycle frequency was defined as nonstandard. The earlier process of integration by means of high-voltage transmission, centralization of control, and standardization had taken more than a decade, but with the coming of the Grid, this remarkable achievement became a 1,400-mile inconsistency.

The Central Electricity Board decided on the uniform three-phase 50-cycle standard instead of NESCO's 40 cycles because 50 cycles was the

[99] "Report of W. W. Monroe of Sargent & Lundy to Samuel Insull on Trip to England and France, 19 June 1922," Merz & McLellan "Reports and Tests" (vol. 1/1.31, Folio 21, Doc. # 35/109), Tyne and Wear County Council Archives. On NESCO power stations in the 1920s, see also N.E.S. Co., Ltd., *Electricity Supply on the North-East Coast* (1924?), pp. 5–6, 28 (a booklet published by NESCO); "The North Tees Power Station," *Engineering* 117 (1924): 753–55 and 118 (1924): 57–59 (reprint in Merz & McLellan Company Archives).

European standard (except in Italy); because most of the existing British polyphase power plants outside NESCO operated at 50 cycles; and because British manufacturers wanted a market outside England for the equipment they would design and supply in quantity for the Grid.[100] The decision entailed serious consequences, for it involved conversion of hundreds of turbogenerators, hundreds of thousands of motors, and almost half a million consumers from other frequencies.

In the 1926 legislation an effort was made to soften the blow for NESCO and its customers by providing for a government loan to accomplish the conversion. The entire electric supply industry would share the repayment cost; each utility, or authorized undertaking, would be required to make an annual payment over forty years on the basis of its revenues. Projections and analyses estimated the total cost of frequency conversion to be £10.5 million. Detailed studies, however, revealed a far greater problem than the one for which Parliament provided. By 1930 the estimate for the conversion in northeast England alone amounted to a gross of £9 million. Because of the escalation, action was delayed and alternatives to integrating the already well-integrated northeast into the large national system were explored.

Electrical suppliers and the large industrial consumers in Merz's northeast opposed the changeover for a number of reasons. The government had arranged funding for the conversion, but had not provided compensation for the disruption of equipment and facilities while the work was under way. Moreover, the northeast system had a high load factor, and the argument could be made that increased economies resulting from incorporation into a larger system would be small, if any. In a report on standardization of frequency prepared by NESCO and associated companies, a member of the Merz & McLellan firm advised the Central Electricity Generating Board not to proceed with conversion.[101]

Events on a larger scale, however, relieved the tension. After the Central Electricity Board found that the choice was between conversion or isolation of the northeast coastal region, it petitioned the government to allocate several million pounds in unemployment grants to the project. This would at least mean that the country's suppliers would not be burdened by an intolerable debt. The industrial northeast was feeling the depression, and the changeover was seen as a palliative, highly rational, and in the long run an economically desirable project. The funds were appropriated.

Confronted by the problems of conversion and the loss of autonomy through inclusion in the Grid, the designers and operators of NESCO could still look back on an unusually interesting and, for England, paradoxical history. From a historical perspective, the principal explanation for the evolution of NESCO into a regional system was political and social. The geographical, economic, and organizational factors were favorable, but that combination (with minor variations) could be found elsewhere in England.

[100] Merz, "Electric Power Distribution," pp. 262–63.
[101] The cost of frequency standardization was estimated in Minutes of the Central Electricity Board, 10 January 1930, "Standardization of Frequency"; and Merz & McLellan, *Report on the Standardization of Frequency on the System of the Newcastle upon Tyne Electric Supply Company, Ltd., and Associated Companies* (January 1928); both in the Electricity Council Archives, London. See also Great Britain, Ministry of Transport, *Report of the Committee Appointed to Review the National Problem of the Supply of Electrical Energy* (London: HMSO, 1926).

What was unique in the Newcastle and Tyneside region was the ability of the Merzes and their associates to put aside the frustrating contradictions that were extant elsewhere in England. In London, the conflict between the parties favoring the centralization of London's government and those advocating its decentralization stymied the development of a centralized London power system. In the large urban-industrial centers of the Midlands and the Liverpool-Manchester region, the forces of municipal socialism contended with those of private ownership, and as a result, the government-owned utilities were confined mostly to the political jurisdiction of the municipality, and the private power companies, with their franchises for large areas, were denied access to the urban districts. NESCO, by contrast, was a privately owned urban utility that metamorphosed into a privately owned power company that crossed the parochial political boundaries. The explanation lies to a large extent in the political, financial, and industrial network, or community, presided over, or participated in, by the kin and friends of the Quaker Merzes. This social and technological network tended to see power—both political and electric—as the Germans and Americans saw it: as a means to technological ends.

CHAPTER XV

Epilogue

THIS study ends with the year 1930. The regional power systems, including those owned by private utilities, government agencies, and mixed private and government enterprises, had matured. After 1930, changes in these systems became less qualitative and more regular and predictable. The reader who is disappointed that the study does not extend to the present should consider that the analysis of a coherent, formative period in the history of power systems is likely to be more meaningful than an effort to extend the study to a so-called present, which is past by the time the study is published.

The model of system formation and growth used as the mode of organization for this study allowed the author to develop and coordinate a number of subthemes. These may prove relevant not only to the history of power systems but to history focused on technology generally. The subthemes were in most instances related to questions often asked about technological systems and about the history of technology in general. For instance, the nature of the inventive act was explored under the rubric of invention and development. Of special interest in this regard was the creativity of Thomas Edison—the motivations, methods, and sociological circumstances pertaining to his invention of a system. Also important were the contributions made by Edison's co-workers. Edison was seen inventing by analogy and organizing his co-workers and his laboratory facilities as a mirror image of the system being invented. The technology-transfer stage of the history raised questions about adaptation. The Edison direct-current system was invented and developed for a specific area of New York City; in order for it to be transferred to other places and societal contexts, such as London and Berlin, changes had to be made in the system. Because failure in technological endeavors is so rarely discussed and because it reveals important and sometimes intractable problems, attention was directed to failures resulting from inappropriate technology transfer, especially in London.

Historians are now interested in the interaction of technology and politics. In this study the interaction was explored in the context of three major Western industrial cities. In Chicago, technology dominated politics; in London, the reverse was true; and in pre–World War I Berlin there was coordination of political and technological power. The reverse salient–critical problems analysis of technological change provided an opportunity to discuss the mechanism, or internal dynamic, of technological evolution. The nature of the resolution of the conflict that developed in the 1890s between technological systems permitted exploration of the art of compromise and nonviolent change. The conflict also revealed that technical problems are sometimes in essence institutional and value conflicts. Another subtheme articulated within the context of the overarching model was technology as an improver on natural circumstances. Natural resources in the Sierras of California were seen being transformed into economic goods by engineers and entrepreneurs using high-voltage transmission.

One question—Is technology autonomous?—was the theme of Chapter XI, "War and Acquired Characteristics," but it was explored throughout the study. To draw together the various threads and to explore and propose at least a partial answer to the question of autonomy, we must consider the internal technical forces that facilitated growth as well as the external, nontechnical, cultural forces that helped shape the electric supply systems. Throughout this study it has been necessary to reach out beyond the technology, outside the history of technical things, to explain the style of the various systems, so obviously the technological systems were not simplistically autonomous, or free of the influence of nontechnical factors. The evolving power systems were not, metaphorically speaking, driverless vehicles carrying society to destinations unknown and perhaps undesired. The systems did, however, have an internal drive and an increasing momentum. The continuous emergence of reverse salients and the ongoing solution of critical problems by inventors, engineers, and entrepreneurs provided this. The external, cultural factors that gave control and direction were systematically summarized in Chapter XIV. The style of each system was found to be based on entrepreneurial drive and decisions, economic principles, legislative constraints or supports, institutional structures, historical contingencies, and geographical factors, both human and natural. All but the natural geographical factors involved the actions of society, which were sometimes purposeful, sometimes inadvertent.

The cultural forces influencing the systems stemmed from the societies within which the systems grew. These societies were of various kinds according to time and place. The early systems were subject mostly to urban influences. In the 1920s the influences stemmed from regional circumstances and the social organizations associated with them—for instance, states in the United States, counties in England, and provinces in Germany. As was explained in Chapter XIII, the cultural forces varied from society to society, but there were also forces that transcended local or regional characteristics. These were mostly economic in nature. These economic forces in turn manifested the values that transcended time and place and pertained to Western society, or at least to the United States, Germany, and England. The values were those of a cost-accounting, capitalistic civilization.

If a would-be Darwin of the technological world is looking for laws analogous to the environmental forces that operate in the world of natural selection, the economic principles of load factor and economic mix are likely candidates. In the history of supply systems, these embodied the values of a culture that was capitalistic, a culture where interest on capital was calculated to ascertain the cost of goods and services. Because electric power systems were capital intensive, interest was of paramount importance. The cost of capital was calculated by the utilities operating electric power systems irrespective of the form of ownership—private, public, or mixed private and public. In a culture that did not calculate capital cost—the medieval Western civilization for instance—electric light and power systems would have grown differently.

Owners and managers of utilities, unlike the owners and managers of railroads, steel mills, automobile factories, and many other large-scale technological enterprises, were not diverted by harassing labor problems from a close analysis of, and emphasis on, capital cost. In the sources pertaining to the problems of the electric power systems, rarely was more than a passing reference to labor costs encountered. Instead, the emphasis was on load factor. Load factor structured the decision-making environment. Probably for this reason the utility engineers and managers were the first to clearly articulate and thoroughly exploit the implications of load factor. The railroads had a load-factor problem in their need to utilize rolling stock more fully, but load factor was a concept articulated by the electric power utilities.

How did close attention to load factor cause system growth? The essence of the answer is diversity, the diversity of human geography, which is often a correlate of the size of an area. As this study has shown, utility managers like Insull strove in a purposeful way to expand the territory of their utilities. The objective was not simply size, as crude explanations for the large scale of modern technology and business insist, but expansion to encompass the diversity of loads that brought a fuller round-the-clock utilization of generating equipment. A utility manager with a peak load caused by rush-hour use of electrified streetcars soon learned that it was not in his interest simply to expand the traction load. Instead, the utility reached out like a tree in a dark forest stretching its limbs into the sustaining sunlight. When sustenance for the load-hungry utility with a traction peak was the night-shift operation of a chemical plant, the system's distribution lines reached out in that direction. System builders knew that the diversity of load that allowed load management, a resulting improvement in load factor, and a lowering of unit capital cost was likely to be found in a large geographical area where the population engaged in a wide variety of energy-consuming activities. This study has made this clear in a number of instances.

Economic mix also enhanced system expansion. This was explained using examples like the spread of the high-voltage transmission lines of RWE, the Ruhr-based utility, into the lignite fields around Cologne and the water-power regions of the Alps. Power plants in these diverse regions complemented each other economically. Again, expansion was not simply an aggressive drive for undifferentiated size; it was a purposeful move to lower the cost of energy. Whereas load-factor considerations led utilities to exploit the diversity of human geography, economic mix dictated expansion to exploit the diversity of natural geography. The imperatives of fuel economy and load factor often interacted because load diversity was encountered along the transmission lines extending to new sources of energy.

The decisions that were made to improve load factor and economic mix shaped the growing electric supply systems in their cost-accounting settings. Only rarely were these principles violated, either by private undertakings or government-owned utilities. London was a salient exception. London's local government and the forces it represented were successful in containing electric utilities within the boundaries of local-government jurisdiction and thereby restricted them to small-scale technology and limited diversity. This was the price, literally and figuratively, that Londoners paid for placing a higher value on the traditional power of local government than on the lower cost of electric power.

The managers, engineers, and owners who applied the principles of load factor and economic mix were also motivated by the conviction that each customer should at least pay the cost of the electricity delivered. The methods of calculating this cost were complex, but with rare exception the variations in price for peak load and low load, as well as for small and large customers, were not intended to subsidize. When Gifford Pinchot and Morris Cooke argued that the small rural consumer should be subsidized, the private utilities objected. Such a policy was considered by them to be a matter of social reform, not the business of private enterprise. Oskar von Miller in Bavaria advocated supplying all of the people with the benefits of electricity, not only those who were able to pay. Generally, however, before 1930, supply was limited to those who could afford to pay. The owners and managers of utilities, whether government or private, usually viewed any other policy as uneconomical and irrational.

As a result of the emphasis on economic factors, decision making in the utility industry was relatively straightforward compared to what it would have been if engineers and managers had felt obliged—or indeed had been obligated—to fulfill social needs despite costs. In the fifty-year history of electric power systems considered here, influential voices did not call for the cross-subsidized supply of light to hospitals, prisons, and other public-welfare institutions; the argument was not made that the streets and houses of the slums should be lit first because lighting would have a more dramatic impact on the quality of life there; and demands that electricity be used first in workplaces where the physical burdens were heavy were not expressed. If these social values had been stressed, the decision-making process would have been more complex and electric power systems would have developed different shapes. In essence, the map of supply would not have been a map of transmission and distribution to industry, traction, and the economic classes able to pay for luxury lighting. Electric power systems before 1930 were mostly artifacts that manifested culture.

In the decade after 1930, interest in introducing a more complex set of values into the design of electric power systems increased. In the United States, planners of the Tennessee Valley Authority stressed that the TVA system would not simply supply electric power; it would also endeavor to meet such social objectives as providing labor-saving appliances for a depressed rural population and making the rivers of an economically backward region navigable. Other social objectives included soil reclamation through the manufacture of fertilizers (using cheap electricity) and the prevention of soil erosion through the control of flood waters and runoff. In England, following the establishment of the National Grid, there was increased emphasis on supplying the needs of the less affluent classes by means of, for example, the widespread sale and use of electric heaters. As a matter of fact, by means of aggressive sales and favorable pricing, electric heating was cultivated to such an extent that the system's load curves were greatly distorted.

Economics was not an absolute determinant of the growth and shape of electric power systems before 1930, application of the principles of load factor, economic mix, and pricing on the basis of cost notwithstanding. There were exceptions, as has been noted. Also, there were innumerable variations in the way in which economic principles were applied. Therefore, the economic factor should be considered deterministic rather than determining, and the result a soft determinism. During the half-century of history considered here, the soft determinism of economics was reinforced by another deterministic influence. As the power systems grew in size, they gathered momentum. As was pointed out in a number of sections of this study, power systems encompassed a technical core of components as well as institutional components, most notably utilities. Occasionally, regulatory and law-making bodies were subject to the control of the system builders and managers and thus became part of the system. In addition, electrical manufacturing enterprises were systematically related to the utilities. Even educational institutions were coordinated. Such encompassing systems should be labeled sociotechnical systems rather than technological systems. These sociotechnical systems had high momentum, force, and direction because of their institutionally structured nature, heavy capital investments, supportive legislation, and the commitment of know-how and experience. This momentum was a conservative force reacting against abrupt changes in the line of development. Because of the conservative momentum, rarely were radical inventions, technical or social, introduced. Radical innovations would have changed the rate and direction of growth. After "the battle of the currents," the system builders' efforts were usually directed to increasing the size of systems incrementally, but not to changing their direction to fulfill radically different economic or social goals.

In sum, it is difficult to change the direction of large electric power systems—and perhaps that of large sociotechnical systems in general—but such systems are not autonomous. Those who seek to control and direct them must acknowledge the fact that systems are evolving cultural artifacts rather than isolated technologies. As cultural artifacts, they reflect the past as well as the present. Attempting to reform technology without systematically taking into account the shaping context and the intricacies of internal dynamics may well be futile. If only the technical components of a system are changed, they may snap back into their earlier shape like charged particles in a strong electromagnetic field. The field also must be attended to; values may need to be changed, institutions reformed, or legislation recast.

Index

Marindin Committee, 238, 241–42, 247
Markgrafenstrasse central station, 73, 186
Martin, John, 270–71, 272–74, 276
Maschinenfabrik Oerlikon, 131; in Frankfort exhibition, 133
Massachusetts Electrical Engineering Co. of Boston, 386. See also Stone & Webster Co.
Massachusetts Institute of Technology (M.I.T.), 144, 145–46, 155–57; electrical engineering at, 151–53; network analyzer of, 376–77
Materials handling, 43, 413, 415–16
Mather, T., 154
Mauerstrasse central station, 73, 183
Maxim, Hiram, 100
Mayer, Wilhelm, 313, 314
Menlo Park laboratory, 23–25. See also Edison, Thomas Alva
Merger: of manufacturers, 163–64; of utilities, 276, 430, 431, 440
Mershon, Ralph, 161–63
Merz, Charles H., 237, 249–50, 449–51; assesses London electric supply system, 228; and British Grid, 350, 352–53, 355; in conflict with John Snell, 357; as expert parliamentary witness, 453; and William McLellan, 451–52; and NESCO, 448; and politics, 204–5; and power bill, 251–56; and regional character of technology, 454; and reorganization of electric supply, 319–20; and standard cycles, 129; on technological style, 405; in World War I, 290
Merz, John Theodore, 248, 249, 250, 448–49, 451; influence of, on son, 455
Merz & McLellan, 249, 256, 356, 357, 452; as consulting engineering firm, 385, 386; and the Grid, 458–59; and NESCO, 449; and turbines, 453–54
Metropolitan Supply Co., 235, 360
Middle West Utilities Co., 204, 392, 402
Miesbach, Germany, 335
Miller, Oskar von, 264, 334–37, 464; as consulting engineer, 385, 386; at Frankfort exhibition, 131; and interconnection of utilities, 317; at Paris exhibition, 51, 66; and Emil Rathenau, 67; report of, on German electric supply, 315. See also Bayernwerk; Walchenseewerk
Ministry of Reconstruction, United Kingdom. See Haldane-Merz Report
Minshall, T. H., 251, 253
Mitchell, S. Z., 393, 395, 396; on diversity principle, 399; and Lehigh Navigation Electric Co., 437, 439
Mitten, Thomas E., 326–28
Moabit power station, 195
Momentum, technological, 15–16, 140–41, 465
Monroe, W. S., 457
Morgan, Arthur E., 301
Morgan, J. P., 387
Morrill Act of 1862 (U.S.), and engineering education, 143
Motor: direct-current, 82; polyphase,

109–11, 455–56. See also Dolivo-Dobrowolsky, M.; Industrialization, and electrification; Load, power and light; Power, stationary; Tesla, Nikola
Munich International Electrical Exhibition (1882), 335
Municipal socialism, in Britain, 59–60, 255, 261
Murray, William S., 296, 297
Muscle Shoals hydroelectric plant, 287, 293–95

Nashville Light & Power Co., 387
National-Bank für Deutschland, 67
National City Bank of New York, 428
National Civic Federation, 207
National Defense Act of 1916 (U.S.), 287, 293
National Electric Light Assoc. (NELA), 174, 207
National Power & Light Co., 441
National Socialists (National Socialist German Workers' party), 315
Neptune Bank central station, 451, 456
Nernst, Walther, 166
Network analyzer, 376
Nevada City plant, 271
Newcastle & District Co., 447
Newcastle Corp., 447
Newcastle upon Tyne Electric Supply Co. (NESCO), 249, 290, 357, 358, 448; and characteristics of region, 445–46; and the Grid, 458–59; and industrial power load, 455; organizational structure of, 449; and politics, 447–48; technological style of, 404, 407
Niagara Falls power project, 135–39
Niagara Falls power station, 135, 137–39, 264–65
Niagara Hudson Power Corp., 400
Noble, Sir Anthony, 250
Norris, Henry, 154
Northcote, A. S., 250
North Eastern Electric Supply Co. See Newcastle upon Tyne Electric Supply Co.
Northern Central Gas Co., 439
North Tees power station, 444, 457
Northumberland County Gas & Electric Co., 440
N. W. Harris & Co., 274

Oakland Gas Light & Heat Co., 276
Oakland Transit Co., 274
Oberspree power station, 192
Ogden Gas Co. case, 206
Ohio Brass Co., 268, 375
Ohm's law, in Edison's work, 34–37
Oliven, Oskar, 317–18
Operator, 173
Operator and Electrical World, 173

Pacific Gas & Electric Co., 276–80, 379
Pacific Gas & Electric Investment Co., 277
Pacific Light & Power Co., 281

Page, Archibald, 354, 355–56, 360–61
Page, Charles Grafton, 86
Pantaleoni, Guido, 102
Panter, T. A., testimony of, for Giant Power bill, 311
Parallel distribution. See Distribution, parallel
Pardee, Calvin, 435
Paris International Electrical Exhibition (1881), 66, 335
Parochialism: of local authorities in London, 233–34, 248–49, 253–54; and utility growth, 318, 323, 324, 338
Parsons, Charles, 319; as head of Committee on Electrical Trades after the War, 320; and Newcastle & District Co., 447; and steam turbines, 211, 454
Parsons Co., 232
Parsons Committee. See Committee on Electrical Trades after the War
Passavant, Dr. Herman, 200
Patents, 84, 91–93, 94; Edison, 32, 67, 68, 71, 83; Edwards and Normandy, 92; Gaulard and Gibbs, 91–92; Haselwander, 118; Jablochkoff, 92; Tesla, 115
Pearl Street central station, 40–45, 81. See also Edison, Thomas Alva
Peck, Charles F., and financial backing for Nikola Tesla, 115
Peek, F. W., Jr., 380–84
Pelton, Lester, 263
Pelton Co., 268
Pennsylvania Electric Association, 298, 443
Pennsylvania Lighting Co., 439, 440
Pennsylvania–New Jersey Interconnection (PNJ), 325, 331–34, 401, 440
Pennsylvania Power & Light Co. (PP&L), 326, 330, 440; compared with RWE, 428–29; and Electric Bond & Share Co., 401; and rural electrification, 442–43; technological style of, 404, 407
Pennsylvania State Council of Farm Organizations, 443
Pennsylvania Water & Power Co., 326
Penrose, Charles, 308–9, 312
Peoples Gas Light & Coke Co., 206
Perry, John, 150
Peterson, Waldemar, 348
Petroleum, as power-plant fuel, 278
Pfalzwerk AG, 336–37
Philadelphia Electric Co., 299; and Conowingo project, 325–31; versus Cooke, 300–301; as PNJ dispatcher, 332. See also Pennsylvania–New Jersey Interconnection
Philadelphia Electric Power Co., 326
Philadelphia Rapid Transit Co., 326–28
Physikalisch-Technische Reichsanstalt, 178
Piesteritz, Germany, 288–89
Pinchot, Gifford, 296, 301, 312, 464; and Conowingo project, 328; and Giant Power, 297–98, 301–5
Pine Grove power station, 440
Politics, and technology, 318–20, 352–53, 447–48, 461